Geophysical Monograph Series

183 **Carbon Sequestration and Its Role in the Global Carbon Cycle** *Brian J. McPherson and Eric T. Sundquist (Eds.)*
184 **Carbon Cycling in Northern Peatlands** *Andrew J. Baird, Lisa R. Belyea, Xavier Comas, A. S. Reeve, and Lee D. Slater (Eds.)*
185 **Indian Ocean Biogeochemical Processes and Ecological Variability** *Jerry D. Wiggert, Raleigh R. Hood, S. Wajih A. Naqvi, Kenneth H. Brink, and Sharon L. Smith (Eds.)*
186 **Amazonia and Global Change** *Michael Keller, Mercedes Bustamante, John Gash, and Pedro Silva Dias (Eds.)*
187 **Surface Ocean–Lower Atmosphere Processes** *Corinne Le Quèrè and Eric S. Saltzman (Eds.)*
188 **Diversity of Hydrothermal Systems on Slow Spreading Ocean Ridges** *Peter A. Rona, Colin W. Devey, Jérôme Dyment, and Bramley J. Murton (Eds.)*
189 **Climate Dynamics: Why Does Climate Vary?** *De-Zheng Sun and Frank Bryan (Eds.)*
190 **The Stratosphere: Dynamics, Transport, and Chemistry** *L. M. Polvani, A. H. Sobel, and D. W. Waugh (Eds.)*
191 **Rainfall: State of the Science** *Firat Y. Testik and Mekonnen Gebremichael (Eds.)*
192 **Antarctic Subglacial Aquatic Environments** *Martin J. Siegert, Mahlon C. Kennicut II, and Robert A. Bindschadler*
193 **Abrupt Climate Change: Mechanisms, Patterns, and Impacts** *Harunur Rashid, Leonid Polyak, and Ellen Mosley-Thompson (Eds.)*
194 **Stream Restoration in Dynamic Fluvial Systems: Scientific Approaches, Analyses, and Tools** *Andrew Simon, Sean J. Bennett, and Janine M. Castro (Eds.)*
195 **Monitoring and Modeling the Deepwater Horizon Oil Spill: A Record-Breaking Enterprise** *Yonggang Liu, Amy MacFadyen, Zhen-Gang Ji, and Robert H. Weisberg (Eds.)*
196 **Extreme Events and Natural Hazards: The Complexity Perspective** *A. Surjalal Sharma, Armin Bunde, Vijay P. Dimri, and Daniel N. Baker (Eds.)*
197 **Auroral Phenomenology and Magnetospheric Processes: Earth and Other Planets** *Andreas Keiling, Eric Donovan, Fran Bagenal, and Tomas Karlsson (Eds.)*
198 **Climates, Landscapes, and Civilizations** *Liviu Giosan, Dorian Q. Fuller, Kathleen Nicoll, Rowan K. Flad, and Peter D. Clift (Eds.)*
199 **Dynamics of the Earth's Radiation Belts and Inner Magnetosphere** *Danny Summers, Ian R. Mann, Daniel N. Baker, and Michael Schulz (Eds.)*
200 **Lagrangian Modeling of the Atmosphere** *John Lin (Ed.)*
201 **Modeling the Ionosphere-Thermosphere** *Jospeh D. Huba, Robert W. Schunk, and George V Khazanov (Eds.)*
202 **The Mediterranean Sea: Temporal Variability and Spatial Patterns** *Gian Luca Eusebi Borzelli, Miroslav GaCiC, Piero Lionello, and Paola Malanotte-Rizzoli (Eds.)*
203 **Future Earth - Advancing Civic Understanding of the Anthropocene** *Diana Dalbotten, Gillian Roehrig, and Patrick Hamilton (Eds.)*
204 **The Galápagos: A Natural Laboratory for the Earth Sciences** *Karen S. Harpp, Eric Mittelstaedt, Noémi d'Ozouville, and David W. Graham (Eds.)*
205 **Modeling Atmospheric and Oceanic Flows: Insights from Laboratory Experiments and Numerical Simulations** *Thomas von Larcher, and Paul D. Williams (Eds.)*
206 **Remote Sensing of the Terrestrial Water Cycle** *Lakshmi (Eds.)*
207 **Magnetotails in the Solar System** *Andreas Keiling, Caitríona Jackman, and Peter Delamere (Eds.)*
208 **Hawaiian Volcanoes: From Source to Surface** *Rebecca Carey, Valerie Cayol, Michael Poland and Dominique Weis (Eds.)*
209 **Sea Ice: Physics, Mechanics, and Remote Sensing** *Mohammed Shokr and Nirmal Sinha (Eds.)*
210 **Fluid Dynamics in Complex Fractured-Porous Systems** *Boris Faybishenko, Sally M. Benson, and John E. Gale (Eds.)*
211 **Subduction Dynamics: From Mantle Flow to Mega Disasters** *Gabriele Morra, David A. Yuen, Scott King, Sang Mook Lee, and Seth Stein (Eds.)*
212 **The Early Earth: Accretion and Differentiation** *James Badro and Michael Walter (Eds.)*
213 **Global Vegetation Dynamics: Concepts and Applications in the MC1 Model** *Dominique Bachelet and David Turner (Eds.)*
214 **Extreme Events: Observations, Modeling and Economics** *Mario Chavez, Michael Ghil, and Jaime Urrutia-Fucugauchi (Eds.)*
215 **Auroral Dynamics and Space Weather** *Yongliang Zhang and Larry Paxton (Eds.)*
216 **Low-Frequency Waves in Space Plasmas** *Andreas Keiling, Dong-Hun Lee, and Valery Nakariakov (Eds.)*
217 **Deep Earth: Physics and Chemistry of the Lower Mantle and Core** *Hidenori Terasaki and Rebecca A. Fischer (Eds.)*
218 **Integrated Imaging of the Earth: Theory and Applications** *Max Moorkamp, Peter G. Lelievre, Niklas Linde, and Amir Khan (Eds.)*
219 **Plate Boundaries and Natural Hazards** *Joao Duarte and Wouter Schellart (Eds.)*
220 **Ionospheric Space Weather: Longitude and Hemispheric Dependences and Lower Atmosphere Forcing** *Timothy Fuller-Rowell, Endawoke Yizengaw, Patricia H. Doherty, and Sunanda Basu (Eds.)*
221 **Terrestrial Water Cycle and Climate Change: Natural and Human-Induced Impacts** *Qiuhong Tang and Taikan Oki (Eds.)*
222 **Magnetosphere-Ionosphere Coupling in the Solar System** *Charles R. Chappell, Robert W. Schunk, Peter M. Banks, James L. Burch, and Richard M. Thorne (Eds.)*
223 **Natural Hazard Uncertainty Assessment: Modeling and Decision Support** *Karin Riley, Peter Webley, and Matthew Thompson (Eds.)*

Geophysical Monograph 224

Hydrodynamics of Time-Periodic Groundwater Flow

Diffusion Waves in Porous Media

Joe S. Depner
Todd C. Rasmussen

This Work is a co-publication between the American Geophysical Union and John Wiley & Sons, Inc.

AGU
American Geophysical Union

WILEY

This Work is a co-publication between the American Geophysical Union and John Wiley & Sons, Inc.

Published under the aegis of the AGU Publications Committee

Published by John Wiley & Sons, Inc., Hoboken, New Jersey
Published simultaneously in Canada

Library of Congress Cataloging-in-Publication Data

9781119133940

Cover image: The image presents the water-level response in an aquifer due to periodic excitation of four groundwater wells at different amplitudes and frequencies.

Printed in the United States of America

10 9 8 7 6 5 4 3 2 1

CONTENTS

PREFACE

Goal and Purpose

Our goal in writing this book is to present a clear and accessible mathematical introduction to the basic theory of time-periodic groundwater flow. Understanding the basic theory is essential for those who seek a comprehensive knowledge of groundwater hydraulics and groundwater hydrology. In addition, the basic theory has an aesthetic beauty that readers can learn to appreciate and thereby enjoy.

Intended Audience

We intend this book to be used primarily for self-directed study by advanced undergraduate and graduate students and by working scientists and engineers in the earth and environmental sciences. This book is suitable for well-prepared readers who either (a) are new to the field of periodic groundwater flow and seek a formal introduction to the theory or (b) were introduced to the field in the distant past and wish to renew their knowledge and enrich their understanding. Additionally, we hope that this book will be a useful resource for educators.

The mathematical framework for time-periodic groundwater flow is structurally equivalent to that of time-periodic diffusion. Therefore, some of the theory presented in this book may be relevant to time-periodic phenomena encountered in fields other than groundwater flow, like electrical conduction, thermal conduction, and molecular diffusion. Consequently, we expect that students and professionals in these other fields also will find parts of this book useful.

Prerequisites

We assume that the reader has completed university courses in multivariable calculus, linear algebra, and subsurface fluid dynamics (e.g., groundwater hydraulics). Also, the reader should have a *basic* familiarity with complex variables, Fourier series, and partial differential equations (PDEs). Readers do not need to know contour integration in the complex plane or Green functions.

Approach

Our development is quantitative. We emphasize problem definition and problem understanding, rather than problem solution techniques, because we believe the former are fundamental prerequisites of the latter and because solution techniques have been described exhaustively by other authors (e.g., *Carslaw and Jaeger* [1986], *Özişik* [1989], *Hermance* [1998], *Bruggeman* [1999], *Mandelis* [2001]).

Much of the information presented here could be gleaned from reading articles in peer-reviewed scientific publications such as those listed in the Bibliography. However, one would have to read many such articles, which typically present only terse descriptions of the mathematical development. This book is more explicit to accommodate the needs of those who are new to the field of periodic groundwater flow. It shows more intermediate steps so that readers can follow the logic of the development, understand the mathematical context, and recognize the limitations of the approach.

Scope

Assumptions

The scope of this book is limited to time-periodic flows of homogeneous fluids through fully saturated, elastically deformable, porous media in which Darcy's law is satisfied. Within this scope, we have attempted to present the basic theory in a general form so that the results are widely applicable. To that end we make the following basic assumptions, among others:

- The relevant space domain is N-dimensional, where N can be 1, 2, or 3.
- The porous medium is macroscopically nonhomogeneous (i.e., spatially nonuniform) with respect to material hydrologic properties. That is, the medium's hydraulic conductivity and specific storage are functions of the space coordinates.
- In multidimensional cases, the porous medium generally is nonuniformly anisotropic with respect to hydraulic conductivity. That is, the problem under consideration cannot be transformed to one in which the anisotropic medium is replaced by an equivalent, macroscopically isotropic one simply by linearly transforming the space coordinates.
- The periodic component of the forcing need not be strictly periodic; it may be *almost periodic* (see Section 1.1 for a discussion of relevant terminology).

With the exception of some illustrative examples and exercises, which we clearly identify, we adhere to these assumptions throughout this book.

Organization

This book consists of the following parts:

- Part I (Introduction, Chapter 1) introduces basic terminology, proposes criteria for the classification of time-periodic forcing, and lists potential areas of application for the theory of time-periodic groundwater flow.
- Part II (Problem Definition, Chapters 2–8) describes the conceptual, mathematical basis of periodic groundwater flow within the framework of the classical boundary-value problem (BVP). It lays the foundation for subsequent parts.
- Part III (Elementary Examples, Chapters 9–13) presents examples of elementary solutions of the complex-variable form of the space BVP.
- Part IV (Essential Concepts, Chapters 14 and 15) explores some basic concepts of periodic flow, such as attenuation, delay, and local time variation of the specific discharge.
- Part V (Stationary Points, Chapters 16–18) examines the existence and nature of stationary points of the hydraulic head amplitude and phase functions and their relation to flow stagnation.
- Part VI (Wave Propagation, Chapters 19–22) presents a conceptualization of periodic groundwater flow as propagation of spatially attenuated (damped), traveling diffusion waves, i.e., harmonic, hydraulic head waves.
- Part VII (Energy Transport, Chapters 23–25) explores the transport of fluid mechanical energy by periodic groundwater flow under isothermal conditions.
- Part VIII (Conclusion, Chapter 26) briefly summarizes the results obtained in the preceding chapters, unresolved issues, and limitations of the book.

Suggested Use

The chapters are meant to be read sequentially within each part.

We believe that all readers should begin by studying Parts I and II. This material forms the core of the subject and is prerequisite for learning about the more advanced topics presented in subsequent parts. After studying Part II, readers should at least browse Parts III and IV to familiarize themselves with their scope. The reader's subsequent course of action depends on individual preference. Those who have both the interest and sufficient time should read all of the remaining parts sequentially. Those who are pressed for time or whose interests are more limited may study a combination of Parts V–VII. Lastly, all readers will want to read Part VIII.

We have embedded more than 360 exercises (see List of Exercises) in the text and included the solutions for nearly all. Each exercise is numbered and accompanied by a title that briefly summarizes its topic. The exercises are intended to reinforce the ideas presented in the text and in many cases are essential elements of the theoretical development. Exercises typically emphasize abstract reasoning, requiring symbolic manipulation rather than numerical computation. Believing that most readers will be more familiar with the material in the earlier parts than that in the later parts, we have placed more exercises in the later parts. Ideally, less advanced readers should attempt to complete every exercise they encounter. More advanced readers and those pressed for time should, at minimum, carefully read each exercise and the accompanying solution to maintain the flow of the presentation.

Usability

The electronic version of this book employs the following features for reader convenience:

- All book components (parts, chapters, sections, subsections, appendices) listed in the Contents are digitally "bookmarked." This allows the reader to navigate to the beginning of any such component, from any point in the book, via hyperlink. To activate a bookmark hyperlink, click on the corresponding label in the bookmark's navigation panel.
- In-line hyperlinks are used extensively. These include references to the following items:
  - Worldwide website uniform resource locators (URLs)
  - Specific book components: parts, chapters, sections, subsections, appendices, etc.
  - Specific content features: equations, examples, exercises, figures, notes, tables
  - Page references in the keyword index

Hyperlinked references, both in the table of contents and elsewhere, appear as blue-colored text. To activate an in-line hyperlink, click on the corresponding blue-colored text.

- Important terms appear in *italics* to draw the reader's attention.

• Examples, exercises, and notes appear with translucent shaded backgrounds colored blue violet, yellow, and green, respectively, to help the reader quickly recognize them.

Website and Contacts

Readers are invited to help improve the quality of this book by reporting errors and suggesting changes. For details, visit the companion website at hydrology.uga.edu/periodic/.

Joe S. Depner
Seattle, Washington

Todd C. Rasmussen
Athens, Georgia

NOTATION

Latin Symbols

Symbol	Description
A	Coefficient of cosine term in Fourier series; real component of complex amplitude; dimensionless coefficient in frequency response function for one-dimensional flow
$\mathbf{A}$	Coefficient matrix of linear differential operator
adj	Adjugate matrix
Arctan	Principal value of inverse-tangent function
arctan	Arctangent (inverse-tangent) function
arg	Argument of complex number
B	Coefficient of sine term in Fourier series; imaginary component of complex amplitude; dimensionless coefficient in frequency response function for one-dimensional flow
$\mathbf{b}$	Coefficient vector of linear differential operator
BC	Boundary condition
ber_ν, bei_ν	Kelvin functions, b-type, order ν ($\nu \in \mathbb{R}$)
BVP	Boundary value problem
$\mathbf{C}$	Matrix of linear transformation of space coordinates
c	Constant; propagation speed of harmonic, traveling wave
$\mathbf{c}$	Eigenvector of FRF for uniform-gradient flow in exponential media
c_h	Coefficient of hydraulic head term in boundary condition equation; propagation speed of hydraulic head constituent wave
$c^{(n)}$	Propagation speed of nth-component wave (n integer, $n > 0$)
c_q	Coefficient of specific-discharge term in boundary condition equation
const	Constant
cos	Cosine function
cosh	Hyperbolic cosine function
1D	One dimensional
2D	Two dimensional
3D	Three dimensional
d	Ordinary differentiation operator
dB	Decibel(s)
det	Determinant of matrix
D	N-dimensional space domain
$D_0(\omega_m)$	Zero set of hydraulic head constituent amplitude $M_\mathrm{h}(\mathbf{x}; \omega_m)$
$D_+(\omega_m)$	Cozero set of hydraulic head constituent amplitude $M_\mathrm{h}(\mathbf{x}; \omega_m)$
E	Fluid mechanical energy
e	Void ratio
e	Euler's number (also, Napier's constant), the mathematical constant, $e = 2.7182818\ldots$
$\mathbf{e}_m$	Unit basis vector for Cartesian coordinate x_m ($m = 1, 2, 3$)
exp	Exponentiation operator, i.e., $\exp(\,) = e^{(\,)}$
F	Complex amplitude
$F^{(n)}$	Frequency response eigenfunction
FRE	Frequency response eigenfunction
FRF	Frequency response function
g	Acceleration of gravity
G	Space domain
$G(r)$, $G(x)$	Frequency response function

Symbol	Description
G_{D}	Dimensionless frequency response function
$G^{(n)}$	Frequency response eigenfunction
$\mathbf{H}[f(\mathbf{x})]$	Hessian matrix of the function $f(\mathbf{x})$
H	Complex-valued harmonic constituent of periodic transient component of hydraulic head
h	Hydraulic head
h'	Harmonic constituent of periodic transient component of hydraulic head
$\mathbf{I}$	Identity tensor
I_0	Modified Bessel function of the first kind, order zero
IBVP	Initial boundary value problem
IC	Initial condition
Im	Imaginary part of complex number
i	Imaginary unit (i.e., $i = \sqrt{-1}$)
i_0	Modified spherical Bessel function of the first kind, order zero
$(\,)_{\mathrm{Ip}}$	In-phase component of harmonic constituent
$\mathbf{J}$	Fluid mechanical energy flux density (vector)
$\mathbf{K}$	Hydraulic conductivity (tensor) of porous medium
K	Hydraulic conductivity (scalar) of porous medium
K_0	Modified Bessel function of the second kind, order zero
$K_{-1/2}$	Modified Bessel function of the second kind, order $-1/2$
k	Wave number
$k^{(n)}$	Wave number for nth-component wave (n integer, $n > 0$)
$\mathbf{k}$	Wave vector
$\mathbf{k}^{(n)}$	Wave vector for nth-component wave (n integer, $n > 0$)
k_0	Modified spherical Bessel function of the second kind, order zero
ker_ν, kei_ν	Kelvin functions, k-type, order ν ($\nu \in \mathbb{R}$)
kg	Kilogram
L	Length (dimension)
$L^{(n)}$	Penetration depth for nth-component wave (n integer, $n > 0$)
$L[\,]$	Linear differential operator
$L_{\mathrm{h}}[\,]$	Linear differential operator
$L_{\mathrm{M}}[\,]$	Linear differential operator
$L_\theta[\,]$	Linear differential operator
LHS	Left-hand side (of equation or inequality)
l	Length of one-dimensional space domain
lim	Limit
ln	Natural (base-e) logarithm
$\log_{10}$	Base-10 logarithm
M	Mass (dimension)
M_{h}	Amplitude function for hydraulic head harmonic constituent
$M^{(n)}$	Amplitude function for nth-component wave (n integer, $n > 0$)
M_{u}	Amplitude function for source term harmonic constituent
M_ψ	Amplitude function for boundary condition harmonic constituent
M_ν	Kelvin modulus function, b-type, order ν ($\nu \in \mathbb{R}$)
m	Harmonic constituent index (m integer, $m > 0$)
m	Meter
max	Maximum value
$\max_t$	Maximum value with respect to time
ME	Mechanical energy
min	Minimum value
N	Newton

Symbol	Description
N	Dimension of space domain ($N = 1, 2, 3$)
$N(\omega_m)$	Number of component waves for the mth-harmonic constituent (N integer, $N > 0$)
N_ν	Kelvin modulus function, k-type, order ν
$\hat{\mathbf{n}}$	Unit vector outwardly perpendicular to space domain boundary
$(\ )^{(n)}$	Component wave index (n integer, $n > 0$)
O	Order of
ODE	Ordinary differential equation
p	Fluid pressure
$p_m^{(n)}$	Dimensionless exponent of FRF for power law media (m, n integer; $m, n > 0$)
PDE	Partial differential equation
$\mathbf{Q}$	Complex-valued, harmonic constituent (vector) of periodic transient component of groundwater specific discharge
Q	Quality factor
Q_0	Complex amplitude of point source
ΔQ_0	Complex amplitude of line source or plane source
$\mathbf{q}$	Specific discharge (vector)
$\mathbf{q}'$	Harmonic constituent (vector) of periodic transient component of groundwater specific discharge
$(\ )_{\mathrm{Qu}}$	Quadrature component of harmonic constituent
$\mathbb{R}$	Set of all real numbers
R	Dimensionless radial space coordinate
rad	Radian
r_{D}	Dimensionless envelope of specific-discharge constituent
Re	Real part of complex number
RHS	Right-hand side (of equation or inequality)
S	Control surface
s	Second
s	Wave travel distance
$s^{(n)}$	Travel distance for component wave (n integer, $n > 0$)
$\hat{\mathbf{s}}_{\mathrm{h}}$	Unit ray path vector of hydraulic head constituent wave
$\hat{\mathbf{s}}^{(n)}$	Unit ray path vector of nth-component wave (n integer, $n > 0$)
sech	Hyperbolic secant function
sign	Sign function
sin	Sine function
sinh	Hyperbolic sine function
S_{s}	Specific storage of porous medium
T	Travel time; fluid temperature
t	Time (dimension or independent variable)
T_m	Period of mth-harmonic constituent (m integer, $m > 0$)
$T^{(n)}$	Travel time of nth-component wave (n integer, $n > 0$)
tr	Trace of matrix
$(\ )^{\mathrm{T}}$	Matrix transpose
U	Complex-valued harmonic constituent of periodic transient component of groundwater volumetric source strength
u'	Harmonic constituent of periodic transient component of groundwater volumetric source strength
$\hat{\mathbf{u}}$	Vector of unit length pointing in the direction of $\mathbf{K}\nabla M_{\mathrm{h}}$
URL	Uniform resource locator
V	Control volume
$\mathbf{v}$	Nominal seepage velocity; phase velocity of traveling wave
$\mathbf{v}_{\mathrm{h}}$	Phase velocity of hydraulic head constituent wave
$\hat{\mathbf{v}}$	Vector of unit length pointing in the direction of $\mathbf{K}\nabla\theta_{\mathrm{h}}$

Symbol	Description
$\mathbf{v}^{(n)}$	Phase velocity of *n*th-component wave (*n* integer, $n > 0$)
$W[\ ]$	Linear differential operator for boundary condition equation
X	Dimensionless space coordinate
$\mathbf{x}$	*N*-dimensional space coordinate vector
$\hat{\mathbf{x}}$	Unit basis vector for Cartesian space coordinate x (also, $\mathbf{e}_1$)
x	Cartesian space coordinate (also, x_1)
x_m	Cartesian space coordinate ($m = 1, 2, 3$)
Y_h	Logarithm of amplitude of hydraulic head harmonic constituent
$\mathbb{Z}$	Set of all integers
z	Complex variable

Greek Symbols

Symbol	Description
α	Bulk compressibility of porous medium
α_m	Constituent parameter [see equation (5.31)] (*m* integer, $m > 0$)
β	Coefficient of order-zero term in generalized wave equation
β_w	Isothermal compressibility of liquid water
Γ	Boundary of space domain
Γ_h	Subset of the boundary for which the pertinent boundary condition is of the Dirichlet type
Γ_hq	Subset of the boundary for which the pertinent boundary condition is of the Robin type
$\Gamma_m^{(n)}$	Dimensionless eigenvalue (*m*, *n* integer; $m, n > 0$)
$\Gamma^{(n)}$	Ray path geometric divergence for *n*th-component wave (*n* integer, $n > 0$)
Γ_q	Subset of the boundary for which the pertinent boundary condition is of the Neumann type
$\Gamma_+(\omega_m)$	Boundary of subregion $D_+(\omega_m)$
γ	Coefficient of order-one term in generalized wave equation
δ_h	Spatial attenuation scale for hydraulic head harmonic constituent
δ_{mn}	Kronecker delta (sometimes referred to as *Kronecker's delta*) (*m*, *n* integer; $m, n > 0$)
$\delta^{(n)}$	Spatial attenuation scale for *n*th-component wave (*n* integer, $n > 0$)
ϵ	Fluid mechanical energy density
ϵ_{ijk}	Three-dimensional Levi-Civita symbol
ζ_h	Space derivative of hydraulic head harmonic constituent phase function in one-dimensional flow
$\zeta^{(n)}$	Space derivative of *n*th-component wave phase function in one-dimensional flow (*n* integer, $n > 0$)
η	Local, per-volume, fluid mechanical energy dissipation rate
η_m	Dimensionless parameter used for plotting dimensionless FRF solutions (*m* integer, $m > 0$)
θ_D	Phase function for dimensionless hydraulic head frequency response function
θ_h	Phase function for hydraulic head harmonic constituent
$\theta^{(n)}$	Phase function for *n*th-component wave (*n* integer, $n > 0$)
θ_u	Phase function for source term harmonic constituent
θ_ψ	Phase function for boundary condition harmonic constituent
θ_ν	Kelvin phase function, *b*-type, order ν ($\nu \in \mathbb{R}$)
κ	Hydraulic diffusivity
Λ	Logarithm of bulk compressibility
$\Lambda_m^{(1)}, \Lambda_m^{(2)}$	Dimensionless eigenvalues for *m*th constituent (*m* integer, $m > 0$)
λ	Wavelength
$\lambda_m, \lambda_m^{(1)}, \lambda_m^{(2)}$	Eigenvalues for *m*th constituent (*m* integer, $m > 0$)
$\lambda^{(n)}$	Local wavelength of *n*th-component wave (*n* integer, $n > 0$)
μ	Reciprocal length scale for spatial variation of natural logarithm of hydraulic conductivity or specific storage

Symbol	Description
$\boldsymbol{\mu}$	Gradient (vector) of natural logarithm of hydraulic conductivity or specific storage
ν	Dimensionless length scale of hydraulic conductivity or specific storage; order of Bessel function or Kelvin function
ξ	Wave phase
ξ_{h}	Wave phase for hydraulic head constituent wave
$\xi^{(n)}$	Wave phase for nth-component wave (n integer, $n > 0$)
π	The mathematical constant, $\pi = 3.14159265\ldots$
ρ_{s}	Volumetric mass density of solid phase of porous medium
ρ_{w}	Volumetric mass density of groundwater
Σ	Summation
σ	Local, per-volume rate of delivery of fluid mechanical energy by internal source(s)
σ_{e}	Effective stress
ϕ	Initial-value function for hydraulic head; porosity
ϕ_ν	Kelvin phase function, k-type, order ν ($\nu \in \mathbb{R}$)
ϕ_{e}	Effective porosity
Ψ	Complex-valued harmonic constituent of periodic transient component of boundary value function
ψ	Boundary value function
ψ'	Harmonic constituent of periodic transient component of boundary value function
ω	Angular frequency
ω_m	Angular frequency of mth-harmonic constituent (m integer, $m > 0$)

Other Symbols

Symbol	Description
$\bar{}$	(bar accent) Steady component
$\tilde{}$	(tilde accent) Transient component
$\hat{}$	(hat accent) Nonperiodic transient component
$\mathring{}$	(ring accent) Time-periodic transient component
$\breve{}$	(breve accent) Transformed variable
$*$	(asterisk) Complex conjugate
$\mathbf{0}$	Zero vector
$(\,)^{-1}$	Reciprocal; matrix inverse
$\int dv$	Integration with respect to the variable v
∞	Infinity
∂	Partial differentiation operator
∇	N-dimensional spatial gradient operator (vector)
$\nabla\cdot$	N-dimensional spatial divergence operator (scalar)
∇^2	N-dimensional Laplacian operator (scalar)
∇^4	Fourth-order, N-dimensional Laplacian operator (scalar)
$\sqrt{}$	Positive square root
$=$	Is equal to
$\approx$	Is approximately equal to
$<$	Is less than
$\ll$	Is much less than
$>$	Is greater than
$\gg$	Is much greater than
$\equiv$	Is defined as
$\in$	Is an element of the set
$\subset$	Is a subset of the set

Symbol	Description
$\cap$	Intersection of sets
$\cup$	Union of sets
$\varnothing$	Empty set
$\forall$	For all
:	Such that
:	Matrix double inner product
|	Such that
| |	Absolute value of real number; modulus of complex number; magnitude (L^2-norm) of n-dimensional vector
$\cdot$	Inner (scalar) product on finite-dimensional vector space
$\times$	Cross (vector) product
$\rightarrow$	Approaches
$\Rightarrow$	Implies
$\langle\ \rangle$	Inner product on infinite-dimensional vector space; time average
$\{\ \}$	Set

ACKNOWLEDGMENTS

The authors thank the following people:

- Gary Streile, for thoroughly reviewing draft versions of Chapters 1–3. His helpful comments led to significant improvements in those and other chapters.
- Several anonymous reviewers for their helpful suggestions.
- Hans Weinberger, for his permission to excerpt from the book *Maximum Principles in Differential Equations* [see *Protter and Weinberger*, 1999].
- Richard Koch, and many others too numerous to mention individually here, who have contributed to the development and maintenance of TeXShop (www.texshop.org)—a noncommercial TeX previewer for Mac OS X. Prepublication drafts of this document were typeset using TeXShop.

Part I
Introduction

1
Introduction

Abstractly, one can consider time-periodic groundwater flow to be the response of a physical system to a stimulus or excitation. The groundwater system consists of a subsurface porous medium and a resident pore fluid such as fresh or saline water, oil, air, or natural gas. The stimulus is some kind of time-periodic forcing, and the response is the time-periodic variation of hydraulic head and specific discharge. Thus, a possible alternative title for this book is "Introduction to the Theory of Periodically Forced Groundwater Systems."

1.1. TERMINOLOGY

To clearly articulate ideas about periodic flow, it will be useful to first clarify and standardize some basic related terminology.

For this book, we define *functions of practical interest* as those mathematical functions that are capable of representing real physical phenomena. We will assume that every function of practical interest is either periodic or aperiodic.

If $f(t)$ is a *periodic* function of time, then there exists a nonzero real number (period) T for which

$$f(t+T) = f(t) \quad \forall\, t \in \mathbb{R}.$$

Thus, periodicity is a type of global translational symmetry. Synonyms for the term *periodic* include the terms *fully periodic*, *purely periodic*, and *strictly periodic*.

We can represent every periodic function of time as the sum of one or more distinct *harmonic constituents* (also *frequency components* or *modes*), each of which is a purely sinusoidal function of time, wherein the constituent frequencies are rational multiples of one another. This description coincides with the classical Fourier series representation of a periodic function.

An *aperiodic* function (also *nonperiodic* function) is any function that is not periodic. Aperiodic functions include an important class of functions that are closely related to the periodic functions—the *almost-periodic* functions. An almost-periodic function (also, *quasiperiodic* function) is a function composed of (i.e., formed by summing) two or more harmonic constituents, at least two of which have frequencies that are not rational multiples of one another. A simple example of an almost-periodic function having two distinct harmonic constituents is

$$f(t) = 3\sin(2t) + 7\cos(3\sqrt{2}t) \quad \forall\, t \in \mathbb{R}.$$

Equivalently, an almost-periodic function is an aperiodic function that we can represent as the sum of two or more periodic functions and thus as a generalized Fourier series.

Throughout this book we generally use the term *periodic function* to represent that broad class of functions that includes both the strictly periodic functions and the almost-periodic functions. What these functions have in common is that we can represent both types by generalized Fourier series. Similarly, we use the term *nonperiodic function* to represent the class of functions that are neither strictly periodic nor almost periodic.

In the literature readers may encounter numerous terms that have meanings related to concepts of periodicity. For instance, some authors use the terms *cyclic* (or *cyclical*) and *rhythmic* as synonyms for *periodic*. In some contexts the terms *oscillating* (or *oscillatory*) and *undulating* also can have meanings similar to that of the term *periodic*; in other contexts these terms might be used to describe types of variation that are more irregular. Similarly, the term *fluctuating* is commonly used to describe variations that are less regular or less predictable than those described by the term *periodic*; in fact, the term *fluctuating* is frequently used to describe variations controlled by random processes.

Hydrodynamics of Time-Periodic Groundwater Flow: Diffusion Waves in Porous Media, Geophysical Monograph 224, First Edition. Joe S. Depner and Todd C. Rasmussen.

Other terms that may reflect temporally periodic flow include *alternating* (e.g., *alternating flow*; see *Stewart et al.* [1961]), *pulsatile* (also, *pulsating*, *pulsed*, *pulsing*), and *reciprocating*.

To distinguish between periodicity in time and periodicity in space, some authors use the terms *time periodic* (also *temporally periodic* or *steady periodic*) and *space periodic* (or *spatially periodic*). In the literature, the term *periodic media* generally refers to (porous) media that are spatially periodic with respect to material properties.

1.2. PERIODIC FORCING

Time-periodic flow occurs in a groundwater system only if the system undergoes periodic forcing. Based on system geometry, periodic forcing can be of two basic types, which can occur either alone or in combination. In *boundary forcing*, a system boundary is subject to time-periodic conditions. An example is the time variation of hydraulic head at an aquifer's seaward boundary. In *internal forcing*, an internal water source/sink is time periodic. An example is time-periodic water injection/pumping at an injection well.

We can classify periodic forcing of groundwater systems according to various additional criteria as well. In summary, criteria for the classification of periodic forcing include the following:

Origin Natural versus artificial.

Frequency High frequency (short period) versus low frequency (long period).

Geometry Boundary versus internal.

Periodicity Purely periodic versus almost periodic.

Other Hydraulic versus nonhydraulic (e.g., periodic water pressurization at a vertical boundary versus periodic mechanical loading on a horizontal upper boundary).

This book addresses both purely periodic and almost-periodic forcing. This choice is largely a matter of convenience—we can represent the time behavior of both types mathematically using general trigonometric series.

1.3. POTENTIAL AREAS OF APPLICATION

The following is a summary of potential areas of application for the theory of time-periodic groundwater flow. While this summary is broad, it is not comprehensive; there likely are additional applications, and we expect the number of applications to grow with time.

Atmospheric Pressure Natural. Diurnal, annual (seasonal), etc. Mechanical effect of barometric forcing on the upper surface of the capillary fringe, on the upper surface of a confining unit, or on the free surface within a well. Here the theory is used both to understand aquifer response to atmospheric forcing and to investigate aquifer hydrologic properties. Examples: *Furbish* [1991], *Hanson* [1980], *Hobbs and Fourie* [2000], *Merritt* [2004], *Neeper* [2001, 2002, 2003], *Rasmussen and Crawford* [1997], *Rinehart* [1972], *Ritzi et al.* [1991], *Rojstaczer* [1988], *Rojstaczer and Agnew* [1989], *Rojstaczer and Riley* [1990, 1992], *Seo* [2001], *Toll and Rasmussen* [2007], *van der Kamp and Gale* [1983], *Weeks* [1979]. Recently the theory has been used to assess the effectiveness of subsurface energy resource exploitation efforts [e.g., *Burbey and Zhang*, 2010].

Infiltration/Recharge Mass flow (hydraulic) effect at recharge boundaries. Here the theory is used to model the effects of periodic recharge cycles on groundwater systems.

- **Artificial Recharge** Seasonal and other cycles. Examples: *Latinopoulos* [1984, 1985].
- **Natural Infiltration/Recharge** Associated with seasonal cycles of precipitation and evapotranspiration. Examples: *Latinopoulos* [1984], *Maddock and Vionnet* [1998], *Rasmussen and Mote* [2007].

Plant Water Uptake/Transpiration Mass flow effect at or near the water table. Seasonal and diurnal cycles. Here the theory is used for modeling the interaction of the biosphere with groundwater systems. Examples: *Butler et al.* [2007], *Kruseman and de Ridder* [2000], *Lautz* [2008a,b].

Tides Natural. Multiple periods, from semidiurnal to monthly and longer. Both hydraulic and mechanical effects.

- **Earth Tides** The theory is used to infer aquifer and petroleum reservoir physical properties [e.g., *Bredehoeft*, 1967; *Chang and Firoozabadi*, 2000; *Cutillo and Bredehoeft*, 2011; *Hsieh et al.*, 1987, 1988; *Kümpel et al.*, 1999; *Marine*, 1975; *Morland and Donaldson*, 1984; *Narasimhan et al.* 1984; *Ritzi et al.*, 1991], to assess the effectiveness of subsurface energy resource exploitation efforts [e.g., *Burbey and Zhang*, 2010], and to understand geyser eruption timing [e.g., *Rinehart*, 1972].
- **Ocean Tides** The theory is used to infer aquifer hydraulic properties [e.g., *Carr and van der Kamp*, 1969; *Erskine*, 1991; *Ferris*, 1951; *Jacob*, 1950; *Jha et al.*, 2008; *Trefry and Bekele*, 2004; *Trefry and Johnston*, 1998], to correct nonsinusoidal hydraulic test results for tidal influence [e.g., *Chapuis et al.*, 2006; *Trefry and Johnston*, 1998], to assess groundwater fluxes in coastal aquifers [e.g., *Serfes*, 1991], and to assess groundwater–surface water fluxes in coastal environments [e.g., *Burnett et al.*, 2006; *Taniguchi*, 2002].

Sinusoidal Hydraulic Tests Artificial. Variable period(s). The theory is used to design and interpret the results of tests to infer material hydraulic properties.

Field (Pumping) Tests Hydraulic effect at face of well. The testing is conducted in situ. Examples: *Black and Kipp* [1981], *Cardiff et al.* [2013], *Hvorslev* [1951], *Mehnert et al.* [1999], *Rasmussen et al.* [2003], *Renner and Messar* [2006].

Laboratory Tests Hydraulic effect at opposite faces of material sample. The testing is conducted on material samples in the laboratory. Examples: *Adachi and Detournay* [1997], *Bernabé et al.* [2006], *Fischer* [1992], *Kranz et al.* [1990], *Rigord et al.* [1993], *Song and Renner* [2006, 2007].

Periodic Groundwater Pumping/Injection Artificial. Variable period(s). Hydraulic effect at face of well(s). Here the theory could be used to design subsurface environmental remediation systems [e.g., *Zawadzki et al.*, 2002], to design aquifer recharge systems [e.g., *Latinopoulos*, 1984, 1985], or to evaluate hydraulic connectivity in functioning geothermal well fields [e.g., *Becker and Guiltinan*, 2010; *Yano et al.*, 2000].

This list represents only a sample of the available literature.

1.4. CHAPTER SUMMARY

In this chapter we introduced basic terminology on the time behavior of periodically forced groundwater systems and proposed criteria for the classification of time-periodic forcing. We also briefly listed some potential areas of application for the theory of time-periodic groundwater flow. The list illustrates that time-periodic groundwater flow is relevant across multiple fields:

Earth sciences: geophysics, groundwater hydrology and hydrogeology, and oceanography.

Engineering fields: civil (environmental and geotechnical) and energy (geothermal and petroleum).

Part II
Problem Definition

In this part we describe the conceptual, mathematical basis of time-periodic groundwater flow within the framework of the classical boundary value problem (BVP). This part lays the foundation for subsequent parts.

Hydrodynamics of Time-Periodic Groundwater Flow: Diffusion Waves in Porous Media, Geophysical Monograph 224,
First Edition. Joe S. Depner and Todd C. Rasmussen.

2

Initial Boundary Value Problem for Hydraulic Head

Consider a confined and fully saturated groundwater system defined on a particular space domain. We wish to define an *initial boundary value problem* (IBVP) to formally describe the physical behavior of this system. The IBVP consists of the following elements:

- Space-time domain
- Governing equation
- Initial condition (IC)
- Boundary condition (BC)
- Other parameters

The following sections discuss each of these elements.

2.1. SPACE-TIME DOMAIN

We define the *time domain* of the BVP as

$$t \in \mathbb{R},\ t > 0,$$

where t is time and $\mathbb{R}$ denotes the set of real numbers.

We will assume that the *space domain* of the BVP, which we will denote D, is an open, connected set in N-dimensional space, where N can be 1, 2, or 3 depending on the particular situation. Let $\mathbf{x}$ denote the N-dimensional vector of space coordinates. Then the space coordinates of points in D satisfy

$$\mathbf{x} \in D.$$

Let Γ denote the (closed) set of points that lie on the boundary of the space domain D. We call Γ the *domain boundary*. The space coordinates of the boundary points satisfy

$$\mathbf{x} \in \Gamma.$$

We can think of D as the N-dimensional region enclosed by the boundary Γ. The domain and its boundary are mutually disjoint sets:

$$D \cap \Gamma = \varnothing.$$

We assume that we can represent the geometry of the space domain equivalently by specifying a dimensionless vector field, $\hat{\mathbf{n}}(\mathbf{x})$, that satisfies the following conditions:

$\hat{\mathbf{n}}(\mathbf{x})$ is directed outwardly perpendicular to the boundary at all points that lie on the boundary and

$\hat{\mathbf{n}}(\mathbf{x})$ vanishes at all points that do not lie on the boundary.

Thus, $\hat{\mathbf{n}}$ is N dimensional. In addition, $\hat{\mathbf{n}}$ is normalized so that it satisfies

$$|\hat{\mathbf{n}}(\mathbf{x})| = \begin{cases} 1, & \mathbf{x} \in \Gamma \\ 0, & \mathbf{x} \notin \Gamma . \end{cases}$$

This definition and notation are convenient because they allow us to compactly represent the geometry of the space domain using the parameter $\hat{\mathbf{n}}(\mathbf{x})$, which is also convenient for the expression of a flux boundary condition (see Section 2.4).

2.2. GOVERNING EQUATION

We assume that groundwater generally is homogeneous and therefore incompressible except insofar as its low but nonzero compressibility contributes to the storage capacity of porous media. Then consideration of groundwater mass conservation leads to the following continuity equation [see *Freeze and Cherry*, 1979]:

$$-\nabla \cdot \mathbf{q}^{(\mathrm{r})}(\mathbf{x}, t) + u(\mathbf{x}, t) = S_{\mathrm{S}}(\mathbf{x}) \frac{\partial h(\mathbf{x}, t)}{\partial t} \quad \forall\, t > 0,\ \mathbf{x} \in D \tag{2.1}$$

Hydrodynamics of Time-Periodic Groundwater Flow: Diffusion Waves in Porous Media, Geophysical Monograph 224,
First Edition. Joe S. Depner and Todd C. Rasmussen.

where

$\nabla\cdot$ denotes the N-dimensional divergence operator (dimensions: L^{-1});

$h(\mathbf{x}, t)$ is the *hydraulic head* as a function of position and time (dimensions: L);

$\mathbf{q}^{(r)}(\mathbf{x}, t)$ is the *specific discharge* (vector), as observed in a frame moving with the solid matrix of the deforming porous medium, as a function of position and time (dimensions: $L\,t^{-1}$);

$u(\mathbf{x}, t)$ is the volumetric strength of internal water *sources/sinks* as a function of position and time (dimensions: t^{-1}); and

$S_s(\mathbf{x})$ is the volumetric *specific storage* of the porous medium as a function of position (dimensions: L^{-1}). Section 2.5.2 discusses the specific storage.

We define the hydraulic head as

$$h(\mathbf{x}, t) \equiv x_3 + \int_{p_0}^{p} \frac{du}{\rho_w(u)g} \tag{2.2}$$

where $\rho_w(p)$ denotes water density as a function of pressure, p_0 denotes a reference pressure, and g is the acceleration of gravity. Consequently, variations in hydraulic head are related to variations in elevation and pressure as

$$dh = dx_3 + \frac{dp}{\rho_w(p)g}. \tag{2.3}$$

The constitutive relation known as *Darcy's law* describes the relationship between the specific discharge and the *hydraulic gradient* (∇h). The generalized Darcy's law is

$$\mathbf{q}^{(r)}(\mathbf{x}, t) = -\mathbf{K}(\mathbf{x})\nabla h(\mathbf{x}, t) \tag{2.4}$$

where

$\mathbf{K}(\mathbf{x})$ is the medium's hydraulic conductivity (tensor) as a function of position (dimensions: Lt^{-1}) and

∇ is the gradient operator (dimensions: L^{-1}).

See *Bear* [1972] for a discussion of the generalized Darcy's law. The hydraulic gradient is a dimensionless, vector function of position and time. Section 2.5.1 discusses the hydraulic conductivity.

Substituting the right-hand side (RHS) of (2.4) for $\mathbf{q}^{(r)}$ in (2.1) yields the equation governing the transient flow of homogeneous groundwater through a fully saturated, elastically deformable, porous medium:

$$\nabla\cdot\left[\mathbf{K}(\mathbf{x})\nabla h(\mathbf{x}, t)\right] + u(\mathbf{x}, t) = S_s(\mathbf{x})\frac{\partial h(\mathbf{x}, t)}{\partial t} \quad \forall\, t > 0,\ \mathbf{x} \in D. \tag{2.5}$$

The *groundwater flow equation* (2.5) is the *governing equation* for hydraulic head. All of the quantities in (2.5) are real valued. Equation (2.5) is linear in h and its derivatives and is a second-order, nonhomogeneous partial differential equation (PDE). Its coefficients ($\mathbf{K}$ and S_s) are spatially variable but time invariant. We can write this equation more compactly as

$$L_h\left[h(\mathbf{x}, t)\right] + u(\mathbf{x}, t) = 0 \quad \forall\, t > 0,\ \mathbf{x} \in D \tag{2.6}$$

where we define $L_h[\ ]$, the homogeneous, linear, second-order, differential operator, as

$$L_h[f] \equiv \nabla\cdot\left[\mathbf{K}(\mathbf{x})\nabla f\right] - S_s(\mathbf{x})\frac{\partial f}{\partial t}. \tag{2.7}$$

2.3. INITIAL CONDITION

We assume that the hydraulic head satisfies the following *IC equation*:

$$h(\mathbf{x}, 0) = \phi(\mathbf{x}) \quad \forall\, \mathbf{x} \in D \tag{2.8}$$

where the *initial-value function*, $\phi(\mathbf{x})$, is known (specified) at every point in the space domain. The IC equation (2.8) is linear in h, nonhomogeneous, and with constant coefficient [i.e., the implied "1" immediately preceding the h on the left-hand side (LHS) of the equation].

2.4. BOUNDARY CONDITIONS

We assume that the hydraulic head satisfies the following general *mixed BC equation*:

$$c_h(\mathbf{x})h(\mathbf{x}, t) + c_q(\mathbf{x})\left[\mathbf{K}(\mathbf{x})\nabla h(\mathbf{x}, t)\right]\cdot\hat{\mathbf{n}}(\mathbf{x}) = \psi(\mathbf{x}, t) \quad \forall\, t > 0,\ \mathbf{x} \in \Gamma \tag{2.9}$$

where

The *BC coefficient functions* $c_h(\mathbf{x})$ and $c_q(\mathbf{x})$ and the *boundary value function* $\psi(\mathbf{x}t)$ are known (specified) at every point on the boundary;

$c_h(\mathbf{x})$ has dimensions L^{-1}, $c_q(\mathbf{x})$ has dimensions $L^{-1}t$, and $\psi(\mathbf{x}, t)$ is dimensionless; and

all of the quantities in the BC equation are real valued.

The BC equation (2.9) is linear in h and its derivatives, nonhomogeneous, and with coefficients (c_h, c_q, $\mathbf{K}$, and $\hat{\mathbf{n}}$) that are spatially variable but time invariant.

The mixed BC formulation (2.9) can accommodate problems for which the hydraulic head satisfies any one of the following three particular types of boundary condition at different locations on the boundary:

- specified hydraulic head (also *Dirichlet*, *essential*, or *first type*);
- specified volumetric flux (also *Neumann*, *natural*, or *second type*); and
- impedance (also *Robin* or *third type*).

Numerous types of boundary conditions occur in the definition of groundwater flow BVPs, but these three probably are the simplest and most commonly used types. Other notable types of boundary conditions include *Cauchy* (i.e., spatially coincident and simultaneous application of Dirichlet and Neumann conditions) and *moving* boundary conditions, including nonlinear moving boundary conditions. The mixed BC formulation (2.9) is not capable of representing Cauchy or moving boundary conditions. However, it is sufficiently general to describe the boundary conditions for a wide variety of groundwater flow problems, and it is linear in the hydraulic head h and its derivatives.

We assume that we can partition the domain boundary into three mutually disjoint subsets; i.e.,

$$\Gamma = \Gamma_{\mathrm{h}} \cup \Gamma_{\mathrm{q}} \cup \Gamma_{\mathrm{hq}}$$

where

$$\begin{aligned} \Gamma_{\mathrm{h}} \cap \Gamma_{\mathrm{q}} &= \varnothing \\ \Gamma_{\mathrm{h}} \cap \Gamma_{\mathrm{hq}} &= \varnothing \\ \Gamma_{\mathrm{q}} \cap \Gamma_{\mathrm{hq}} &= \varnothing \end{aligned}$$

and where the coefficient functions satisfy

$$\mathbf{x} \in \Gamma_{\mathrm{h}} \quad \begin{cases} c_{\mathrm{h}}(\mathbf{x}) \neq 0 \\ c_{\mathrm{q}}(\mathbf{x}) = 0 \end{cases} \tag{2.10a}$$

$$\mathbf{x} \in \Gamma_{\mathrm{q}} \quad \begin{cases} c_{\mathrm{h}}(\mathbf{x}) = 0 \\ c_{\mathrm{q}}(\mathbf{x}) \neq 0 \end{cases} \tag{2.10b}$$

$$\mathbf{x} \in \Gamma_{\mathrm{hq}} \quad \begin{cases} c_{\mathrm{h}}(\mathbf{x}) \neq 0 \\ c_{\mathrm{q}}(\mathbf{x}) \neq 0. \end{cases} \tag{2.10c}$$

Notice that no boundary points exist for which both c_{h} and c_{q} vanish.

Considering each boundary subset (i.e., Γ_{h}, Γ_{q}, and Γ_{hq}) separately leads to the following explicit form for the boundary condition:

$$\begin{aligned} h(\mathbf{x}, t) &= \frac{\psi(\mathbf{x}, t)}{c_{\mathrm{h}}(\mathbf{x})} \quad \forall\, \mathbf{x} \in \Gamma_{\mathrm{h}} \\ \left[\mathbf{K}(\mathbf{x})\nabla h(\mathbf{x}, t)\right] \cdot \hat{\mathbf{n}}(\mathbf{x}) &= \frac{\psi(\mathbf{x}, t)}{c_{\mathrm{q}}(\mathbf{x})} \quad \forall\, \mathbf{x} \in \Gamma_{\mathrm{q}} \\ c_{\mathrm{h}}(\mathbf{x})h(\mathbf{x}, t) + c_{\mathrm{q}}(\mathbf{x})\left[\mathbf{K}(\mathbf{x})\nabla h(\mathbf{x}, t)\right] \cdot \hat{\mathbf{n}}(\mathbf{x}) &= \psi(\mathbf{x}, t) \quad \forall\, \mathbf{x} \in \Gamma_{\mathrm{hq}} \\ &\forall\, t > 0. \end{aligned} \tag{2.11}$$

Thus, the BC equation (2.9) simultaneously represents the following:

- Dirichlet boundary condition on subset Γ_{h};
- Neumann boundary condition on subset Γ_{q}; and
- Robin boundary condition on subset Γ_{hq}.

We can express the BC equation (2.9) even more compactly as

$$W[h(\mathbf{x}, t)] = \psi(\mathbf{x}, t) \quad \forall\, t > 0,\ \mathbf{x} \in \Gamma \tag{2.12}$$

where we define $W[\ ]$ as the homogeneous, linear, first-order, differential operator

$$W[f] \equiv c_{\mathrm{h}}(\mathbf{x})f + c_{\mathrm{q}}(\mathbf{x})\left[\mathbf{K}(\mathbf{x})\nabla f\right] \cdot \hat{\mathbf{n}}(\mathbf{x}). \tag{2.13}$$

2.5. OTHER PARAMETERS

To apply the governing, IC, and BC equations to any particular problem, all of the quantities (variables and constants) in those equations, both known and unknown, must be defined. In the case of our hydraulic head IBVP, we assume that the only unknown quantity is the hydraulic head h; we assume that all other quantities are known. In the context of the IBVP, we refer to the known quantities as *parameters*. The parameters are listed below, with the space domain on which each must be specified:

- Geometry of the space domain, represented by $\hat{\mathbf{n}}(\mathbf{x})\ \forall\, \mathbf{x} \in D \cup \Gamma$
- Material hydrologic properties of the porous medium:

Hydraulic conductivity, $\mathbf{K}(\mathbf{x})$, $\forall\, \mathbf{x} \in D$
Specific storage, $S_{\mathrm{s}}(\mathbf{x})$, $\forall\, \mathbf{x} \in D$

- Internal source/sink strength, $u(\mathbf{x}, t)$, $\forall\, t > 0,\ \mathbf{x} \in D$
- Initial-value function, $\phi(\mathbf{x})$, $\forall\, \mathbf{x} \in D$
- BC coefficient functions, $c_{\mathrm{h}}(\mathbf{x})$ and $c_{\mathrm{q}}(\mathbf{x})$, $\forall\, \mathbf{x} \in \Gamma$
- Boundary value function, $\psi(\mathbf{x}, t)$, $\forall\, t > 0,\ \mathbf{x} \in \Gamma$

For this particular IBVP all of the parameters are space or space-time fields.

2.5.1. Hydraulic Conductivity

For the remainder of this book we assume that $\mathbf{K}$ is symmetric and positive definite throughout the space domain. *Nye* [1957] and *Bear* [1972] present theoretical arguments for the symmetry of $\mathbf{K}$ based on the Onsager-Casimir reciprocal relations [e.g., see *Onsager*, 1931a, b, and *Casimir*, 1945]. Note 2.1 presents rationale for the positive definiteness of $\mathbf{K}$.

Note 2.1 Positive Definiteness of Hydraulic Conductivity Tensor

We expect that for real porous media whenever the hydraulic gradient is nonzero the component of the specific discharge in the direction of the fluid driving force (i.e., opposite the hydraulic gradient) is positive:

$$-\nabla h(\mathbf{x}, t) \cdot \mathbf{q}(\mathbf{x}, t) > 0 \quad \forall\, |\nabla h(\mathbf{x}, t)| > 0.$$

That is, the specific discharge is not perpendicular to, or opposite, $-\nabla h$. Using Darcy's law (2.4) we can express this as

$$\nabla h(\mathbf{x}, t) \cdot [\mathbf{K}(\mathbf{x})\nabla h(\mathbf{x}, t)] > 0 \quad \forall\, |\nabla h(\mathbf{x}, t)| > 0.$$

This must be satisfied for all nonzero hydraulic gradient values, so we require the following:

$$\mathbf{u} \cdot [\mathbf{K}(\mathbf{x})\mathbf{u}] > 0 \quad \forall\, |\mathbf{u}| > 0 \quad \forall\, \mathbf{x} \in D. \tag{2.14}$$

Using Einstein notation, we can also write this as

$$u_m K_{mn}(\mathbf{x}) u_n > 0 \quad \forall\, |\mathbf{u}| > 0 \quad \forall\, \mathbf{x} \in D. \tag{2.15}$$

Inequalities (2.14) and (2.15) are mathematically equivalent; both state that the hydraulic conductivity tensor **K** is positive definite throughout the space domain. △

Because **K** is real valued, symmetric, and positive definite, it has real-valued, positive eigenvalues. Hence, **K** is nonsingular. The inverse of **K** is the *reciprocal hydraulic conductivity tensor*, $\mathbf{K}^{-1}$.

2.5.2. Specific Storage

Freeze and Cherry [1979] give the specific storage as

$$S_s = \rho_w g(\alpha + \phi\beta_w)$$

where
α is the *bulk compressibility* of the porous medium;
ϕ is the porosity; and
β_w is the isothermal *compressibility of water*.

If we allow for the possibility that the bulk compressibility and/or the porosity depend on position, then the specific storage generally depends on position.

For the remainder of this book we assume that $S_s(\mathbf{x}) > 0$ throughout the space domain. The rationale for this assumption is as follows. If fluid flow occurs at any point $\mathbf{x}$ in D, then both the solid skeleton of the porous medium and the resident fluid (i.e., groundwater) make positive, albeit possibly very small, relative contributions to the specific storage at $\mathbf{x}$ via their respective bulk compressibilities.

2.6. SPECIAL CASES

2.6.1. Ideal Media

An *ideal medium* is one that is homogeneous and isotropic with respect to hydraulic conductivity and homogeneous with respect to specific storage. We represent an ideal medium mathematically as follows:

$$\mathbf{K}(\mathbf{x}) = K\mathbf{I} \quad (\nabla K = \mathbf{0},\ K > 0) \tag{2.16a}$$

$$\nabla S_s(\mathbf{x}) = \mathbf{0} \tag{2.16b}$$

$$\forall\, \mathbf{x} \in D.$$

Here **I** denotes the $N \times N$ *identity tensor*, i.e.,

$$I_{mp} \equiv \delta_{mp}$$

where δ_{mp} is the *Kronecker delta*:

$$\delta_{mp} \equiv \begin{cases} 1, & m = p \\ 0, & m \neq p \end{cases} \quad \forall\, m, p \in \mathbb{Z} \tag{2.17}$$

and both $1 \le m \le N$ and $1 \le p \le N$.

Consequently, the *hydraulic diffusivity*, which we define as

$$\kappa \equiv \frac{K}{S_s} \tag{2.18}$$

is a real, positive constant.

2.6.2. Source-Free Domains

When the space domain is free of internal sources and sinks, we have

$$u(\mathbf{x}, t) = 0 \quad \forall\, t > 0,\ \mathbf{x} \in D.$$

2.7. CHAPTER SUMMARY

In this chapter we derived a general IBVP to describe the transient flow of homogeneous groundwater through a multidimensional, nonhomogeneous, anisotropic, elastically deformable, saturated porous medium whose domain boundary is motionless. We described the major elements of the IBVP, including the space-time domain, governing equation, initial condition, boundary conditions, and other parameters. The IBVP governing and BC equations are jointly linear in the following variables:

- hydraulic head, $h(\mathbf{x}, t)$, and its derivatives;
- internal source/sink strength, $u(\mathbf{x}, t)$;
- initial-value function, $\phi(\mathbf{x})$; and
- boundary-value function, $\psi(\mathbf{x}, t)$.

3
Hydraulic Head Components and Their IBVPs

In Chapter 2 we described an IBVP for hydraulic head h in a saturated medium with stationary boundaries, and we showed that the IBVP is jointly linear with respect to the hydraulic head and its derivatives, the source/sink strength u, the initial value ϕ, and the boundary value ψ. In this chapter, we exploit this linearity to resolve the hydraulic head into multiple components based on their general behavior over time, and we describe the general IBVPs corresponding to these components.

3.1. STEADY AND TRANSIENT COMPONENTS

Assume that we can resolve the hydraulic head, source/sink strength, and boundary value, respectively, into steady and transient components as

$$h(\mathbf{x}, t) = \bar{h}(\mathbf{x}) + \tilde{h}(\mathbf{x}, t) \qquad (3.1a)$$

$$u(\mathbf{x}, t) = \bar{u}(\mathbf{x}) + \tilde{u}(\mathbf{x}, t) \qquad (3.1b)$$

$$\psi(\mathbf{x}, t) = \bar{\psi}(\mathbf{x}) + \tilde{\psi}(\mathbf{x}, t). \qquad (3.1c)$$

Here the bar (i.e., $\bar{\ }$) and tilde (i.e., $\tilde{\ }$) accents denote the steady and transient components, respectively. The steady components are those for which there is no time variation:

$$\frac{\partial \bar{h}}{\partial t} = 0 \qquad (3.2a)$$

$$\frac{\partial \bar{u}}{\partial t} = 0 \qquad (3.2b)$$

$$\frac{\partial \bar{\psi}}{\partial t} = 0 \qquad (3.2c)$$

$$\forall\, t > 0.$$

We also assume that we can resolve the initial value into steady and transient components:

$$\phi(\mathbf{x}) = \bar{h}(\mathbf{x}) + \tilde{\phi}(\mathbf{x}).$$

3.2. STEADY AND TRANSIENT HYDRAULIC HEAD BVPs

We assume that the steady and transient hydraulic head components independently satisfy corresponding BVPs, which we describe below. Appendix A.1 presents the rationale for this assumption. The steady component of the hydraulic head satisfies the following BVP:

$$L_{\mathrm{h}}\left[\bar{h}(\mathbf{x})\right] + \bar{u}(\mathbf{x}) = 0 \quad \forall\, \mathbf{x} \in D \qquad (3.3a)$$

$$W\left[\bar{h}(\mathbf{x})\right] = \bar{\psi}(\mathbf{x}) \quad \forall\, \mathbf{x} \in \Gamma \qquad (3.3b)$$

where the operators L_{h} and W were defined by (2.7) and (2.13), respectively. This BVP has no corresponding IC. The transient component of the hydraulic head satisfies the following IBVP:

$$L_{\mathrm{h}}\left[\tilde{h}(\mathbf{x}, t)\right] + \tilde{u}(\mathbf{x}, t) = 0 \quad \forall\, t > 0,\ \mathbf{x} \in D \qquad (3.4a)$$

$$\tilde{h}(\mathbf{x}, 0) = \tilde{\phi}(\mathbf{x}) \quad \forall\, \mathbf{x} \in D \qquad (3.4b)$$

$$W\left[\tilde{h}(\mathbf{x}, t)\right] = \tilde{\psi}(\mathbf{x}, t) \quad \forall\, t > 0,\ \mathbf{x} \in \Gamma. \qquad (3.4c)$$

3.3. NONPERIODIC AND PERIODIC TRANSIENT COMPONENTS

We will assume that we can resolve the transient components of the hydraulic head, source/sink strength, and

Hydrodynamics of Time-Periodic Groundwater Flow: Diffusion Waves in Porous Media, Geophysical Monograph 224,
First Edition. Joe S. Depner and Todd C. Rasmussen.

boundary value, respectively, into components that are nonperiodic and time periodic as

$$\tilde{h}(\mathbf{x}, t) = \hat{h}(\mathbf{x}, t) + \mathring{h}(\mathbf{x}, t) \tag{3.5a}$$

$$\tilde{u}(\mathbf{x}, t) = \hat{u}(\mathbf{x}, t) + \mathring{u}(\mathbf{x}, t) \tag{3.5b}$$

$$\tilde{\psi}(\mathbf{x}, t) = \hat{\psi}(\mathbf{x}, t) + \mathring{\psi}(\mathbf{x}, t). \tag{3.5c}$$

Here the hat (i.e., ^) and ring (i.e., °) accents denote the nonperiodic transient and time-periodic transient components, respectively. Each time-periodic component represents that part of the transient component that we can represent globally in time by a trigonometric series. Conversely, each nonperiodic transient component represents that part of the transient component that we cannot represent globally in time by a trigonometric series.

Although the nonperiodic transient components may be oscillatory in time, they are neither strictly periodic nor almost periodic. For example, the nonperiodic transient component might be represented by a sum of functions that decay exponentially with time, possibly at different rates. Such functions may include exponentially decaying (damped) sinusoids.

In contrast, the time-periodic transient components are either strictly periodic or almost periodic in time. Although they do not converge to fixed values as time approaches infinity, they do have finite upper and lower bounds because the alternative is not physically meaningful.

We also assume that we can resolve the initial value for the transient component into nonperiodic and time-periodic components:

$$\tilde{\phi}(\mathbf{x}) = \hat{\phi}(\mathbf{x}) + \mathring{\phi}(\mathbf{x}). \tag{3.6}$$

A *warning*. Terminology is not uniform across different scientific and engineering disciplines. In some analogous contexts, such as electrical conduction in continuous media, authors sometimes use other terms (e.g., *steady state* or *steady periodic*) as synonyms for the term *periodic transient*.

3.4. NONPERIODIC AND PERIODIC TRANSIENT HYDRAULIC HEAD BVPs

We assume that the nonperiodic and time-periodic transient hydraulic head components satisfy corresponding BVPs, which we describe below. Appendix A.2 presents the rationale for this assumption. The nonperiodic transient component satisfies the following IBVP:

$$L_{\mathrm{h}}\left[\hat{h}(\mathbf{x}, t)\right] + \hat{u}(\mathbf{x}, t) = 0 \quad \forall\, t > 0,\ \mathbf{x} \in D$$
$$\hat{h}(\mathbf{x}, 0) = \hat{\phi}(\mathbf{x}) \quad \forall\, \mathbf{x} \in D$$
$$W\left[\hat{h}(\mathbf{x}, t)\right] = \hat{\psi}(\mathbf{x}, t) \quad \forall\, t > 0,\ \mathbf{x} \in \Gamma.$$

The time-periodic transient component satisfies the following IBVP equations:

$$L_{\mathrm{h}}\left[\mathring{h}(\mathbf{x}, t)\right] + \mathring{u}(\mathbf{x}, t) = 0 \quad \forall\, t > 0,\ \mathbf{x} \in D \tag{3.7a}$$

$$\mathring{h}(\mathbf{x}, 0) = \mathring{\phi}(\mathbf{x}) \quad \forall\, \mathbf{x} \in D \tag{3.7b}$$

$$W\left[\mathring{h}(\mathbf{x}, t)\right] = \mathring{\psi}(\mathbf{x}, t) \quad \forall\, t > 0,\ \mathbf{x} \in \Gamma. \tag{3.7c}$$

3.5. CHAPTER SUMMARY

In this chapter we showed how, conceptually, one could use superposition to resolve the hydraulic head, source/sink strength, initial value, and boundary value into steady, nonperiodic transient, and time-periodic transient components. Combining those results leads to the following decompositions:

$$\begin{aligned} h(\mathbf{x}, t) &= \bar{h}(\mathbf{x}) + \hat{h}(\mathbf{x}, t) + \mathring{h}(\mathbf{x}, t) \\ u(\mathbf{x}, t) &= \bar{u}(\mathbf{x}) + \hat{u}(\mathbf{x}, t) + \mathring{u}(\mathbf{x}, t) \\ \phi(\mathbf{x}) &= \bar{h}(\mathbf{x}) + \hat{\phi}(\mathbf{x}) + \mathring{\phi}(\mathbf{x}) \\ \psi(\mathbf{x}, t) &= \bar{\psi}(\mathbf{x}) + \hat{\psi}(\mathbf{x}, t) + \mathring{\psi}(\mathbf{x}, t). \end{aligned} \tag{3.8}$$

Similarly, we can resolve the hydraulic gradient and specific-discharge vectors into their corresponding steady, nonperiodic transient, and time-periodic transient components:

$$\begin{aligned} \nabla h(\mathbf{x}, t) &= \nabla \bar{h}(\mathbf{x}) + \nabla \hat{h}(\mathbf{x}, t) + \nabla \mathring{h}(\mathbf{x}, t) \\ \mathbf{q}(\mathbf{x}, t) &= \bar{\mathbf{q}}(\mathbf{x}) + \hat{\mathbf{q}}(\mathbf{x}, t) + \mathring{\mathbf{q}}(\mathbf{x}, t) \end{aligned} \tag{3.9}$$

where

$$\bar{\mathbf{q}}(\mathbf{x}) = -\mathbf{K}(\mathbf{x}) \nabla \bar{h}(\mathbf{x}) \tag{3.10a}$$

$$\hat{\mathbf{q}}(\mathbf{x}, t) = -\mathbf{K}(\mathbf{x}) \nabla \hat{h}(\mathbf{x}, t) \tag{3.10b}$$

$$\mathring{\mathbf{q}}(\mathbf{x}, t) = -\mathbf{K}(\mathbf{x}) \nabla \mathring{h}(\mathbf{x}, t). \tag{3.10c}$$

We also defined a BVP corresponding to each of the steady, nonperiodic transient, and time-periodic transient hydraulic head components. While in principle we can represent each of the time-periodic transient components of hydraulic head, source/sink strength, and boundary value globally in time by a nontrivial trigonometric series, this is not true of the corresponding nonperiodic transient and steady components.

4
Periodic Transient Components

In Chapter 3 we saw that the periodic transient components of hydraulic head, source/sink strength, and boundary value are periodic or almost periodic in time, so that we can represent them globally in time by trigonometric series (i.e., sums of harmonic constituents). In this chapter we give expressions for the trigonometric series and the harmonic constituents. For each constituent we describe the relevant parameters (frequency, amplitude, and phase) and the relationships between them. We also express the hydraulic gradient and specific-discharge vector fields as trigonometric series and relate their harmonic constituents to those of the hydraulic head.

4.1. TRIGONOMETRIC SERIES REPRESENTATION

We assume that we can represent the time-periodic transient components of hydraulic head, source strength, and boundary value globally in time by trigonometric series as

$$\mathring{h}(\mathbf{x}, t) = \sum_{m=1}^{\infty} h'(\mathbf{x}, t; \omega_m) \quad \forall\, \mathbf{x} \in D \tag{4.1a}$$

$$\mathring{u}(\mathbf{x}, t) = \sum_{m=1}^{\infty} u'(\mathbf{x}, t; \omega_m) \quad \forall\, \mathbf{x} \in D \tag{4.1b}$$

$$\mathring{\psi}(\mathbf{x}, t) = \sum_{m=1}^{\infty} \psi'(\mathbf{x}, t; \omega_m) \quad \forall\, \mathbf{x} \in \Gamma \tag{4.1c}$$

$$\forall\, t > 0$$

where the corresponding *harmonic constituents* are

$$h'(\mathbf{x}, t; \omega_m) = A_{\mathrm{h}}(\mathbf{x}; \omega_m)\cos(\omega_m t) - B_{\mathrm{h}}(\mathbf{x}; \omega_m)\sin(\omega_m t) \tag{4.2a}$$

$$u'(\mathbf{x}, t; \omega_m) = A_{\mathrm{u}}(\mathbf{x}; \omega_m)\cos(\omega_m t) - B_{\mathrm{u}}(\mathbf{x}; \omega_m)\sin(\omega_m t) \tag{4.2b}$$

$$\psi'(\mathbf{x}, t; \omega_m) = A_{\psi}(\mathbf{x}; \omega_m)\cos(\omega_m t) - B_{\psi}(\mathbf{x}; \omega_m)\sin(\omega_m t) \tag{4.2c}$$

$$m = 1, 2, 3, \ldots.$$

We refer to (4.2) as the *rectangular form* for the harmonic constituents. We define the harmonic constituent parameters as follows:

m is the harmonic constituent *index* (integer) (dimensionless) and

ω_m is the *angular frequency* (also *circular frequency*) of the mth harmonic constituent (i.e., the *constituent frequency*) (dimensions: t^{-1}, units: rad s^{-1}).

We can express the harmonic constituents in *polar form* also, as

$$h'(\mathbf{x}, t; \omega_m) = M_{\mathrm{h}}(\mathbf{x}; \omega_m)\cos\left[\omega_m t + \theta_{\mathrm{h}}(\mathbf{x}; \omega_m)\right] \tag{4.3a}$$

$$u'(\mathbf{x}, t; \omega_m) = M_{\mathrm{u}}(\mathbf{x}; \omega_m)\cos\left[\omega_m t + \theta_{\mathrm{u}}(\mathbf{x}; \omega_m)\right] \tag{4.3b}$$

$$\psi'(\mathbf{x}, t; \omega_m) = M_{\psi}(\mathbf{x}; \omega_m)\cos\left[\omega_m t + \theta_{\psi}(\mathbf{x}; \omega_m)\right] \tag{4.3c}$$

$$m = 1, 2, 3, \ldots$$

where

$M(\mathbf{x}; \omega_m)$ is the *amplitude* (also *amplitude function*) of the mth harmonic constituent as a function of position:
$M_{\mathrm{h}}(\mathbf{x}; \omega_m)$ has dimensions L,
$M_{\mathrm{u}}(\mathbf{x}; \omega_m)$ has dimensions t^{-1}, and
$M_{\psi}(\mathbf{x}; \omega_m)$ is dimensionless;

$\theta(\mathbf{x}; \omega_m)$ is the *phase* (also *phase function*) of the mth harmonic constituent as a function of position (dimensionless, units: rad); and

$\cos\left[\omega_m t + \theta(\mathbf{x}; \omega_m)\right]$ is the corresponding *unit-amplitude constituent* of the mth harmonic constituent as a function of position (dimensionless).

Hydrodynamics of Time-Periodic Groundwater Flow: Diffusion Waves in Porous Media, Geophysical Monograph 224,
First Edition. Joe S. Depner and Todd C. Rasmussen.

The A and B coefficient functions are related to the amplitude and phase functions as

$$A_h(\mathbf{x};\omega_m) = M_h(\mathbf{x};\omega_m)\cos\theta_h(\mathbf{x};\omega_m) \tag{4.4a}$$

$$A_u(\mathbf{x};\omega_m) = M_u(\mathbf{x};\omega_m)\cos\theta_u(\mathbf{x};\omega_m) \tag{4.4b}$$

$$A_\psi(\mathbf{x};\omega_m) = M_\psi(\mathbf{x};\omega_m)\cos\theta_\psi(\mathbf{x};\omega_m) \tag{4.4c}$$

$$m = 1, 2, 3, \ldots$$

and

$$B_h(\mathbf{x};\omega_m) = M_h(\mathbf{x};\omega_m)\sin\theta_h(\mathbf{x};\omega_m) \tag{4.5a}$$

$$B_u(\mathbf{x};\omega_m) = M_u(\mathbf{x};\omega_m)\sin\theta_u(\mathbf{x};\omega_m) \tag{4.5b}$$

$$B_\psi(\mathbf{x};\omega_m) = M_\psi(\mathbf{x};\omega_m)\sin\theta_\psi(\mathbf{x};\omega_m) \tag{4.5c}$$

$$m = 1, 2, 3, \ldots.$$

Equations (4.4) and (4.5) follow from (4.2), (4.3), and trigonometric identity (B.2a).

Solving (4.4) and (4.5) for the amplitudes yields

$$M_h(\mathbf{x};\omega_m) = \sqrt{A_h^2(\mathbf{x};\omega_m) + B_h^2(\mathbf{x};\omega_m)} \tag{4.6a}$$

$$M_u(\mathbf{x};\omega_m) = \sqrt{A_u^2(\mathbf{x};\omega_m) + B_u^2(\mathbf{x};\omega_m)} \tag{4.6b}$$

$$M_\psi(\mathbf{x};\omega_m) = \sqrt{A_\psi^2(\mathbf{x};\omega_m) + B_\psi^2(\mathbf{x};\omega_m)} \tag{4.6c}$$

$$m = 1, 2, 3, \ldots.$$

Equations (4.4) and (4.5) yield

$$\tan\theta_h(\mathbf{x};\omega_m) = \frac{B_h(\mathbf{x};\omega_m)}{A_h(\mathbf{x};\omega_m)}, \quad |A_h(\mathbf{x};\omega_m)| > 0 \tag{4.7a}$$

$$\tan\theta_u(\mathbf{x};\omega_m) = \frac{B_u(\mathbf{x};\omega_m)}{A_u(\mathbf{x};\omega_m)}, \quad |A_u(\mathbf{x};\omega_m)| > 0 \tag{4.7b}$$

$$\tan\theta_\psi(\mathbf{x};\omega_m) = \frac{B_\psi(\mathbf{x};\omega_m)}{A_\psi(\mathbf{x};\omega_m)}, \quad |A_\psi(\mathbf{x};\omega_m)| > 0 \tag{4.7c}$$

$$m = 1, 2, 3, \ldots.$$

Exercise 4.1 Invariance under Time Shift: Rectangular Form Coefficients

Problem: Consider a uniform translation of the time coordinate:

$$\breve{t} \equiv t + \Delta t \tag{4.8}$$

where Δt is a constant time shift and the breve accent (i.e., "˘") over a variable indicates the variable is expressed relative to the transformed time coordinate, $\breve{t}$. Derive expressions showing how the rectangular form amplitude functions, A_h and B_h, transform under a uniform time shift. Are A_h and B_h invariant under a uniform time shift?

Solution: Solving (4.8) for t and substituting the result for t in (4.2a) lead to

$$h'(\mathbf{x},t;\omega_m) = A_h(\mathbf{x};\omega_m)\cos\left[\omega_m(\breve{t}-\Delta t)\right] - B_h(\mathbf{x};\omega_m)\sin\left[\omega_m(\breve{t}-\Delta t)\right].$$

Using trigonometric identities (B.2b) and (B.2d), we can write this as

$$\begin{aligned} h'(\mathbf{x},t;\omega_m) = {} & A_h(\mathbf{x};\omega_m)\left[\cos(\omega_m\breve{t})\cos(\omega_m\Delta t) + \sin(\omega_m\breve{t})\sin(\omega_m\Delta t)\right] \\ & - B_h(\mathbf{x};\omega_m)\left[\sin(\omega_m\breve{t})\cos(\omega_m\Delta t) - \cos(\omega_m\breve{t})\sin(\omega_m\Delta t)\right]. \end{aligned}$$

Grouping terms gives

$$\begin{aligned} h'(\mathbf{x},t;\omega_m) = {} & \left[A_h(\mathbf{x};\omega_m)\cos(\omega_m\Delta t) + B_h(\mathbf{x};\omega_m)\sin(\omega_m\Delta t)\right]\cos(\omega_m\breve{t}) \\ & + \left[A_h(\mathbf{x};\omega_m)\sin(\omega_m\Delta t) - B_h(\mathbf{x};\omega_m)\cos(\omega_m\Delta t)\right]\sin(\omega_m\breve{t}). \end{aligned}$$

Noting that $h'(\mathbf{x},t;\omega_m) = \breve{h}'(\mathbf{x},\breve{t};\omega_m)$, we can express this as

$$\breve{h}'(\mathbf{x},\breve{t};\omega_m) = \breve{A}_h(\mathbf{x};\omega_m)\cos(\omega_m\breve{t}) - \breve{B}_h(\mathbf{x};\omega_m)\sin(\omega_m\breve{t}). \tag{4.9}$$

where we define the transformed amplitudes as

$$\breve{A}_h(\mathbf{x};\omega_m) \equiv A_h(\mathbf{x};\omega_m)\cos(\omega_m\Delta t) + B_h(\mathbf{x};\omega_m)\sin(\omega_m\Delta t) \tag{4.10a}$$

$$\breve{B}_h(\mathbf{x};\omega_m) \equiv -A_h(\mathbf{x};\omega_m)\sin(\omega_m\Delta t) + B_h(\mathbf{x};\omega_m)\cos(\omega_m\Delta t). \tag{4.10b}$$

Thus, the rectangular form amplitudes A_h and B_h are *not* invariant under a time shift. □

Exercise 4.2 Invariance under Uniform Time Shift: Amplitude Function

Problem: Derive an equation that shows how the amplitude function, M_h, transforms under a uniform time shift. Is M_h invariant under a uniform time shift?

Solution: The amplitude function for the transformed harmonic constituent (4.9) is [compare equation (4.6a)]

$$\breve{M}_h(\mathbf{x};\omega_m) = \sqrt{\breve{A}_h^2(\mathbf{x};\omega_m) + \breve{B}_h^2(\mathbf{x};\omega_m)}.$$

Substituting the RHSs of (4.10) for the corresponding terms in the previous equation and using trigonometric identity (B.1) to simplify the resulting expression yield

$$\begin{aligned}\breve{M}_{\mathrm{h}}(\mathbf{x};\omega_m) &= \sqrt{A_{\mathrm{h}}^2(\mathbf{x};\omega_m) + B_{\mathrm{h}}^2(\mathbf{x};\omega_m)} \\ &= M_{\mathrm{h}}(\mathbf{x};\omega_m). \qquad (4.11)\end{aligned}$$

Thus, the amplitude function M_{h} is invariant under a uniform time shift. □

Exercise 4.3 Invariance under Uniform Time Shift: Phase Function

Problem: Derive an equation that shows how the phase function, θ_{h}, transforms under a uniform time shift. Is θ_{h} invariant under a uniform time shift?

Solution: Substituting the RHSs of (4.4a) and (4.5a) for A_{h} and B_{h}, respectively, in (4.10) yields

$$\breve{A}_{\mathrm{h}}(\mathbf{x};\omega_m) = M_{\mathrm{h}}\big[\cos\theta_{\mathrm{h}}\cos(\omega_m\Delta t) + \sin\theta_{\mathrm{h}}\sin(\omega_m\Delta t)\big]$$
$$\breve{B}_{\mathrm{h}}(\mathbf{x};\omega_m) = M_{\mathrm{h}}\big[-\cos\theta_{\mathrm{h}}\sin(\omega_m\Delta t) + \sin\theta_{\mathrm{h}}\cos(\omega_m\Delta t)\big].$$

Using trigonometric identities (B.2b) and (B.2d), respectively, we can write these as

$$\breve{A}_{\mathrm{h}}(\mathbf{x};\omega_m) = M_{\mathrm{h}}(\mathbf{x};\omega_m)\cos\left[\theta_{\mathrm{h}}(\mathbf{x};\omega_m) - \omega_m\Delta t\right] \quad (4.12a)$$
$$\breve{B}_{\mathrm{h}}(\mathbf{x};\omega_m) = M_{\mathrm{h}}(\mathbf{x};\omega_m)\sin\left[\theta_{\mathrm{h}}(\mathbf{x};\omega_m) - \omega_m\Delta t\right]. \quad (4.12b)$$

We also require

$$\breve{A}_{\mathrm{h}}(\mathbf{x};\omega_m) = \breve{M}_{\mathrm{h}}(\mathbf{x};\omega_m)\cos\breve{\theta}_{\mathrm{h}}(\mathbf{x};\omega_m) \quad (4.13a)$$
$$\breve{B}_{\mathrm{h}}(\mathbf{x};\omega_m) = \breve{M}_{\mathrm{h}}(\mathbf{x};\omega_m)\sin\breve{\theta}_{\mathrm{h}}(\mathbf{x};\omega_m). \quad (4.13b)$$

Combining (4.12), (4.13), and (4.11) gives

$$\cos\breve{\theta}_{\mathrm{h}}(\mathbf{x};\omega_m) = \cos\left[\theta_{\mathrm{h}}(\mathbf{x};\omega_m) - \omega_m\Delta t\right]$$
$$\sin\breve{\theta}_{\mathrm{h}}(\mathbf{x};\omega_m) = \sin\left[\theta_{\mathrm{h}}(\mathbf{x};\omega_m) - \omega_m\Delta t\right].$$

Simultaneous solution of these requires

$$\breve{\theta}_{\mathrm{h}}(\mathbf{x};\omega_m) = \theta_{\mathrm{h}}(\mathbf{x};\omega_m) - \omega_m\Delta t + n2\pi \text{ rad}, \quad n \in \mathbb{Z}.$$

To ensure that the phase is single valued, we must fix the value of n; we arbitrarily set $n = 0$, giving

$$\breve{\theta}_{\mathrm{h}}(\mathbf{x};\omega_m) = \theta_{\mathrm{h}}(\mathbf{x};\omega_m) - \omega_m\Delta t. \quad (4.14)$$

Thus, the phase function θ_{h} is *not* invariant under a uniform time shift. □

Exercises 4.1, 4.2, and 4.3 illustrate an important difference between the polar and rectangular forms with regard to uniform time shifts. With the polar form, the amplitude is invariant and the phase undergoes a shift of $-\omega_m\Delta t$. The effect of a uniform time shift on the rectangular form amplitudes A_{h} and B_{h} is less intuitive.

4.2. HARMONIC CONSTITUENT PARAMETERS

4.2.1. Amplitude

The harmonic constituent amplitudes are, by definition, real valued and nonnegative:

$$M_{\mathrm{h}}(\mathbf{x};\omega_m) \geq 0 \quad (4.15a)$$
$$M_{\mathrm{u}}(\mathbf{x};\omega_m) \geq 0 \quad (4.15b)$$
$$M_{\psi}(\mathbf{x};\omega_m) \geq 0 \quad (4.15c)$$
$$m = 1, 2, 3, \ldots.$$

This assumption does not reduce the generality of our results, because we can represent any harmonic constituent with a negative amplitude by an equivalent constituent with a positive amplitude and a phase shift of π rad:

$$-M\cos\alpha = M\cos(\alpha + \pi).$$

This is based on the angle-sum relation for the cosine (see trigonometric identity B.2a with $\beta = \pi$ rad).

Any time-independent terms (i.e., harmonic constituents for which $\omega_m = 0$) in the trigonometric series in (4.1) vanish,

$$M_{\mathrm{h}}(\mathbf{x};0) = 0 \quad \forall\, \mathbf{x} \in D \quad (4.16a)$$
$$M_{\mathrm{u}}(\mathbf{x};0) = 0 \quad \forall\, \mathbf{x} \in D \quad (4.16b)$$
$$M_{\psi}(\mathbf{x};0) = 0 \quad \forall\, \mathbf{x} \in \Gamma \quad (4.16c)$$

because, by definition, such terms would contribute to the corresponding steady component rather than the transient component.

4.2.2. Phase

Consider a general phase function, θ, which could represent θ_{h}, θ_{u}, or θ_{ψ}. The phase function is real valued at every point in its domain. We will assume that, with few exceptions, the phase function is spatially continuous on its domain. In addition, we expect that the phase function is bounded on any finite space domain; in contrast, on an infinite domain it may be unbounded.

The phase function is undefined at the *zero points* of the corresponding amplitude function (i.e., at those points where $A = B = 0$); hence, the domain of the phase function does not include the zero points of the amplitude function. That is, the *domain* of the phase function is discontinuous at such points. Strictly, the phase function is neither continuous nor discontinuous at these points.

Because the inverse-tangent function (arctan) is multiple valued, equations (4.7) by themselves are insufficient to unambiguously define the phase functions on their respective domains. Generally one must also impose the requirement that the phase function is spatially continuous on its domain. Then each of equations (4.7) leads to an equation of the form

$$\theta(\mathbf{x};\omega_m) = \begin{cases} \left(2n - \frac{1}{2}\right)\pi \text{ rad}, & A(\mathbf{x};\omega_m) = 0, \\ & B(\mathbf{x};\omega_m) < 0 \\ \text{Arctan}\frac{B(\mathbf{x};\omega_m)}{A(\mathbf{x};\omega_m)} + n\pi \text{ rad}, & |A(\mathbf{x};\omega_m)| > 0 \\ \left(2n + \frac{1}{2}\right)\pi \text{ rad}, & A(\mathbf{x};\omega_m) = 0, \\ & B(\mathbf{x};\omega_m) > 0 \\ \text{undefined}, & A(\mathbf{x};\omega_m) \\ & = B(\mathbf{x};\omega_m) = 0 \end{cases} \quad (4.17)$$

$$n \in \mathbb{Z}; \quad m = 1, 2, 3, \ldots$$

where the value of n is determined by the requirement that the phase function is spatially continuous and arctan denotes the principal-value inverse-tangent function, i.e.,

$$\frac{-\pi}{2} \text{ rad} < \text{Arctan}\, y < \frac{\pi}{2} \text{ rad}.$$

4.2.3. Frequency

We assume that the angular frequencies are unique:

$$|\omega_m - \omega_n| > 0 \quad \forall\, |m - n| > 0 \quad (4.18)$$

where m and n are positive integers. This assumption does not reduce the generality of our results because we can represent the sum of any two constituents having identical frequencies (and arbitrary amplitudes and phases) as a single equivalent constituent having the same frequency:

$$M_1 \cos(\omega_m t + \alpha) + M_2 \cos(\omega_m t + \beta) = M \cos(\omega_m t + \gamma)$$

where

$$M^2 = M_1^2 + M_2^2 + 2M_1 M_2 \cos(\alpha - \beta)$$

$$\tan\gamma = \frac{M_1 \sin\alpha + M_2 \sin\beta}{M_1 \cos\alpha + M_2 \cos\beta}.$$

We also assume that the angular frequencies are positive:

$$\omega_m > 0 \quad \forall m.$$

This assumption does not reduce the generality of our results, because any constituent with a negative frequency is equivalent to a constituent with a positive frequency and a corresponding phase reversal:

$$\cos(-\omega_m t + \alpha) = \cos(\omega_m t - \alpha).$$

For convenience, we assign numeric values for the indices according to the relative numeric values of the corresponding frequencies:

$$\omega_m > \omega_n \quad \forall\, m > n. \quad (4.19)$$

In summary, the frequencies satisfy

$$0 < \omega_1 < \omega_2 < \omega_3 < \cdots$$

which is obtained by combining (4.18)–(4.19). Otherwise, the values of the constituent frequencies (ω_m) are not explicitly restricted. However, for extremely high and extremely low frequencies, those assumptions on which our physical conceptual model is based, whether explicitly or implicitly, may not be valid. For example, for periods on the order of thousands or millions of years (i.e., geological time scales), geomechanical and geochemical processes can profoundly alter the physical characteristics of a groundwater system, changing both the material physical properties and the domain geometry. The conceptual model described here is inadequate to handle such long-term temporal variations.

4.3. VECTOR HARMONIC CONSTITUENTS

The time-periodic transient hydraulic head is resolved into harmonic constituents, and the hydraulic gradient and specific-discharge vectors are linear in the hydraulic head and its derivatives. Therefore, we can resolve these vectors into harmonic constituents:

$$\nabla\mathring{h}(\mathbf{x}, t) = \sum_{m=1}^{\infty} \nabla h'(\mathbf{x}, t; \omega_m) \quad (4.20)$$

$$\mathring{\mathbf{q}}(\mathbf{x}, t) = \sum_{m=1}^{\infty} \mathbf{q}'(\mathbf{x}, t; \omega_m) \quad (4.21)$$

where each specific-discharge harmonic constituent is related to the corresponding hydraulic head harmonic constituent via Darcy's law:

$$\mathbf{q}'(\mathbf{x}, t; \omega_m) = -\mathbf{K}(\mathbf{x})\nabla h'(\mathbf{x}, t; \omega_m) \quad (m = 1, 2, 3, \ldots). \quad (4.22)$$

Exercise 4.4 Vector Harmonic Constituents—Rectangular Form

Problem: Express the hydraulic gradient and specific-discharge vector harmonic constituents in terms of the rectangular form coefficients [i.e., $A_h(\mathbf{x};\omega_m)$ and $B_h(\mathbf{x};\omega_m)$].

Solution: Taking the gradient of (4.2a) gives

$$\nabla h'(\mathbf{x}, t; \omega_m) = \nabla A_h(\mathbf{x};\omega_m)\cos(\omega_m t) - \nabla B_h(\mathbf{x};\omega_m)\sin(\omega_m t) \quad (4.23)$$

$$m = 1, 2, 3, \ldots.$$

Substituting the RHS of (4.23) for $\nabla h'$ in (4.22) gives

$$\begin{aligned}\mathbf{q}'(\mathbf{x}, t; \omega_m) &= -\mathbf{K}(\mathbf{x})\nabla A_{\mathrm{h}}(\mathbf{x}; \omega_m)\cos(\omega_m t) \\ &\quad + \mathbf{K}(\mathbf{x})\nabla B_{\mathrm{h}}(\mathbf{x}; \omega_m)\sin(\omega_m t) \qquad (4.24)\\ m &= 1, 2, 3, \ldots .\end{aligned}$$

□

Exercise 4.5 Vector Harmonic Constituents—Polar Form

Problem: Express the hydraulic gradient and specific-discharge vector harmonic constituents in terms of the polar form coefficient functions [i.e., the hydraulic head amplitude and phase functions $M_{\mathrm{h}}(\mathbf{x}; \omega_m)$ and $\theta_{\mathrm{h}}(\mathbf{x}; \omega_m)$].

Solution: Taking the gradient of (4.3a) gives

$$\begin{aligned}\nabla h'(\mathbf{x}, t; \omega_m) &= \nabla M_{\mathrm{h}}(\mathbf{x}; \omega_m)\cos\left[\omega_m t + \theta_{\mathrm{h}}(\mathbf{x}; \omega_m)\right] \\ &\quad - M_{\mathrm{h}}(\mathbf{x}; \omega_m)\nabla\theta_{\mathrm{h}}(\mathbf{x}; \omega_m) \\ &\quad \times \sin\left[\omega_m t + \theta_{\mathrm{h}}(\mathbf{x}; \omega_m)\right] \\ m &= 1, 2, 3, \ldots . \qquad (4.25)\end{aligned}$$

Substituting the RHS of (4.25) for $\nabla h'$ in (4.22) yields

$$\begin{aligned}\mathbf{q}'(\mathbf{x}, t; \omega_m) &= -\mathbf{K}(\mathbf{x})\nabla M_{\mathrm{h}}(\mathbf{x}; \omega_m)\cos\left[\omega_m t + \theta_{\mathrm{h}}(\mathbf{x}; \omega_m)\right] \\ &\quad + M_{\mathrm{h}}(\mathbf{x}; \omega_m)\mathbf{K}(\mathbf{x})\nabla\theta_{\mathrm{h}}(\mathbf{x}; \omega_m) \\ &\quad \times \sin\left[\omega_m t + \theta_{\mathrm{h}}(\mathbf{x}; \omega_m)\right] \\ m &= 1, 2, 3, \ldots . \qquad (4.26)\end{aligned}$$

□

4.4. CHAPTER SUMMARY

In this chapter we expressed each of the time-periodic transient components of hydraulic head, source strength, and boundary value as a sum of time-harmonic constituents (i.e., as a trigonometric series) whose amplitudes and phases are functions of position and whose frequencies are constants. We described the harmonic constituent parameters and inferred basic relations between them. We expressed the time-periodic transient components of the hydraulic gradient and specific-discharge vectors as sums of vector harmonic constituents that are directly related to the hydraulic head harmonic constituents via the gradient operator and Darcy's law.

5
BVP for Harmonic Constituents

In this chapter we show that each term of the trigonometric series for the time-periodic transient hydraulic head satisfies a component BVP, and we derive various forms of that BVP.

5.1. RECTANGULAR FORM OF SPACE BVP

Substituting the RHSs of (4.1) for the corresponding terms in the IBVP equations (3.7) yields

$$\sum_{m=1}^{\infty}\left\{L_{\mathrm{h}}\left[h'(\mathbf{x},t;\omega_m)\right]+u'(\mathbf{x},t;\omega_m)\right\}=0\quad\forall\, t>0,\ \mathbf{x}\in D \tag{5.1a}$$

$$\sum_{m=1}^{\infty}h'(\mathbf{x},0;\omega_m)=\mathring{\phi}(\mathbf{x})\quad\forall\,\mathbf{x}\in D \tag{5.1b}$$

$$\sum_{m=1}^{\infty}\left\{W\left[h'(\mathbf{x},t;\omega_m)\right]-\psi'(\mathbf{x},t;\omega_m)\right\}=0\quad\forall\, t>0,\ \mathbf{x}\in\Gamma. \tag{5.1c}$$

The IC equation (5.1b) is time invariant, while the other two equations, (5.1a) and (5.1c), are time variant. The LHSs of (5.1a) and (5.1c) are trigonometric series.

5.1.1. Derivation

Define the *inner product* of two functions $f(t)$ and $g(t)$ as the time average of their product:

$$\left\langle f(t),g(t)\right\rangle\equiv\lim_{T\to\infty}\frac{1}{T}\int_{t}^{t+T}f(v)g(v)\,dv. \tag{5.2}$$

Forming the inner products of the LHS of the governing equation (5.1a) with the sinusoids $\cos(\omega_m t)$ and $\sin(\omega_m t)$, respectively, gives

$$\left\langle\sum_{n=1}^{\infty}\left\{L_{\mathrm{h}}\left[h'(\mathbf{x},t;\omega_n)\right]+u'(\mathbf{x},t;\omega_n)\right\},\cos(\omega_m t)\right\rangle=0 \tag{5.3a}$$

$$\left\langle\sum_{n=1}^{\infty}\left\{L_{\mathrm{h}}\left[h'(\mathbf{x},t;\omega_n)\right]+u'(\mathbf{x},t;\omega_n)\right\},\sin(\omega_m t)\right\rangle=0 \tag{5.3b}$$

$$\forall\,\mathbf{x}\in D;\quad m=1,2,3,\ldots.$$

Similarly, forming the inner products of the LHS of the BC equation (5.1c) with the sinusoids $\cos(\omega_m t)$ and $\sin(\omega_m t)$, respectively, gives

$$\left\langle\sum_{n=1}^{\infty}\left\{W\left[h'(\mathbf{x},t;\omega_n)\right]-\psi'(\mathbf{x},t;\omega_n)\right\},\cos(\omega_m t)\right\rangle=0 \tag{5.4a}$$

$$\left\langle\sum_{n=1}^{\infty}\left\{W\left[h'(\mathbf{x},t;\omega_n)\right]-\psi'(\mathbf{x},t;\omega_n)\right\},\sin(\omega_m t)\right\rangle=0 \tag{5.4b}$$

$$\forall\,\mathbf{x}\in\Gamma;\quad m=1,2,3,\ldots.$$

Changing the order of the summation and the time averaging in (5.3) and (5.4), respectively, gives

$$\sum_{n=1}^{\infty}\left\langle\left\{L_{\mathrm{h}}\left[h'(\mathbf{x},t;\omega_n)\right]+u'(\mathbf{x},t;\omega_n)\right\},\cos(\omega_m t)\right\rangle=0\quad\forall\,\mathbf{x}\in D \tag{5.5a}$$

$$\sum_{n=1}^{\infty}\left\langle\left\{L_{\mathrm{h}}\left[h'(\mathbf{x},t;\omega_n)\right]+u'(\mathbf{x},t;\omega_n)\right\},\sin(\omega_m t)\right\rangle=0\quad\forall\,\mathbf{x}\in D \tag{5.5b}$$

Hydrodynamics of Time-Periodic Groundwater Flow: Diffusion Waves in Porous Media, Geophysical Monograph 224, First Edition. Joe S. Depner and Todd C. Rasmussen.

$$\sum_{n=1}^{\infty}\Big\langle\Big\{W\big[h'(\mathbf{x},t;\omega_n)\big] - \psi'(\mathbf{x},t;\omega_n)\Big\},\cos(\omega_m t)\Big\rangle = 0 \quad \forall\,\mathbf{x}\in\Gamma \tag{5.6a}$$

$$\sum_{n=1}^{\infty}\Big\langle\Big\{W\big[h'(\mathbf{x},t;\omega_n)\big] - \psi'(\mathbf{x},t;\omega_n)\Big\},\sin(\omega_m t)\Big\rangle = 0 \quad \forall\,\mathbf{x}\in\Gamma \tag{5.6b}$$

$$m = 1,2,3,\ldots.$$

Expressing the quantities within the curved braces in terms of the rectangular forms for the corresponding harmonic constituents (4.2) gives

$$\begin{aligned}
L_{\mathrm{h}}\big[h'(\mathbf{x},t;\omega_n)\big] + u'(\mathbf{x},t;\omega_n)
&= \Big\{\nabla\cdot\big[\mathbf{K}\nabla A_{\mathrm{h}}(\mathbf{x};\omega_n)\big] + \omega_n S_{\mathrm{s}} B_{\mathrm{h}}(\mathbf{x};\omega_n) \\
&\quad + A_{\mathrm{u}}(\mathbf{x};\omega_n)\Big\}\cos(\omega_n t) \\
&\quad + \Big\{-\nabla\cdot\big[\mathbf{K}\nabla B_{\mathrm{h}}(\mathbf{x};\omega_n)\big] + \omega_n S_{\mathrm{s}} A_{\mathrm{h}}(\mathbf{x};\omega_n) \\
&\quad - B_{\mathrm{u}}(\mathbf{x};\omega_n)\Big\}\sin(\omega_n t)
\end{aligned}$$

$$\forall\,\mathbf{x}\in D,\ t>0;\quad n=1,2,3,\ldots \tag{5.7a}$$

$$\begin{aligned}
W\big[h'(\mathbf{x},t;\omega_n)\big] - \psi'(\mathbf{x},t;\omega_n)
&= \Big\{c_{\mathrm{h}} A_{\mathrm{h}}(\mathbf{x};\omega_n) + c_{\mathrm{q}}\big[\mathbf{K}\nabla A_{\mathrm{h}}(\mathbf{x};\omega_n)\big]\cdot\hat{\mathbf{n}}(\mathbf{x}) \\
&\quad - A_{\psi}(\mathbf{x};\omega_n)\Big\}\cos(\omega_n t) \\
&\quad - \Big\{c_{\mathrm{h}} B_{\mathrm{h}}(\mathbf{x};\omega_n) + c_{\mathrm{q}}\big[\mathbf{K}\nabla B_{\mathrm{h}}(\mathbf{x};\omega_n)\big]\cdot\hat{\mathbf{n}}(\mathbf{x}) \\
&\quad - B_{\psi}(\mathbf{x};\omega_n)\Big\}\sin(\omega_n t)
\end{aligned}$$

$$\forall\,\mathbf{x}\in\Gamma,\ t>0;\quad n=1,2,3,\ldots. \tag{5.7b}$$

Forming the inner products of these equations with the sinusoids $\cos(\omega_m t)$ and $\sin(\omega_m t)$ and using the orthogonality relations (B.12) yields

$$\begin{aligned}
\Big\langle\Big\{L_{\mathrm{h}}\big[h'(\mathbf{x},t;\omega_n)\big] + u'(\mathbf{x},t;\omega_n)\Big\},\cos(\omega_m t)\Big\rangle
&= \frac{\delta_{mn}}{2}\Big\{\nabla\cdot\big[\mathbf{K}\nabla A_{\mathrm{h}}(\mathbf{x};\omega_m)\big] \\
&\quad + \omega_m S_{\mathrm{s}} B_{\mathrm{h}}(\mathbf{x};\omega_m) + A_{\mathrm{u}}(\mathbf{x};\omega_m)\Big\}
\end{aligned} \tag{5.8a}$$

$$\begin{aligned}
\Big\langle\Big\{L_{\mathrm{h}}\big[h'(\mathbf{x},t;\omega_n)\big] + u'(\mathbf{x},t;\omega_n)\Big\},\sin(\omega_m t)\Big\rangle
&= \frac{\delta_{mn}}{2}\Big\{-\nabla\cdot\big[\mathbf{K}\nabla B_{\mathrm{h}}(\mathbf{x};\omega_m)\big] \\
&\quad + \omega_m S_{\mathrm{s}} A_{\mathrm{h}}(\mathbf{x};\omega_m) - B_{\mathrm{u}}(\mathbf{x};\omega_m)\Big\}
\end{aligned} \tag{5.8b}$$

$$\begin{aligned}
\Big\langle\Big\{W\big[h'(\mathbf{x},t;\omega_n)\big] - \psi'(\mathbf{x},t;\omega_n)\Big\},\cos(\omega_m t)\Big\rangle
&= \frac{\delta_{mn}}{2}\Big\{c_{\mathrm{h}} A_{\mathrm{h}}(\mathbf{x};\omega_m) + c_{\mathrm{q}}\big[\mathbf{K}\nabla A_{\mathrm{h}}(\mathbf{x};\omega_m)\big] \\
&\quad \cdot\hat{\mathbf{n}}(\mathbf{x}) - A_{\psi}(\mathbf{x};\omega_m)\Big\}
\end{aligned} \tag{5.9a}$$

$$\begin{aligned}
\Big\langle\Big\{W\big[h'(\mathbf{x},t;\omega_n)\big] - \psi'(\mathbf{x},t;\omega_n)\Big\},\sin(\omega_m t)\Big\rangle
&= -\frac{\delta_{mn}}{2}\Big\{c_{\mathrm{h}} B_{\mathrm{h}}(\mathbf{x};\omega_m) + c_{\mathrm{q}}\big[\mathbf{K}\nabla B_{\mathrm{h}}(\mathbf{x};\omega_m)\big] \\
&\quad \cdot\hat{\mathbf{n}}(\mathbf{x}) - B_{\psi}(\mathbf{x};\omega_m)\Big\}.
\end{aligned} \tag{5.9b}$$

Substituting the RHSs of (5.8) for the corresponding terms of the series in (5.5) and simplifying gives

$$\nabla\cdot\big[\mathbf{K}(\mathbf{x})\nabla A_{\mathrm{h}}(\mathbf{x};\omega_m)\big] + \omega_m S_{\mathrm{s}}(\mathbf{x}) B_{\mathrm{h}}(\mathbf{x};\omega_m) + A_{\mathrm{u}}(\mathbf{x};\omega_m) = 0 \tag{5.10a}$$

$$\nabla\cdot\big[\mathbf{K}(\mathbf{x})\nabla B_{\mathrm{h}}(\mathbf{x};\omega_m)\big] - \omega_m S_{\mathrm{s}}(\mathbf{x}) A_{\mathrm{h}}(\mathbf{x};\omega_m) + B_{\mathrm{u}}(\mathbf{x};\omega_m) = 0 \tag{5.10b}$$

$$\forall\,\mathbf{x}\in D;\quad m=1,2,3,\ldots.$$

Similarly, substituting the RHSs of (5.9) for the corresponding terms of the series in (5.6) and simplifying gives

$$c_{\mathrm{h}}(\mathbf{x}) A_{\mathrm{h}}(\mathbf{x};\omega_m) + c_{\mathrm{q}}(\mathbf{x})\big[\mathbf{K}(\mathbf{x})\nabla A_{\mathrm{h}}(\mathbf{x};\omega_m)\big]\cdot\hat{\mathbf{n}}(\mathbf{x}) = A_{\psi}(\mathbf{x};\omega_m) \tag{5.11a}$$

$$c_{\mathrm{h}}(\mathbf{x}) B_{\mathrm{h}}(\mathbf{x};\omega_m) + c_{\mathrm{q}}(\mathbf{x})\big[\mathbf{K}(\mathbf{x})\nabla B_{\mathrm{h}}(\mathbf{x};\omega_m)\big]\cdot\hat{\mathbf{n}}(\mathbf{x}) = B_{\psi}(\mathbf{x};\omega_m) \tag{5.11b}$$

$$\forall\,\mathbf{x}\in\Gamma;\quad m=1,2,3,\ldots.$$

Using (4.2a) to express the IC equation (5.1b) in terms of the constituent parameters and rearranging yields

$$\sum_{m=1}^{\infty} A_{\mathrm{h}}(\mathbf{x};\omega_m) - \mathring{\phi}(\mathbf{x}) = 0 \quad \forall\,\mathbf{x}\in D.$$

Forming the inner products of this equation with the sinusoids $\cos(\omega_m t)$ and $\sin(\omega_m t)$, respectively, produces

$$\Big\langle\sum_{n=1}^{\infty}\Big\{A_{\mathrm{h}}(\mathbf{x};\omega_n) - \mathring{\phi}(\mathbf{x})\Big\},\cos(\omega_m t)\Big\rangle = 0 \tag{5.12a}$$

$$\Big\langle\sum_{n=1}^{\infty}\Big\{A_{\mathrm{h}}(\mathbf{x};\omega_n) - \mathring{\phi}(\mathbf{x})\Big\},\sin(\omega_m t)\Big\rangle = 0 \tag{5.12b}$$

$$\forall\,\mathbf{x}\in D;\quad m=1,2,3,\ldots.$$

Changing the order of the summation and the time averaging gives

$$\sum_{n=1}^{\infty}\left\langle\left\{A_{\rm h}(\mathbf{x};\omega_n)-\mathring{\phi}(\mathbf{x})\right\},\cos(\omega_m t)\right\rangle=0 \qquad (5.13a)$$

$$\sum_{n=1}^{\infty}\left\langle\left\{A_{\rm h}(\mathbf{x};\omega_n)-\mathring{\phi}(\mathbf{x})\right\},\sin(\omega_m t)\right\rangle=0 \qquad (5.13b)$$

$$\forall\,\mathbf{x}\in D;\quad m=1,2,3,\ldots.$$

Evaluating each of these gives the identity 0 = 0—a trivial result. This effectively eliminates the IC (5.1b) from the BVP for harmonic constituents.

Regarding (5.10) and (5.11), we observe the following:

- The only independent variables are the space coordinates; time does not appear.
- For each harmonic constituent (i.e., each value of m), we have the following:

The unknowns are the dependent variables (space fields) $A_{\rm h}(\mathbf{x};\omega_m)$ and $B_{\rm h}(\mathbf{x};\omega_m)$.
Equations (5.10) are governing equations.
Equations (5.11) are BC equations.

For a given harmonic constituent (i.e., a given value of m), (5.10) and (5.11) constitute two equations in each of the two unknowns, $A_{\rm h}(\mathbf{x};\omega_m)$ and $B_{\rm h}(\mathbf{x};\omega_m)$. We assume that these equations, together with the parameters, are sufficient to fully determine the two unknowns. That is, the IC equation (5.1.1) is redundant, so we neglect it from this point forward. Therefore, (5.10) and (5.11), together with the specification of material hydrologic properties and space domain geometry, constitute a BVP for the space fields $A_{\rm h}(\mathbf{x};\omega_m)$ and $B_{\rm h}(\mathbf{x};\omega_m)$; we call it the *rectangular form of the space BVP*.

The governing equations (5.10) are coupled, second-order, nonhomogeneous, linear PDEs with space-variable coefficients. In Section 5.1.2 we derive mathematically equivalent, uncoupled forms of these equations.

Exercise 5.1 Spatially Uniform Rectangular Form Coefficients

Problem: If both $A_{\rm h}(\mathbf{x};\omega_m)$ and $B_{\rm h}(\mathbf{x};\omega_m)$ are spatially uniform, then what are their values? What constraints do these impose on the corresponding source distribution?

Solution: Substituting $\nabla A_{\rm h}(\mathbf{x};\omega_m)=\nabla B_{\rm h}(\mathbf{x};\omega_m)=\mathbf{0}$ in (5.10) and rearranging gives

$$B_{\rm h}(\mathbf{x};\omega_m)=-\frac{A_{\rm u}(\mathbf{x};\omega_m)}{\omega_m S_{\rm s}(\mathbf{x})} \qquad (5.14a)$$

$$A_{\rm h}(\mathbf{x};\omega_m)=\frac{B_{\rm u}(\mathbf{x};\omega_m)}{\omega_m S_{\rm s}(\mathbf{x})} \qquad (5.14b)$$

$$\forall\,\mathbf{x}\in D.$$

By assumption, the RHSs of (5.14) are constant. This requires the coefficient functions for the source-sink distribution to be proportional to the specific storage:

$$A_{\rm u}(\mathbf{x};\omega_m)=a(\omega_m)S_{\rm s}(\mathbf{x})$$

$$B_{\rm u}(\mathbf{x};\omega_m)=b(\omega_m)S_{\rm s}(\mathbf{x})$$

$$\forall\,\mathbf{x}\in D$$

where, for fixed ω_m, $a(\omega_m)$ and $b(\omega_m)$ are real-valued constants. □

Exercise 5.2 Spatially Uniform Rectangular Form Coefficients

Problem: Consider time-periodic flow in a source-free domain. Show that if $A_{\rm h}(\mathbf{x};\omega_m)$ is spatially uniform, then both $A_{\rm h}(\mathbf{x};\omega_m)$ and $B_{\rm h}(\mathbf{x};\omega_m)$ vanish everywhere. Similarly for $B_{\rm h}(\mathbf{x};\omega_m)$.

Solution: Use (5.10) with $A_{\rm u}(\mathbf{x};\omega_m)=B_{\rm u}(\mathbf{x};\omega_m)=0$ $\forall\mathbf{x}\in D$. □

Exercise 5.3 Spatially Uniform Rectangular Form Coefficients

Problem: Derive general expressions for both $A_{\rm h}(\mathbf{x};\omega_m)$ and $B_{\rm h}(\mathbf{x};\omega_m)$ in the special case where $B_{\rm h}(\mathbf{x};\omega_m)$ is spatially uniform.

Solution: Substituting $\nabla B_{\rm h}(\mathbf{x};\omega_m)=\mathbf{0}$ in (5.10b) and solving for $A_{\rm h}(\mathbf{x};\omega_m)$ yield

$$A_{\rm h}(\mathbf{x};\omega_m)=\frac{B_{\rm u}(\mathbf{x};\omega_m)}{\omega_m S_{\rm s}(\mathbf{x})}. \qquad (5.15a)$$

Substituting the RHS of this equation for $A_{\rm h}(\mathbf{x};\omega_m)$ in (5.10a) and solving for $B_{\rm h}(\mathbf{x};\omega_m)$ yield

$$B_{\rm h}(\mathbf{x};\omega_m)=\frac{-1}{\omega_m S_{\rm s}(\mathbf{x})}\times\left(\nabla\cdot\left\{\mathbf{K}(\mathbf{x})\nabla\left[\frac{B_{\rm u}(\mathbf{x};\omega_m)}{\omega_m S_{\rm s}(\mathbf{x})}\right]\right\}+A_{\rm u}(\mathbf{x};\omega_m)\right). \qquad (5.15b)$$

□

5.1.2. Uncoupling the Governing Equations

In this section we derive uncoupled forms of the governing equations for the rectangular form of the space BVP. We begin with the governing equations (5.10). For each value of m, these represent a pair of equations in which $A_{\rm h}(\mathbf{x};\omega_m)$ and $B_{\rm h}(\mathbf{x};\omega_m)$ are the unknowns. The two equations are said to be *coupled* because both of the unknowns appear in each equation. To maximize the readability of the derivation, we will use the following notation:

$$
\begin{aligned}
A_h &= A_h(\mathbf{x};\omega_m)\\
B_h &= B_h(\mathbf{x};\omega_m)\\
A_u &= A_u(\mathbf{x};\omega_m)\\
B_u &= B_u(\mathbf{x};\omega_m)\\
\mathbf{K} &= \mathbf{K}(\mathbf{x})\\
S_s &= S_s(\mathbf{x}).
\end{aligned} \tag{5.16}
$$

Solving (5.10a) and (5.10b) for B_h and A_h, respectively, gives

$$B_h = -(\omega_m S_s)^{-1}\left[\nabla\cdot(\mathbf{K}\nabla A_h) + A_u\right] \tag{5.17a}$$

$$A_h = (\omega_m S_s)^{-1}\left[\nabla\cdot(\mathbf{K}\nabla B_h) + B_u\right]. \tag{5.17b}$$

Taking the gradients of (5.17) yields

$$\nabla B_h = -\omega_m^{-1}\nabla\left[S_s^{-1}\nabla\cdot(\mathbf{K}\nabla A_h) + S_s^{-1}A_u\right] \tag{5.18a}$$

$$\nabla A_h = \omega_m^{-1}\nabla\left[S_s^{-1}\nabla\cdot(\mathbf{K}\nabla B_h) + S_s^{-1}B_u\right]. \tag{5.18b}$$

Premultiplying (5.18) by $\mathbf{K}$ and then taking the divergence of the products lead to

$$\nabla\cdot(\mathbf{K}\nabla B_h) = -\omega_m^{-1}\nabla\cdot\left\{\mathbf{K}\nabla\left[S_s^{-1}\nabla\cdot(\mathbf{K}\nabla A_h) + S_s^{-1}A_u\right]\right\} \tag{5.19a}$$

$$\nabla\cdot(\mathbf{K}\nabla A_h) = -\omega_m^{-1}\nabla\cdot\left\{\mathbf{K}\nabla\left[S_s^{-1}\nabla\cdot(\mathbf{K}\nabla B_h) + S_s^{-1}B_u\right]\right\}. \tag{5.19b}$$

Substituting the RHSs of (5.19) for the corresponding divergence terms in (5.10), multiplying by ω_m, and rearranging give

$$
\begin{aligned}
&\nabla\cdot\left\{\mathbf{K}\nabla\left[S_s^{-1}\nabla\cdot(\mathbf{K}\nabla B_h)\right]\right\} + \omega_m^2 S_s B_h\\
&\quad + \nabla\cdot\left[\mathbf{K}\nabla\left(S_s^{-1}B_u\right)\right] + \omega_m A_u = 0
\end{aligned} \tag{5.20a}
$$

$$
\begin{aligned}
&\nabla\cdot\left\{\mathbf{K}\nabla\left[S_s^{-1}\nabla\cdot(\mathbf{K}\nabla A_h)\right]\right\} + \omega_m^2 S_s A_h\\
&\quad + \nabla\cdot\left[\mathbf{K}\nabla\left(S_s^{-1}A_u\right)\right] + \omega_m B_u = 0.
\end{aligned} \tag{5.20b}
$$

Changing the order of the two equations and reverting to explicit notation produce

$$
\begin{aligned}
&\nabla\cdot\left\{\mathbf{K}(\mathbf{x})\nabla\left(S_s^{-1}(\mathbf{x})\nabla\cdot[\mathbf{K}(\mathbf{x})\nabla A_h(\mathbf{x};\omega_m)]\right)\right\}\\
&\quad + \omega_m^2 S_s(\mathbf{x})A_h(\mathbf{x};\omega_m)\\
&\quad + \nabla\cdot\left\{\mathbf{K}(\mathbf{x})\nabla\left(S_s^{-1}(\mathbf{x})A_u(\mathbf{x};\omega_m)\right)\right\} + \omega_m B_u(\mathbf{x};\omega_m) = 0
\end{aligned} \tag{5.21a}
$$

$$
\begin{aligned}
&\nabla\cdot\left\{\mathbf{K}(\mathbf{x})\nabla\left(S_s^{-1}(\mathbf{x})\nabla\cdot[\mathbf{K}(\mathbf{x})\nabla B_h(\mathbf{x};\omega_m)]\right)\right\}\\
&\quad + \omega_m^2 S_s(\mathbf{x})B_h(\mathbf{x};\omega_m)\\
&\quad + \nabla\cdot\left\{\mathbf{K}(\mathbf{x})\nabla\left(S_s^{-1}(\mathbf{x})B_u(\mathbf{x};\omega_m)\right)\right\}\\
&\quad + \omega_m A_u(\mathbf{x};\omega_m) = 0
\end{aligned} \tag{5.21b}
$$

$$\forall\,\mathbf{x}\in D;\quad m = 1,2,3,\ldots.$$

Equations (5.21) are uncoupled, fourth-order, nonhomogeneous, linear PDEs with space-variable coefficients. Recall that the corresponding BC equations (5.11) also are uncoupled:

$$
\begin{aligned}
&c_h(\mathbf{x})A_h(\mathbf{x};\omega_m) + c_q(\mathbf{x})\left[\mathbf{K}(\mathbf{x})\nabla A_h(\mathbf{x};\omega_m)\right]\\
&\quad\cdot\hat{\mathbf{n}}(\mathbf{x}) = A_\psi(\mathbf{x};\omega_m)
\end{aligned} \tag{5.22a}
$$

$$
\begin{aligned}
&c_h(\mathbf{x})B_h(\mathbf{x};\omega_m) + c_q(\mathbf{x})\left[\mathbf{K}(\mathbf{x})\nabla B_h(\mathbf{x};\omega_m)\right]\\
&\quad\cdot\hat{\mathbf{n}}(\mathbf{x}) = B_\psi(\mathbf{x};\omega_m)
\end{aligned} \tag{5.22b}
$$

$$\forall\,\mathbf{x}\in\Gamma;\quad m = 1,2,3,\ldots.$$

Presumably one could solve (5.21a) for $A_h(\mathbf{x};\omega_m)$ subject to the corresponding BC equation (5.22a) and then use (5.17a) to determine $B_h(\mathbf{x};\omega_m)$. Alternatively, one could solve (5.21b) for $B_h(\mathbf{x};\omega_m)$ and then use (5.17b) to determine $A_h(\mathbf{x};\omega_m)$.

5.2. SPACE-TIME BVP

5.2.1. Rectangular Form

Multiplying (5.10a) by $\cos(\omega_m t)$ and (5.10b) by $\sin(\omega_m t)$ and rearranging yield

$$
\begin{aligned}
&\nabla\cdot\left[\mathbf{K}(\mathbf{x})\nabla A_h(\mathbf{x};\omega_m)\cos(\omega_m t)\right] + \omega_m S_s(\mathbf{x})B_h(\mathbf{x};\omega_m)\\
&\quad\times\cos(\omega_m t) + A_u(\mathbf{x};\omega_m)\cos(\omega_m t) = 0
\end{aligned} \tag{5.23a}
$$

$$
\begin{aligned}
&\nabla\cdot\left[\mathbf{K}(\mathbf{x})\nabla B_h(\mathbf{x};\omega_m)\sin(\omega_m t)\right] - \omega_m S_s(\mathbf{x})A_h(\mathbf{x};\omega_m)\\
&\quad\times\sin(\omega_m t) + B_u(\mathbf{x};\omega_m)\sin(\omega_m t) = 0
\end{aligned} \tag{5.23b}
$$

$$\forall\,t>0,\ \mathbf{x}\in D;\quad m = 1,2,3,\ldots.$$

Similarly, multiplying (5.11a) by $\cos(\omega_m t)$ and (5.11b) by $\sin(\omega_m t)$ and rearranging yield

$$
\begin{aligned}
&c_h(\mathbf{x})A_h(\mathbf{x};\omega_m)\cos(\omega_m t) + c_q(\mathbf{x})\left[\mathbf{K}(\mathbf{x})\nabla A_h(\mathbf{x};\omega_m)\right.\\
&\quad\left.\times\cos(\omega_m t)\right]\cdot\hat{\mathbf{n}}(\mathbf{x}) = A_\psi(\mathbf{x};\omega_m)\cos(\omega_m t)
\end{aligned} \tag{5.24a}
$$

$$
\begin{aligned}
&c_h(\mathbf{x})B_h(\mathbf{x};\omega_m)\sin(\omega_m t) + c_q(\mathbf{x})\left[\mathbf{K}(\mathbf{x})\nabla B_h(\mathbf{x};\omega_m)\right.\\
&\quad\left.\times\sin(\omega_m t)\right]\cdot\hat{\mathbf{n}}(\mathbf{x}) = B_\psi(\mathbf{x};\omega_m)\sin(\omega_m t)
\end{aligned} \tag{5.24b}
$$

$$\forall\,t>0,\ \mathbf{x}\in\Gamma;\quad m = 1,2,3,\ldots.$$

Equations (5.23) and (5.24) are the governing and BC equations, respectively, for the *rectangular form of the space-time BVP* for the hydraulic head harmonic constituents.

5.2.2. General Form

Subtracting (5.23b) from (5.23a) and (5.24b) from (5.24a) and simplifying give

$$\nabla \cdot \left[\mathbf{K}(\mathbf{x})\nabla h'(\mathbf{x}, t; \omega_m)\right] + u'(\mathbf{x}, t; \omega_m) = S_s(\mathbf{x})\frac{\partial h'(\mathbf{x}, t; \omega_m)}{\partial t} \quad \forall\, \mathbf{x} \in D \tag{5.25a}$$

$$c_h(\mathbf{x})h'(\mathbf{x}, t; \omega_m) + c_q(\mathbf{x})\left[\mathbf{K}(\mathbf{x})\nabla h'(\mathbf{x}, t; \omega_m)\right] \cdot \hat{\mathbf{n}}(\mathbf{x}) = \psi'(\mathbf{x}, t; \omega_m) \quad \forall\, \mathbf{x} \in \Gamma \tag{5.25b}$$

$$\forall\, t > 0; \quad m = 1, 2, 3, \ldots.$$

Using the linear operator notation introduced in (2.7), we can express (5.25) compactly as

$$L_h\left[h'(\mathbf{x}, t; \omega_m)\right] = -u'(\mathbf{x}, t; \omega_m) \quad \forall\, \mathbf{x} \in D \tag{5.26a}$$

$$W\left[h'(\mathbf{x}, t; \omega_m)\right] = \psi'(\mathbf{x}, t; \omega_m) \quad \forall\, \mathbf{x} \in \Gamma \tag{5.26b}$$

$$\forall\, t > 0; \quad m = 1, 2, 3, \ldots.$$

Equations (5.26) are the governing and BC equations, respectively, for the *general form of the space-time BVP* for the hydraulic head harmonic constituents.

The goal of Exercises 5.4–5.7 is to derive the governing equations for the polar form of the space BVP.

Exercise 5.4 Part 1 of 4: Governing Equation for Polar Form Space BVP

Problem: Show that

$$\nabla h'(\mathbf{x}, t; \omega_m) = \cos\left[\omega_m t + \theta_h(\mathbf{x}; \omega_m)\right]\nabla M_h(\mathbf{x}; \omega_m) - \sin\left[\omega_m t + \theta_h(\mathbf{x}; \omega_m)\right]M_h(\mathbf{x}; \omega_m)\nabla\theta_h(\mathbf{x}; \omega_m). \tag{5.27}$$

□

Exercise 5.5 Part 2 of 4: Governing Equation for Polar Form Space BVP

Problem: Using (5.27) and the symmetry of **K**, show that

$$\begin{aligned}\nabla \cdot \left[\mathbf{K}(\mathbf{x})\nabla h'(\mathbf{x}, t; \omega_m)\right] &= \cos\left[\omega_m t + \theta_h(\mathbf{x}; \omega_m)\right] \\ &\times \Big\{\nabla \cdot \left[\mathbf{K}(\mathbf{x})\nabla M_h(\mathbf{x}; \omega_m)\right] - M_h(\mathbf{x}; \omega_m) \\ &\times \left[\mathbf{K}(\mathbf{x})\nabla\theta_h(\mathbf{x}; \omega_m)\right] \cdot \nabla\theta_h(\mathbf{x}; \omega_m)\Big\} \\ &- \sin\left[\omega_m t + \theta_h(\mathbf{x}; \omega_m)\right] \\ &\times \Big\{2\left[\mathbf{K}(\mathbf{x})\nabla M_h(\mathbf{x}; \omega_m)\right] \cdot \nabla\theta_h(\mathbf{x}; \omega_m) \\ &+ M_h(\mathbf{x}; \omega_m)\nabla \cdot \left[\mathbf{K}(\mathbf{x})\nabla\theta_h(\mathbf{x}; \omega_m)\right]\Big\}.\end{aligned}$$

□

Exercise 5.6 Part 3 of 4: Governing Equation for Polar Form Space BVP

Problem: Show that

$$S_s(\mathbf{x})\frac{\partial h'(\mathbf{x}, t; \omega_m)}{\partial t} = -\omega_m S_s(\mathbf{x})M_h(\mathbf{x}; \omega_m) \times \sin\left[\omega_m t + \theta_h(\mathbf{x}; \omega_m)\right].$$

□

Exercise 5.7 Part 4 of 4: Governing Equation for Polar Form Space BVP

Problem: Using trigonometric identity (B.2a), show that

$$\begin{aligned}u'(\mathbf{x}, t; \omega_m) &= M_h(\mathbf{x}; \omega_m) \\ &\times \Big\{\cos\left[\omega_m t + \theta_h(\mathbf{x}; \omega_m)\right]\cos\left[\theta_u(\mathbf{x}; \omega_m) - \theta_h(\mathbf{x}; \omega_m)\right] \\ &- \sin\left[\omega_m t + \theta_h(\mathbf{x}; \omega_m)\right]\sin\left[\theta_u(\mathbf{x}; \omega_m) - \theta_h(\mathbf{x}; \omega_m)\right]\Big\}.\end{aligned} \tag{5.28}$$

□

Exercise 5.8 Spatially Uniform Specific-Discharge Constituent

Problem: Consider time-periodic flow in a source-free domain. Show that the governing equation for the general form of the space-time BVP does not admit solutions for which the specific-discharge harmonic constituent, $\mathbf{q}'$, is spatially uniform.

Solution: Substituting $u' = 0$ in (5.25a) and then combining the result with (4.22) yield

$$\nabla \cdot \mathbf{q}'(\mathbf{x}, t; \omega_m) = S_s(\mathbf{x})\frac{\partial h'(\mathbf{x}, t; \omega_m)}{\partial t} \times (\forall\, t > 0,\ \mathbf{x} \in D; \quad m = 1, 2, 3, \ldots).$$

For periodic flow, the RHS is nonzero except at two times within every constituent period. If $\mathbf{q}'$ is spatially uniform, then the LHS vanishes for all times. This is a contradiction, so the equation generally is not satisfied. Thus, the general form of the space-time BVP does not admit solutions for which the specific-discharge harmonic constituent is spatially uniform. □

5.3. SPECIAL CASE: IDEAL MEDIA

Throughout this section we assume that the porous medium is ideal (see Section 2.6.1). We also assume that the domain is source free:

$$A_u(\mathbf{x}; \omega_m) = B_u(\mathbf{x}; \omega_m) = 0 \quad \forall\, \mathbf{x} \in D \quad (m = 1, 2, 3, \ldots). \tag{5.29}$$

5.3.1. Governing and BC Equations

Substituting the RHSs of (2.16) and (5.29) for the corresponding quantities in (5.21), dividing by S_s, and then simplifying yield the homogeneous, Helmholtz-type equations:

$$\nabla^4 A_h(\mathbf{x}; \omega_m) + \alpha_m^2 A_h(\mathbf{x}; \omega_m) = 0 \tag{5.30a}$$

$$\nabla^4 B_h(\mathbf{x}; \omega_m) + \alpha_m^2 B_h(\mathbf{x}; \omega_m) = 0 \tag{5.30b}$$

$$\forall\, \mathbf{x} \in D; \quad m = 1, 2, 3, \ldots$$

where we define the real-valued constants α_m as

$$\alpha_m \equiv \frac{\omega_m S_s}{K} \tag{5.31}$$

(dimensions: L^{-2}) and the *biharmonic operator* (also, *bi-Laplacian* or *fourth-order Laplacian*) as

$$\nabla^4(\,) \equiv \nabla^2[\nabla^2(\,)].$$

Substituting the RHS of (2.16a) for **K** in (5.22) and simplifying yield the BC equations

$$c_h(\mathbf{x})A_h(\mathbf{x};\omega_m) + c_q(\mathbf{x})K\nabla A_h(\mathbf{x};\omega_m)\cdot\mathbf{n}(\mathbf{x}) = A_\psi(\mathbf{x};\omega_m) \tag{5.32a}$$

$$c_h(\mathbf{x})B_h(\mathbf{x};\omega_m) + c_q(\mathbf{x})K\nabla B_h(\mathbf{x};\omega_m)\cdot\mathbf{n}(\mathbf{x}) = B_\psi(\mathbf{x};\omega_m) \tag{5.32b}$$

$$\forall\, \mathbf{x} \in \Gamma; \quad m = 1, 2, 3, \ldots.$$

5.3.2. Example: One Dimension

In the 1D case the (uncoupled) governing equations (5.30) become

$$\frac{d^4}{dx^4}A_h(x;\omega_m) + \alpha_m^2 A_h(x;\omega_m) = 0 \tag{5.33a}$$

$$\frac{d^4}{dx^4}B_h(x;\omega_m) + \alpha_m^2 B_h(x;\omega_m) = 0 \tag{5.33b}$$

$$\forall\, \mathbf{x} \in D; \quad m = 1, 2, 3, \ldots.$$

Consider the following trial solution for (5.33a):

$$f(x;\omega_m) = e^{g_m x} \tag{5.34}$$

where g_m is a complex-valued, nonzero constant (dimensions: L^{-1}).

Exercise 5.9 Trial Functions

Problem: Show that if the constant g_m is appropriately chosen, then the trial function $f(x;\omega_m)$ is a solution of (5.33a). Also, write the corresponding characteristic equation.

Solution: $g_m^4 + \alpha_m^2 = 0.$ □

Exercise 5.10 Finding Eigenvalues

Problem: Find the eigenvalues (i.e., solve the characteristic equation; see Exercise 5.9).

Solution

$$g_m^{(1)} = +\left(\frac{1+i}{\sqrt{2}}\right)\sqrt{\alpha_m}$$

$$g_m^{(2)} = +\left(\frac{1-i}{\sqrt{2}}\right)\sqrt{\alpha_m}$$

$$g_m^{(3)} = -\left(\frac{1+i}{\sqrt{2}}\right)\sqrt{\alpha_m}$$

$$g_m^{(4)} = -\left(\frac{1-i}{\sqrt{2}}\right)\sqrt{\alpha_m}$$

where i denotes the *imaginary unit* (i.e., $i \equiv \sqrt{-1}$). Note that $g_m^{(1)}$ and $g_m^{(2)}$ are complex conjugate, as are $g_m^{(3)}$ and $g_m^{(4)}$. □

We can write the corresponding eigenfunctions as

$$f^{(n)}(x;\omega_m) = \exp\left[g_m^{(n)}x\right] \quad (n = 1, 2, 3, 4).$$

The governing equation (5.33a) is linear, so we can form the solution by superposition:

$$A_h(x;\omega_m) = \sum_{n=1}^{4} c_m^{(n)} f^{(n)}(x;\omega_m) \tag{5.35}$$

where the $c_m^{(n)}$ ($n = 1, 2, 3, 4$) coefficients generally are complex valued. Similarly, the general solution of (5.33b) is

$$B_h(x;\omega_m) = \sum_{n=1}^{4} d_m^{(n)} f^{(n)}(x;\omega_m) \tag{5.36}$$

where the $d_m^{(n)}$ ($n = 1, 2, 3, 4$) coefficients generally are complex valued.

Exercise 5.11 Eigenfunctions Occur in Conjugate Pairs

Problem: Show that the eigenfunctions are conjugate pairs [i.e., that $f^{(1)}(x;\omega_m)$ and $f^{(2)}(x;\omega_m)$ are conjugate and that $f^{(3)}(x;\omega_m)$ and $f^{(4)}(x;\omega_m)$ are conjugate]. □

Exercise 5.12 Conditions on Coefficients

Problem: The solution $A_h(x;\omega_m)$ must be real valued. What restrictions does this impose on the coefficients $c_m^{(n)}$?

Solution The coefficients $c_m^{(1)}$ and $c_m^{(2)}$ must be conjugate, as must $c_m^{(3)}$ and $c_m^{(4)}$. □

Exercise 5.13 Form of Solution

Problem: Show that by imposing the conditions determined in Exercise 5.12 the solution (5.35) becomes

$$\begin{aligned} A_h(x;\omega_m) = {} & \exp\left(\sqrt{\frac{\alpha_m}{2}}x\right)\left[A_m^{(1)}\cos\left(\sqrt{\frac{\alpha_m}{2}}x\right)\right. \\ & \left. + A_m^{(2)}\sin\left(\sqrt{\frac{\alpha_m}{2}}x\right)\right] \\ & + \exp\left(-\sqrt{\frac{\alpha_m}{2}}x\right)\left[A_m^{(3)}\cos\left(\sqrt{\frac{\alpha_m}{2}}x\right)\right. \\ & \left. + A_m^{(4)}\sin\left(\sqrt{\frac{\alpha_m}{2}}x\right)\right] \end{aligned} \tag{5.37}$$

where the $A_m^{(n)}$ $(n = 1, 2, 3, 4)$ are real-valued constants. Also, express these constants in terms of the coefficients $c_m^{(n)}$.

Solution

$$A_m^{(1)} = c_m^{(1)} + c_m^{(2)} \tag{5.38a}$$

$$A_m^{(2)} = \mathrm{i}\big(c_m^{(1)} - c_m^{(2)}\big) \tag{5.38b}$$

$$A_m^{(3)} = c_m^{(3)} + c_m^{(4)} \tag{5.38c}$$

$$A_m^{(4)} = -\mathrm{i}\big(c_m^{(3)} - c_m^{(4)}\big) \tag{5.38d}$$

□

Exercise 5.14 Form of Solution

Problem: Show that one can write the solution (5.37) as

$$A_{\rm h}(x;\omega_m) = a_m^{(1)} \exp\left(\sqrt{\frac{\alpha_m}{2}}x\right) \cos\left(\sqrt{\frac{\alpha_m}{2}}x + \mu_m^{(1)}\right) + a_m^{(2)} \exp\left(-\sqrt{\frac{\alpha_m}{2}}x\right) \cos\left(\sqrt{\frac{\alpha_m}{2}}x + \mu_m^{(2)}\right) \tag{5.39}$$

where $a_m^{(1)}$, $a_m^{(2)}$, $\mu_m^{(1)}$, and $\mu_m^{(2)}$ are real-valued constants. Also, express these constants in terms of the coefficients $c_m^{(n)}$.

Solution

$$a_m^{(1)} = \sqrt{\big(A_m^{(1)}\big)^2 + \big(A_m^{(2)}\big)^2} = 2\sqrt{c_m^{(1)} c_m^{(2)}} \tag{5.40a}$$

$$a_m^{(2)} = \sqrt{\big(A_m^{(3)}\big)^2 + \big(A_m^{(4)}\big)^2} = 2\sqrt{c_m^{(3)} c_m^{(4)}} \tag{5.40b}$$

$$\mu_m^{(1)} = -\arctan\frac{A_m^{(2)}}{A_m^{(1)}} = \arctan\frac{i\big(c_m^{(2)} - c_m^{(1)}\big)}{c_m^{(1)} + c_m^{(2)}} \tag{5.40c}$$

$$\mu_m^{(2)} = -\arctan\frac{A_m^{(4)}}{A_m^{(3)}} = \arctan\frac{i\big(c_m^{(3)} - c_m^{(4)}\big)}{c_m^{(3)} + c_m^{(4)}}. \tag{5.40d}$$

□

The solution (5.39) is the sum of two modulated sinusoids whose amplitudes increase or decrease exponentially with distance.

Regardless of whether one uses (5.35), (5.37), or (5.39) to express the general solution for $A_{\rm h}(x;\omega_m)$, the expression involves four undetermined real-valued constants. Similar results hold for the solution $B_{\rm h}(x;\omega_m)$, so there are a total of eight undetermined real-valued constants for each harmonic constituent. In the following exercises we show how one can use the coupled governing equations to effectively eliminate four of these unknowns.

Exercise 5.15 Coupled Governing Equations: Special Case

Problem: Starting with the *coupled* governing equation (5.10), derive the equivalent forms for flow in a source-free domain filled with an ideal medium in N dimensions. Also write the equations for the case $N = 1$.

Solution: In N dimensions the equations take the form

$$\nabla^2 A_{\rm h}(\mathbf{x};\omega_m) + \alpha_m B_{\rm h}(\mathbf{x};\omega_m) = 0 \tag{5.41a}$$

$$\nabla^2 B_{\rm h}(\mathbf{x};\omega_m) - \alpha_m A_{\rm h}(\mathbf{x};\omega_m) = 0 \tag{5.41b}$$

$$\forall\, \mathbf{x} \in D; \quad m = 1, 2, 3, \ldots.$$

For $N = 1$ these become

$$\frac{d^2}{dx^2} A_{\rm h}(x;\omega_m) + \alpha_m B_{\rm h}(x;\omega_m) = 0 \tag{5.42a}$$

$$\frac{d^2}{dx^2} B_{\rm h}(x;\omega_m) - \alpha_m A_{\rm h}(x;\omega_m) = 0 \tag{5.42b}$$

$$\forall\, x \in D; \quad m = 1, 2, 3, \ldots.$$

□

Exercise 5.16 Relations between Coefficients

Problem: Substitute the RHS of (5.35) for $A_{\rm h}(x;\omega_m)$ and the RHS of (5.36) for $B_{\rm h}(x;\omega_m)$ in (5.42) and eliminate $f^{(n)}(x;\omega_m)$ to obtain the following relations between the $c_m^{(n)}$ and $d_m^{(n)}$ coefficients:

$$d_m^{(n)} \big(g_m^{(n)}\big)^2 = \alpha_m c_m^{(n)} \tag{5.43}$$

$$n = 1, 2, 3, 4; \quad m = 1, 2, 3, \ldots.$$

□

Each of these is easily solved for either $c_m^{(n)}$ or $d_m^{(n)}$. Thus, using the coupled governing equations we have effectively reduced the number of unknown coefficients from eight to four for each harmonic constituent.

In the 1D case the BC equations (5.11) become

$$c_{\rm h}(0) A_{\rm h}(0;\omega_m) - c_{\rm q}(0) K \left[\frac{d}{dx} A_{\rm h}(x;\omega_m)\right]\Bigg|_{x=0} = A_\psi(0;\omega_m) \tag{5.44a}$$

$$c_{\rm h}(l) A_{\rm h}(l;\omega_m) + c_{\rm q}(l) K \left[\frac{d}{dx} A_{\rm h}(x;\omega_m)\right]\Bigg|_{x=l} = A_\psi(l;\omega_m) \tag{5.44b}$$

$$c_{\rm h}(0) B_{\rm h}(0;\omega_m) - c_{\rm q}(0) K \left[\frac{d}{dx} B_{\rm h}(x;\omega_m)\right]\Bigg|_{x=0} = B_\psi(0;\omega_m) \tag{5.44c}$$

$$c_{\rm h}(l) B_{\rm h}(l;\omega_m) + c_{\rm q}(l) K \left[\frac{d}{dx} B_{\rm h}(x;\omega_m)\right]\Bigg|_{x=l} = B_\psi(l;\omega_m) \tag{5.44d}$$

$$m = 1, 2, 3, \ldots$$

where we have assumed that the two endpoints of the domain have the coordinates $x = 0$ and $x = l$. That is, each of the BC equations (5.11) actually represents two separate equations— one for each endpoint. This gives a total of four equations, which is sufficient to solve for the four remaining unknown constants.

5.3.3. From Fourth Order to Second Order

We can reduce the order of (5.30) from four to two by transforming the unknown dependent variables. We will demonstrate the procedure for (5.30a). Our goal is to transform the fourth-order equation (5.30a) into the following second-order equation:

$$\nabla^2 C_\mathrm{h}(\mathbf{x};\omega_m) - \gamma_m C_\mathrm{h}(\mathbf{x};\omega_m) = 0 \tag{5.45}$$

where C_h is a transformed dependent variable that we define as

$$C_\mathrm{h}(\mathbf{x};\omega_m) \equiv \nabla^2 A_\mathrm{h}(\mathbf{x};\omega_m) + \eta_m A_\mathrm{h}(\mathbf{x};\omega_m) \tag{5.46}$$

and γ_m and η_m are constants. To determine the values of the constants γ_m and η_m, substitute the RHS of (5.46) for C_h in (5.45). After simplifying, we obtain

$$\begin{aligned}\nabla^4 A_\mathrm{h}(\mathbf{x};\omega_m) + (\eta_m - \gamma_m)\nabla^2 A_\mathrm{h}(\mathbf{x};\omega_m)\\ - \gamma_m\eta_m A_\mathrm{h}(\mathbf{x};\omega_m) = 0.\end{aligned} \tag{5.47}$$

Equating the LHSs of (5.47) and (5.30a) yields

$$\eta_m - \gamma_m = 0 \tag{5.48a}$$

$$-\gamma_m\,\eta_m = \alpha_m^2. \tag{5.48b}$$

The solutions of (5.48) are the two eigenvalues

$$\eta_m = \gamma_m = \pm i\alpha_m.$$

Substituting the RHS for η_m in (5.46) gives the two corresponding eigenfunctions:

$$C_\mathrm{h}^+(\mathbf{x};\omega_m) = \nabla^2 A_\mathrm{h}(\mathbf{x};\omega_m) + i\alpha_m A_\mathrm{h}(\mathbf{x};\omega_m) \tag{5.49a}$$

$$C_\mathrm{h}^-(\mathbf{x};\omega_m) = \nabla^2 A_\mathrm{h}(\mathbf{x};\omega_m) - i\alpha_m A_\mathrm{h}(\mathbf{x};\omega_m). \tag{5.49b}$$

These are generally complex valued. Solving the system (5.49) for A_h yields

$$A_\mathrm{h}(\mathbf{x};\omega_m) = \frac{\mathrm{i}}{2\alpha_m}\left[C_\mathrm{h}^-(\mathbf{x};\omega_m) - C_\mathrm{h}^+(\mathbf{x};\omega_m)\right]. \tag{5.50}$$

A general procedure for solution in this case is as follows. First, solve the following Helmholtz-type equations to obtain the eigenfunctions:

$$\nabla^2 C_\mathrm{h}^+(\mathbf{x};\omega_m) - i\alpha_m C_\mathrm{h}^+(\mathbf{x};\omega_m) = 0$$

$$\nabla^2 C_\mathrm{h}^-(\mathbf{x};\omega_m) + i\alpha_m C_\mathrm{h}^-(\mathbf{x};\omega_m) = 0.$$

Then use (5.50) to determine $A_\mathrm{h}(\mathbf{x};\omega_m)$, applying the BC equation (5.11a) to determine any unknown constants. Finally, use (5.17a) to determine $B_\mathrm{h}(\mathbf{x};\omega_m)$.

5.4. CHAPTER SUMMARY

In this chapter we derived two general types of BVP for the hydraulic head harmonic constituents—a space BVP and a space-time BVP. We expressed the space BVP equations in rectangular form, and we expressed the space-time BVP equations in two forms—rectangular and general. We found that these various BVPs are mathematically equivalent.

6

Polar Form of Space BVP

In this chapter we derive the polar form of the space BVP from the corresponding rectangular form.

6.1. ASSUMPTIONS

With the exceptions noted below, throughout this chapter we assume that the following *conditions* are satisfied in the space domain (D) and on its boundary (Γ):

1. The elements of $\mathbf{K}$ are C^1 continuous (i.e., their first partial derivatives with respect to the space coordinates exist and are continuous).
2. The hydraulic head amplitude and phase functions are C^2 continuous in space (i.e., their second partial derivatives with respect to the space coordinates exist and are continuous).

Recall that the hydraulic head phase function, $\theta_h(\mathbf{x};\omega_m)$, is undefined at those points (if any) where the amplitude function, $M_h(\mathbf{x};\omega_m)$, is zero; the phase cannot be continuous at such points because they are not within its domain. In such cases we partition the space domain into two sets—one consisting of those points for which the amplitude is zero and the other those points for which it is positive:

$$D = D_o(\omega_m) \cup D_+(\omega_m) \tag{6.1}$$

where

$$D_o(\omega_m) = \{\mathbf{x} | M_h(\mathbf{x};\omega_m) = 0\} \cap D$$
$$D_+(\omega_m) = \{\mathbf{x} | M_h(\mathbf{x};\omega_m) > 0\} \cap D.$$

We refer to the sets $D_o(\omega_m)$ and $D_+(\omega_m)$ as the *zero set* and the *cozero set*, respectively, for the amplitude function $M_h(\mathbf{x};\omega_m)$. The partition generally depends on the constituent frequency because each harmonic constituent has its own space BVP with unique boundary conditions and unique source distribution. We then treat the two sets separately. For instance, we assume that *condition* 2 is valid on the set $D_+(\omega_m)$, but we make no such assumption for the set $D_o(\omega_m)$.

6.2. GOVERNING EQUATIONS

6.2.1. Derivation

First, consider those points in the domain at which both the hydraulic head and source amplitudes are positive. Substituting the RHSs of (4.4a) and (4.5a) for the corresponding divergence terms in the rectangular forms of the governing equations (5.10) gives

$$\nabla \cdot \left[\mathbf{K}\nabla A_h\right] = \nabla \cdot \left[\mathbf{K}\nabla(M_h \cos\theta_h)\right] \tag{6.2a}$$
$$\nabla \cdot \left[\mathbf{K}\nabla B_h\right] = \nabla \cdot \left[\mathbf{K}\nabla(M_h \sin\theta_h)\right]. \tag{6.2b}$$

Using vector calculus to evaluate the RHSs then yields, respectively,

$$\begin{aligned}\nabla \cdot \left[\mathbf{K}\nabla A_h\right] = &-M_h \cos\theta_h(\mathbf{K}\nabla\theta_h)\cdot\nabla\theta_h \\ &- 2\sin\theta_h(\mathbf{K}\nabla M_h)\cdot\nabla\theta_h \\ &- M_h \sin\theta_h \nabla\cdot(\mathbf{K}\nabla\theta_h) \\ &+ \cos\theta_h\nabla\cdot(\mathbf{K}\nabla M_h)\end{aligned} \tag{6.3a}$$

$$\begin{aligned}\nabla \cdot \left[\mathbf{K}\nabla B_h\right] = &-M_h \sin\theta_h(\mathbf{K}\nabla\theta_h)\cdot\nabla\theta_h \\ &+ 2\cos\theta_h(\mathbf{K}\nabla M_h)\cdot\nabla\theta_h \\ &+ M_h \cos\theta_h \nabla\cdot(\mathbf{K}\nabla\theta_h) \\ &+ \sin\theta_h\nabla\cdot(\mathbf{K}\nabla M_h),\end{aligned} \tag{6.3b}$$

where we have exploited the symmetry of $\mathbf{K}$. Substituting the expressions on the RHSs of (4.4), (4.5), and (6.3) for

Hydrodynamics of Time-Periodic Groundwater Flow: Diffusion Waves in Porous Media, Geophysical Monograph 224, First Edition. Joe S. Depner and Todd C. Rasmussen.

the corresponding terms in the rectangular forms of the governing equations (5.10) and simplifying give

$$\begin{aligned}&- M_h \cos\theta_h (\mathbf{K}\nabla\theta_h)\cdot\nabla\theta_h - 2\sin\theta_h(\mathbf{K}\nabla M_h)\cdot\nabla\theta_h\\&- M_h \sin\theta_h \nabla\cdot(\mathbf{K}\nabla\theta_h) + \cos\theta_h\nabla\cdot(\mathbf{K}\nabla M_h)\\&+ \omega_m S_s M_h \sin\theta_h + M_u\cos\theta_u = 0\end{aligned} \tag{6.4a}$$

$$\begin{aligned}&- M_h \sin\theta_h (\mathbf{K}\nabla\theta_h)\cdot\nabla\theta_h + 2\cos\theta_h(\mathbf{K}\nabla M_h)\cdot\nabla\theta_h\\&+ M_h \cos\theta_h \nabla\cdot(\mathbf{K}\nabla\theta_h) + \sin\theta_h\nabla\cdot(\mathbf{K}\nabla M_h)\\&- \omega_m S_s M_h \cos\theta_h + M_u\sin\theta_u = 0.\end{aligned} \tag{6.4b}$$

Multiplying (6.4a) and (6.4b) by $-\sin\theta_h$ and $\cos\theta_h$, respectively, adding the resulting equations, and simplifying give

$$\begin{aligned}&M_h\nabla\cdot(\mathbf{K}\nabla\theta_h) + 2(\mathbf{K}\nabla M_h)\cdot\nabla\theta_h - \omega_m S_s M_h\\&\quad + M_u\sin(\theta_u - \theta_h) = 0\end{aligned} \tag{6.5}$$

where we have used both the symmetry of **K** and trigonometric identity (B.2d). Multiplying by M_h and combining the first two terms on the LHS into a single divergence term lead to the compact form

$$\nabla\cdot(M_h^2\mathbf{K}\nabla\theta_h) - \omega_m S_s M_h^2 + M_u M_h\sin(\theta_u - \theta_h) = 0. \tag{6.6}$$

Multiplying (6.4a) and (6.4b) by $\cos\theta_h$ and $\sin\theta_h$, respectively, adding the resulting equations, and simpify-ing give

$$\nabla\cdot(\mathbf{K}\nabla M_h) - M_h(\mathbf{K}\nabla\theta_h)\cdot\nabla\theta_h + M_u\cos(\theta_u - \theta_h) = 0 \tag{6.7}$$

where we have used trigonometric identity (B.2b). Since **K** is positive definite and symmetric, it possesses a positive-definite, symmetric square root, $\mathbf{K}^{1/2}$. Thus, we can write (6.7) as

$$\nabla\cdot(\mathbf{K}\nabla M_h) - M_h|\mathbf{K}^{1/2}\nabla\theta_h|^2 + M_u\cos(\theta_u - \theta_h) = 0. \tag{6.8}$$

Exercise 6.1 Algebra of Inner Products

Problem: Show that

$$(\mathbf{Ku})\cdot\mathbf{v} = (\mathbf{K}^{1/2}\mathbf{u})\cdot(\mathbf{K}^{1/2}\mathbf{v})$$

where

$$\mathbf{K}^{1/2}\mathbf{K}^{1/2} = \mathbf{K}$$

and **u** and **v** are column N-vectors.

Solution

$$\begin{aligned}(\mathbf{Ku})\cdot\mathbf{v} &= (\mathbf{Ku})^T\mathbf{v}\\&= \mathbf{u}^T\mathbf{K}^T\mathbf{v}\\&= \mathbf{u}^T\mathbf{K}\mathbf{v} (\text{because } \mathbf{K} \text{ is symmetric})\\&= \mathbf{u}^T\mathbf{K}^{1/2}\mathbf{K}^{1/2}\mathbf{v}\\&\quad\times(\mathbf{K}^{1/2} \text{exists because } \mathbf{K} \text{ is positive}\\&\quad\text{definite and symmetric})\\&= \mathbf{u}^T(\mathbf{K}^{1/2})^T\mathbf{K}^{1/2}\mathbf{v}\\&= (\mathbf{K}^{1/2}\mathbf{u})^T(\mathbf{K}^{1/2}\mathbf{v})\\&= (\mathbf{K}^{1/2}\mathbf{u})\cdot(\mathbf{K}^{1/2}\mathbf{v}).\end{aligned}$$

□

In summary, (6.6) and (6.8) are written below using explicit notation:

$$\begin{aligned}&\nabla\cdot\left[\mathbf{K}(\mathbf{x})\nabla M_h(\mathbf{x};\omega_m)\right] - M_h(\mathbf{x};\omega_m)\left|\mathbf{K}^{1/2}(\mathbf{x})\nabla\theta_h(\mathbf{x};\omega_m)\right|^2\\&\quad + M_u(\mathbf{x};\omega_m)\cos\left[\theta_u(\mathbf{x};\omega_m) - \theta_h(\mathbf{x};\omega_m)\right] = 0\end{aligned} \tag{6.9a}$$

$$\begin{aligned}&\nabla\cdot\left[M_h^2(\mathbf{x};\omega_m)\mathbf{K}(\mathbf{x})\nabla\theta_h(\mathbf{x};\omega_m)\right] - \omega_m S_s(\mathbf{x})M_h^2(\mathbf{x};\omega_m)\\&\quad + M_u(\mathbf{x};\omega_m)M_h(\mathbf{x};\omega_m)\sin\left[\theta_u(\mathbf{x};\omega_m) - \theta_h(\mathbf{x};\omega_m)\right] = 0\end{aligned} \tag{6.9b}$$

$$M_u(\mathbf{x};\omega_m) > 0; \quad \forall\, \mathbf{x}\in D_+(\omega_m); \quad m = 1, 2, 3, \ldots.$$

Next, consider those points in the space domain at which the hydraulic head amplitude is positive and the source amplitude is zero. Recall that at such points the source phase function is undefined. The relevant equations are obtained as follows. First, substitute zero for A_u and B_u in the governing equations (5.10). Then follow a procedure similar to that described above to yield

$$\begin{aligned}&\nabla\cdot\left[\mathbf{K}(\mathbf{x})\nabla M_h(\mathbf{x};\omega_m)\right]\\&\quad - M_h(\mathbf{x};\omega_m)\left|\mathbf{K}^{1/2}(\mathbf{x})\nabla\theta_h(\mathbf{x};\omega_m)\right|^2 = 0\end{aligned} \tag{6.10a}$$

$$\nabla\cdot\left[M_h^2(\mathbf{x};\omega_m)\mathbf{K}(\mathbf{x})\nabla\theta_h(\mathbf{x};\omega_m)\right] - \omega_m S_s(\mathbf{x})M_h^2(\mathbf{x};\omega_m) = 0 \tag{6.10b}$$

$$M_u(\mathbf{x};\omega_m) > 0; \quad \forall\, \mathbf{x}\in D_+(\omega_m); \quad m = 1, 2, 3, \ldots.$$

For the sake of convenience, we implicitly represent both sets of governing equations (6.9) and (6.10) compactly as

$$\begin{aligned}&\nabla\cdot\left[\mathbf{K}(\mathbf{x})\nabla M_h(\mathbf{x};\omega_m)\right] - M_h(\mathbf{x};\omega_m)\left|\mathbf{K}^{1/2}(\mathbf{x})\nabla\theta_h(\mathbf{x};\omega_m)\right|^2\\&\quad + M_u(\mathbf{x};\omega_m)\cos\left[\theta_u(\mathbf{x};\omega_m) - \theta_h(\mathbf{x};\omega_m)\right] = 0\end{aligned} \tag{6.11a}$$

$$\begin{aligned}&\nabla\cdot\left[M_h^2(\mathbf{x};\omega_m)\mathbf{K}(\mathbf{x})\nabla\theta_h(\mathbf{x};\omega_m)\right] - \omega_m S_s(\mathbf{x})M_h^2(\mathbf{x};\omega_m)\\&\quad + M_u(\mathbf{x};\omega_m)M_h(\mathbf{x};\omega_m)\sin\left[\theta_u(\mathbf{x};\omega_m) - \theta_h(\mathbf{x};\omega_m)\right] = 0\end{aligned} \tag{6.11b}$$

$$\forall\, \mathbf{x}\in D_+(\omega_m); \quad m = 1, 2, 3, \ldots$$

where it is understood that at those points (if any) where the source amplitude is zero the terms in (6.11) that are multiplied by the source amplitude vanish, even though technically the source phase function is undefined at such points. This convention is purely an abbreviation; it allows us to write one set of governing equations rather than two sets.

The governing equations (6.11) are coupled, second-order, *nonlinear* PDEs with space-variable coefficients. How, or even if, we can derive equivalent uncoupled forms is not immediately evident.

Exercise 6.2 Governing Equation for Polar Form Space BVP

Problem: Beginning with governing equation (5.25a) for the general form of the space-time BVP and using the results obtained in Exercises 5.4–5.7, derive the governing equation (6.11) for the polar form of the space BVP. □

6.2.2. Properties

This section explores the basic properties of the polar form of the governing equations (6.11) through a series of exercises, none of which is especially difficult. We recommend that readers complete, or at least study, all of the exercises in this section.

Exercise 6.3 Properties of Governing Equations: Linearity

Problem: Show that the governing equations (6.11) are linear in the hydraulic head amplitude.

Solution: Suppose that $M_{\mathrm{h}}^{(1)}(\mathbf{x};\omega_m)$ and $M_{\mathrm{h}}^{(2)}(\mathbf{x};\omega_m)$ are two distinct solutions of the governing equations. Define the linear combination

$$M_{\mathrm{h}}(\mathbf{x};\omega_m) \equiv c^{(1)}M_{\mathrm{h}}^{(1)}(\mathbf{x};\omega_m) + c^{(2)}M_{\mathrm{h}}^{(2)}(\mathbf{x};\omega_m) \quad (6.12)$$

where $c^{(1)}$ and $c^{(2)}$ are real-valued constants and $c^{(1)} + c^{(2)} = 1$. Substituting the RHS of (6.12) for the hydraulic head amplitude in (6.11) and using the fact that both $M_{\mathrm{h}}^{(1)}(\mathbf{x};\omega_m)$ and $M_{\mathrm{h}}^{(2)}(\mathbf{x};\omega_m)$ are solutions of the governing equations twice yield the identity $0 = 0$. Thus, the governing equations are linear in the hydraulic head amplitude. □

Exercise 6.4 Properties of Governing Equations: Linearity

Problem: Are the governing equations (6.11) linear or nonlinear in the hydraulic head phase? Why? A formal proof is not necessary; instead, very briefly discuss the terms in each equation that support your conclusion. Does your conclusion change if the medium is free of sources/sinks?

Solution: The terms $[\mathbf{K}\nabla\theta_{\mathrm{h}}] \cdot \nabla\theta_{\mathrm{h}}$ and $\cos[\theta_{\mathrm{u}} - \theta_{\mathrm{h}}]$ of (6.11a) and the term $\sin[\theta_{\mathrm{u}} - \theta_{\mathrm{h}}]$ of (6.11b) are nonlinear in the phase. In the case where the medium is source free, the source terms vanish, but these account for only two of the three nonlinear terms. Thus, while (6.11b) reduces to a form that is linear in the phase, (6.11a) reduces to a form that is nonlinear in the phase. Therefore, the equations are nonlinear in the hydraulic head phase, even in the case where the domain is source free. □

Exercise 6.5 Properties of Governing Equations: Uniform Phase Shift

Problem: Consider a frequency-dependent, spatially uniform phase shift:

$$\breve{\theta}(\mathbf{x};\omega_m) = \theta(\mathbf{x};\omega_m) - \theta_0(\omega_m) \quad (6.13)$$

where $\nabla\theta_0(\omega_m) = \mathbf{0}$. Show that the shifted hydraulic head phase $\breve{\theta}_{\mathrm{h}}(\mathbf{x};\omega_m)$ satisfies the governing equations (6.11).

Solution: The transformed (shifted) phase variables are

$$\breve{\theta}_{\mathrm{h}}(\mathbf{x};\omega_m) \equiv \theta_{\mathrm{h}}(\mathbf{x};\omega_m) - \theta_0(\omega_m) \quad (6.14a)$$

$$\breve{\theta}_{\mathrm{u}}(\mathbf{x};\omega_m) \equiv \theta_{\mathrm{u}}(\mathbf{x};\omega_m) - \theta_0(\omega_m). \quad (6.14b)$$

Solving (6.14) for the corresponding untransformed phases, substituting the results for the corresponding terms in the governing equations (6.11) and simplifying yield

$$\nabla \cdot \left[\mathbf{K}(\mathbf{x})\nabla M_{\mathrm{h}}(\mathbf{x};\omega_m)\right] - M_{\mathrm{h}}(\mathbf{x};\omega_m)\left|\mathbf{K}^{1/2}(\mathbf{x})\nabla\breve{\theta}_{\mathrm{h}}(\mathbf{x};\omega_m)\right|^2 + M_{\mathrm{u}}(\mathbf{x};\omega_m)\cos\left[\breve{\theta}_{\mathrm{u}}(\mathbf{x};\omega_m) - \breve{\theta}_{\mathrm{h}}(\mathbf{x};\omega_m)\right] = 0 \quad (6.15a)$$

$$\nabla \cdot \left[M_{\mathrm{h}}^2(\mathbf{x};\omega_m)\mathbf{K}(\mathbf{x})\nabla\breve{\theta}_{\mathrm{h}}(\mathbf{x};\omega_m)\right] - \omega_m S_{\mathrm{s}}(\mathbf{x})M_{\mathrm{h}}^2(\mathbf{x};\omega_m) + M_{\mathrm{u}}(\mathbf{x};\omega_m)M_{\mathrm{h}}(\mathbf{x};\omega_m)\sin\left[\breve{\theta}_{\mathrm{u}}(\mathbf{x};\omega_m) - \breve{\theta}_{\mathrm{h}}(\mathbf{x};\omega_m)\right] = 0 \quad (6.15b)$$

$\forall\, \mathbf{x} \in D_{+}(\omega_m); \quad m = 1, 2, 3, \ldots.$

These are structurally identical to the original, untransformed equations (6.11). □

Exercise 6.6 Properties of Governing Equations: Amplitude Shift

Problem: In Exercise 6.5 it was shown that if the phase is transformed by a frequency-dependent, spatially uniform shift, then the transformed phase satisfies the governing equations (6.11), even though the equations are nonlinear in the phase. Is this true of the amplitude as well? Why or why not?

Solution: No, it is not true of the amplitude. Although the governing equations are nonlinear in the phase, the phase appears in them only as derivatives or differences (i.e., $\theta_{\mathrm{u}} - \theta_{\mathrm{h}}$), both of which are invariant to uniform translations. In contrast, the hydraulic head amplitude appears not only in gradient and divergence terms but also in undifferentiated and undifferenced in some terms. □

Exercise 6.7 Properties of Governing Equations: Amplitude Scaling

Problem: Consider a frequency-dependent, spatially uniform amplitude scaling:

$$\check{M}(\mathbf{x};\omega_m) \equiv \frac{M(\mathbf{x};\omega_m)}{M_0(\omega_m)} \tag{6.16}$$

where $M_0(\omega_m) > 0$ and $\nabla M_0(\omega_m) = \mathbf{0}$. Show that the scaled hydraulic head amplitude $\check{M}_\mathrm{h}(\mathbf{x};\omega_m)$ satisfies the governing equations (6.11).

Solution: The transformed (scaled) amplitude variables are

$$\check{M}_\mathrm{h}(\mathbf{x};\omega_m) = \frac{M_\mathrm{h}(\mathbf{x};\omega_m)}{M_0(\omega_m)} \tag{6.17a}$$

$$\check{M}_\mathrm{u}(\mathbf{x};\omega_m) = \frac{M_\mathrm{u}(\mathbf{x};\omega_m)}{M_0(\omega_m)}. \tag{6.17b}$$

Solving (6.17) for the corresponding untransformed amplitudes, substituting the results for the corresponding terms in the governing equations (6.11), and simplifying yield

$$\begin{aligned} &\nabla\cdot\left[\mathbf{K}(\mathbf{x})\nabla\check{M}_\mathrm{h}(\mathbf{x};\omega_m)\right] \\ &\quad - \check{M}_\mathrm{h}(\mathbf{x};\omega_m)\left|\mathbf{K}^{1/2}(\mathbf{x})\nabla\theta_\mathrm{h}(\mathbf{x};\omega_m)\right|^2 \\ &\quad + \check{M}_\mathrm{u}(\mathbf{x};\omega_m)\cos\left[\theta_\mathrm{u}(\mathbf{x};\omega_m)-\theta_\mathrm{h}(\mathbf{x};\omega_m)\right] = 0 \end{aligned} \tag{6.18a}$$

$$\begin{aligned} &\nabla\cdot\left[\check{M}_\mathrm{h}^2(\mathbf{x};\omega_m)\mathbf{K}(\mathbf{x})\nabla\theta_\mathrm{h}(\mathbf{x};\omega_m)\right] - \omega_m S_\mathrm{s}(\mathbf{x})\check{M}_\mathrm{h}^2(\mathbf{x};\omega_m) \\ &\quad + \check{M}_\mathrm{u}(\mathbf{x};\omega_m)\check{M}_\mathrm{h}(\mathbf{x};\omega_m) \\ &\quad \sin\left[\theta_\mathrm{u}(\mathbf{x};\omega_m)-\theta_\mathrm{h}(\mathbf{x};\omega_m)\right] = 0 \end{aligned} \tag{6.18b}$$

$$\forall\, \mathbf{x} \in D_+(\omega_m); \quad m = 1,2,3,\ldots.$$

These are structurally identical to the original, untransformed equations (6.11). □

Exercise 6.8 Properties of Governing Equations: Phase Scaling

Problem: In Exercise 6.7 it was shown that if the amplitude is transformed by a frequency-dependent, spatially uniform scaling, then the transformed amplitude satisfies the governing equations (6.11). Is this true of the phase as well? Why or why not?

Solution: No, it is not true of the phase, because although the governing equations are linear in the amplitudes M_h and M_u, they are nonlinear in the phases θ_h and θ_u. □

6.2.3. Special Case: One-Dimensional (1D) Flow

In this section we uncouple the polar form governing equations (6.11) in the special case where the domain is source free and the flow is 1D. Throughout this section we will assume that the material hydrologic properties vary with one space coordinate (x) and that both the hydraulic gradient and the specific discharge are parallel to the x axis.

Exercise 6.9 Polar Form Governing Equations in One Dimension

Problem: Consider time-periodic flow in a source-free, 1D, generally nonhomogeneous, porous medium. In this case we have

$$M_\mathrm{h}(\mathbf{x};\omega_m) = M_\mathrm{h}(x;\omega_m) \tag{6.19a}$$

$$M_\mathrm{u}(\mathbf{x};\omega_m) = 0 \tag{6.19b}$$

$$\theta_\mathrm{h}(\mathbf{x};\omega_m) = \theta_\mathrm{h}(x;\omega_m) \tag{6.19c}$$

and

$$\mathbf{K}(\mathbf{x}) = K(x)\mathbf{I} \tag{6.20a}$$

$$S_\mathrm{s}(\mathbf{x}) = S_\mathrm{s}(x) \tag{6.20b}$$

where $K(x) > 0$ and $S_\mathrm{s}(x) > 0$. Write the 1D equivalents of the polar form governing equations (6.11) for this case.

Solution: Substituting the RHSs of (6.19) and (6.20) for the corresponding terms in the governing equations (6.11) gives

$$\begin{aligned} &\frac{d}{dx}\left[K(x)\frac{d}{dx}M_\mathrm{h}(x;\omega_m)\right] \\ &\quad - M_\mathrm{h}(x;\omega_m)K(x)\left|\frac{d}{dx}\theta_\mathrm{h}(x;\omega_m)\right|^2 = 0 \end{aligned} \tag{6.21a}$$

$$\begin{aligned} &\frac{d}{dx}\left[M_\mathrm{h}^2(x;\omega_m)K(x)\frac{d}{dx}\theta_\mathrm{h}(x;\omega_m)\right] \\ &\quad - \omega_m S_\mathrm{s}(x)M_\mathrm{h}^2(x;\omega_m) = 0 \end{aligned} \tag{6.21b}$$

$$\forall\, x \in D_+(\omega_m); \quad m = 1,2,3,\ldots.$$

□

Define

$$\zeta_\mathrm{h}(x;\omega_m) \equiv \frac{d}{dx}\theta_\mathrm{h}(x;\omega_m). \tag{6.22}$$

For the remainder of this section we will consider only those flows for which

$$|\zeta_\mathrm{h}(x;\omega_m)| > 0 \quad \forall\, x \in D_+(\omega_m).$$

Assuming $\zeta_\mathrm{h}(x;\omega_m)$ is a continuous function of x, this implies that either (a) $\zeta_\mathrm{h}(x;\omega_m) > 0$ everywhere in $D_+(\omega_m)$ or (b) $\zeta_\mathrm{h}(x;\omega_m) < 0$ everywhere in $D_+(\omega_m)$. We discuss this assumption in the context of wave propagation in

Section 21.2.1. Solving (6.21a) and (6.21b) for ζ_h and M_h, respectively, gives

$$\zeta_\mathrm{h}(x;\omega_m) = \pm\sqrt{\frac{1}{M_\mathrm{h}}\left[\frac{d^2M_\mathrm{h}}{dx^2} + \left(\frac{d\ln K}{dx}\right)\frac{dM_\mathrm{h}}{dx}\right]} \tag{6.23a}$$

$$\frac{M_\mathrm{h}(x;\omega_m)}{M_\mathrm{h}(x_0;\omega_m)} = \sqrt{\frac{K(x_0)\zeta_\mathrm{h}(x_0;\omega_m)}{K(x)\zeta_\mathrm{h}(x;\omega_m)}} \times \exp\left(\frac{\omega_m}{2}\int\limits_{x_0}^{x}\frac{S_\mathrm{s}(v)}{K(v)\zeta_\mathrm{h}(v;\omega_m)}dv\right) \tag{6.23b}$$

$\forall\, x \in D_+(\omega_m)$.

Exercise 6.10 Derivation: Polar Form Governing Equations in One Dimension

Problem: Consider time-periodic flow in a source-free, 1D, nonhomogeneous, porous medium. Manipulate the governing equation (6.21a) to derive (6.23a).

Solution: Substituting the LHS of (6.22) for $d\theta_\mathrm{h}/dx$ in (6.21a) and then solving for ζ_h give

$$\zeta_\mathrm{h}(x;\omega_m) = \pm\sqrt{\frac{1}{KM_\mathrm{h}}\frac{d}{dx}\left(K\frac{dM_\mathrm{h}}{dx}\right)}.$$

Expanding the derivative and then simplifying yield (6.23a). □

Exercise 6.11 Derivation: Polar Form Governing Equations in One Dimension

Problem: Consider time-periodic flow in a source-free, 1D, generally nonhomogeneous, porous medium. Integrate the governing equation (6.21b) to derive (6.23b).

Solution: Substituting the RHS of (6.22) for the product of the hydraulic conductivity and the derivative of the phase function in the governing equation (6.21b) and rearranging yield

$$\frac{d}{dx}\left[M_\mathrm{h}^2(x;\omega_m)K(x)\zeta_\mathrm{h}(x;\omega_m)\right] = \omega_m S_\mathrm{s}(x)M_\mathrm{h}^2(x;\omega_m).$$

Dividing by $M_\mathrm{h}^2K\zeta_\mathrm{h}$, and then substituting $x = v$, gives

$$\frac{\frac{d}{dv}\left[M_\mathrm{h}^2(v;\omega_m)K(v)\zeta_\mathrm{h}(v;\omega_m)\right]}{M_\mathrm{h}^2(v;\omega_m)K(v)\zeta_\mathrm{h}(v;\omega_m)} = \omega_m\frac{S_\mathrm{s}(v)}{K(v)\zeta_\mathrm{h}(v;\omega_m)}.$$

Integrating with respect to v, from $v = x_0$ to $v = x$, and incorporating the assumed properties of the function $\zeta_\mathrm{h}(x;\omega_m)$ produce

$$\left\{\ln\left[M_\mathrm{h}^2(v;\omega_m)K(v)|\zeta_\mathrm{h}(v;\omega_m)|\right]\right\}\Bigg|_{x_0}^{x} = \omega_m\int\limits_{x_0}^{x}\frac{S_\mathrm{s}(v)}{K(v)\zeta_\mathrm{h}(v;\omega_m)}dv.$$

Evaluating the LHS at the indicated limits leads to

$$\ln\frac{M_\mathrm{h}^2(x;\omega_m)K(x)\zeta_\mathrm{h}(x;\omega_m)}{M_\mathrm{h}^2(x_0;\omega_m)K(x_0)\zeta_\mathrm{h}(x_0;\omega_m)} = \omega_m\int\limits_{x_0}^{x}\frac{S_\mathrm{s}(v)}{K(v)\zeta_\mathrm{h}(v;\omega_m)}dv.$$

Exponentiating and then taking the positive square root give

$$\frac{M_\mathrm{h}(x;\omega_m)}{M_\mathrm{h}(x_0;\omega_m)}\sqrt{\frac{K(x)\zeta_\mathrm{h}(x;\omega_m)}{K(x_0)\zeta_\mathrm{h}(x_0;\omega_m)}} = \exp\left(\frac{\omega_m}{2}\int\limits_{x_0}^{x}\frac{S_\mathrm{s}(v)}{K(v)\zeta_\mathrm{h}(v;\omega_m)}dv\right).$$

Solving for $M_\mathrm{h}(x;\omega_m)/M_\mathrm{h}(x_0;\omega_m)$ then yields (6.23b). □

Substituting the RHS of (6.23a) for ζ_h in (6.21b) and the RHS of (6.23b) for M_h in (6.21a), respectively, leads to

$$\left(\frac{\omega_m S_\mathrm{s}}{K}\right)^2 + 2\frac{d}{dx}\left(\frac{\omega_m S_\mathrm{s}}{K}\right)\zeta_\mathrm{h} - \left[\left(\frac{d\ln K}{dx}\right)^2 + 2\frac{d}{dx}\left(\frac{d\ln K}{dx}\right)\right]\zeta_\mathrm{h}^2 - 4\zeta_\mathrm{h}^4 - 4\left(\frac{\omega_m S_\mathrm{s}}{K}\right)\frac{d\zeta_\mathrm{h}}{dx} + 3\left(\frac{d\zeta_\mathrm{h}}{dx}\right)^2 - 2\zeta_\mathrm{h}\frac{d^2\zeta_\mathrm{h}}{dx^2} = 0 \tag{6.24a}$$

$$\left(\frac{d\ln K}{dx}\right)M_\mathrm{h}^{3/2}\sqrt{\frac{d^2M_\mathrm{h}}{dx^2} + \left(\frac{d\ln K}{dx}\right)\frac{dM_\mathrm{h}}{dx}} + \frac{d}{dx}\left[M_\mathrm{h}^{3/2}\sqrt{\frac{d^2M_\mathrm{h}}{dx^2} + \left(\frac{d\ln K}{dx}\right)\frac{dM_\mathrm{h}}{dx}}\right] - c\left(\frac{\omega_m S_\mathrm{s}}{K}\right)M_\mathrm{h}^2 = 0 \tag{6.24b}$$

$\forall\, x \in D_+(\omega_m)$

where

$$c(x;\omega_m) \equiv \mathrm{sign}\left[\zeta_\mathrm{h}(x;\omega_m)\right].$$

Equations (6.24) are nonlinear, ordinary differential equations (ODEs) with variable coefficients. Equation (6.24a) is second order in ζ_h, which makes it third order in θ_h, and (6.24b) is third order in M_h. Notice that we have expressed the coefficients in both equations in terms of $\omega_m S_\mathrm{s}/K$ and $d\ln K/(dx)$ and their space derivatives.

Exercise 6.12 Derivation: Polar Form Governing Equations in One Dimension

Problem: Consider time-periodic flow in a 1D, source-free domain for which the space derivative of the phase function is nonzero everywhere. Derive expressions for dM_h/dx and d^2M_h/dx^2 by differentiating the RHS of (6.23b).

Solution: Differentiating the RHS of (6.23b) with respect to x gives

$$\frac{d}{dx}M_\mathrm{h}(x;\omega_m) = \frac{1}{2}M_\mathrm{h}\left[\left(\frac{\omega_m S_\mathrm{s}}{K}\right)\frac{1}{\zeta_\mathrm{h}} - \frac{d\ln K}{dx} - \frac{1}{\zeta_\mathrm{h}}\frac{d\zeta_\mathrm{h}}{dx}\right] \tag{6.25}$$

where $K = K(x)$, $M_\mathrm{h} = M_\mathrm{h}(x;\omega_m)$, and $\zeta_\mathrm{h} = \zeta_\mathrm{h}(x;\omega_m)$. Differentiating (6.25) with respect to x yields

$$\frac{d^2}{dx^2}M_\mathrm{h}(x;\omega_m) = \frac{1}{2}M_\mathrm{h}\left[\frac{1}{2}\left(\frac{\omega_m S_\mathrm{s}}{K}\frac{1}{\zeta_\mathrm{h}} - \frac{d\ln K}{dx} - \frac{1}{\zeta_\mathrm{h}}\frac{d\zeta_\mathrm{h}}{dx}\right)^2 + \frac{d}{dx}\left(\frac{\omega_m S_\mathrm{s}}{K\zeta_\mathrm{h}}\right) - \frac{d^2\ln K}{dx^2} - \frac{d}{dx}\left(\frac{1}{\zeta_\mathrm{h}}\frac{d\zeta_\mathrm{h}}{dx}\right)\right].$$

Expanding the RHS and then grouping terms produce

$$\begin{aligned}\frac{d^2}{dx^2}M_\mathrm{h}(x;\omega_m) &= \frac{1}{2}M_\mathrm{h}\left[\frac{1}{2}\left(\frac{\omega_m S_\mathrm{s}}{K}\right)^2\frac{1}{\zeta_\mathrm{h}^2} + \frac{1}{2}\left(\frac{d\ln K}{dx}\right)^2 + \frac{3}{2}\frac{1}{\zeta_\mathrm{h}^2}\left(\frac{d\zeta_\mathrm{h}}{dx}\right)^2\right.\\ &\quad - \left(\frac{\omega_m S_\mathrm{s}}{K}\right)\left(\frac{d\ln K}{dx}\right)\frac{1}{\zeta_\mathrm{h}} + \left(\frac{d\ln K}{dx}\right)\frac{1}{\zeta_\mathrm{h}}\frac{d\zeta_\mathrm{h}}{dx}\\ &\quad - 2\left(\frac{\omega_m S_\mathrm{s}}{K}\right)\frac{1}{\zeta_\mathrm{h}^2}\frac{d\zeta_\mathrm{h}}{dx} + \frac{d}{dx}\left(\frac{\omega_m S_\mathrm{s}}{K}\right)\frac{1}{\zeta_\mathrm{h}}\\ &\quad\left. - \frac{d^2\ln K}{dx^2} - \frac{1}{\zeta_\mathrm{h}}\frac{d^2\zeta_\mathrm{h}}{dx^2}\right].\end{aligned} \tag{6.26}$$

□

Exercise 6.13 Derivation: Polar Form Governing Equations in One Dimension

Problem: Consider time-periodic flow in a 1D, source-free domain for which the space derivative of the phase function is nonzero everywhere. Verify (6.24a) to your own satisfaction. Start with the governing equation (6.21a).

Solution: First, substitute the RHSs of (6.23b), (6.25), and (6.26), respectively, for M_h, dM_h/dx, and d^2M_h/dx^2 in (6.21a). Then multiply by $4\zeta_\mathrm{h}^2/M_\mathrm{h}$ and group terms to obtain (6.24a). □

Exercise 6.14 Derivation: Polar Form Governing Equations in One Dimension

Problem: Consider time-periodic flow in a 1D, source-free domain for which the space derivative of the phase function is nonzero everywhere. Verify (6.24b) to your own satisfaction. Start with the governing equation (6.21b).

Solution: First, substitute the RHS of (6.23a) for $d\theta_\mathrm{h}/dx$ in (6.21b). Then manipulate the result to obtain (6.24b). □

We can summarize the results of this section as follows. If for some particular harmonic constituent (i.e., ω_m fixed) we have $|\zeta_\mathrm{h}(x;\omega_m)| > 0$ throughout the 1D space domain $D_+(\omega_m)$, then the set of uncoupled governing equations (6.24) is mathematically equivalent to the set of governing equations (6.21).

6.3. BC EQUATIONS

First, consider those points on the boundary at which both the hydraulic head and boundary value amplitudes are positive. Substituting the expressions on the RHSs of (4.3) for the corresponding terms in the rectangular forms of the BC equations (5.11) and simplifying give

$$c_\mathrm{h}M_\mathrm{h}\cos\theta_\mathrm{h} + c_\mathrm{q}\big[\mathbf{K}\nabla(M_\mathrm{h}\cos\theta_\mathrm{h})\big]\cdot\hat{\mathbf{n}} = M_\psi\cos\theta_\psi \tag{6.27a}$$

$$c_\mathrm{h}M_\mathrm{h}\sin\theta_\mathrm{h} + c_\mathrm{q}\big[\mathbf{K}\nabla(M_\mathrm{h}\sin\theta_\mathrm{h})\big]\cdot\hat{\mathbf{n}} = M_\psi\sin\theta_\psi. \tag{6.27b}$$

Following a procedure similar to that used previously for the governing equations, we obtain

$$c_\mathrm{q}(\mathbf{x})M_\mathrm{h}(\mathbf{x};\omega_m)\big[\mathbf{K}(\mathbf{x})\nabla\theta_\mathrm{h}(\mathbf{x};\omega_m)\big]\cdot\hat{\mathbf{n}}(\mathbf{x}) = M_\psi(\mathbf{x};\omega_m)\sin\left[\theta_\psi(\mathbf{x};\omega_m) - \theta_\mathrm{h}(\mathbf{x};\omega_m)\right] \tag{6.28a}$$

$$c_\mathrm{h}(\mathbf{x})M_\mathrm{h}(\mathbf{x};\omega_m) + c_\mathrm{q}(\mathbf{x})\big[\mathbf{K}(\mathbf{x})\nabla M_\mathrm{h}(\mathbf{x};\omega_m)\big]\cdot\hat{\mathbf{n}}(\mathbf{x}) = M_\psi(\mathbf{x};\omega_m)\cos\left[\theta_\psi(\mathbf{x};\omega_m) - \theta_\mathrm{h}(\mathbf{x};\omega_m)\right] \tag{6.28b}$$

$$M_\psi(\mathbf{x};\omega_m) > 0;\quad \forall\,\mathbf{x}\in\Gamma_+(\omega_m);\quad m = 1,2,3,\ldots$$

where $\Gamma_+(\omega_m)$ denotes the boundary of the domain $D_+(\omega_m)$.

Next, consider those points on the boundary at which the hydraulic head amplitude is positive and the boundary value amplitude is zero. Substituting zero for A_ψ and B_ψ in the rectangular forms of the BC equations (5.11) and following a procedure similar to that described above yield

$$c_\mathrm{q}(\mathbf{x})M_\mathrm{h}(\mathbf{x};\omega_m)\big[\mathbf{K}(\mathbf{x})\nabla\theta_\mathrm{h}(\mathbf{x};\omega_m)\big]\cdot\hat{\mathbf{n}}(\mathbf{x}) = 0 \tag{6.29a}$$

$$c_\mathrm{h}(\mathbf{x})M_\mathrm{h}(\mathbf{x};\omega_m) + c_\mathrm{q}(\mathbf{x})\big[\mathbf{K}(\mathbf{x})\nabla M_\mathrm{h}(\mathbf{x};\omega_m)\big]\cdot\hat{\mathbf{n}}(\mathbf{x}) = 0 \tag{6.29b}$$

$$M_\psi(\mathbf{x};\omega_m) > 0;\quad \forall\,\mathbf{x}\in\Gamma_+(\omega_m);\quad m = 1,2,3,\ldots.$$

For the sake of convenience, we implicitly represent both sets of BC equations (6.28) and (6.29) compactly as

$$c_{\mathrm{q}}(\mathbf{x})M_{\mathrm{h}}(\mathbf{x};\omega_m)\big[\mathbf{K}(\mathbf{x})\nabla\theta_{\mathrm{h}}(\mathbf{x};\omega_m)\big]\cdot\hat{\mathbf{n}}(\mathbf{x}) = M_{\psi}(\mathbf{x};\omega_m)\sin\big[\theta_{\psi}(\mathbf{x};\omega_m)-\theta_{\mathrm{h}}(\mathbf{x};\omega_m)\big] \quad (6.30a)$$

$$c_{\mathrm{h}}(\mathbf{x})M_{\mathrm{h}}(\mathbf{x};\omega_m)+c_{\mathrm{q}}(\mathbf{x})\big[\mathbf{K}(\mathbf{x})\nabla M_{\mathrm{h}}(\mathbf{x};\omega_m)\big]\cdot\hat{\mathbf{n}}(\mathbf{x}) = M_{\psi}(\mathbf{x};\omega_m)\cos\big[\theta_{\psi}(\mathbf{x};\omega_m)-\theta_{\mathrm{h}}(\mathbf{x};\omega_m)\big] \quad (6.30b)$$

$$\forall\, \mathbf{x}\in\Gamma_{+}(\omega_m);\quad m=1,2,3,\ldots$$

where it is understood that at those points (if any) where the boundary value amplitude is zero, the RHSs of (6.30) vanish. This convention is purely an abbreviation; it allows us to write one set of governing equations rather than two sets. The BC equations (6.30), like the governing equations (6.11), are coupled and nonlinear.

Exercise 6.15 Boundary Condition for Polar Form of Space BVP

Problem: Beginning with the BC equation (5.25b) for the general form of the space-time BVP and using trigonometric identity (B.2a), derive the BC equation (6.30) for the polar form of the space BVP. □

Exercise 6.16 Polar Form Boundary Conditions: Uniform Phase Shift

Problem: Consider the general definition of a frequency-dependent, spatially uniform phase shift given by (6.13). Show that the shifted hydraulic head phase $\breve{\theta}_{\mathrm{h}}(\mathbf{x};\omega_m)$ satisfies the BC equations (6.30). *Hint*: Follow a procedure similar to that used in Exercise 6.5. Use the following definition for the shifted phase of the source distribution:

$$\breve{\theta}_{\psi}(\mathbf{x};\omega_m)=\theta_{\psi}(\mathbf{x};\omega_m)-\theta_0(\omega_m).$$ □

Exercise 6.17 Polar Form Boundary Conditions: Amplitude Scaling

Problem: Consider the general definition of a frequency-dependent, spatially uniform amplitude scaling given by (6.16). Following a procedure similar to that used in Exercise 6.7, show that the scaled hydraulic head amplitude $\breve{M}_{\mathrm{h}}(\mathbf{x};\omega_m)$ satisfies the BC equations (6.30). Use the following definition for the scaled amplitude of the source distribution:

$$\breve{M}_{\psi}(\mathbf{x};\omega_m)=\frac{M_{\psi}(\mathbf{x};\omega_m)}{M_0(\omega_m)}.$$ □

The governing equations (6.11) and the BC equations (6.30), together with the specification of material hydrologic properties and space domain geometry, constitute the *polar form of the space BVP*. Like the rectangular form, the polar form is mathematically equivalent to the general form of the space-time BVP (5.26).

6.4. CHAPTER SUMMARY

In this chapter we derived the polar form of the space BVP for hydraulic head harmonic constituents from the corresponding rectangular form. We described how the polar forms of the governing and BC equations are invariant to frequency-dependent, spatially uniform transformations of the dependent variables—multiplicative scaling of the amplitude function and shifting of the phase function. We also showed that for time-periodic flow in a source-free, 1D medium, under some conditions it is possible to uncouple the polar form governing equations.

7
Complex-Variable Form of Space BVP

In previous chapters we derived two basic forms of the space BVP for hydraulic head harmonic constituents—the rectangular form and the polar form. In this chapter we derive yet another form of the space BVP using complex variables.

7.1. FREQUENCY RESPONSE FUNCTIONS

Recall the equations for the general form of the space-time BVP for the hydraulic head harmonic constituents (5.26):

$$L_h\left[h'(\mathbf{x}, t; \omega_m)\right] + u'(\mathbf{x}, t; \omega_m) = 0 \quad \forall\, \mathbf{x} \in D \qquad (7.1a)$$

$$W\left[h'(\mathbf{x}, t; \omega_m)\right] - \psi'(\mathbf{x}, t; \omega_m) = 0 \quad \forall\, \mathbf{x} \in \Gamma \qquad (7.1b)$$

$$\forall\, t > 0; \quad m = 1, 2, 3, \ldots .$$

Each harmonic constituent (4.3) is real valued, and we express each as the real part of a product of a *complex amplitude* (F) and a *unit-amplitude complex exponential* function of time ($e^{i\omega_m t}$) as

$$h'(\mathbf{x}, t; \omega_m) = \mathrm{Re}\left[F_h(\mathbf{x}; \omega_m) e^{i\omega_m t}\right] \qquad (7.2a)$$

$$u'(\mathbf{x}, t; \omega_m) = \mathrm{Re}\left[F_u(\mathbf{x}; \omega_m) e^{i\omega_m t}\right] \qquad (7.2b)$$

$$\psi'(\mathbf{x}, t; \omega_m) = \mathrm{Re}\left[F_\psi(\mathbf{x}; \omega_m) e^{i\omega_m t}\right] \qquad (7.2c)$$

where i is the *imaginary unit* (i.e., $i \equiv \sqrt{-1}$) and we define the complex amplitudes as

$$F_h(\mathbf{x}; \omega_m) \equiv M_h(\mathbf{x}; \omega_m) e^{i\theta_h(\mathbf{x};\omega_m)} \qquad (7.3a)$$

$$F_u(\mathbf{x}; \omega_m) \equiv M_u(\mathbf{x}; \omega_m) e^{i\theta_u(\mathbf{x};\omega_m)} \qquad (7.3b)$$

$$F_\psi(\mathbf{x}; \omega_m) \equiv M_\psi(\mathbf{x}; \omega_m) e^{i\theta_\psi(\mathbf{x};\omega_m)}. \qquad (7.3c)$$

We refer to equations (7.2) as the *complex form* for the harmonic constituents.

We refer to the complex amplitudes F_h, F_u, and F_ψ, respectively, as the hydraulic head, source, and boundary value *frequency response functions (FRFs)*. The hydraulic head FRF measures the sinusoidal hydraulic head response of the groundwater system to sinusoidal source and boundary inputs (u' and ψ') of frequency ω at the point with space coordinates $\mathbf{x}$.

The complex amplitudes F convey, in compact notation, information about both the real amplitudes M and the phases θ of the corresponding harmonic constituents. The real amplitudes are the moduli of the corresponding complex amplitudes:

$$M_h(\mathbf{x}; \omega_m) = |F_h(\mathbf{x}; \omega_m)| \qquad (7.4a)$$

$$M_u(\mathbf{x}; \omega_m) = |F_u(\mathbf{x}; \omega_m)| \qquad (7.4b)$$

$$M_\psi(\mathbf{x}; \omega_m) = |F_\psi(\mathbf{x}; \omega_m)|. \qquad (7.4c)$$

The phases are the arguments of the corresponding complex amplitudes:

$$\theta_h(\mathbf{x}; \omega_m) = \arg F_h(\mathbf{x}; \omega_m) \qquad (7.5a)$$

$$\theta_u(\mathbf{x}; \omega_m) = \arg F_u(\mathbf{x}; \omega_m) \qquad (7.5b)$$

$$\theta_\psi(\mathbf{x}; \omega_m) = \arg F_\psi(\mathbf{x}; \omega_m). \qquad (7.5c)$$

The modulus M_h and phase θ_h of the complex-valued hydraulic head FRF give the hydraulic head *amplitude response function* and *phase response function*, respectively, both of which are real valued. We will refer to these as the *amplitude function* and *phase function*, respectively.

We can express the complex amplitudes in rectangular form as

$$F_h(\mathbf{x}; \omega_m) = A_h(\mathbf{x}; \omega_m) + iB_h(\mathbf{x}; \omega_m) \qquad (7.6a)$$

$$F_u(\mathbf{x}; \omega_m) = A_u(\mathbf{x}; \omega_m) + iB_h(\mathbf{x}; \omega_m) \qquad (7.6b)$$

$$F_\psi(\mathbf{x}; \omega_m) = A_\psi(\mathbf{x}; \omega_m) + iB_\psi(\mathbf{x}; \omega_m) \qquad (7.6c)$$

Hydrodynamics of Time-Periodic Groundwater Flow: Diffusion Waves in Porous Media, Geophysical Monograph 224,
First Edition. Joe S. Depner and Todd C. Rasmussen.

where the corresponding *rectangular components* (i.e., the A and B coefficient functions) are real valued and represent the real and imaginary parts, respectively, of the corresponding complex amplitudes:

$$\begin{aligned} A_{\rm h}(\mathbf{x};\omega_m) &= \operatorname{Re} F_{\rm h}(\mathbf{x};\omega_m) \\ A_{\rm u}(\mathbf{x};\omega_m) &= \operatorname{Re} F_{\rm u}(\mathbf{x};\omega_m) \\ A_{\psi}(\mathbf{x};\omega_m) &= \operatorname{Re} F_{\psi}(\mathbf{x};\omega_m) \end{aligned} \tag{7.7a}$$

$$\begin{aligned} B_{\rm h}(\mathbf{x};\omega_m) &= \operatorname{Im} F_{\rm h}(\mathbf{x};\omega_m) \\ B_{\rm u}(\mathbf{x};\omega_m) &= \operatorname{Im} F_{\rm u}(\mathbf{x};\omega_m) \\ B_{\psi}(\mathbf{x};\omega_m) &= \operatorname{Im} F_{\psi}(\mathbf{x};\omega_m). \end{aligned} \tag{7.7b}$$

7.2. SPACE BVP FOR HYDRAULIC HEAD FRF

In this section we derive the equations of the space BVP for the hydraulic head FRF. First we establish a few basic results in the following exercises.

Exercise 7.1 Commutativity of Real-Part and Linear Operators

Problem: Suppose L is a linear operator with real-valued coefficients and $Z(\mathbf{x}, t; \omega_m)$ is a complex-valued function. Show that

$$L\big[\operatorname{Re} Z(\mathbf{x}, t;\omega_m)\big] = \operatorname{Re} L\big[Z(\mathbf{x}, t;\omega_m)\big]. \tag{7.8}$$

Solution: Define

$$\begin{aligned} X(\mathbf{x}, t;\omega_m) &\equiv \operatorname{Re} Z(\mathbf{x}, t;\omega_m) \\ Y(\mathbf{x}, t;\omega_m) &\equiv \operatorname{Im} Z(\mathbf{x}, t;\omega_m). \end{aligned}$$

Then

$$L\big[\operatorname{Re} Z(\mathbf{x}, t;\omega_m)\big] = L\big[X(\mathbf{x}, t;\omega_m)\big].$$

Assume X and Y are real valued and the coefficients of L are real valued. Therefore, both $L[X]$ and $L[Y]$ are real valued. Consequently,

$$\begin{aligned} L\big[\operatorname{Re} Z(\mathbf{x}, t;\omega_m)\big] &= \operatorname{Re}\Big\{L\big[X(\mathbf{x}, t;\omega_m)\big]\Big\} \\ &= \operatorname{Re}\Big\{L\big[X(\mathbf{x}, t;\omega_m)\big]\Big\} \\ &\quad + \operatorname{Re}\Big\{iL\big[Y(\mathbf{x}, t;\omega_m)\big]\Big\}. \end{aligned}$$

Grouping terms on the RHS gives

$$L\big[\operatorname{Re} Z(\mathbf{x}, t;\omega_m)\big] = \operatorname{Re}\Big\{L\big[X(\mathbf{x}, t;\omega_m)\big] + iL\big[Y(\mathbf{x}, t;\omega_m)\big]\Big\}.$$

Because the operator L is linear with real coefficients, we can write this as

$$\begin{aligned} L\big[\operatorname{Re} Z(\mathbf{x}, t;\omega_m)\big] &= \operatorname{Re}\Big\{L\big[X(\mathbf{x}, t;\omega_m) + iY(\mathbf{x}, t;\omega_m)\big]\Big\} \\ &= \operatorname{Re} L\big[Z(\mathbf{x}, t;\omega_m)\big]. \end{aligned} \tag{7.9}$$

□

Exercise 7.2 Properties of Complex Sinusoids

Problem: Suppose $F(\mathbf{x};\omega_m)$ is a complex-valued function that is independent of time. Show that

$$\operatorname{Re}\Big[F(\mathbf{x};\omega_m)e^{i\omega_m t}\Big] = 0 \quad \forall\, t > 0 \tag{7.10}$$

implies

$$F(\mathbf{x};\omega_m) = 0. \tag{7.11}$$

Solution: Define

$$\begin{aligned} A(\mathbf{x};\omega_m) &\equiv \operatorname{Re} F(\mathbf{x};\omega_m) \\ B(\mathbf{x};\omega_m) &\equiv \operatorname{Im} F(\mathbf{x};\omega_m). \end{aligned}$$

Substituting $F = A + iB$ and $e^{i\omega_m t} = \cos(\omega_m t) + i\sin(\omega_m t)$ in (7.10), expanding the product, and then taking the real part yield

$$A(\mathbf{x};\omega_m)\cos(\omega_m t) - B(\mathbf{x};\omega_m)\sin(\omega_m t) = 0 \quad \forall\, t > 0. \tag{7.12}$$

Multiplying this equation by $\cos(\omega_m t)$ and then integrating the product with respect to t from $t = 0$ to $t = 2\pi\ \text{rad}/\omega_m$ give $A(\mathbf{x};\omega_m) = 0$. Repeating this procedure with $\sin(\omega_m t)$ rather than $\cos(\omega_m t)$ gives $B(\mathbf{x};\omega_m) = 0$. Together these results prove (7.11). □

Substituting the RHSs of (7.2a), (7.2b), and (7.2c) for h', u', and ψ', respectively, in (5.26) gives

$$\begin{aligned} &L_{\rm h}\Big\{\operatorname{Re}\Big[F_{\rm h}(\mathbf{x};\omega_m)e^{i\omega_m t}\Big]\Big\} \\ &\quad + \operatorname{Re}\Big[F_{\rm u}(\mathbf{x};\omega_m)e^{i\omega_m t}\Big] = 0 \quad \forall\, \mathbf{x} \in D \qquad (7.13) \\ &W\Big\{\operatorname{Re}\Big[F_{\rm h}(\mathbf{x};\omega_m)e^{i\omega_m t}\Big]\Big\} \\ &\quad - \operatorname{Re}\Big[F_{\psi}(\mathbf{x};\omega_m)e^{i\omega_m t}\Big] = 0 \quad \forall\, \mathbf{x} \in \Gamma \qquad (7.14) \\ &\forall\, t > 0; \quad m = 1, 2, 3, \ldots. \end{aligned}$$

The operators $L_{\rm h}$ and W are linear with real-valued coefficients [see equations (2.7) and (2.13), respectively], so we can use the result of Exercise 7.1 to rewrite these as

$$\begin{aligned} &\operatorname{Re}\Big\{L_{\rm h}\Big[F_{\rm h}(\mathbf{x};\omega_m)e^{i\omega_m t}\Big]\Big\} \\ &\quad + \operatorname{Re}\Big[F_{\rm u}(\mathbf{x};\omega_m)e^{i\omega_m t}\Big] = 0 \quad \forall\, \mathbf{x} \in D \qquad (7.15a) \\ &\operatorname{Re}\Big\{W\Big[F_{\rm h}(\mathbf{x};\omega_m)e^{i\omega_m t}\Big]\Big\} \\ &\quad - \operatorname{Re}\Big[F_{\psi}(\mathbf{x};\omega_m)e^{i\omega_m t}\Big] = 0 \quad \forall\, \mathbf{x} \in \Gamma \qquad (7.15b) \\ &\forall\, t > 0; \quad m = 1, 2, 3, \ldots. \end{aligned}$$

Grouping terms then leads to

$$\mathrm{Re}\left\{L_{\mathrm{h}}\left[F_{\mathrm{h}}(\mathbf{x};\omega_m)e^{i\omega_m t}\right] + F_{\mathrm{u}}(\mathbf{x};\omega_m)ei\omega_m t\right\} = 0 \quad \forall\, \mathbf{x} \in D \tag{7.16a}$$

$$\mathrm{Re}\left\{W\left[F_{\mathrm{h}}(\mathbf{x};\omega_m)e^{i\omega_m t}\right] - F_{\psi}(\mathbf{x};\omega_m)e^{i\omega_m t}\right\} = 0 \quad \forall\, \mathbf{x} \in \Gamma \tag{7.16b}$$

$$\forall\, t > 0; \quad m = 1, 2, 3, \ldots .$$

Writing the result of the operation $L_{\mathrm{h}}[F_{\mathrm{h}}]$ [see equation (2.7)] explicitly, noting that the operator W is time invariant [see equation (2.13)], and then factoring $e^{i\omega_m t}$ in both equations yield

$$\mathrm{Re}\left[\left\{\nabla \cdot \left[\mathbf{K}(\mathbf{x})\nabla F_{\mathrm{h}}(\mathbf{x};\omega_m)\right] - i\omega_m S_{\mathrm{s}} F_{\mathrm{h}}(\mathbf{x};\omega_m) + F_{\mathrm{u}}(\mathbf{x};\omega_m)\right\}e^{i\omega_m t}\right] = 0 \quad \forall\, \mathbf{x} \in D$$

$$\mathrm{Re}\left[\left\{W\left[F_{\mathrm{h}}(\mathbf{x};\omega_m)\right] - F_{\psi}(\mathbf{x};\omega_m)\right\}e^{i\omega_m t}\right] = 0 \quad \forall\, \mathbf{x} \in \Gamma$$

$$\forall\, t > 0; \quad m = 1, 2, 3, \ldots .$$

The quantities within the curved braces are time invariant. Applying the result of Exercise 7.2 to each of these and then moving the nonhomogeneous terms to the RHSs yield

$$\nabla \cdot \left[\mathbf{K}(\mathbf{x})\nabla F_{\mathrm{h}}(\mathbf{x};\omega_m)\right] - i\omega_m S_{\mathrm{s}}(\mathbf{x}) F_{\mathrm{h}}(\mathbf{x};\omega_m) = -F_{\mathrm{u}}(\mathbf{x};\omega_m) \quad \forall\, \mathbf{x} \in D \tag{7.17a}$$

$$W\left[F_{\mathrm{h}}(\mathbf{x};\omega_m)\right] = F_{\psi}(\mathbf{x};\omega_m) \quad \forall\, \mathbf{x} \in \Gamma \tag{7.17b}$$

$$m = 1, 2, 3, \ldots .$$

Equations (7.17a) and (7.17b), respectively, are the governing and BC equations for the hydraulic head FRF, $F_{\mathrm{h}}(\mathbf{x};\omega_m)$. Together with the specification of material hydrologic properties and space domain geometry, they represent a space BVP for the FRF. We call it the *complex-variable form of the space BVP*. Like the rectangular and polar forms of the space BVP, the complex-variable form is mathematically equivalent to the general form of the space-time BVP [see equations (5.26)].

Exercise 7.3 Classification of Complex-Form Governing Equation

Problem: Is the complex-variable form of the governing equation (7.17a) elliptic, hyperbolic, or parabolic? Why?

Solution: The second-derivative terms on the LHS of (7.17a) sum to

$$\mathbf{K}(\mathbf{x}) : \mathbf{H}\left[F_{\mathrm{h}}(\mathbf{x};\omega_m)\right] = \sum_{n=1}^{N}\sum_{p=1}^{N} K_{np}(\mathbf{x})\frac{\partial^2 F_{\mathrm{h}}(\mathbf{x};\omega_m)}{\partial x_n \partial x_p}$$

where $\mathbf{H}[F_{\mathrm{h}}]$ denotes the *Hessian matrix* of the function F_{h} (see Appendix C). The coefficient matrix for the second-derivative terms (i.e., $\mathbf{K}$) is positive definite, so the equation is elliptic. □

We can use the complex-variable form of the space BVP as a starting point in deriving the rectangular and polar forms of the space BVP, as the following exercises illustrate.

Exercise 7.4 Derivation of Governing Equations

Problem: Consider the rectangular form of the space BVP for the hydraulic head harmonic constituents. Derive the governing equations (5.10) from the corresponding governing equation (7.17a) of the complex-variable form of the space BVP. *Hint*: First, substitute the RHSs of (7.6) for the corresponding complex amplitudes in the governing equation (7.17a); then take the real and imaginary parts. □

Exercise 7.5 Derivation of BC Equations

Problem: Consider the rectangular form of the space BVP for the hydraulic head harmonic constituents. Derive the BC equations (5.11) from the corresponding BC equation (7.17b) in the complex-variable form of the space BVP. *Hint*: First, substitute the RHSs of (7.6) for the corresponding complex amplitudes in the BC equation (7.17b); then take the real and imaginary parts. □

Exercise 7.6 Derivation of Governing Equations

Problem: Consider the polar form of the space BVP for the hydraulic head harmonic constituents. Derive the governing equations (6.11) from the corresponding governing equation (7.17a) of the FRF space BVP. *Hint*: First, substitute the RHSs of (7.6) for the corresponding complex amplitudes in the governing equation (7.17a); then take the real and imaginary parts. □

Exercise 7.7 Derivation of BC Equations

Problem: Consider the polar form of the space BVP for the hydraulic head harmonic constituents. Derive the BC equations (6.30) from the corresponding BC equation (7.17b) of the FRF space BVP. *Hint*: First, substitute the RHSs of (7.6) for the corresponding complex amplitudes in the BC equation (7.17b); then take the real and imaginary parts. □

7.3. COMPLEX-VALUED HARMONIC CONSTITUENTS

Define the *complex-valued harmonic constituents* of hydraulic head, source strength, and boundary value, respectively, as

$$H(\mathbf{x}, t; \omega_m) \equiv F_h(\mathbf{x}; \omega_m)e^{i\omega_m t} \quad (7.18a)$$

$$U(\mathbf{x}, t; \omega_m) \equiv F_u(\mathbf{x}; \omega_m)e^{i\omega_m t} \quad (7.18b)$$

$$\Psi(\mathbf{x}, t; \omega_m) \equiv F_\psi(\mathbf{x}; \omega_m)e^{i\omega_m t}. \quad (7.18c)$$

Combining (7.2) and (7.18) leads to the following relations between the real-valued and complex-valued harmonic constituents:

$$h'(\mathbf{x}, t; \omega_m) = \operatorname{Re} H(\mathbf{x}, t; \omega_m) \quad (7.19a)$$

$$u'(\mathbf{x}, t; \omega_m) = \operatorname{Re} U(\mathbf{x}, t; \omega_m) \quad (7.19b)$$

$$\psi'(\mathbf{x}, t; \omega_m) = \operatorname{Re} \Psi(\mathbf{x}, t; \omega_m). \quad (7.19c)$$

We can express the complex-valued harmonic constituents as the following linear combinations:

$$H(\mathbf{x}, t; \omega_m) = h'(\mathbf{x}, t; \omega_m) + ih'\left(\mathbf{x}, t - \frac{\pi \text{ rad}}{2\omega_m}; \omega_m\right) \quad (7.20a)$$

$$U(\mathbf{x}, t; \omega_m) = u'(\mathbf{x}, t; \omega_m) + iu'\left(\mathbf{x}, t - \frac{\pi \text{ rad}}{2\omega_m}; \omega_m\right) \quad (7.20b)$$

$$\Psi(\mathbf{x}, t; \omega_m) = \psi'(\mathbf{x}, t; \omega_m) + i\psi'\left(\mathbf{x}, t - \frac{\pi \text{ rad}}{2\omega_m}; \omega_m\right). \quad (7.20c)$$

Each complex-valued harmonic constituent is the sum of the corresponding real-valued constituent and a purely imaginary-valued, time-shifted constituent. The real part is equal to, and therefore *in phase* with, the corresponding real harmonic constituent. In contrast, the imaginary part is a sinusoid whose phase differs from that of the real part by $\pi/2$ rad, so the imaginary part is in *quadrature* with the real part. See Section 14.4 for relevant discussion.

Exercise 7.8 Complex-Valued Harmonic Constituents

Problem: Use exponential notation to express the complex-valued harmonic constituents H, U, and Ψ compactly in terms of their corresponding real amplitude and phase functions.

Solution: Substituting the RHSs of (7.3) for the corresponding FRFs in (7.18) and grouping exponential terms give

$$H(\mathbf{x}, t; \omega_m) = M_h(\mathbf{x}; \omega_m) \exp\left\{i\left[\omega_m t + \theta_h(\mathbf{x}; \omega_m)\right]\right\} \quad (7.21a)$$

$$U(\mathbf{x}, t; \omega_m) = M_u(\mathbf{x}; \omega_m) \exp\left\{i\left[\omega_m t + \theta_u(\mathbf{x}; \omega_m)\right]\right\} \quad (7.21b)$$

$$\Psi(\mathbf{x}, t; \omega_m) = M_\psi(\mathbf{x}; \omega_m) \exp\left\{i\left[\omega_m t + \theta_\psi(\mathbf{x}; \omega_m)\right]\right\}. \quad (7.21c)$$

□

Multiplying (7.17) by $e^{i\omega_m t}$ produces

$$\nabla \cdot \left[\mathbf{K}(\mathbf{x})\nabla F_h e^{i\omega_m t}\right] - i\omega_m S_s(\mathbf{x}) F_h e^{i\omega_m t} = -F_u e^{i\omega_m t} \quad \forall\, \mathbf{x} \in D \quad (7.22a)$$

$$W\left[F_h e^{i\omega_m t}\right] = F_\psi e^{i\omega_m t} \quad \forall\, \mathbf{x} \in \Gamma \quad (7.22b)$$

$$\forall\, t > 0; \quad m = 1, 2, 3, \ldots.$$

Using the definitions of the complex-valued harmonic constituents (7.18) allows us to write these as

$$\nabla \cdot \left[\mathbf{K}(\mathbf{x})\nabla H(\mathbf{x}, t; \omega_m)\right] - i\omega_m S_s(\mathbf{x}) H(\mathbf{x}, t; \omega_m) = -U(\mathbf{x}, t; \omega_m) \quad \forall\, \mathbf{x} \in D \quad (7.23a)$$

$$W\left[H(\mathbf{x}, t; \omega_m)\right] = \Psi(\mathbf{x}, t; \omega_m) \quad \forall\, \mathbf{x} \in \Gamma \quad (7.23b)$$

$$\forall\, t > 0; \quad m = 1, 2, 3, \ldots.$$

Thus, the complex-valued, hydraulic head harmonic constituent $H(\mathbf{x}, t; \omega_m)$ satisfies BVP equations similar to those of the space BVP for the hydraulic head FRF (7.17). Using the definitions in (2.7) and (7.18a) to rewrite (7.23a), we obtain

$$L_h\left[H(\mathbf{x}, t; \omega_m)\right] + U(\mathbf{x}, t; \omega_m) = 0 \quad \forall\, \mathbf{x} \in D \quad (7.24a)$$

$$W\left[H(\mathbf{x}, t; \omega_m)\right] = \Psi(\mathbf{x}, t; \omega_m) \quad \forall\, \mathbf{x} \in \Gamma \quad (7.24b)$$

$$\forall\, t > 0; \quad m = 1, 2, 3, \ldots.$$

Therefore, the complex-valued, hydraulic head harmonic constituent $H(\mathbf{x}, t; \omega_m)$ also satisfies BVP equations similar to those of the space-time BVP for the hydraulic head harmonic constituent (5.26).

Exercise 7.9 Equivalence of Governing Equations

Problem: Show that the governing equation (7.24a) in the space-time BVP for the complex-valued, hydraulic head harmonic constituent is equivalent to the governing equations (5.10) in the rectangular form space BVP.

Solution: First rewrite (7.24a) in the form of (7.23a). Then, substituting the RHSs of (7.6a) and (7.6b) for F_h and F_u, respectively, in (7.18a) and (7.18b) gives

$$H(\mathbf{x}, t; \omega_m) \equiv \left[A_h(\mathbf{x}; \omega_m) + iB_h(\mathbf{x}; \omega_m)\right]e^{i\omega_m t}$$

$$U(\mathbf{x}, t; \omega_m) \equiv \left[A_u(\mathbf{x}; \omega_m) + iB_u(\mathbf{x}; \omega_m)\right]e^{i\omega_m t}.$$

Substituting the RHSs for the corresponding complex constituents in (7.23a) and then multiplying the resulting equation by $e^{-i\omega_m t}$ produce

$$\nabla \cdot \left[\mathbf{K}(\mathbf{x})\nabla A_h(\mathbf{x}; \omega_m)\right] + i\nabla \cdot \left[\mathbf{K}(\mathbf{x})\nabla B_h(\mathbf{x}; \omega_m)\right] + \omega_m S_s(\mathbf{x}) B_h(\mathbf{x}; \omega_m) - i\omega_m S_s(\mathbf{x}) A_h(\mathbf{x}; \omega_m) = -A_u(\mathbf{x}; \omega_m) - iB_u(\mathbf{x}; \omega_m).$$

Taking the real and imaginary parts, respectively, yields (5.10a) and (5.10b). □

Exercise 7.10 Equivalence of BC Equations

Problem: Show that the BC equation (7.24b) in the space-time BVP for the complex-valued, hydraulic head harmonic constituent is equivalent to the BC equations (5.11) in the rectangular form space BVP.

Solution: Substituting the RHSs of (7.6a) and (7.6c) for F_h and F_ψ, respectively, in (7.18a) and (7.18c) gives

$$H(\mathbf{x},t;\omega_m) \equiv \left[A_h(\mathbf{x};\omega_m) + iB_h(\mathbf{x};\omega_m)\right]e^{i\omega_m t}$$
$$\Psi(\mathbf{x},t;\omega_m) \equiv \left[A_\psi(\mathbf{x};\omega_m) + iB_\psi(\mathbf{x};\omega_m)\right]e^{i\omega_m t}.$$

Substituting the RHSs for the corresponding complex constituents in (7.24b), using (2.13) to expand the W operator, and then multiplying the resulting equation by $e^{-i\omega_m t}$ produce

$$\begin{aligned} &c_h(\mathbf{x})A_h(\mathbf{x};\omega_m)] + ic_h(\mathbf{x})B_h(\mathbf{x};\omega_m) \\ &\quad + c_q(\mathbf{x})\left[\mathbf{K}(\mathbf{x})\nabla A_h(\mathbf{x};\omega_m)\right]\cdot\hat{\mathbf{n}}(\mathbf{x}) \\ &\quad + ic_q(\mathbf{x})\left[\mathbf{K}(\mathbf{x})\nabla B_h(\mathbf{x};\omega_m)\right]\cdot\hat{\mathbf{n}}(\mathbf{x}) \\ &\quad = A_\psi(\mathbf{x};\omega_m) + iB_\psi(\mathbf{x};\omega_m). \end{aligned}$$

Taking the real and imaginary parts, respectively, yields (5.11). □

7.4. HYDRAULIC GRADIENT AND SPECIFIC-DISCHARGE FRFs

Define the complex-valued specific-discharge harmonic constituent (vector) as

$$\mathbf{Q}(\mathbf{x},t;\omega_m) \equiv \mathbf{q}'(\mathbf{x},t;\omega_m) + i\mathbf{q}'\left(\mathbf{x}, t - \frac{\pi\ \text{rad}}{2\omega_m};\omega_m\right) \quad (m = 1,2,3,\ldots)$$

where $\mathbf{q}'$ is the specific-discharge harmonic constituent (see Section 4.3). Then $\mathbf{Q}$ satisfies Darcy's law,

$$\mathbf{Q}(\mathbf{x},t;\omega_m) = -\mathbf{K}(\mathbf{x})\nabla H(\mathbf{x},t;\omega_m) \quad (m = 1,2,3,\ldots)$$

where H is the complex-valued hydraulic head harmonic constituent (7.18a). We can express the complex-valued hydraulic gradient and specific-discharge constituents as products of complex-amplitude vectors and a unit-modulus complex exponential of time:

$$\begin{aligned} \nabla H(\mathbf{x},t;\omega_m) &= \mathbf{F}_{\nabla h}(\mathbf{x};\omega_m)e^{i\omega_m t} \\ \mathbf{Q}(\mathbf{x},t;\omega_m) &= \mathbf{F}_q(\mathbf{x};\omega_m)e^{i\omega_m t} \\ m &= 1,2,3,\ldots \end{aligned}$$

where we define the complex-amplitude vectors as

$$\begin{aligned} \mathbf{F}_{\nabla h}(\mathbf{x};\omega_m) &\equiv \nabla F_h(\mathbf{x};\omega_m) \\ \mathbf{F}_q(\mathbf{x};\omega_m) &\equiv -\mathbf{K}(\mathbf{x})\nabla F_h(\mathbf{x};\omega_m) \\ m &= 1,2,3,\ldots. \end{aligned}$$

We refer to $\mathbf{F}_{\nabla h}(\mathbf{x};\omega_m)$ and $\mathbf{F}_q(\mathbf{x};\omega_m)$, respectively, as the hydraulic gradient and specific-discharge *vector FRFs*.

We can express each component of the FRF vectors as the product of a positive, real amplitude and a unit-modulus complex exponential of the corresponding phase:

$$\begin{aligned} F_{\nabla h,1}(\mathbf{x};\omega_m) &= M_{\nabla h,1}(\mathbf{x};\omega_m)\exp\left[i\theta_{\nabla h,1}(\mathbf{x};\omega_m)\right] \\ F_{\nabla h,2}(\mathbf{x};\omega_m) &= M_{\nabla h,2}(\mathbf{x};\omega_m)\exp\left[i\theta_{\nabla h,2}(\mathbf{x};\omega_m)\right] \\ F_{\nabla h,3}(\mathbf{x};\omega_m) &= M_{\nabla h,3}(\mathbf{x};\omega_m)\exp\left[i\theta_{\nabla h,3}(\mathbf{x};\omega_m)\right] \\ m &= 1,2,3,\ldots \end{aligned}$$

$$\begin{aligned} F_{q,1}(\mathbf{x};\omega_m) &= M_{q,1}(\mathbf{x};\omega_m)\exp\left[i\theta_{q,1}(\mathbf{x};\omega_m)\right] \\ F_{q,2}(\mathbf{x};\omega_m) &= M_{q,2}(\mathbf{x};\omega_m)\exp\left[i\theta_{q,2}(\mathbf{x};\omega_m)\right] \\ F_{q,3}(\mathbf{x};\omega_m) &= M_{q,3}(\mathbf{x};\omega_m)\exp\left[i\theta_{q,3}(\mathbf{x};\omega_m)\right] \\ m &= 1,2,3,\ldots. \end{aligned}$$

For each complex-vector component, the modulus and argument give the amplitude response function and phase response function, respectively. For the hydraulic gradient components these are

$$\begin{aligned} |F_{\nabla h,1}(\mathbf{x};\omega_m)| &= M_{\nabla h,1}(\mathbf{x};\omega_m) \\ |F_{\nabla h,2}(\mathbf{x};\omega_m)| &= M_{\nabla h,2}(\mathbf{x};\omega_m) \\ |F_{\nabla h,3}(\mathbf{x};\omega_m)| &= M_{\nabla h,3}(\mathbf{x};\omega_m) \\ m &= 1,2,3,\ldots \end{aligned}$$

$$\begin{aligned} \arg F_{\nabla h,1}(\mathbf{x};\omega_m) &= \theta_{\nabla h,1}(\mathbf{x};\omega_m) \\ \arg F_{\nabla h,2}(\mathbf{x};\omega_m) &= \theta_{\nabla h,2}(\mathbf{x};\omega_m) \\ \arg F_{\nabla h,3}(\mathbf{x};\omega_m) &= \theta_{\nabla h,3}(\mathbf{x};\omega_m) \\ m &= 1,2,3,\ldots. \end{aligned}$$

For the specific-discharge components these are

$$\begin{aligned} |F_{q,1}(\mathbf{x};\omega_m)| &= M_{q,1}(\mathbf{x};\omega_m) \\ |F_{q,2}(\mathbf{x};\omega_m)| &= M_{q,2}(\mathbf{x};\omega_m) \\ |F_{q,3}(\mathbf{x};\omega_m)| &= M_{q,3}(\mathbf{x};\omega_m) \\ m &= 1,2,3,\ldots \end{aligned}$$

$$\arg F_{q,1}(\mathbf{x};\omega_m) = \theta_{q,1}(\mathbf{x};\omega_m)$$
$$\arg F_{q,2}(\mathbf{x};\omega_m) = \theta_{q,2}(\mathbf{x};\omega_m)$$
$$\arg F_{q,3}(\mathbf{x};\omega_m) = \theta_{q,3}(\mathbf{x};\omega_m)$$
$$m = 1, 2, 3, \ldots .$$

7.5. CHAPTER SUMMARY

In this chapter we defined complex-valued frequency response functions for the harmonic constituents of hydraulic head, source strength, and boundary value. Then we derived a space BVP that governs the hydraulic head FRF. This BVP is mathematically equivalent to the space-time BVP for the real-valued, hydraulic head harmonic constituent. We presented exercises to illustrate that the complex-variable form of the space BVP is mathematically equivalent to the rectangular and polar forms derived in previous chapters.

We also defined complex-valued harmonic constituents of hydraulic head, source strength, and boundary value. Then we showed that the complex-valued, hydraulic head harmonic constituent satisfies BVPs analogous to (a) the general space-time BVP for the real-valued, hydraulic head harmonic constituent and (b) the complex form of the space BVP that governs the hydraulic head FRF.

Finally, we derived vector expressions for the FRFs of the corresponding hydraulic gradient and specific-discharge vectors.

8
Comparison of Space BVP Forms

8.1. INTRODUCTION

In Part II, we developed the following three basic forms of the space BVP for hydraulic head harmonic constituents:

- Rectangular form
- Polar form
- Complex-variable form

These forms are distinct yet mathematically equivalent. In this chapter we briefly summarize their characteristics and compare their strengths and weaknesses (see Table 8.1).

Each of the basic forms of the space BVP for hydraulic head harmonic constituents consists of the following specifications:

- Governing equation(s)
- BC equation(s)
- Spatial distributions of material hydrologic properties (i.e., hydraulic conductivity and specific storage)
- Space domain geometry [i.e., $\hat{\mathbf{n}}(\mathbf{x})$; see Section 2.1]

However, the three basic forms use the same specifications of hydraulic properties and domain geometry, so for the remainder of this chapter we will discuss the equations only.

No particular form is superior to the others for all applications. Rather, each form has unique characteristics that render it more suitable than the others for analysis or computation within a limited range of circumstances. Additionally, in any given situation personal preference may play a role in the selection of a particular form.

8.2. THREE MAIN FORMS

In all three of the main forms, the governing equations generally are second-order partial differential equations with space-variable coefficients.

8.2.1. Rectangular Form

The equations for the rectangular form of the space BVP consist of the governing equations (5.10) and the BC equations (5.11).

The strengths of the rectangular form are as follows:

- The corresponding unknowns are real valued.
- Its equations are linear in the unknowns and their derivatives.
- One can analytically uncouple its two governing equations (see Section 5.1.2), although doing so increases their order from 2 to 4. Compare the polar form, whose governing equations cannot be uncoupled except in special cases.
- Its BC equations are uncoupled.

Therefore, the rectangular form may be the preferred form for computation.

8.2.2. Polar Form

The equations for the polar form of the space BVP consist of the governing equations (6.11) and the BC equations (6.30).

The strengths of the polar form are as follows:

- The corresponding unknowns are real valued.
- It describes the space BVP in terms of the amplitude and the phase, whose meanings are physically intuitive.
- It clearly illustrates the nonlinear relationship between the amplitude and the phase.
- It is convenient for theoretical analyses of the following:

Time variation of the specific-discharge harmonic constituent (see Chapter 15)
Amplitude and phase extrema (see Chapter 17)
Diffusion wave propagation (see Chapter 19)
Mechanical energy transport (see Chapter 25)

Hydrodynamics of Time-Periodic Groundwater Flow: Diffusion Waves in Porous Media, Geophysical Monograph 224,
First Edition. Joe S. Depner and Todd C. Rasmussen.

Table 8.1 Comparison of forms of the Space BVP.

	Unknown(s)		Equations			
Form	Symbols	Type	Governing	BC	Linearity	Derivation
Rectangular	A_h, B_h	Real	2	2	Linear	Chapter 5
Polar	M_h, θ_h	Real	2	2	Nonlinear	Chapter 6
Complex	F_h	Complex	1	1	Linear	Chapter 7

- One can convert solutions expressed in polar form to complex-variable form, and vice versa, straightforwardly.

The first two of these points make the polar form useful for graphical illustration of FRFs.

With few exceptions, it is not possible to uncouple the governing equations for the polar form of the space BVP. One such exception involves time-periodic flow in a source-free, 1D medium, over which the amplitude function varies monotonically with distance. Even then, the uncoupled equations are third-order, nonlinear ODEs. For details see Section 6.2.3.

The polar form of the space BVP is impractical for solution for the following reasons. First, the governing equations for the polar form are coupled and nonlinear in the dependent variables and their derivatives. Second, in some problems the hydraulic head amplitude vanishes at one or more points in the space domain or on its boundary. At such points the hydraulic head phase is undefined and hence the polar forms of the governing or BC equations are invalid. This requires redefining the problem so that neither the space domain nor its boundary includes points for which the amplitude is zero. In addition, at such points the phase may be spatially discontinuous. Therefore, the polar form is not the preferred form for solution by analytical or numerical means.

8.2.3. Complex-Variable Form

The equations for the complex-variable form of the space BVP are equations 7.17.

The strengths of the complex-variable form are as follows:

- It is more compact than the polar and rectangular forms (two equations versus four equations).
- Its equations are linear in the dependent variables and their derivatives.
- One can convert solutions expressed in complex-variable form to polar form, and vice versa, straightforwardly.

These characteristics make the complex-variable form very convenient for symbolic manipulation and mathematical analysis and therefore the preferred form for analytical solution.

A disadvantage of the complex-variable form is that it seems mathematically more abstract than the real-variable (rectangular and polar) forms, especially for those who are unaccustomed to working with complex-variable BVPs. This is a psychological obstacle that one can overcome through experience.

8.3. OTHER FORMS

It may be possible to derive forms of the space BVP other than the three basic forms presented here, and such forms may have analytical or computational advantages under particular circumstances.

Part III
Elementary Examples

In this part we explore examples of elementary solutions of the complex-variable form of the space BVP.

The analytical solutions presented in this part serve multiple purposes, including the following:

- They illustrate the basic theory presented in Part II.
- In subsequent parts of this book we use them to illustrate advanced concepts such as stationary points, wave propagation, and energy transport.
- Generally, analytical solutions are useful for evaluating computational algorithms and numerical models.

In addition to the general assumptions of Part II (see Chapters 2 and 3), throughout this part we make the following specific assumptions:

- The medium is isotropic with respect to hydraulic conductivity:

$$\mathrm{K}(x) = K(x)\mathrm{I}$$

where I is the 3×3 identity tensor. We call $K(\mathrm{x})$ the *scalar hydraulic conductivity*.

- With a few exceptions, the space domain is source free:

$$F_u(x; \omega_m) = 0 \;\; \forall x \in D \;\; (m = 1, 2, 3, \ldots).$$

Exceptions include time-periodic flows involving point, line, and plane sources, but in such cases the solutions presented are valid only in the source-free portions of the space domains.

- The hydraulic head FRF consists of a single harmonic constituent, ω_m (m fixed). This is not a severe restriction because one can compose a more complicated FRF by superposing those FRFs corresponding to multiple, distinct harmonic constituents.

Readers interested in other examples of analytical solutions of time-periodic groundwater flow should consult the literature. For example, *Black and Kipp* [1981] and *Townley* [1995] present basic solutions for homogeneous media, while *Monachesi and Guarracino* [2011] and *Trefry* [1999] discuss solutions for nonhomogeneous media.

Hydrodynamics of Time-Periodic Groundwater Flow: Diffusion Waves in Porous Media, Geophysical Monograph 224,
First Edition. Joe S. Depner and Todd C. Rasmussen.

9
Examples: 1D Flow in Ideal Media

In this chapter we explore elementary examples of FRFs for time-periodic, 1D flow in ideal media.

9.1. ASSUMPTIONS

In addition to the assumptions listed in the introduction to Part III, throughout this chapter we assume the porous medium is ideal (see Section 2.6.1).

We also assume that the boundary conditions are such that the flow is purely 1D, so that we can represent the hydraulic head FRF as a function of a single Cartesian space coordinate (x):

$$F_{\mathrm{h}}(\mathbf{x}; \omega_m) = G(x; \omega_m). \tag{9.1}$$

9.2. GOVERNING EQUATION AND GENERAL SOLUTION

Exercise 9.1 Governing Equation for 1D Flow in Homogeneous Medium

Problem: Consider time-periodic, 1D flow in a source-free domain filled with a homogeneous, porous medium. Write the simplified governing equation for the hydraulic head FRF for this particular case using the parameter α_m.

Solution: Substituting the RHS of (9.1) for the FRF in the governing equation (12.1a) and evaluating the Laplacian (in rectangular coordinates) yield

$$\frac{d^2G}{dx^2} - i\alpha_m G = 0 \tag{9.2}$$

where (5.31) defines the real-valued constants α_m. □

The governing equation (9.2) is a linear, second-order, homogeneous ODE with constant coefficients. It is valid only in the source-free portion of the domain.

Exercise 9.2 Trial Solution for 1D Flow in Homogeneous Medium

Problem: Regarding the governing equation (9.2), consider the trial solution

$$G(x; \omega_m) = e^{\lambda_m x} \tag{9.3}$$

where λ_m is an undetermined constant (eigenvalue). Show that the trial function satisfies the governing equation by deriving the characteristic equation and then solving it to find the eigenvalue(s). How many distinct eigenvalues are there?

Solution: Substituting the RHS of (9.3) for G in (9.2), carrying out the differentiation, and then dividing by G lead to the characteristic equation

$$\lambda_m^2 - i\alpha_m = 0$$

which is quadratic in λ_m. Using the quadratic formula to solve it gives

$$\lambda_m = \pm\sqrt{i\alpha_m} \tag{9.4}$$

where $\sqrt{z}$ denotes the principal square root of z (see Section D.2.1). There are two distinct eigenvalues. □

We define the eigenvalues as

$$\lambda_m^{(n)} \equiv (-1)^n \sqrt{i\alpha_m} \quad (n = 1, 2) \tag{9.5}$$

where $\sqrt{i} = (1 + i)/\sqrt{2}$, and the corresponding *frequency response eigenfunctions* (FREs) as

$$G^{(n)}(x; \omega_m) = e^{\lambda_m^{(n)} x} \quad (n = 1, 2). \tag{9.6}$$

These are linearly independent, so we can construct a general solution to the homogeneous governing equation (9.2) by superposition as

Hydrodynamics of Time-Periodic Groundwater Flow: Diffusion Waves in Porous Media, Geophysical Monograph 224, First Edition. Joe S. Depner and Todd C. Rasmussen.

$$G(x;\omega_m) = a_m^{(1)} e^{\lambda_m^{(1)} x} + a_m^{(2)} e^{-\lambda_m^{(1)} x} \tag{9.7}$$

where the $a_m^{(n)}$ are complex constants (dimensions: L) to be determined by the boundary conditions.

Example 9.1 1D Flow in Homogeneous Medium: Zero-Frequency Limit
In Exercise 9.2 we developed a trial solution based on the assumption that $\omega_m > 0$, and we will continue to make this assumption throughout most of this chapter. However, if we want to get some idea of how the FRF behaves for very small frequencies, we also need to consider the special case where ω_m approaches zero. If the limit

$$\lim_{\omega \to 0} G(x;\omega)$$

exists, then we call this the *zero-frequency* limit. Knowing the zero-frequency limit FRF is useful because it gives us an endpoint, so to speak, for the range of behaviors possible for the FRF.

Substituting $\omega_m = 0$ in (5.31) gives $\alpha_m = 0$. Substituting $\alpha_m = 0$ in the governing equation (9.2) leads to

$$\frac{d^2}{dx^2} G(x;\omega_m) = 0.$$

Integrating twice gives

$$G(x;\omega_m) = c_m x + d_m \tag{9.8}$$

where c_m and d_m are complex constants to be determined by the boundary conditions. Equation (9.8) shows that, subject to the assumptions of this chapter, when the low-frequency limit FRF exists it is a linear function of the space coordinate. ◊

9.3. FINITE DOMAIN

In this section we examine time-periodic, 1D flow in a homogeneous domain of finite length (l).

9.3.1. Dirichlet Conditions at Both Boundaries

In this section we consider problems in which the FRF at the left and right boundaries are known (Dirichlet conditions):

$$G(0;\omega_m) = \text{known (constant for fixed } \omega_m) \tag{9.9a}$$

$$G(l;\omega_m) = \text{known (constant for fixed } \omega_m). \tag{9.9b}$$

Exercise 9.3 FRF Coefficients

Problem: Suppose the hydraulic head FRF is subject to the boundary conditions (9.9). Express the coefficients $a_m^{(1)}$ and $a_m^{(2)}$ in terms of $G(0;\omega_m)$ and $G(l;\omega_m)$, $\lambda_m^{(1)}$, and l.

Solution: Substituting $x = 0$, $x = l$, and $\lambda_m^{(2)} = -\lambda_m^{(1)}$ in (9.7) yields

$$G(0;\omega_m) = a_m^{(1)} + a_m^{(2)} \tag{9.10a}$$

$$G(l;\omega_m) = a_m^{(1)} e^{\lambda_m^{(1)} l} + a_m^{(2)} e^{-\lambda_m^{(1)} l}. \tag{9.10b}$$

This is a system of two linear, generally nonhomogeneous equations in the two unknowns $a_m^{(1)}$ and $a_m^{(2)}$. Its solution is

$$a_m^{(1)} = \frac{G(l;\omega_m) - G(0;\omega_m) e^{-\lambda_m^{(1)} l}}{2\sinh(\lambda_m^{(1)} l)} \tag{9.11a}$$

$$a_m^{(2)} = \frac{G(0;\omega_m) e^{\lambda_m^{(1)} l} - G(l;\omega_m)}{2\sinh(\lambda_m^{(1)} l)}. \tag{9.11b}$$

□

Consider the case where $|G(0;\omega_m)| > 0$. The BVP solution is given by (9.7), with the coefficients given by (9.11). In this case the FRF depends on five independent variables: x, l, α_m, $G(0;\omega_m)$, and $G(l;\omega_m)$. We can also express the FRF in terms of dimensionless variables. Define the dependent dimensionless variable as

$$G_D(X;\eta_m) \equiv \frac{G(x;\omega_m)}{G(0;\omega_m)} \tag{9.12}$$

and the independent dimensionless variables as

$$G_{DL} \equiv \frac{G(l;\omega_m)}{G(0;\omega_m)}$$
$$X \equiv \frac{x}{l} \tag{9.13}$$
$$\eta_m \equiv \sqrt{\alpha_m}\, l.$$

Rearranging these leads to

$$G(x;\omega_m) = G(0;\omega_m) G_D$$
$$G(l;\omega_m) = G(0;\omega_m) G_{DL}$$
$$x = lX$$
$$\alpha_m = \frac{\eta_m^2}{l^2}.$$

Substituting the RHSs for the corresponding quantities in (9.7) and (9.11) gives

$$G_D(X;\eta_m) = A_m^{(1)} e^{\Lambda_m^{(1)} X} + A_m^{(2)} e^{-\Lambda_m^{(1)} X} \tag{9.14}$$

where we define the dimensionless coefficients as

$$A_m^{(1)} \equiv \frac{G_{DL} - e^{-\Lambda_m^{(1)}}}{2\sinh \Lambda_m^{(1)}} \tag{9.15a}$$

$$A_m^{(2)} \equiv \frac{e^{\Lambda_m^{(1)}} - G_{DL}}{2\sinh \Lambda_m^{(1)}} \tag{9.15b}$$

Table 9.1 FRF cases: 1D, homogeneous media, Dirichlet/Dirichlet BCs

Case ID	G_{DL}	FRF graphs
a-0	0	Figure 9.1
b-0	1	Figure 9.2
c-0	−1	Figure 9.3
d-0	i	Figure 9.4
e-0	0.5	Figure 9.5

and the dimensionless eigenvalue as

$$\Lambda_m^{(1)} \equiv -\eta_m \sqrt{i}. \tag{9.16}$$

Thus, by expressing the FRF in terms of dimensionless variables, we have effectively reduced the total number of independent variables from five to three (X, η_m, and G_{DL}).

To illustrate the results of this section, we constructed the dimensionless FRF solutions for the cases listed in Table 9.1. Each case includes amplitude-versus-distance and phase-versus-distance plots for the three subcases $\eta_m = 0$, $\eta_m = 5$, and $\eta_m = 10$. Technically the subcases corresponding to $\eta_m = 0$ (i.e., zero frequency) are steady state rather than time periodic, but these are included because they illustrate limiting behavior of the amplitude and phase functions. In all of these plots we define the dimensionless phase function as

$$\theta_D(X; \eta_m) \equiv \arg G_D(X; \eta_m).$$

Exercise 9.4 Dimensionless FRF, Zero-Frequency Limit

Problem: Express the dimensionless hydraulic head FRF (9.14) compactly in terms of hyperbolic functions. Then derive an expression for the zero-frequency limit FRF in dimensionless form. Use the expression to evaluate the zero-frequency limit FRFs for the cases listed in Table 9.1. Are your results consistent with the amplitude and phase graphs shown in Figures 9.1–9.5?

Solution: Substituting the RHSs of (9.15) for $A_m^{(1)}$ and $A_m^{(2)}$, respectively, in (9.14) and rearranging yield

$$G_D(X; \eta_m) = \frac{G_{DL} \sinh(\Lambda_m^{(1)} X) - \sinh\left[\Lambda_m^{(1)}(X-1)\right]}{\sinh \Lambda_m^{(1)}}. \tag{9.17}$$

Because η_m is directly proportional to $\Lambda_m^{(1)}$ [see equation (9.16)], we have

$$G_D(X; 0) = \lim_{\Lambda_m^{(1)} \to 0} G_D(X; \eta_m).$$

Taking the limit of the RHS of (9.17) gives

$$\lim_{\Lambda_m^{(1)} \to 0} G_D(X; \eta_m) \to \frac{0}{0}$$

which is an indeterminate form. Using l'Hôpital's rule to evaluate the limit yields

$$G_D(X; 0) = (G_{DL} - 1)X + 1.$$

Applying the RHS to cases a-0, b-0, c-0, d-0, and e-0 yields $G_D(X; 0) = 1 - X$, 1, $1 - 2X$, $(i - 1)X + 1$, and $1 - 0.5X$, respectively. These are consistent with the graphs corresponding to $\eta_m = 0$ in Figures 9.1–9.5. □

Exercise 9.5 Dimensionless FRF

Problem: Suppose $G_{DL} = 0$. Express the dimensionless FRF compactly in terms of hyperbolic functions.

Solution: Substituting $G_{DL} = 0$ in (9.15) gives

$$A_m^{(1)} \equiv \frac{-e^{-\Lambda_m^{(1)}}}{2 \sinh \Lambda_m^{(1)}}$$

$$A_m^{(2)} \equiv \frac{e^{\Lambda_m^{(1)}}}{2 \sinh \Lambda_m^{(1)}}.$$

Substituting the RHSs for $A_m^{(1)}$ and $A_m^{(2)}$, respectively, in (9.14) and rearranging yield

$$G_D(X; \eta_m) = \frac{\sinh\left[\Lambda_m^{(1)}(1-X)\right]}{\sinh \Lambda_m^{(1)}}. \tag{9.18}$$

□

Exercise 9.6 Symmetry of Dimensionless FRF

Problem: Show that if $G_{DL} = 1$ then the dimensionless FRF exhibits the symmetry

$$G_D(1 - X; \eta_m) = G_D(X; \eta_m). \tag{9.19}$$

That is, show that the dimensionless FRF is symmetric about the point $X = 0.5$ (see Figure 9.2).

Solution: Substituting $G_{DL} = 1$ in (9.15) gives

$$A_m^{(1)} \equiv \frac{1 - e^{-\Lambda_m^{(1)}}}{2 \sinh \Lambda_m^{(1)}}$$

$$A_m^{(2)} \equiv \frac{e^{\Lambda_m^{(1)}} - 1}{2 \sinh \Lambda_m^{(1)}}.$$

Substituting the RHSs for $A_m^{(1)}$ and $A_m^{(2)}$, respectively, in (9.14) and rearranging yield

$$G_D(X; \eta_m) = \frac{\sinh(\Lambda_m^{(1)} X) + \sinh\left[\Lambda_m^{(1)}(1-X)\right]}{\sinh \Lambda_m^{(1)}}.$$

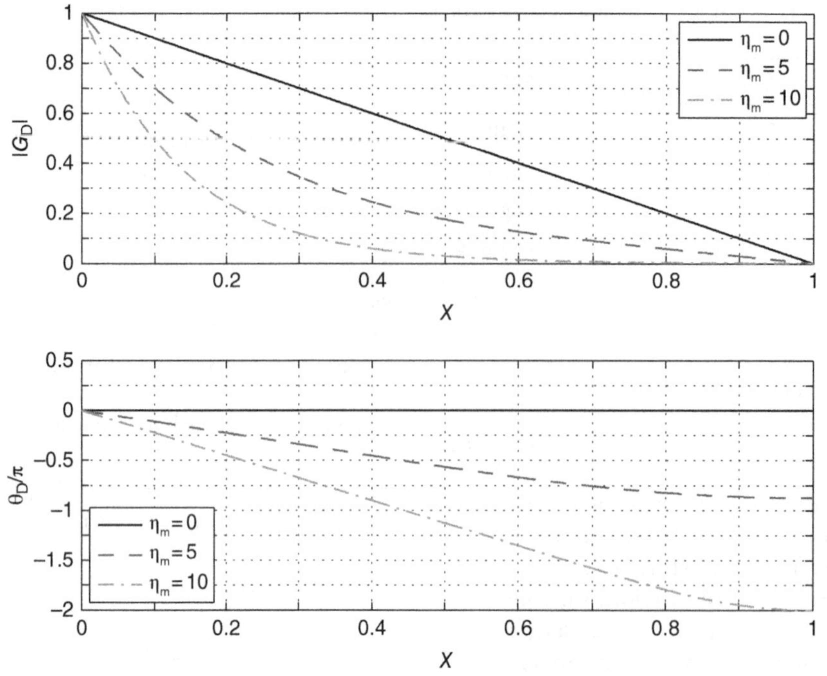

Figure 9.1 Case a-0 FRF (context: see Table 9.1).

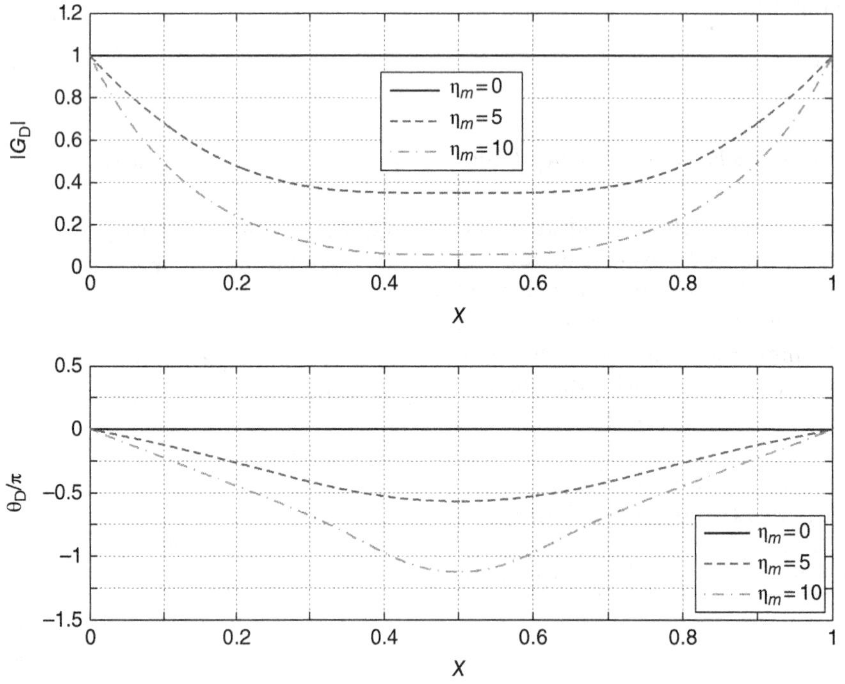

Figure 9.2 Case b-0 FRF (context: see Table 9.1).

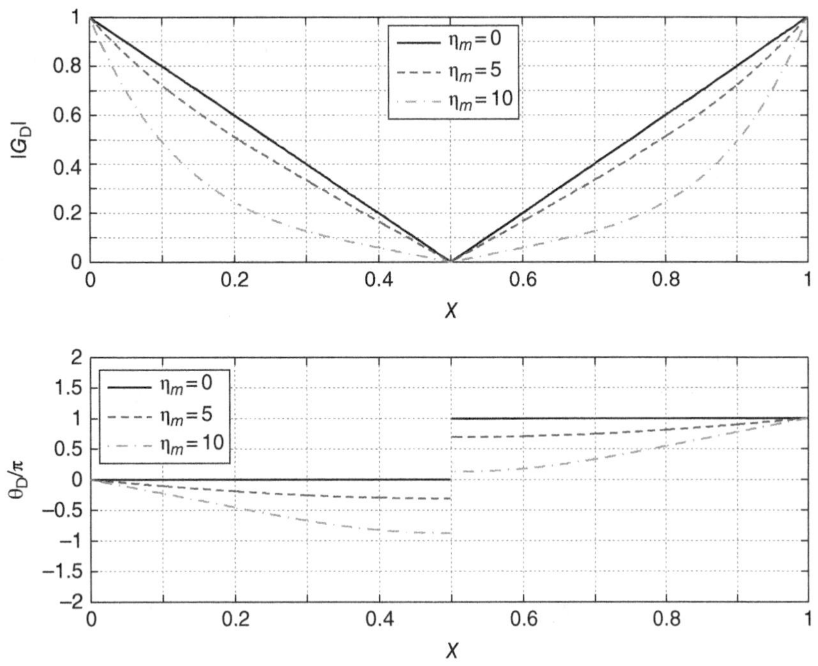

Figure 9.3 Case c-0 FRF (context: see Table 9.1).

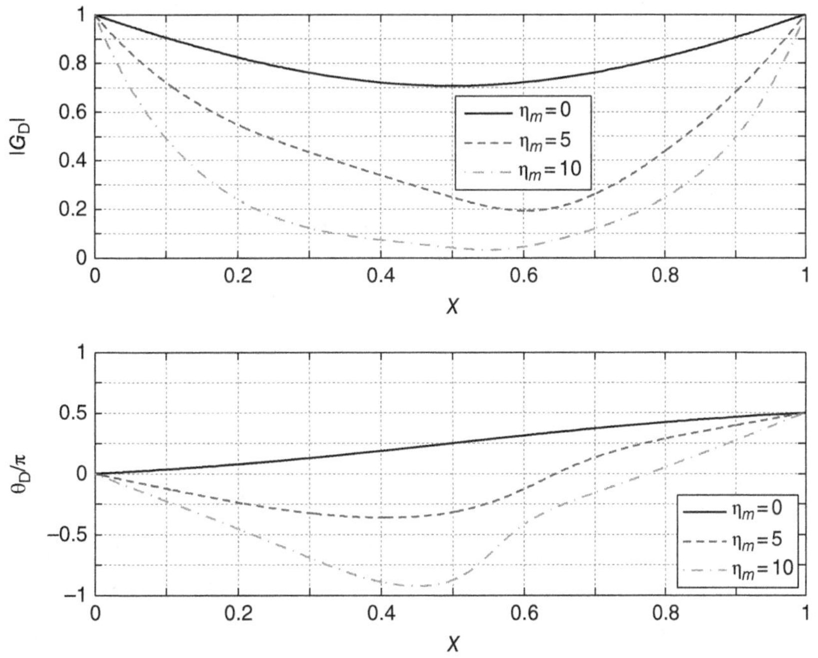

Figure 9.4 Case d-0 FRF (context: see Table 9.1).

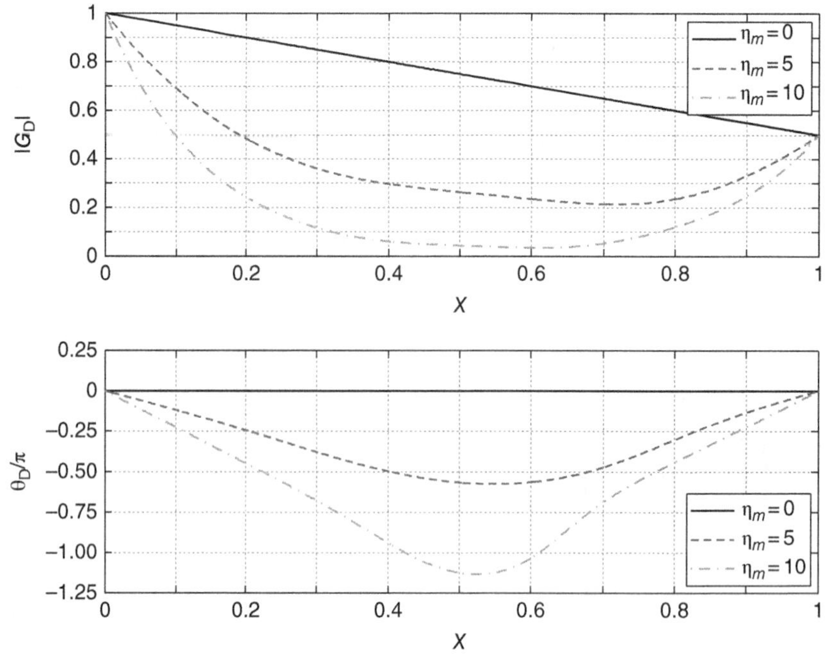

Figure 9.5 Case e-0 FRF (context: see Table 9.1).

Substituting $X = 1 - X$ in this equation

$$G_{\mathrm{D}}(1 - X; \eta_m) = \frac{\sinh\left[\Lambda_m^{(1)}(1 - X)\right] + \sinh(\Lambda_m^{(1)} X)}{\sinh \Lambda_m^{(1)}}. \tag{9.20}$$

The RHSs of the two previous equations are identical, so (9.19) follows. □

Exercise 9.7 Symmetry of Dimensionless FRF

Problem: Show that if $G_{\mathrm{DL}} = -1$ then the dimensionless FRF exhibits the symmetry

$$G_{\mathrm{D}}(1 - X; \eta_m) = -G_{\mathrm{D}}(X; \eta_m). \tag{9.21}$$

That is, show that the dimensionless FRF is antisymmetric about the point $X = 0.5$.

Solution: Substituting $G_{\mathrm{DL}} = -1$ in (9.15) yields

$$A_m^{(1)} \equiv \frac{-1 - e^{-\Lambda_m^{(1)}}}{2\sinh \Lambda_m^{(1)}} \tag{9.22a}$$

$$A_m^{(2)} \equiv \frac{e^{\Lambda_m^{(1)}} + 1}{2\sinh \Lambda_m^{(1)}}. \tag{9.22b}$$

Substituting the RHSs for $A_m^{(1)}$ and $A_m^{(2)}$, respectively, in (9.14) and rearranging yield

$$G_{\mathrm{D}}(X; \eta_m) = \frac{-\sinh(\Lambda_m^{(1)} X) + \sinh\left[\Lambda_m^{(1)}(1 - X)\right]}{\sinh \Lambda_m^{(1)}}. \tag{9.23}$$

Substituting $X = 1 - X$ in this equation gives

$$G_{\mathrm{D}}(1 - X; \eta_m) = \frac{-\sinh\left[\Lambda_m^{(1)}(1 - X)\right] + \sinh(\Lambda_m^{(1)} X)}{\sinh \Lambda_m^{(1)}}. \tag{9.24}$$

The RHSs of (9.23) and (9.24) differ only in sign, effectively proving (9.21). □

Exercise 9.8 Symmetry of Dimensionless Amplitude and Phase Functions

Problem: Suppose $G_{\mathrm{DL}} = -1$. What can you say about the symmetry of the corresponding amplitude and phase functions? *Hint*: Use Exercise 9.7.

Solution: Taking the modulus of (9.21) yields

$$|G_{\mathrm{D}}(1 - X; \eta_m)| = |G_{\mathrm{D}}(X; \eta_m)|.$$

That is, the amplitude function is symmetric about the point $X = 0.5$. Taking the argument of (9.21) and noting that $G_{\mathrm{D}}(0; \eta_m) = 0$ and $G_{\mathrm{D}}(1; \eta_m) = \pi$ lead to

$$\theta_{\mathrm{D}}(1 - X; \eta_m) = \begin{cases} \theta_{\mathrm{D}}(X; \eta_m) + \pi \text{ rad}, & 0 \le X < 0.5 \\ \theta_{\mathrm{D}}(X; \eta_m) - \pi \text{ rad}, & 0.5 < X \le 1. \end{cases}$$

That is, the phase function is symmetric about the point $X = 0.5$, except that, in addition, over the interval $0.5 < X \le 1$ the phase function is biased (shifted upward) by π rad (see Figure 9.3). □

9.3.2. Dirichlet/Neumann Conditions at Left/Right Boundaries

In this section we consider problems in which the FRF at the left boundary is known (Dirichlet condition) and its derivative at the right boundary is known (Neumann condition):

$$G(0;\omega_m) = \text{known (constant for fixed } \omega_m) \tag{9.25a}$$

$$\begin{aligned} G_\nabla(l;\omega_m) &\equiv \left[\frac{d}{dx}G(x;\omega_m)\right]\Bigg|_{x=l} \\ &= \text{known (constant for fixed } \omega_m). \end{aligned} \tag{9.25b}$$

Exercise 9.9 FRF Coefficients

Problem: Suppose the hydraulic head FRF is subject to the boundary conditions (9.25). Express the coefficients $a_m^{(1)}$ and $a_m^{(2)}$ in terms of $G(0;\omega_m)$ and $G_\nabla(l;\omega_m)$, $\lambda_m^{(1)}$, and l.

Solution: Differentiating (9.7) with respect to x gives

$$\frac{d}{dx}G(x;\omega_m) = a_m^{(1)}\lambda_m^{(1)}e^{\lambda_m^{(1)}x} - a_m^{(2)}\lambda_m^{(1)}e^{-\lambda_m^{(1)}x}.$$

Substituting $x = 0$ in (9.7) and $x = l$ in the previous equation, respectively, yields

$$G(0;\omega_m) = a_m^{(1)} + a_m^{(2)} \tag{9.26a}$$

$$G_\nabla(l;\omega_m) = a_m^{(1)}\lambda_m^{(1)}e^{\lambda_m^{(1)}l} - a_m^{(2)}\lambda_m^{(1)}e^{-\lambda_m^{(1)}l}. \tag{9.26b}$$

This is a system of two linear, generally nonhomogeneous equations in the two unknowns $a_m^{(1)}$ and $a_m^{(2)}$. Its solution is

$$a_m^{(1)} = \frac{G_\nabla(l;\omega_m) + \lambda_m^{(1)}e^{-\lambda_m^{(1)}l}G(0;\omega_m)}{2\lambda_m^{(1)}\cosh(\lambda_m^{(1)}l)} \tag{9.27a}$$

$$a_m^{(2)} = \frac{\lambda_m^{(1)}e^{\lambda_m^{(1)}l}G(0;\omega_m) - G_\nabla(l;\omega_m)}{2\lambda_m^{(1)}\cosh(\lambda_m^{(1)}l)}. \tag{9.27b}$$

□

Consider the FRF that satisfies boundary conditions (9.25) with $|G(0;\omega_m)| > 0$. The BVP solution is given by (9.7), with the coefficients given by (9.27). In this case the FRF depends on five independent variables: x, l, α_m, $G(0;\omega_m)$, and $G_\nabla(l;\omega_m)$. We can also express the FRF in terms of dimensionless variables. Define the dependent dimensionless variable

$$G_D(X;\eta_m) \equiv \frac{G(x;\omega_m)}{G(0;\omega_m)} \tag{9.28}$$

and the independent dimensionless variables

$$\begin{aligned} G_{D\nabla L} &\equiv \frac{lG_\nabla(l;\omega_m)}{G(0;\omega_m)} \\ X &\equiv \frac{x}{l} \\ \eta_m &\equiv \sqrt{\alpha_m}\, l. \end{aligned} \tag{9.29}$$

Table 9.2 FRF cases: 1D, homogeneous media, Dirichlet/Neumann BCs

Case ID	$G_{D\nabla L}$	FRF graphs
m-0	0	Figure 9.6
n-0	1	Figure 9.7
p-0	−1	Figure 9.8
q-0	i	Figure 9.9

Rearranging these leads to

$$\begin{aligned} G(x;\omega_m) &= G(0;\omega_m)G_D \\ G_\nabla(l;\omega_m) &= \frac{G(0;\omega_m)}{l}G_{D\nabla L} \\ x &= lX \\ \alpha_m &= \frac{\eta_m^2}{l^2}. \end{aligned}$$

Substituting the RHSs for the corresponding quantities in (9.7) and (9.27) leads to

$$G_D(X;\eta_m) = A_m^{(1)}e^{\Lambda_m^{(1)}X} + A_m^{(2)}e^{-\Lambda_m^{(1)}X} \tag{9.30}$$

where we define the dimensionless coefficients as

$$A_m^{(1)} \equiv \frac{G_{D\nabla L} + \Lambda_m^{(1)}e^{-\Lambda_m^{(1)}}}{2\Lambda_m^{(1)}\cosh\Lambda_m^{(1)}} \tag{9.31a}$$

$$A_m^{(2)} \equiv \frac{\Lambda_m^{(1)}e^{\Lambda_m^{(1)}} - G_{D\nabla L}}{2\Lambda_m^{(1)}\cosh\Lambda_m^{(1)}} \tag{9.31b}$$

and the dimensionless eigenvalue as

$$\Lambda_m^{(1)} \equiv -\eta_m\sqrt{i}. \tag{9.32}$$

Thus, by expressing the FRF in terms of dimensionless variables, we have effectively reduced the total number of independent variables from five to three (X, η_m, and $G_{D\nabla L}$).

To illustrate the results of this section, we constructed the dimensionless FRF solutions for the four cases listed in Table 9.2. Each case includes amplitude-versus-distance and phase-versus-distance plots for the three subcases $\eta_m = 0$, $\eta_m = 5$, and $\eta_m = 10$. Technically the subcases corresponding to $\eta_m = 0$ (i.e., zero frequency) are steady state rather than time periodic, but these are included because they illustrate limiting behavior of the amplitude and phase functions. In all of these plots we define the dimensionless phase function as

$$\theta_D(X;\eta_m) \equiv \arg G_D(X;\eta_m).$$

Exercise 9.10 Zero-Frequency Limit of Dimensionless FRF

Problem: Express the dimensionless hydraulic head FRF (9.30) compactly in terms of hyperbolic functions. Then derive an expression for the zero-frequency limit

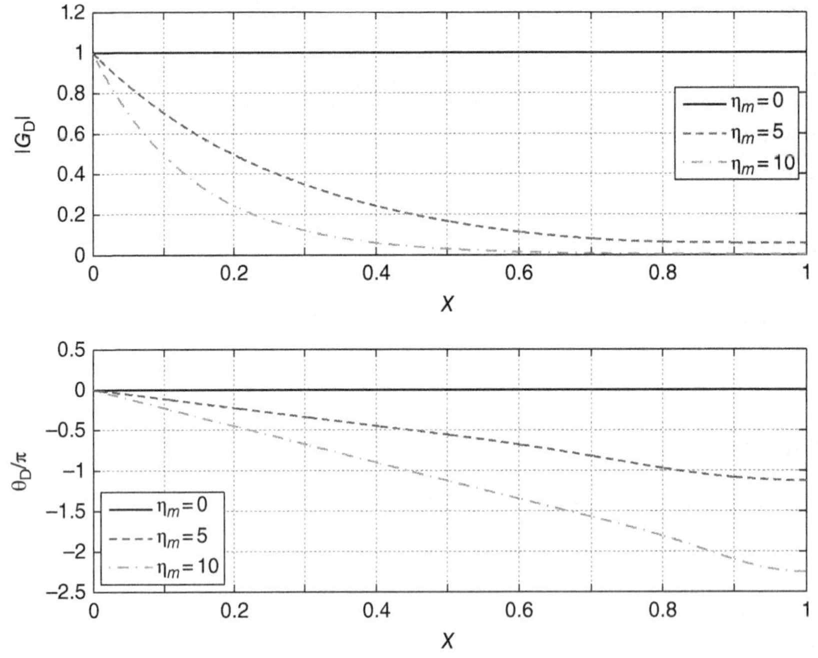

Figure 9.6 Case m-0 FRF (context: see Table 9.2).

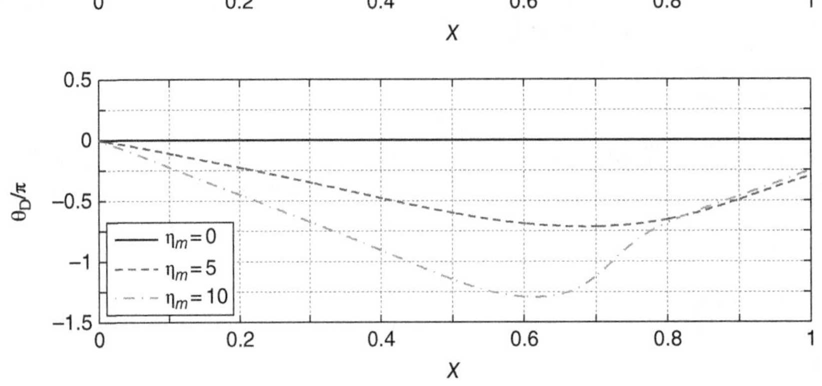

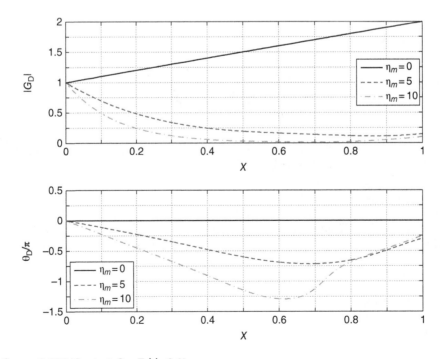

Figure 9.7 Case n-0 FRF (Context: See Table 9.2).

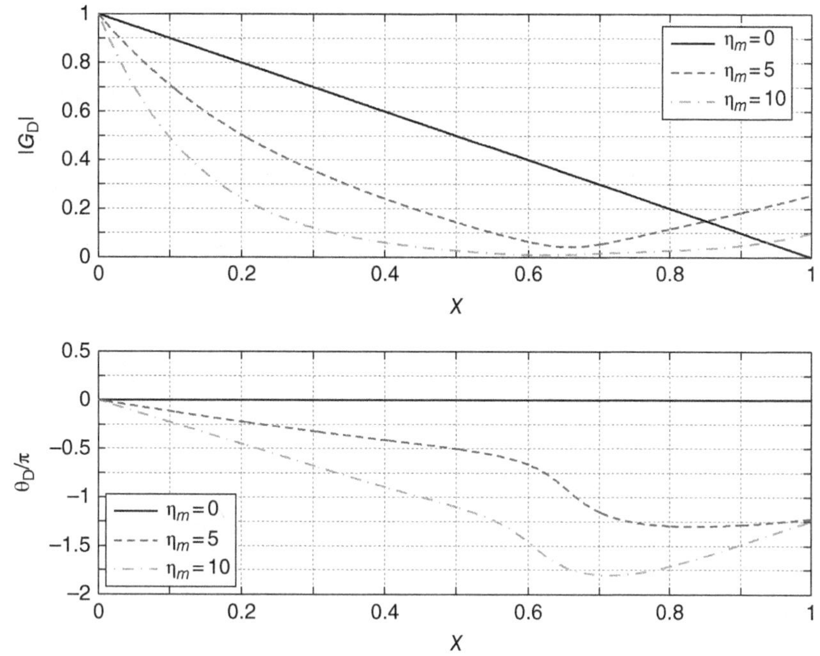

Figure 9.8 Case p-0 FRF (context: see Table 9.2).

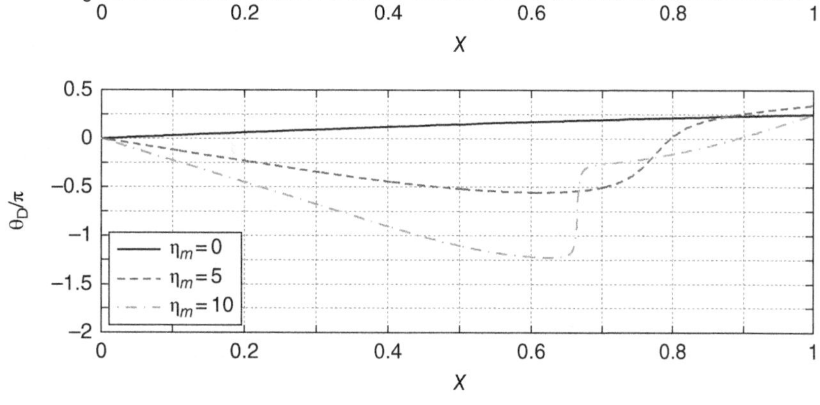

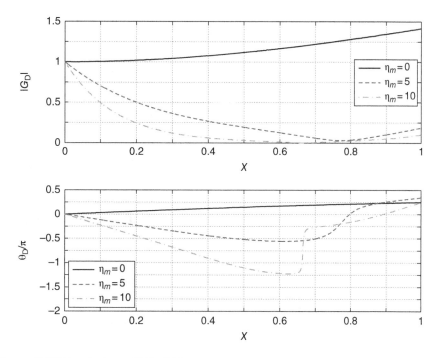

Figure 9.9 Case q-0 FRF (context: see Table 9.2).

FRF in dimensionless form. Use the expression to evaluate the zero-frequency limit FRFs for the cases listed in Table 9.2. Are your results consistent with the amplitude and phase graphs shown in Figures 9.6–9.9?

Solution: Substituting the RHSs of (9.31) for $A_m^{(1)}$ and $A_m^{(2)}$, respectively, in (9.30) and rearranging yield

$$G_D(X;\eta_m) = \frac{G_{D\nabla L}(\Lambda_m^{(1)})^{-1}\sinh(\Lambda_m^{(1)}X) + \cosh\left[\Lambda_m^{(1)}(X-1)\right]}{\cosh\Lambda_m^{(1)}}. \tag{9.33}$$

Because η_m is directly proportional to $\Lambda_m^{(1)}$ [see equation (9.32)], we have

$$G_D(X;0) = \lim_{\Lambda_m^{(1)}\to 0} G_D(X;\eta_m).$$

Taking the limit of the sinh term in the numerator of (9.33) gives

$$\lim_{\Lambda_m^{(1)}\to 0}\frac{\sinh(\Lambda_m^{(1)}X)}{\Lambda_m^{(1)}} \to \frac{0}{0}$$

which is an indeterminate form. Using l'Hôpital's rule to evaluate it leads to

$$\lim_{\Lambda_m^{(1)}\to 0}\frac{\sinh(\Lambda_m^{(1)}X)}{\Lambda_m^{(1)}} = \lim_{\Lambda_m^{(1)}\to 0}\frac{X\cosh(\Lambda_m^{(1)}X)}{1} = X.$$

Using this result to evaluate the zero-frequency limit FRF yields

$$G_D(X;0) = G_{D\nabla L}X + 1.$$

Applying the RHS to cases m-0, n-0, p-0, and q-0 yields $G_D(X;0) = 1$, $X+1$, $1-X$, and $iX+1$, respectively. These are consistent with the graphs corresponding to $\eta_m = 0$ in Figures 9.6–9.9. □

9.3.2.1. No-Flow Boundary

A special type of Neumann boundary condition is the no-flow condition. Consider a time-periodic, 1D flow with a no-flow condition at the right boundary, i.e., $G_\nabla(l;\omega_m) = 0$. What values do the amplitude and phase gradients take at the no-flow boundary? We can write the FRF in terms of its corresponding amplitude and phase functions as

$$G(x;\omega_m) = M_h(x;\omega_m)e^{i\theta_h(x;\omega_m)}.$$

Recall that this expression is only valid at those points where the amplitude is nonzero; at the amplitude function's zero points the phase is undefined (see Section 4.2.2). Differentiating with respect to x, using the product rule, and then simplifying the result give

$$\frac{d}{dx}G(x;\omega_m) = e^{i\theta_h(x;\omega_m)}\left(\frac{dM_h}{dx} + iM_h\frac{d\theta_h}{dx}\right). \tag{9.34}$$

Evaluating this equation at the no-flow boundary (i.e., $x = l$) yields

$$0 = \left.\frac{dM_h}{dx}\right|_{x=l} + iM_h(l;\omega_m)\left.\frac{d\theta_h}{dx}\right|_{x=l}$$

where we have used the fact that the exponential term on the RHS of (9.34) has magnitude 1. Taking the real and imaginary parts, respectively, gives

$$\left.\left[\frac{d}{dx}M_h(x;\omega_m)\right]\right|_{x=l} = 0 \tag{9.35a}$$

$$M_h(l;\omega_m)\left.\left[\frac{d}{dx}\theta_h(x;\omega_m)\right]\right|_{x=l} = 0. \tag{9.35b}$$

The first of these requires that the amplitude gradient vanishes at the no-flow boundary. Assuming the amplitude is nonzero there, the second of these requires that the phase gradient also vanishes at the no-flow boundary.

Exercise 9.11 Expression for FRF

Problem: Suppose the hydraulic head FRF is subject to a Dirichlet condition at the left boundary and a no-flow condition at the right boundary. Express the FRF compactly in terms of $G(0;\omega_m)$, $\lambda_m^{(1)}$, l, and x.

Solution: Substituting $G_\nabla(l;\omega_m) = 0$ in (9.27) gives

$$a_m^{(1)} = \frac{e^{-\lambda_m^{(1)}l}G(0;\omega_m)}{2\cosh(\lambda_m^{(1)}l)} \tag{9.36a}$$

$$a_m^{(2)} = \frac{e^{\lambda_m^{(1)}l}G(0;\omega_m)}{2\cosh(\lambda_m^{(1)}l)}. \tag{9.36b}$$

Substituting the RHSs of (9.36) for $a_m^{(1)}$ and $a_m^{(2)}$ in (9.7) and then simplifying the result yield

$$G(x;\omega_m) = G(0;\omega_m)\frac{\cosh[\lambda_m^{(1)}(x-l)]}{\cosh(\lambda_m^{(1)}l)}.$$

□

9.3.3. Cauchy Conditions at Left Boundary

In this section we briefly consider problems in which the FRF and its derivative with respect to the space coordinate, both evaluated at the left boundary, are known:

$$G(0;\omega_m) = \text{known (constant for fixed } \omega_m) \tag{9.37a}$$

$$G_\nabla(0;\omega_m) \equiv \left.\left[\frac{d}{dx}G(x;\omega_m)\right]\right|_{x=0} = \text{known (constant for fixed } \omega_m). \tag{9.37b}$$

No condition is specified for the right boundary. We refer to (9.37) as *Cauchy boundary conditions*.

Exercise 9.12 Cauchy Boundary Conditions; FRF Coefficients

Problem: Suppose the hydraulic head FRF is subject to the Cauchy boundary conditions (9.37). Express the coefficients $a_m^{(1)}$ and $a_m^{(2)}$ in terms of $G(0;\omega_m)$, $G_\nabla(0;\omega_m)$, and λ_m.

Solution: Substituting $x = 0$ in (9.7) yields

$$G(0;\omega_m) = a_m^{(1)} + a_m^{(2)}. \tag{9.38}$$

Differentiating (9.7) with respect to x and then substituting $x = 0$ in the resulting expression lead to

$$G_\nabla(0;\omega_m) = \lambda_m^{(1)} a_m^{(1)} - \lambda_m^{(1)} a_m^{(2)}. \tag{9.39}$$

Equations (9.38) and (9.39) constitute a linear system of two nonhomogeneous equations in the two unknowns ($a_m^{(1)}$ and $a_m^{(2)}$). The solution is

$$a_m^{(1)} = \frac{1}{2}\left[G(0;\omega_m) + \frac{G_\nabla(0;\omega_m)}{\lambda_m^{(1)}}\right] \tag{9.40a}$$

$$a_m^{(2)} = \frac{1}{2}\left[G(0;\omega_m) - \frac{G_\nabla(0;\omega_m)}{\lambda_m^{(1)}}\right] \tag{9.40b}$$

□

Exercise 9.13 Cauchy Boundary Conditions

Problem: Suppose the hydraulic head FRF is subject to the Cauchy boundary conditions (9.37), with an impermeable boundary at $x = 0$. Express the FRF compactly in terms of $G(0;\omega_m)$, $\lambda_m^{(1)}$, and x.

Solution: Substituting $G_\nabla(0;\omega_m) = 0$ in (9.40) gives

$$a_m^{(1)} = a_m^{(2)} = \frac{1}{2}G(0;\omega_m). \tag{9.41}$$

Substituting these results for $a_m^{(1)}$ and $a_m^{(2)}$ in (9.7) yields

$$G(x;\omega_m) = G(0;\omega_m)\cosh(\lambda_m^{(1)}x).$$

□

Exercise 9.14 Cauchy Boundary Conditions

Problem: Suppose the hydraulic head FRF is subject to the Cauchy boundary conditions (9.37), with $G(0;\omega_m) = 0$. Express the FRF compactly in terms of $G_\nabla(0;\omega_m)$, $\lambda_m^{(1)}$, and x.

Solution: Substituting $G(0;\omega_m) = 0$ in (9.40) yields

$$a_m^{(1)} = \frac{G_\nabla(0;\omega_m)}{2\lambda_m^{(1)}} \tag{9.42a}$$

$$a_m^{(2)} = \frac{-G_\nabla(0;\omega_m)}{2\lambda_m^{(1)}}. \tag{9.42b}$$

Substituting the RHSs of these equations for $a_m^{(1)}$ and $a_m^{(2)}$ in (9.7) and then simplifying the result give

$$G(x;\omega_m) = \frac{G_\nabla(0;\omega_m)}{\lambda_m^{(1)}}\sinh(\lambda_m^{(1)}x).$$

□

Exercise 9.15 Cauchy Boundary Conditions

Problem: Suppose the hydraulic head FRF is subject to the Cauchy boundary conditions (9.37). Does a nontrivial solution exist for which both the FRF and its derivative vanish at the left boundary?

Solution: Substituting $G(0;\omega_m) = G_\nabla(0;\omega_m) = 0$ in (9.40) yields $a_m^{(1)} = a_m^{(2)} = 0$. Alternatively, substituting $G(0;\omega_m) = 0$ in (9.41) or $G_\nabla(0;\omega_m) = 0$ in (9.42) yields the same result. In any case the solution is the trivial one: $G(x;\omega_m) = 0$ for $0 < x < l$. Therefore, for these particular boundary conditions a nontrivial solution does not exist. □

9.4. SEMI-INFINITE DOMAIN

In this section we examine time-periodic, 1D flow in a homogeneous, semi-infinite domain.

9.4.1. Jacob-Ferris Solution

Jacob [1950] and *Ferris* [1951] described a semi-infinite, 1D medium for which one end is periodically forced with a single constituent (i.e., ω_m fixed) and the fluctuations vanish as the distance from the forced end increases without bound. Although neither *Jacob* [1950] nor *Ferris* [1951] formulated the problem in terms of a complex-valued FRF, we will do so here. The boundary conditions are

$$G(0;\omega_m) = \text{known (constant for fixed } \omega_m\text{) and nonzero} \tag{9.43a}$$

$$\lim_{x\to\infty} G(x;\omega_m) = 0. \tag{9.43b}$$

Combining (9.43) with (9.7) and using the eigenvalue definitions in (9.5) give

$$a_m^{(1)} = G(0;\omega_m) \tag{9.44a}$$

$$a_m^{(2)} = 0. \tag{9.44b}$$

Substituting these results for $a_m^{(1)}$ and $a_m^{(2)}$ in (9.7) yields the solution:

$$G(x;\omega_m) = G(0;\omega_m)e^{\lambda_m^{(1)}x}. \tag{9.45}$$

In this book we will refer to (9.45) as the *Jacob-Ferris solution*.

Example 9.2 The *Jacob-Ferris* Solution: Dimensionless FRF

We can express the solution (9.45) in terms of dimensionless variables as

$$G_{\mathrm{D}}(X) = e^{-\sqrt{i}X} \tag{9.46}$$

where

$$G_{\mathrm{D}} \equiv \frac{G(x;\omega_m)}{G(0;\omega_m)} \tag{9.47a}$$

$$X \equiv \sqrt{\alpha_m}x. \tag{9.47b}$$

Figure 9.10 shows the corresponding amplitude and phase functions. We define the dimensionless phase function as

$$\theta_{\mathrm{D}}(X) \equiv \arg G_{\mathrm{D}}(X).$$

Notice that the vertical axis of the amplitude plot is scaled logarithmically. As one moves from the forced boundary into the space domain (i.e., as X increases), the amplitude decreases exponentially and the phase decreases linearly. ◊

Exercise 9.16 Dimensionless FRF for Jacob-Ferris Solution

Problem: Consider the dimensionless FRF corresponding to the Jacob-Ferris solution (see Example 9.2). Derive compact expressions for the dimensionless amplitude and phase functions. What numerical values do your expressions give for the slopes of the $\log_{10}|G_{\mathrm{D}}|$-versus-$X$ and θ-versus-X curves? Are your results consistent with the graphs in Figure 9.10?

Solution: Taking the modulus and argument, respectively, of (9.46) yields

$$\left|G_{\mathrm{D}}(X)\right| = e^{-X/\sqrt{2}} \tag{9.48a}$$

$$\theta_{\mathrm{D}}(X) = \frac{-X}{\sqrt{2}}. \tag{9.48b}$$

Taking the base-10 logarithm of (9.48a) gives

$$\log_{10}\left|G_{\mathrm{D}}(X)\right| = bX$$

where the slope (dimensionless) is

$$b \equiv \frac{-\log_{10} e}{\sqrt{2}} \approx -0.3071.$$

The slope of the line described by (9.48b) is $-1/\sqrt{2} \approx -0.7071$. After scaling by π rad, this becomes $(-\pi\sqrt{2})^{-1} \approx -0.2251$. These numerical values appear to be consistent with the slopes of the graphs shown in Figure 9.10. □

Exercise 9.17 Neumann Boundary Conditions

Problem: Consider time-periodic, 1D flow in a semi-infinite medium, with the hydraulic head FRF subject to the Neumann boundary conditions:

$$\left.\left[\frac{d}{dx}G(x;\omega_m)\right]\right|_{x=0} = G_\nabla(0;\omega_m) = \text{known (constant for fixed } \omega_m) \tag{9.49a}$$

$$\lim_{x\to\infty}\left[\frac{d}{dx}G(x;\omega_m)\right] = 0 \tag{9.49b}$$

where $|G_\nabla(0;\omega_m)| > 0$. Find the hydraulic head FRF, $G(x;\omega_m)$.

Solution: Combining (9.49b) with (9.7) gives $a_m^{(2)} = 0$. Combining this result, (9.49a), and (9.7) gives $a_m^{(1)} = G_\nabla(0;\omega_m)/\lambda_m^{(1)}$. Substituting these results for $a_m^{(1)}$ and $a_m^{(2)}$ in (9.7) yields the solution:

$$G(x;\omega_m) = \frac{G_\nabla(0;\omega_m)}{\lambda_m^{(1)}}e^{\lambda_m^{(1)}x}. \tag{9.50}$$

□

Exercise 9.17 shows the equivalence between the Dirichlet and Neumann BVPs for time-periodic, 1D flow in a semi-infinite medium. Equation (9.50), evaluated at the left boundary ($x = 0$), shows the relationship between $G(0;\omega_m)$ and $G_\nabla(0;\omega_m)$ for this case.

9.4.2. Relation to Finite-Domain Solutions

In this section we show how one can derive the Jacob-Ferris solution from finite-domain solutions by a limit process.

Exercise 9.18 Finite-Domain and Infinite-Domain FRF Solutions

Problem: Consider the problem of time-periodic, 1D flow in a finite-length domain with Dirichlet conditions at both the left and right boundaries. Suppose the length l increases without bound (i.e., $l \to \infty$). What is the limiting form of the solution?

Solution: Substituting the RHSs of (9.11) for $a_m^{(1)}$ and $a_m^{(2)}$ in (9.7) and then taking the limit as l increases without bound lead to (9.45). □

Exercise 9.19 Finite-Domain and Infinite-Domain FRF Solutions

Problem: Consider the problem of time-periodic, 1D flow in a finite-length domain with a Dirichlet condition at the left boundary and a Neumann condition at the right

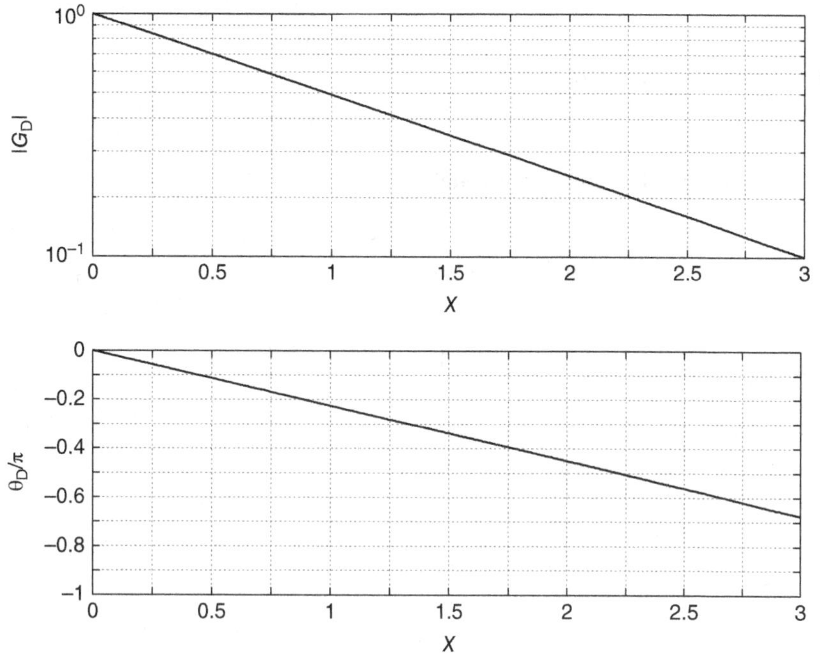

Figure 9.10 FRF for Example 9.2.

boundary. Suppose the length l increases without bound (i.e., $l \to \infty$). What is the limiting form of the solution?

Solution: Substituting the RHSs of (9.27) for $a_m^{(1)}$ and $a_m^{(2)}$ in (9.7) and then taking the limit as l increases without bound lead to (9.45). □

Taken together, Exercises 9.18 and 9.19 show that regardless of whether the right boundary is of the Dirichlet type or the Neumann type, the limiting form of the solution for a semi-infinite domain with Dirichlet condition at the left boundary is given by (9.45).

9.4.3. Relation to Point, Line, and Plane Sources

In this section we show how the Jacob-Ferris solution is related to the solutions corresponding to point, line, and plane sources.

Example 9.3 Rectangular Flow Due to Plane Source
Consider an infinite plane source of uniform strength located at $x = 0$ in a 3D, infinite domain filled with an ideal medium. The boundary conditions are

$$\lim_{D\to 0}\left[\left(-K\frac{dG}{dx}\bigg|_{x=D}\right)-\left(-K\frac{dG}{dx}\bigg|_{x=-D}\right)\right]=\Delta Q_0(\omega_m) \tag{9.51a}$$

$$\lim_{x\to -\infty} G(x;\omega_m) = 0 \tag{9.51b}$$

$$\lim_{x\to \infty} G(x;\omega_m) = 0 \tag{9.51c}$$

where $D > 0$ and $\Delta Q_0(\omega_m)$ denotes the complex amplitude of the plane source [i.e., $|\Delta Q_0(\omega_m)|$ gives the maximum (peak) volumetric rate of recharge per unit area of source] (dimensions: Lt^{-1}). The flow field is symmetric about the source plane, so in effect we only need to find the expression for the FRF on the half space $x > 0$.

Consider the half space $x > 0$. We will assume that the FRF due to the plane source takes the form of (9.7) there. Substituting the RHS of (9.7) for G in the boundary condition (9.51c) yields $a_m^{(2)} = 0$, so

$$G(x;\omega_m) = a_m^{(1)} e^{\lambda_m^{(1)} x}, \quad x > 0.$$

By symmetry, then the solution for the entire domain is

$$G(x;\omega_m) = a_m^{(1)} e^{\lambda_m^{(1)} |x|}, \quad |x| > 0. \tag{9.52}$$

Substituting the RHS for G in the BC (9.51a) yields

$$a_m^{(1)} = \frac{-\Delta Q_0(\omega_m)}{2K\lambda_m^{(1)}}.$$

Substituting the RHS for $a_m^{(1)}$ in (9.52) and then simplifying produce

$$G(x;\omega_m) = \frac{-\Delta Q_0(\omega_m)}{2K\lambda_m^{(1)}} e^{\lambda_m^{(1)}|x|}, \quad |x| > 0. \tag{9.53}$$

The FRF is undefined at $x = 0$. ◊

Exercise 9.20 Boundary Conditions

Problem: Consider Example 9.3. Verify that the FRF solution (9.53) satisfies the boundary conditions (9.51). □

Exercise 9.21 Phase Relationship: Hydraulic Head and Recharge

Problem: Consider Example 9.3. Define the limiting value of the FRF:

$$G_0(\omega_m) \equiv \lim_{x \to 0} G(x; \omega_m). \tag{9.54}$$

In the immediate vicinity of the plane source, what is the phase relationship between the hydraulic head fluctuations and the volumetric recharge fluctuations?

Solution: Substituting the RHS of (9.53) for $G(x; \omega_m)$ in (9.54) and then evaluating the limit give

$$G_0(\omega_m) = \frac{-\Delta Q_0(\omega_m)}{2K\lambda_m^{(1)}}.$$

Substituting the RHS of (9.5) (with $n = 1$) for $\lambda_m^{(1)}$ and simplifying yield

$$G_0(\omega_m) = \frac{\Delta Q_0(\omega_m)}{2K\sqrt{\alpha_m}} e^{-i\pi/4}.$$

That is, the near-source head fluctuations lag the recharge fluctuations by $\pi/4$ rad. □

Example 9.4 Plane Source FRF Derived from Point Source FRF

We can derive the plane source FRF (9.53) by superposition of the point source FRF. Imagine the plane source as an infinite set of identical point sources distributed continuously and uniformly along the x_2x_3 plane. Substitute the plane source amplitude $\Delta Q_0(\omega_m)$ for the point source amplitude $Q_0(\omega_m)$ (because the plane source is distributed continuously over the plane), and

$$\mathbf{x}_0 = [0 \quad u \quad v]^{\mathrm{T}}$$

in (12.16). Then integrate the resulting expression with respect to u and v:

$$G(x; \omega_m) = \frac{\Delta Q_0(\omega_m)}{4\pi K} \int_{-\infty}^{\infty} \int_{-\infty}^{\infty} \frac{\exp\left[-\lambda_m\sqrt{x^2 + (x_2 - u)^2 + (x_3 - v)^2}\right]}{\sqrt{x^2 + (x_2 - u)^2 + (x_3 - v)^2}} \, du \, dv$$

where (12.4) defines λ_m. Substituting $w = u - x_2$ and $y = v - x_3$ gives

$$G(x; \omega_m) = \frac{\Delta Q_0(\omega_m)}{4\pi K} \int_{-\infty}^{\infty} \int_{-\infty}^{\infty} \frac{\exp\left(-\lambda_m\sqrt{x^2 + w^2 + y^2}\right)}{\sqrt{x^2 + w^2 + y^2}} \, dw \, dy.$$

Substituting $s = \sqrt{w^2 + y^2}$ and using polar coordinates give

$$G(x; \omega_m) = \frac{\Delta Q_0(\omega_m)}{4\pi K} \int_0^{2\pi} \int_0^{\infty} \frac{\exp\left(-\lambda_m\sqrt{x^2 + s^2}\right)}{\sqrt{x^2 + s^2}} s \, ds \, d\theta.$$

Making the coordinate transformation $z \equiv \sqrt{x^2 + s^2}$ leads to

$$G(x; \omega_m) = \frac{\Delta Q_0(\omega_m)}{2K} \int_x^{\infty} e^{-\lambda_m z} \, dz.$$

Evaluating the integral leads to

$$G(x; \omega_m) = \frac{\Delta Q_0(\omega_m)}{2K\lambda_m} e^{-\lambda_m x}, \quad x > 0. \tag{9.55}$$

Comparing the definitions of the eigenvalues λ_m and $\lambda_m^{(1)}$ in (12.4) and (9.5), respectively, yields $\lambda_m = -\lambda_m^{(1)}$. Substituting this in (9.55) yields

$$G(x; \omega_m) = \frac{-\Delta Q_0(\omega_m)}{2K\lambda_m^{(1)}} e^{\lambda_m^{(1)} x}. \tag{9.56}$$

Extending the result to the left-half space ($x < 0$) by reflection yields (9.53). ◊

Example 9.5 Plane Source FRF Derived from Line Source FRF

We can also derive the plane source FRF (9.53) by superposition of the line source FRF. Imagine the plane source as an infinite set of identical, vertically oriented line sources distributed continuously and uniformly along the x_2 axis. Substitute the plane source amplitude $\Delta Q_0(\omega_m)$ for the line source amplitude (because the plane source is distributed continuously over the plane) and

$$\mathbf{x}_0 = [0 \quad u \quad x_3]^T$$

in (12.37). Then integrate the resulting expression with respect to u,

$$G(x; \omega_m) = \frac{\Delta Q_0(\omega_m)}{2\pi K} \int_{-\infty}^{\infty} K_0\left(\lambda_m\sqrt{x^2 + (x_2 - u)^2}\right) du$$

where (12.32) defines λ_m. Substituting $v = u - x_2$ gives

$$G(x;\omega_m) = \frac{\Delta Q_0(\omega_m)}{\pi K} \int_0^{\infty} K_0\left(\lambda_m\sqrt{x^2 + v^2}\right) dv$$

where we have used the fact that the integrand is even in v to change the lower limit of integration. Using equation 6.596 3 of *Gradshteyn and Ryzhik* [1980] to evaluate the integral produces

$$G(x;\omega_m) = \frac{\Delta Q_0(\omega_m)}{K} \sqrt{\frac{x}{2\pi\lambda_m}}\, K_{-1/2}\left(\lambda_m x\right)$$

where $K_{-1/2}$ denotes the modified Bessel function of the second kind, order $-1/2$. Using equation 8.469 3 of *Gradshteyn and Ryzhik* [1980] to rewrite the half-integer-order Bessel function in terms of an exponential leads to

$$G(x;\omega_m) = \frac{\Delta Q_0(\omega_m)}{2K\lambda_m} e^{-\lambda_m x}. \tag{9.57}$$

Comparing the definitions of the eigenvalues λ_m and $\lambda_m^{(1)}$ in (12.32) and (9.5), respectively, yields $\lambda_m = -\lambda_m^{(1)}$. Substituting this in (9.57) gives

$$G(x;\omega_m) = \frac{-\Delta Q_0(\omega_m)}{2K\lambda_m^{(1)}} e^{\lambda_m^{(1)} x}.$$

Extending the result to the left-half space ($x < 0$) by reflection yields (9.53). ◊

9.5. DISCUSSION

Additional results that are relevant to time-periodic, 1D flow in homogeneous media are given in Chapter 10, where they are presented within a slightly more general context.

The following observations regard the FRFs presented in this chapter, exclusive of zero-frequency limit FRFs (i.e., those for which $\eta_m = 0$):

- One or two amplitude maxima occurred in every example. These always occurred at one or both of the boundaries; none occurred in the space domain. The same is true of phase maxima.
- One amplitude minimum occurred in every example, either in the space domain or at one of the boundaries (e.g., cases a-0, m-0). The same is true of phase minima.
- In cases b-0, d-0, e-0, n-0, p-0, and q-0, both an amplitude minimum and a phase minimum occurred in the space domain. In case b-0 the amplitude and phase minima coincided, while in the other cases they did not.

In one example (case c-0), a zero-amplitude minimum occurred in the space domain at $X = 0.5$; the phase is discontinuous there. If we view this as two separate flow problems, one for the subregion $0 \le X < 0.5$ and one for the subregion $0.5 < X \le 1$, then the first two observations listed above apply to each of the two subregions in this example as well.

For the examples with Dirichlet conditions at both boundaries (cases a-0, b-0, c-0, d-0, and e-0), the infinite-domain example (see Section 9.4.1), and the example with a Dirichlet condition at one boundary and a no-flow condition at the other boundary (case m-0) and the dimensionless frequencies considered, both the amplitude and phase apparently are decreasing functions of the constituent frequency. This does not hold for the remaining examples with Dirichlet condition at one boundary and Neuman condition at the other (cases n-0, p-0, and q-0).

9.6. CHAPTER SUMMARY

In this chapter we briefly derived several elementary solutions of the complex-variable form of the space BVP for time-periodic, 1D flow in finite and semi-infinite domains filled with ideal media. We also explored, through examples and exercises, relationships between these solutions and the solutions for point, line, and plane sources in multidimensional domains filled with ideal media.

10

Examples: 1D Flow in Exponential Media

In this chapter we extend the 1D solutions for homogeneous media that we explored in Chapter 9 (e.g., 1d flow in ideal media) to a particular type of continuously nonhomogeneous media that we call *exponential media*. These are the simplest nontrivial solutions for time-periodic, 1D flow in continuously nonhomogeneous media of which we are aware.

10.1. ASSUMPTIONS

In addition to the assumptions listed in the introduction to Part III, throughout this chapter we assume the material hydrologic properties vary continuously in space as follows:

- The scalar hydraulic conductivity varies as

$$K(\mathbf{x}) = K(x) = K(0)e^{\mu x} \tag{10.1}$$

where $K(0)$ and μ are real constants, $K(0) > 0$, and x is a Cartesian space coordinate.

- The specific storage varies as

$$S_s(\mathbf{x}) = S_s(x) = S_s(0)e^{\mu x} \tag{10.2}$$

where $S_s(0)$ is a real, positive constant.
Consequently, the hydraulic diffusivity, which (2.18) defines, is a real, positive constant [i.e., $\kappa = K(0)/S_s(0)$]. In this book, we call media that satisfy these assumptions *exponential media*.

We also assume that the boundary conditions are such that the flow is purely 1D, so that we can represent the hydraulic head frequency response function (FRF) as a function of a single Cartesian space coordinate (x):

$$F_h(\mathbf{x};\omega_m) = G(x;\omega_m). \tag{10.3}$$

10.2. GOVERNING EQUATION AND GENERAL SOLUTION

Exercise 10.1 Governing Equation for 1D Flow in Exponential Medium

Problem: Consider time-periodic, 1D flow in a source-free domain filled with an exponential medium. Write the simplified governing equation for the hydraulic head FRF for this particular case using the parameters μ and α_m.

Solution: Substituting the RHSs of (10.1) and (10.2), respectively, for K and S_s, zero for $F_u(\mathbf{x};\omega_m)$, and $G(x;\omega_m)$ for $F_h(\mathbf{x};\omega_m)$ in the governing equation (7.17a) and then simplifying yield

$$\frac{d^2G}{dx^2} + \mu\frac{dG}{dx} - i\alpha_m G = 0 \tag{10.4}$$

where (5.31) defines the real-valued constants α_m. □

The governing equation (10.4) is a linear, second-order, homogeneous ODE with constant coefficients. It differs from (9.2) in that it contains an additional term, $\mu\, dG/dx$.

Exercise 10.2 Trial Solution for 1D Flow in Exponential Medium

Problem: Regarding the governing equation (10.4), consider the trial solution

$$G(x;\omega_m) = e^{\lambda_m x} \tag{10.5}$$

where λ_m is an undetermined constant (eigenvalue). Show that the trial function satisfies the governing equation by deriving the characteristic equation and then solving it to find the eigenvalue(s). How many distinct eigenvalues are there?

Hydrodynamics of Time-Periodic Groundwater Flow: Diffusion Waves in Porous Media, Geophysical Monograph 224, First Edition. Joe S. Depner and Todd C. Rasmussen.

Solution: Substituting the RHS of (10.5) for G in (10.4), carrying out the differentiation, and then dividing by G lead to the characteristic equation

$$\lambda_m^2 + \mu\lambda_m - i\alpha_m = 0$$

which is quadratic in λ_m. Using the quadratic formula to solve it gives

$$\lambda_m = \frac{-\mu \pm \sqrt{\mu^2 + i4\alpha_m}}{2} \tag{10.6}$$

where $\sqrt{z}$ denotes the principal square root of z (see Section D.2.1). There are two distinct eigenvalues. □

We define the eigenvalues as

$$\lambda_m^{(n)} \equiv \frac{-\mu + (-1)^n\sqrt{\mu^2 + i4\alpha_m}}{2} \quad (n = 1, 2) \tag{10.7}$$

and the corresponding FREs as

$$G^{(n)}(x; \omega_m) = e^{\lambda_m^{(n)} x} \quad (n = 1, 2). \tag{10.8}$$

These are linearly independent, so we can construct a general solution to the homogeneous governing equation (10.4) by superposition as

$$G(x; \omega_m) = a_m^{(1)} e^{\lambda_m^{(1)} x} + a_m^{(2)} e^{\lambda_m^{(2)} x} \tag{10.9}$$

where the $a_m^{(n)}$ are complex constants (dimensions: L) to be determined by the boundary conditions.

The general solution (10.9) is identical in form to that for a homogeneous medium (9.7). Additionally, (10.7) indicates that as μ approaches zero, the eigenvalues $\lambda_m^{(n)}$ approach the values corresponding to a homogeneous medium [see equations (9.5)]. Thus, the solution (10.9) is a generalization of that obtained in Section 9.2 for a homogeneous medium.

Exercise 10.3 Eigenvalues—Rectangular Form

Problem: Use the results presented in Section D.2.1 to express the principal square root $\sqrt{\mu^2 + i4\alpha_m}$ in the rectangular form $x + iy$, where x and y are real valued. Then use your result to express the eigenvalues $\lambda_m^{(n)}$ $(n = 1, 2)$ in (10.7) in rectangular form.

Solution: Define

$$z \equiv \mu^2 + i4\alpha_m.$$

Then $z = x + iy$ where $x = \mu^2$ and $y = 4\alpha_m$. Also, $r = \sqrt{\mu^4 + 16\alpha_m^2}$. Substituting these results for r, x, y, and z in (D.6) yields

$$\sqrt{\mu^2 + i4\alpha_m} = \sqrt{\frac{\sqrt{\mu^4 + 16\alpha_m^2} + \mu^2}{2}} + i\sqrt{\frac{\sqrt{\mu^4 + 16\alpha_m^2} - \mu^2}{2}}. \tag{10.10}$$

Substituting the RHS for the radical terms in (10.7) yields

$$\lambda_m^{(n)} = \frac{-\mu}{2} + \frac{(-1)^n}{2}\sqrt{\frac{\sqrt{\mu^4 + 16\alpha_m^2} + \mu^2}{2}} + i\frac{(-1)^n}{2}\sqrt{\frac{\sqrt{\mu^4 + 16\alpha_m^2} - \mu^2}{2}} \tag{10.11}$$

$n = 1, 2.$

□

We deduce the following directly from (10.11):

$$\text{Re}\lambda_m^{(1)} < 0 \tag{10.12a}$$

$$\text{Im}\lambda_m^{(1)} < 0 \tag{10.12b}$$

$$\text{Re}\lambda_m^{(2)} > 0 \tag{10.12c}$$

$$\text{Im}\lambda_m^{(2)} > 0 \tag{10.12d}$$

and

$$\text{Re}\lambda_m^{(1)} + \text{Re}\lambda_m^{(2)} = -\mu \tag{10.13a}$$

$$\text{Im}\lambda_m^{(1)} + \text{Im}\lambda_m^{(2)} = 0. \tag{10.13b}$$

10.3. FINITE DOMAIN

In this section we examine problems of time-periodic, 1D flow in a source-free, finite-length (l) domain filled with an exponential medium.

10.3.1. Dirichlet Conditions at Both Boundaries

In this section we consider problems in which the FRF at the left and right boundaries are known (Dirichlet conditions):

$$G(0; \omega_m) = \text{known (constant for fixed } \omega_m) \tag{10.14a}$$

$$G(l; \omega_m) = \text{known (constant for fixed } \omega_m). \tag{10.14b}$$

Exercise 10.4 FRF Coefficients

Problem: Suppose the hydraulic head FRF is subject to the boundary conditions (10.14). Express the coefficients $a_m^{(1)}$ and $a_m^{(2)}$ in terms of $G(0; \omega_m)$, $G(l; \omega_m)$, $\lambda_m^{(1)}$, $\lambda_m^{(2)}$, and l.

Solution: Substituting $x = 0$ and $x = l$ in (10.9) yields, respectively,

$$G(0; \omega_m) = a_m^{(1)} + a_m^{(2)} \tag{10.15a}$$

$$G(l; \omega_m) = a_m^{(1)} e^{\lambda_m^{(1)} l} + a_m^{(2)} e^{\lambda_m^{(2)} l}. \tag{10.15b}$$

This is a system of two linear, generally nonhomogeneous equations in the two unknowns $a_m^{(1)}$ and $a_m^{(2)}$. Its solution is

$$a_m^{(1)} = \frac{G(l;\omega_m) - G(0;\omega_m)e^{\lambda_m^{(2)} l}}{e^{\lambda_m^{(1)} l} - e^{\lambda_m^{(2)} l}} \tag{10.16a}$$

$$a_m^{(2)} = \frac{G(0;\omega_m)e^{\lambda_m^{(1)} l} - G(l;\omega_m)}{e^{\lambda_m^{(1)} l} - e^{\lambda_m^{(2)} l}}. \tag{10.16b}$$

□

Consider the BVP of Exercise 10.4, where $|G(0;\omega_m)| > 0$. The solution is given by (10.9), with the coefficients given by (10.16). In this case the FRF depends on six independent variables: x, l, μ, α_m, $G(0;\omega_m)$, and $G(l;\omega_m)$. We can also express the FRF in terms of dimensionless variables. Define the dependent dimensionless variable as

$$G_D(X;\eta_m,\nu) \equiv \frac{G(x;\omega_m)}{G(0;\omega_m)} \tag{10.17}$$

and the independent dimensionless variables as

$$G_{DL} \equiv \frac{G(l;\omega_m)}{G(0;\omega_m)} \tag{10.18a}$$

$$X \equiv \frac{x}{l} \tag{10.18b}$$

$$\nu \equiv \mu l \tag{10.18c}$$

$$\eta_m \equiv \sqrt{\alpha_m}\, l. \tag{10.18d}$$

Rearranging these leads to

$$G(x;\omega_m) = G(0;\omega_m)G_D \tag{10.19a}$$

$$G(l;\omega_m) = G(0;\omega_m)G_{DL} \tag{10.19b}$$

$$x = lX \tag{10.19c}$$

$$\mu = \frac{\nu}{l} \tag{10.19d}$$

$$\alpha_m = \frac{\eta_m^2}{l^2}. \tag{10.19e}$$

Substituting the RHSs for the corresponding quantities in (10.9) and (10.16) gives

$$G_D(X;\eta_m,\nu) = A_m^{(1)} e^{\Lambda_m^{(1)} X} + A_m^{(2)} e^{\Lambda_m^{(2)} X} \tag{10.20}$$

where we define the dimensionless coefficients as

$$A_m^{(1)} \equiv \frac{G_{DL} - e^{\Lambda_m^{(2)}}}{e^{\Lambda_m^{(1)}} - e^{\Lambda_m^{(2)}}} \tag{10.21a}$$

$$A_m^{(2)} \equiv \frac{e^{\Lambda_m^{(1)}} - G_{DL}}{e^{\Lambda_m^{(1)}} - e^{\Lambda_m^{(2)}}} \tag{10.21b}$$

and the dimensionless eigenvalues as

$$\Lambda_m^{(n)} \equiv \frac{-\nu + (-1)^n\sqrt{\nu^2 + i4\eta_m^2}}{2}. \tag{10.22}$$

Table 10.1 FRF cases: 1D, nonhomogeneous media, Dirichlet/Dirichlet BCs.

	G_{DL}			
η_m	0	1	-1	i
0	a-1 (Figure 10.1)	b-1 (Figure 10.4)	c-1 (Figure 10.7)	d-1 (Figure 10.10)
5	a-2 (Figure 10.2)	b-2 (Figure 10.5)	c-2 (Figure 10.8)	d-2 (Figure 10.11)
10	a-3 (Figure 10.3)	b-3 (Figure 10.6)	c-3 (Figure 10.9)	d-3 (Figure 10.12)

Thus, by expressing the FRF in terms of dimensionless variables we have effectively reduced the total number of independent variables from six to four (X, ν, η_m, and G_{DL}).

To illustrate the results of this section, we constructed the dimensionless FRF solutions for the 12 cases listed in Table 10.1. Each case includes amplitude-versus-distance and phase-versus-distance plots for the three values $\nu = \ln 0.1$, $\nu = 0$, and $\nu = \ln 10$. Technically the cases corresponding to $\eta_m = 0$ (i.e., the zero-frequency case) are steady state rather than time periodic, but these are included because they illustrate the limiting behavior of the amplitude and phase functions. In all of these plots we define the dimensionless phase function as

$$\theta_D(X;\eta_m,\nu) \equiv \arg G_D(X;\eta_m,\nu).$$

Exercise 10.5 Dimensionless FRF

Problem: Show that if $G(l;\omega_m) = 0$ then

$$G_D(X;\eta_m,\nu) = e^{-\nu X/2}\frac{\sinh[\xi(1-X)]}{\sinh\xi} \tag{10.23}$$

where

$$\xi \equiv \frac{\sqrt{\nu^2 + i4\eta_m^2}}{2}. \tag{10.24}$$

Solution: Using (10.24), we can write (10.22) as

$$\Lambda_m^{(n)} = \frac{-\nu}{2} + (-1)^n\xi \quad (n = 1, 2). \tag{10.25}$$

Substituting zero for G_{DL} and the RHSs of (10.25), respectively, for $\Lambda_m^{(n)}$ in (10.21) and then simplifying give

$$A_m^{(1)} = \frac{-e^{-\xi}}{2\sinh\xi} \tag{10.26a}$$

$$A_m^{(2)} = \frac{e^{\xi}}{2\sinh\xi}. \tag{10.26b}$$

Substituting the RHSs of these equations for $A_m^{(1)}$ and $A_m^{(2)}$ and the RHSs of (10.25) for $\Lambda_m^{\pm}$, respectively, in (10.20) and then simplifying yield (10.23). □

Exercise 10.6 Symmetry of Dimensionless Amplitude and Phase Functions

Problem: Suppose $G_{\mathrm{DL}} = 0$. Consider the amplitude and phase corresponding to the dimensionless FRF G_{D}. As functions of ν, are the amplitude and phase even, odd, or neither?

Solution: Taking the modulus of (10.23) gives the dimensionless amplitude:

$$\left|G_{\mathrm{D}}(X;\eta_m,\nu)\right| = e^{-\nu X/2}\left|\frac{\sinh\left[\xi(1-X)\right]}{\sinh\xi}\right|. \tag{10.27}$$

The first term on the RHS is neither even nor odd in ν. The term ξ is even in ν [see equation (10.24)], so the second term is even in ν. Therefore, the amplitude function is neither even nor odd in ν. Taking the argument of (10.23) yields the phase:

$$\theta_{\mathrm{D}}(X;\eta_m,\nu) = \arg\left[\frac{\sinh\left[\xi(1-X)\right]}{\sinh\xi}\right].$$

The RHS is even in ν. Hence, the phase function is even in ν (see Figures 10.1–10.3). □

Exercise 10.7 Symmetry of Dimensionless FRF

Problem: Show that if $G_{\mathrm{DL}} = 1$ then the dimensionless FRF exhibits the symmetry

$$G_{\mathrm{D}}(X;\eta_m,\nu) = G_{\mathrm{D}}(1-X;\eta_m,-\nu) \tag{10.28}$$

(see Figures 10.4–10.6).

Solution: Define

$$\xi \equiv \frac{\sqrt{\nu^2 + i4\eta_m^2}}{2}.$$

Using this definition we can write (10.22) and (10.21), respectively, as

$$\Lambda_m^{(n)} = \frac{-\nu}{2} + (-1)^n\xi \quad (n = 1, 2)$$

and

$$A_m^{(1)} = \frac{e^{\nu/2} - e^{-\xi}}{2\sinh\xi} \tag{10.29a}$$

$$A_m^{(2)} = \frac{e^{\xi} - e^{\nu/2}}{2\sinh\xi}. \tag{10.29b}$$

Substituting the RHSs of the four previous equations for the corresponding quantities in (10.20) and then simplifying yield

$$G_{\mathrm{D}}(X;\eta_m,\nu) = e^{-\nu X/2}\, g(X;\eta_m,\nu) \tag{10.30}$$

where

$$g(X;\eta_m,\nu) = \frac{\sinh\left[\xi(1-X)\right] + e^{\nu/2}\sinh(\xi X)}{\sinh\xi}.$$

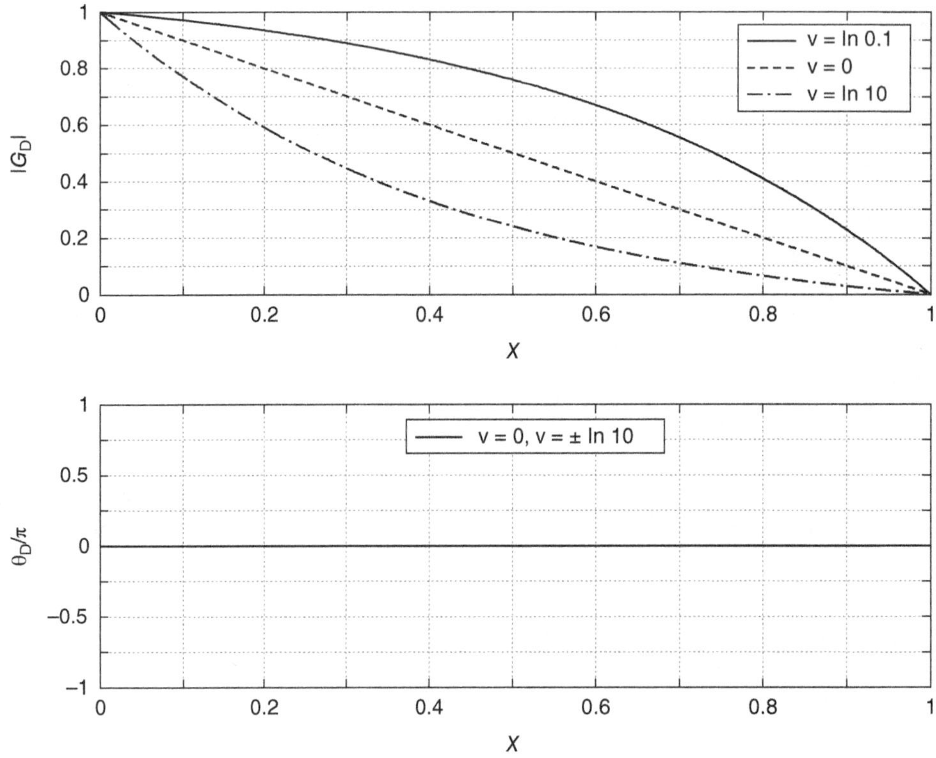

Figure 10.1 Case a-1 FRF (context: see Table 10.1).

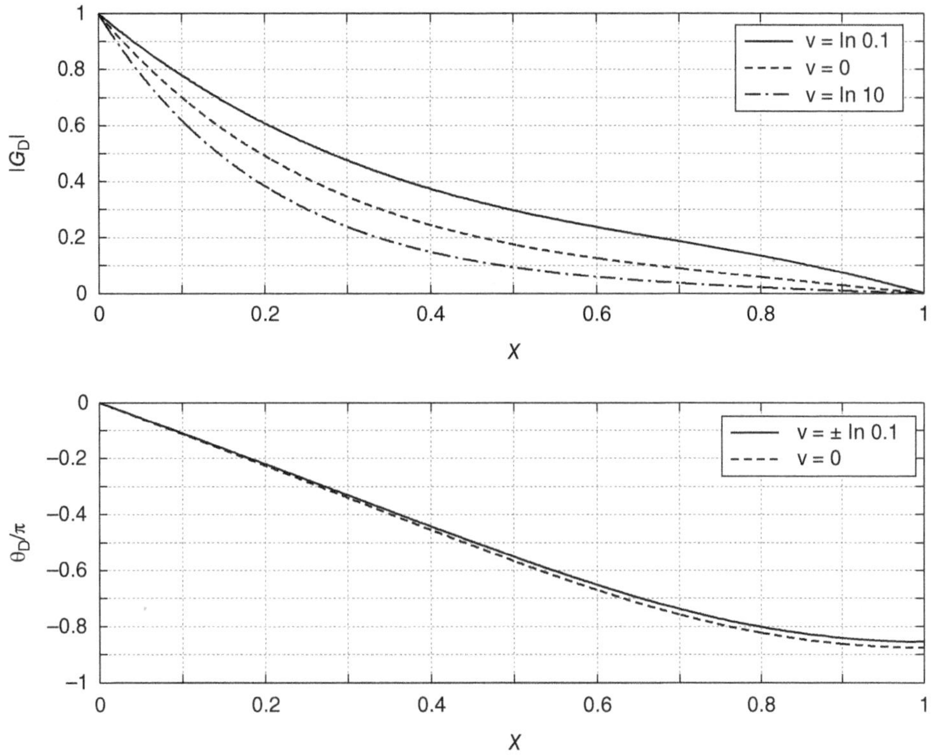

Figure 10.2 Case a-2 FRF (context: see Table 10.1).

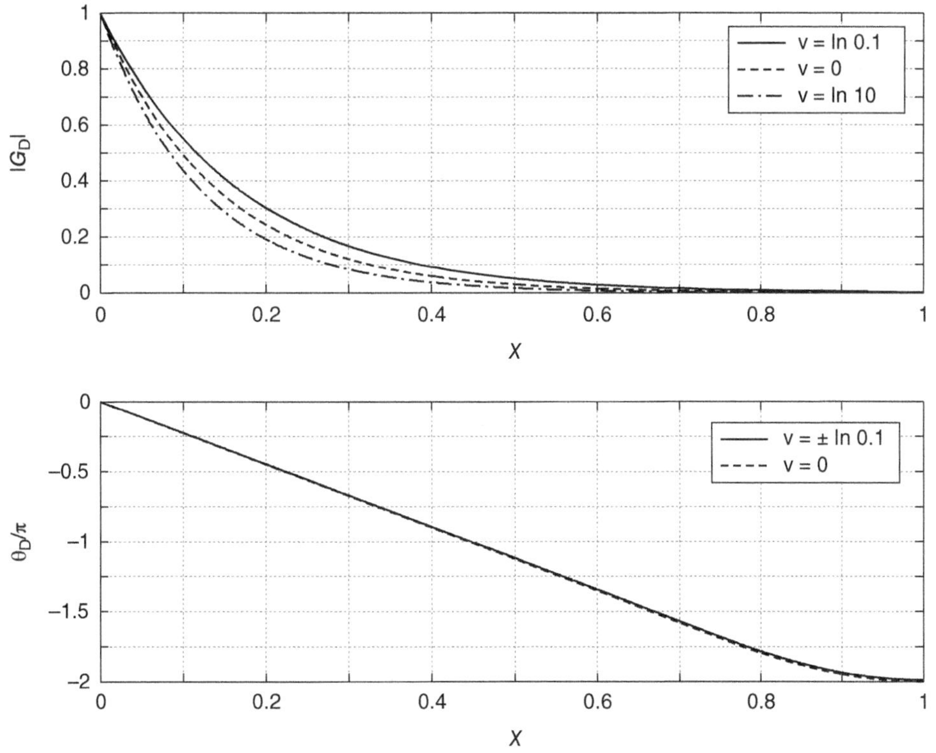

Figure 10.3 Case a-3 FRF (context: see Table 10.1).

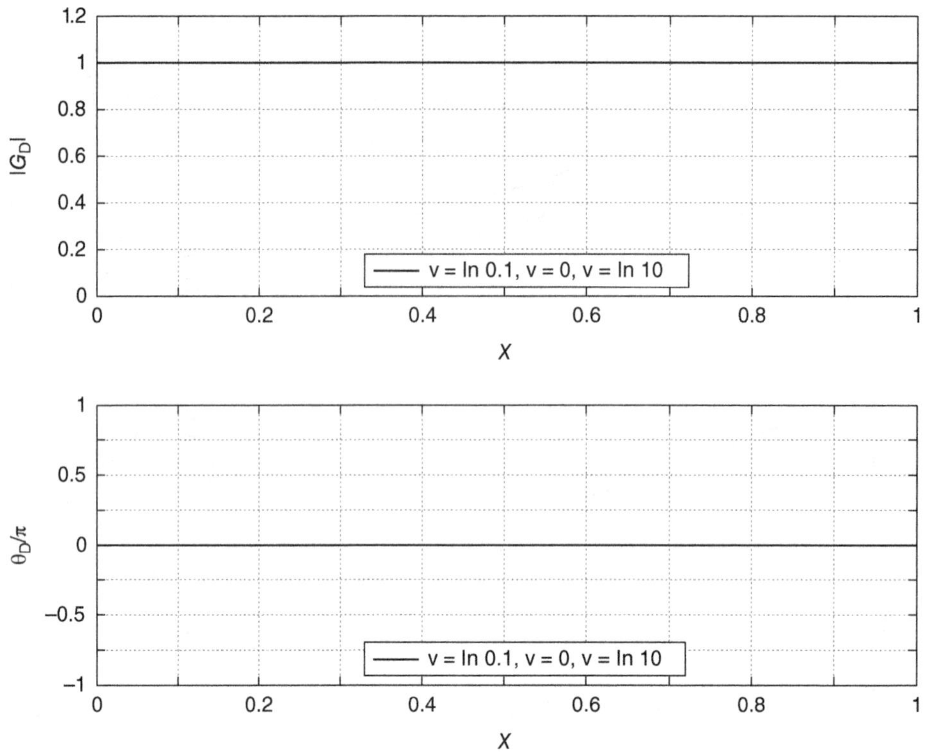

Figure 10.4 Case b-1 FRF (context: see Table 10.1).

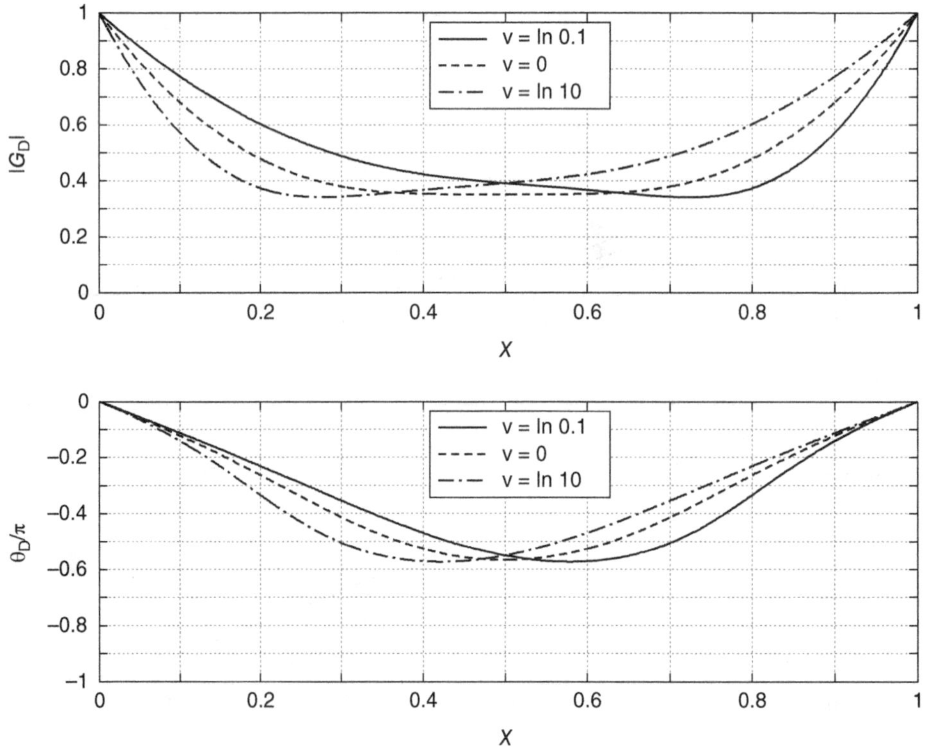

Figure 10.5 Case b-2 FRF (context: see Table 10.1).

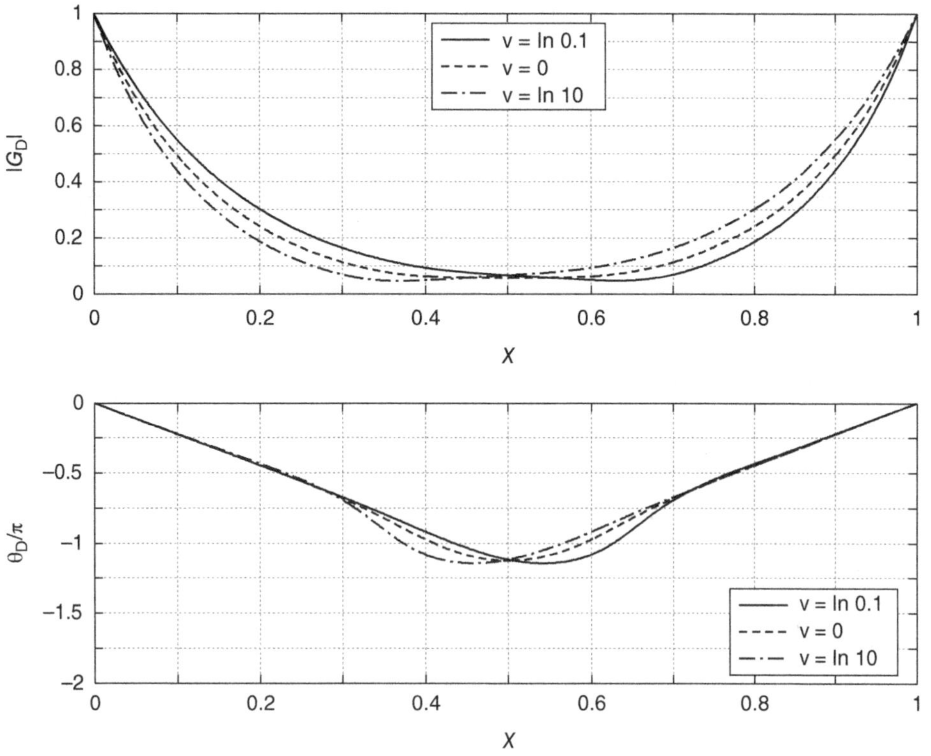

Figure 10.6 Case b-3 FRF (context: see Table 10.1).

Substituting $1-X$ for X and $-\nu$ for ν and then simplifying give

$$G_D(1-X;\eta_m,-\nu) = e^{-\nu X/2}\, g(X;\eta_m,\nu). \qquad (10.31)$$

The RHSs of (10.30) and (10.31) are identical, thus proving (10.28). □

Exercise 10.8 Symmetry of Dimensionless FRF

Problem: Show that if $G_{DL} = -1$ then the dimensionless FRF exhibits the symmetry

$$G_D(X;\eta_m,\nu) = -G_D(1-X;\eta_m,-\nu) \qquad (10.32)$$

(see Figures 10.7–10.9).

Solution: Define

$$\xi \equiv \frac{\sqrt{\nu^2 + i4\eta_m^2}}{2}.$$

Using this definition we can write (10.22) and (10.21), respectively, as

$$\Lambda_m^{(n)} = \frac{-\nu}{2} + (-1)^n \xi \quad (n = 1,2) \qquad (10.33a)$$

and

$$A_m^{(1)} = \frac{-e^{\nu/2} - e^{-\xi}}{2\sinh\xi} \qquad (10.33b)$$

$$A_m^{(2)} = \frac{e^{\xi} + e^{\nu/2}}{2\sinh\xi}. \qquad (10.33c)$$

Substituting the RHSs of the four previous equations for the corresponding quantities in (10.20) and then simplifying yield

$$G_D(X;\eta_m,\nu) = \frac{e^{-\nu X/2}}{\sinh\xi} f(X;\eta_m,\nu) \qquad (10.34)$$

where

$$f(X;\eta_m,\nu) = \sinh(\nu/2 - \xi X) + \sinh\left[\xi(1-X)\right].$$

Substituting $1-X$ for X and $-\nu$ for ν and then simplifying give

$$G_D(1-X;\eta_m,-\nu) = -\frac{e^{-\nu X/2}}{\sinh\xi} f(X;\eta_m,\nu). \qquad (10.35)$$

The RHSs of (10.34) and (10.35) differ only in sign, thus proving (10.32). □

Example 10.1 Dirichlet Conditions: Zero-Frequency Limit—Case 1
We can use the dimensionless form of the FRF to derive the zero-frequency limit FRF. Consider the two cases $\nu \geq 0$ and $\nu < 0$ separately.

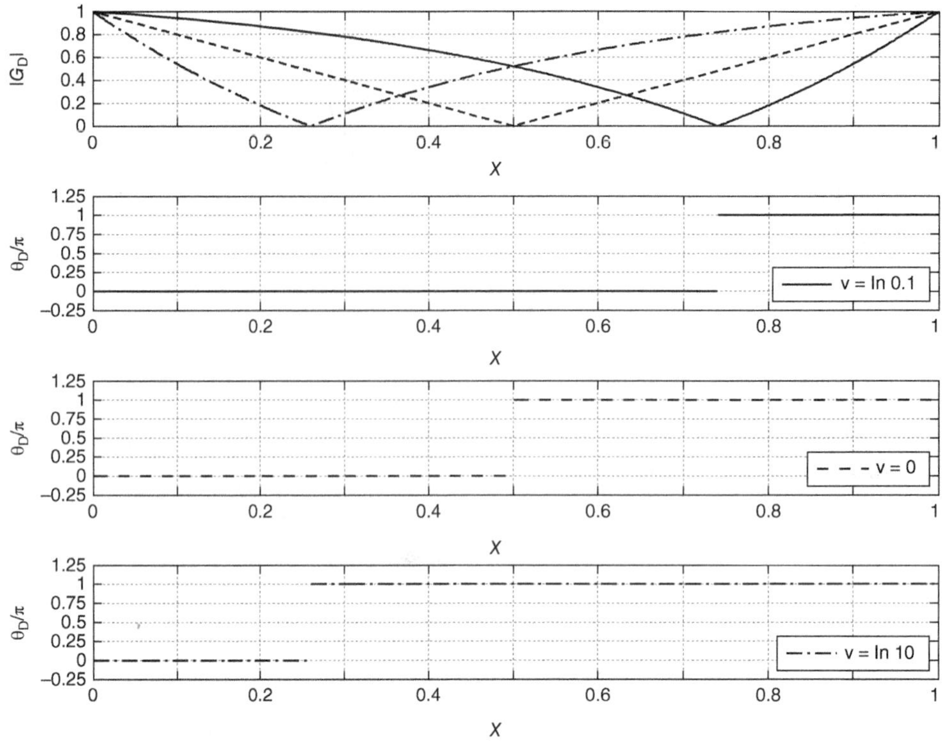

Figure 10.7 Case c-1 FRF (context: see Table 10.1).

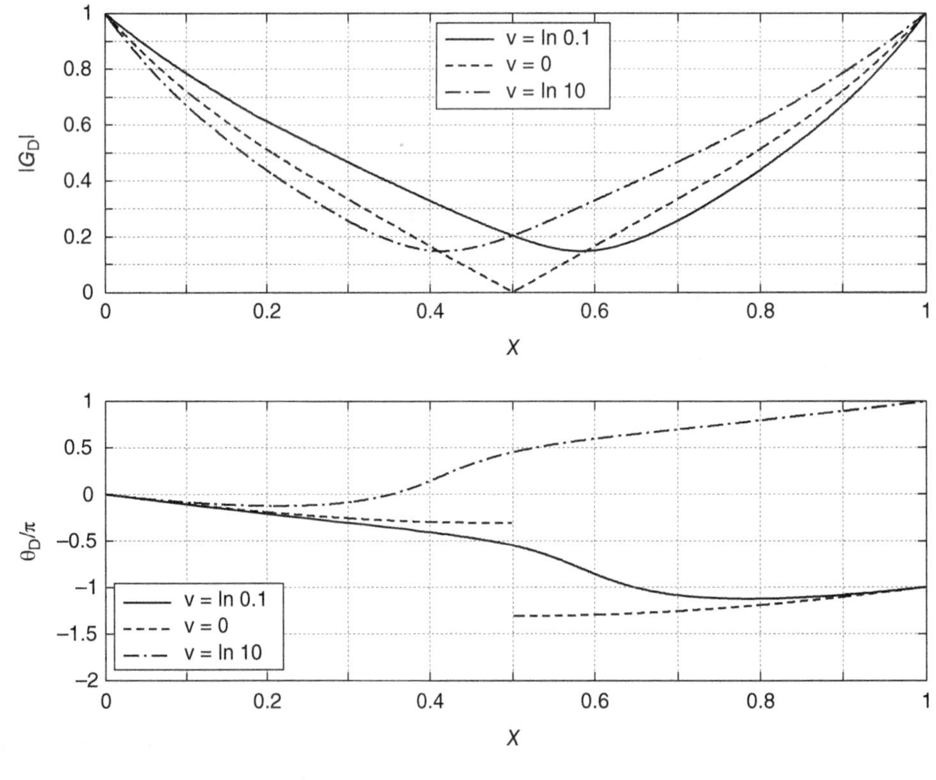

Figure 10.8 Case c-2 FRF (context: see Table 10.1).

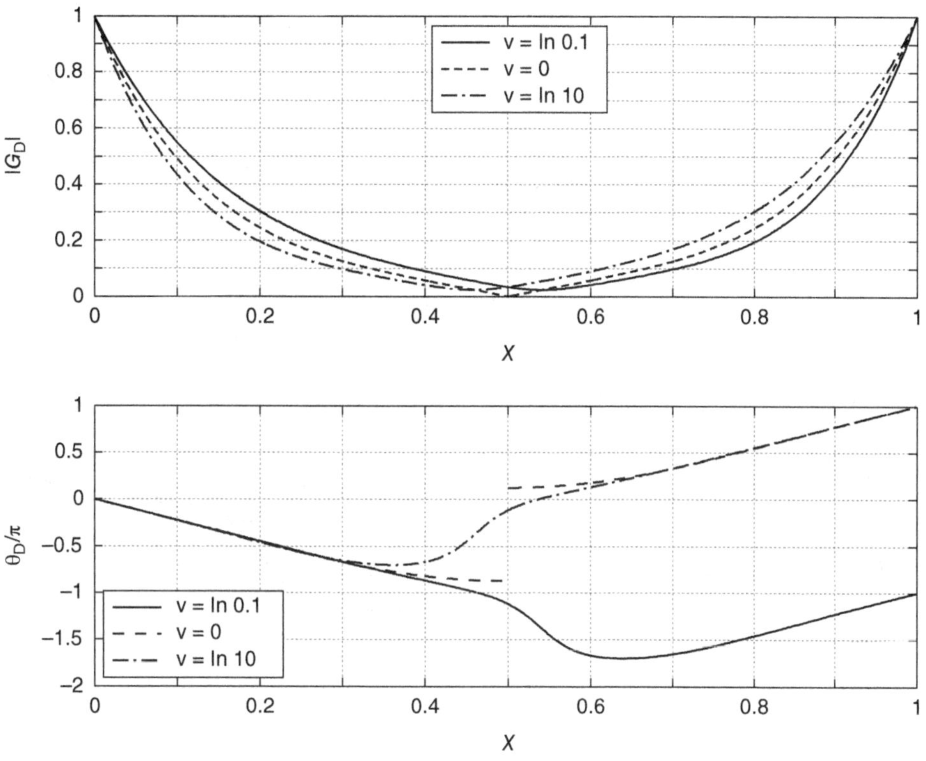

Figure 10.9 Case c-3 FRF (context: see Table 10.1).

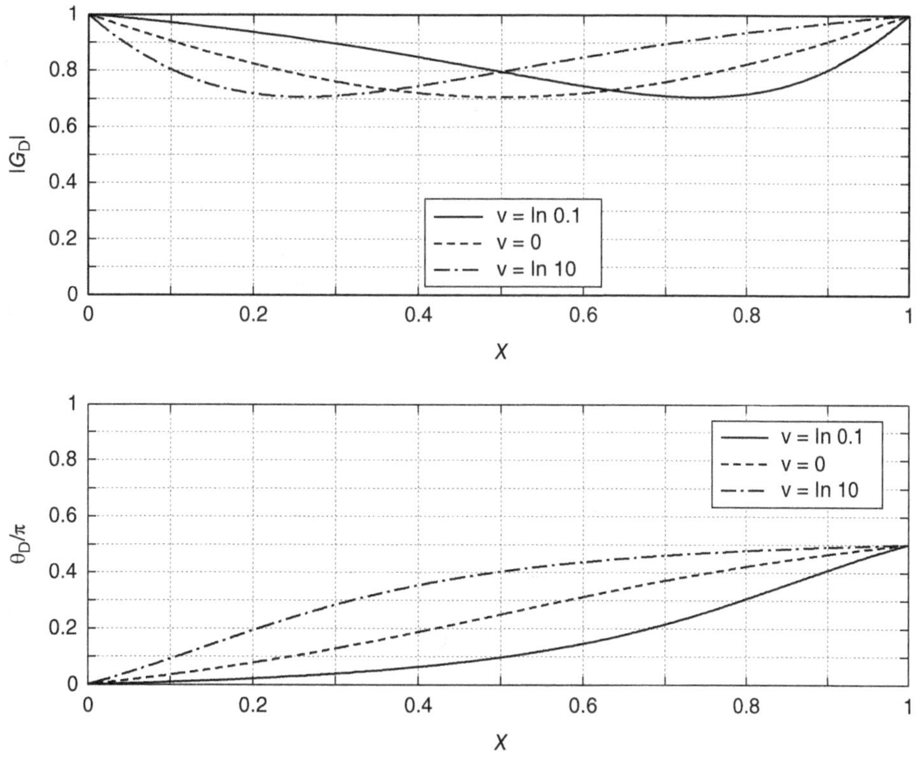

Figure 10.10 Case d-1 FRF (context: see Table 10.1).

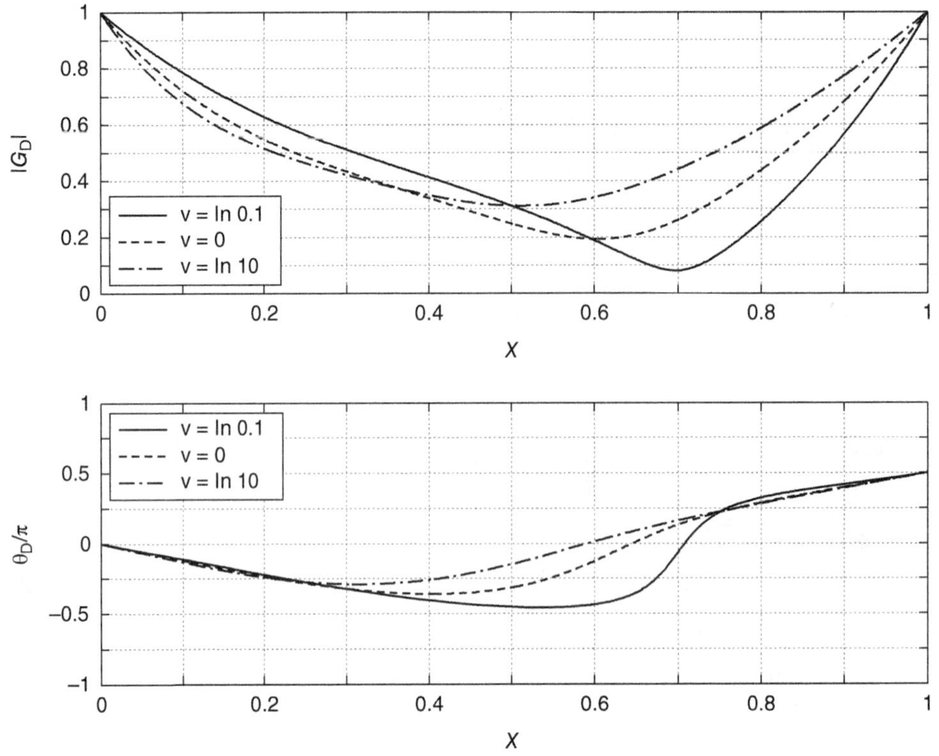

Figure 10.11 Case d-2 FRF (context: see Table 10.1).

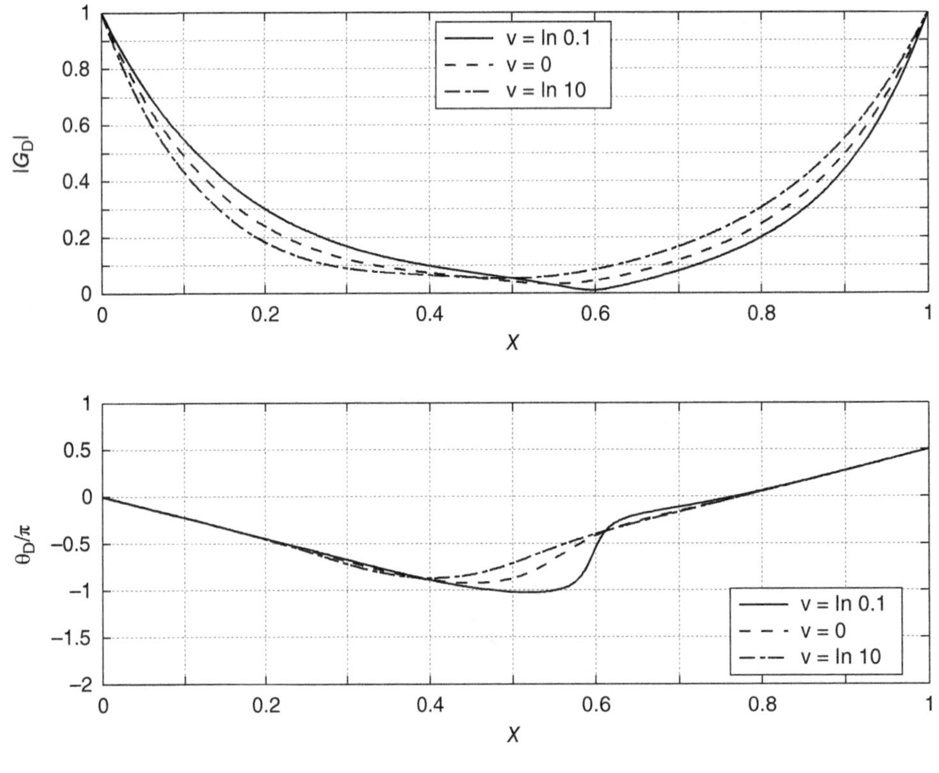

Figure 10.12 Case d-3 FRF (context: see Table 10.1).

Case 1: $\nu \geq 0$: Substituting $\eta_m = 0$ in (10.22) gives

$$\Lambda_m^{(1)} = -\nu \tag{10.36a}$$

$$\Lambda_m^{(2)} = 0. \tag{10.36b}$$

Substituting the RHSs for the corresponding terms on the RHSs of (10.21) gives

$$A_m^{(1)} = \frac{1 - G_{\mathrm{DL}}}{1 - e^{-\nu}} \tag{10.37a}$$

$$A_m^{(2)} = \frac{G_{\mathrm{DL}} - e^{-\nu}}{1 - e^{-\nu}}. \tag{10.37b}$$

Substituting the RHSs of the previous four equations for the corresponding coefficients on the RHS of (10.20) yields

$$G_{\mathrm{D}}(X; 0, \nu) = \frac{(1 - G_{\mathrm{DL}})e^{-\nu X} + G_{\mathrm{DL}} - e^{-\nu}}{1 - \mathrm{e}^{-\nu}}, \quad \nu \geq 0. \tag{10.38}$$

◊

Example 10.2 Dirichlet Conditions: Zero-Frequency Limit—Case 2

Case 2: $\nu < 0$: Substituting $\eta_m = 0$ in (10.22) gives

$$\Lambda_m^{(1)} = 0 \tag{10.39a}$$

$$\Lambda_m^{(2)} = -\nu. \tag{10.39b}$$

Substituting the RHSs for the corresponding quantities on the RHSs of (10.21) gives

$$A_m^{(1)} = \frac{G_{\mathrm{DL}} - e^{-\nu}}{1 - e^{-\nu}} \tag{10.40a}$$

$$A_m^{(2)} = \frac{1 - G_{\mathrm{DL}}}{1 - e^{-\nu}}. \tag{10.40b}$$

Substituting the RHSs of the previous four equations for the corresponding coefficients on the RHS of (10.20) yields

$$G_{\mathrm{D}}(X; 0, \nu) = \frac{(1 - G_{\mathrm{DL}})e^{-\nu X} + G_{\mathrm{DL}} - e^{-\nu}}{1 - e^{-\nu}}, \quad \nu < 0. \tag{10.41}$$

Combining (10.38) and (10.41) leads to the final result:

$$G_{\mathrm{D}}(X; 0, \nu) = \frac{(1 - G_{\mathrm{DL}})e^{-\nu X} + G_{\mathrm{DL}} - e^{-\nu}}{1 - e^{-\nu}}. \tag{10.42}$$

◊

Exercise 10.9 Zero-Frequency Limit of Dimensionless FRF

Problem: What can you infer about the zero-frequency limit FRF (10.42) in the special case where $G_{\mathrm{DL}}(X; 0, \nu) = 0$? Consider the two cases $|\nu| > 0$ and $\nu = 0$ separately.

Solution: Substituting $G_{\mathrm{DL}} = 0$ in (10.42) gives

$$G_{\mathrm{D}}(X; 0, \nu) = \frac{e^{-\nu X} - e^{-\nu}}{1 - e^{-\nu}}. \tag{10.43}$$

When $|\nu| > 0$, evaluation of the RHS is straightforward, but when $\nu = 0$, the RHS yields the indeterminate form $0/0$. Using l'Hôpital's rule then gives

$$\lim_{\nu \to 0} G_{\mathrm{D}}(X; 0, \nu) = 1 - X.$$

The simplicity of this result makes it convenient to use for checking numerical computations. To summarize, in the case where $G_{\mathrm{DL}} = 0$ the zero-frequency limit FRF is

$$G_{\mathrm{D}}(X; 0, \nu) = \begin{cases} \dfrac{e^{-\nu X} - e^{-\nu}}{1 - e^{-\nu}}, & |\nu| > 0 \\ 1 - X, & \nu = 0. \end{cases}$$

For $0 < X < 1$ the FRF G_{D} is real valued and positive, so $|G_{\mathrm{D}}| = G_{\mathrm{D}}$ and the phase of G_{D} is zero. Figure 10.1 shows the amplitude and phase functions. □

Exercise 10.10 Symmetry of Zero-Frequency Limit FRF

Problem: Consider the zero-frequency limit FRF in the special case where $G_{\mathrm{DL}} = 0$ (10.43). Define $G_D(X; \nu_0)$ as the low-frequency FRF corresponding to a particular nonzero value of ν. Show that

$$G_D(X; 0, \nu_0) + G_D(1 - X; 0, -\nu_0) = 1. \tag{10.44}$$

Give a geometric interpretation of this result. *Hint*: Sketch a graph consisting of a single set of X and G_D orthogonal coordinate axes, on which the functions $G_D(X; 0, \nu_0)$ and $G_D(1 - X; 0, -\nu_0)$ are plotted (two curves). The ranges for the abscissa and ordinate should be $0 \leq X \leq 1$ and $0 \leq G_D \leq 1$, respectively.

Solution: [Use algebra to prove that (10.44) is valid]. One possible geometric interpretation is that each point on the graph of $G_D(X; \nu_0)$ versus X is the reflection about the point $(X, G_D) = (0.5, 0.5)$, of the point on the graph of $G_D(1 - X; 0, -\nu_0)$ versus X. There may be other geometric interpretations as well. □

Exercise 10.11 Symmetry of Zero-Frequency Limit FRF

Problem: Show that if $G_{\mathrm{DL}} = i$ then the zero-frequency limit FRF exhibits the symmetry

$$G_{\mathrm{D}}(1 - X; 0, -\nu) = \mathrm{Im}G_{\mathrm{D}}(X; 0, \nu) + i\mathrm{Re}G_{\mathrm{D}}(X; 0, \nu). \tag{10.45}$$

What basic symmetry relations does this imply for the corresponding amplitude and phase?

Solution: Substituting i for G_{DL}, $1 - X$ for X, and $-\nu$ for ν in (10.42) gives

$$G_D(1-X;0,-\nu) = \frac{e^{\nu(1-X)} - e^{\nu}}{1-e^{\nu}} + \frac{i\left[1-e^{\nu(1-X)}\right]}{1-e^{\nu}}.$$

Rearranging then leads to

$$G_D(1-X;0,-\nu) = \frac{(1-e^{-\nu X}) + i(e^{-\nu X} - e^{-\nu})}{1-e^{-\nu}}.$$

Using (10.42) to rewrite the real and imaginary parts then yields (10.45).

Equation (10.45) implies the following symmetry relations for the amplitude and phase:

$$\left|G_D(1-X;0,-\nu)\right| = \left|G_D(X;0,\nu)\right| \tag{10.46a}$$

$$\theta_D(1-X;0,-\nu) = \pi/2 \text{ rad} - \theta_D(X;0,\nu). \tag{10.46b}$$

□

Exercise 10.12 High-Frequency Limit FRF

Problem: Consider the dimensionless form of the FRF, G_D. How does the FRF behave in the limit as the frequency grows very large (i.e., $\eta_m \to \infty$)?

Solution: Evaluating (10.22) in the limit as $\eta_m \to \infty$ gives

$$\begin{aligned} \Lambda_m^{(1)} &\to -\eta_m\sqrt{i} \\ \Lambda_m^{(2)} &\to \eta_m\sqrt{i}. \end{aligned} \tag{10.47a}$$

Substituting the RHSs for the corresponding quantities in (10.21) and evaluating the limit give

$$\begin{aligned} A_m^{(1)} &\to 1 \\ A_m^{(2)} &\to 0. \end{aligned} \tag{10.47b}$$

Substituting the RHSs of the previous four equations for the corresponding coefficients on the RHS of (10.20) and taking the limit as $\eta_m \to \infty$ yield

$$\lim_{\eta_m\to\infty} G_D(X;\eta_m,\nu) = 0 \quad (0 < X < 1). \tag{10.48}$$

□

10.3.2. Dirichlet/Neumann Conditions at Left/Right Boundaries

In this section we consider problems in which the FRF at the left boundary is specified (Dirichlet condition) and the gradient of the FRF at the right boundary is specified (Neumann condition):

$$\begin{aligned} &G(0;\omega_m) \\ &\quad = \text{known (constant for fixed } \omega_m) \end{aligned} \tag{10.49a}$$

$$\begin{aligned} G_\nabla(l;\omega_m) &\equiv \left[\frac{d}{dx}G(x;\omega_m)\right]\Bigg|_{x=l} \\ &= \text{known (constant for fixed } \omega_m). \end{aligned} \tag{10.49b}$$

Exercise 10.13 FRF Coefficients

Problem: Suppose the hydraulic head FRF is subject to the boundary conditions (10.49). Express the coefficients $a_m^{(1)}$ and $a_m^{(2)}$ in terms of $G(0;\omega_m)$ and $G_\nabla(l;\omega_m)$, $\lambda_m^{(1)}$, $\lambda_m^{(2)}$, and l.

Solution: Differentiating (10.9) with respect to x gives

$$\frac{d}{dx}G(x;\omega_m) = a_m^{(1)}\lambda_m^{(1)}e^{\lambda_m^{(1)}x} + a_m^{(2)}\lambda_m^{(2)}e^{\lambda_m^{(2)}x}.$$

Substituting $x = 0$ in (10.9) and $x = l$ in the previous equation, respectively, yields

$$G(0;\omega_m) = a_m^{(1)} + a_m^{(2)} \tag{10.50a}$$

$$G_\nabla(l;\omega_m) = a_m^{(1)}\lambda_m^{(1)}e^{\lambda_m^{(1)}l} + a_m^{(2)}\lambda_m^{(2)}e^{\lambda_m^{(2)}l}. \tag{10.50b}$$

This is a system of two linear, generally nonhomogeneous equations in the two unknowns $a_m^{(1)}$ and $a_m^{(2)}$. Its solution is

$$a_m^{(1)} = \frac{G_\nabla(l;\omega_m) - G(0;\omega_m)\lambda_m^{(2)}e^{\lambda_m^{(2)}l}}{\lambda_m^{(1)}e^{\lambda_m^{(1)}l} - \lambda_m^{(2)}e^{\lambda_m^{(2)}l}} \tag{10.51a}$$

$$a_m^{(2)} = \frac{G(0;\omega_m)\lambda_m^{(1)}e^{\lambda_m^{(1)}l} - G_\nabla(l;\omega_m)}{\lambda_m^{(1)}e^{\lambda_m^{(1)}l} - \lambda_m^{(2)}e^{\lambda_m^{(2)}l}}. \tag{10.51b}$$

□

Consider the FRF that satisfies boundary conditions (10.49) with $|G(0;\omega_m)| > 0$. The solution is given by (10.9) with the coefficients given by (10.51). In this case the FRF depends on six independent variables: x, l, μ, α_m, $G(0;\omega_m)$, and $G_\nabla(l;\omega_m)$. We can also express the FRF in terms of dimensionless variables. Define the dependent dimensionless variable

$$G_D(X;\eta_m,\nu) \equiv \frac{G(x;\omega_m)}{G(0;\omega_m)}$$

and the independent dimensionless variables

$$G_{D\nabla L} \equiv \frac{lG_\nabla(l;\omega_m)}{G(0;\omega_m)} \tag{10.52a}$$

$$X \equiv \frac{x}{l} \tag{10.52b}$$

$$\nu \equiv \mu l \tag{10.52c}$$

$$\eta_m \equiv \sqrt{\alpha_m}\, l. \tag{10.52d}$$

Rearranging these leads to

$$G(x;\omega_m) = G(0;\omega_m)G_D \tag{10.53a}$$

$$G_\nabla(l;\omega_m) = \frac{G(0;\omega_m)}{l}G_{D\nabla L} \tag{10.53b}$$

$$x = lX \tag{10.53c}$$

Table 10.2 FRF Cases: 1D, nonhomogeneous media, Dirichlet/Neuman BCs.

η_m	$G_{D\nabla L}$			
	0	1	−1	i
0	m-1 (Figure 10.13)	n-1 (Figure 10.16)	p-1 (Figure 10.19)	q-1 (Figure 10.22)
5	m-2 (Figure 10.14)	n-2 (Figure 10.17)	p-2 (Figure 10.20)	q-2 (Figure 10.23)
10	m-3 (Figure 10.15)	n-3 (Figure 10.18)	p-3 (Figure 10.21)	q-3 (Figure 10.24)

$$\mu = \frac{\nu}{l} \tag{10.53d}$$

$$\alpha_m = \frac{\eta_m^2}{l^2}. \tag{10.53e}$$

Substituting the RHSs for the corresponding quantities in (10.9) and (10.51) leads to

$$G_D(X;\eta_m,\nu) = A_m^{(1)} e^{\Lambda_m^{(1)} X} + A_m^{(2)} e^{\Lambda_m^{(2)} X} \tag{10.54}$$

where we define the dimensionless coefficients as

$$A_m^{(1)} \equiv \frac{G_{D\nabla L} - \Lambda_m^{(2)} e^{\Lambda_m^{(2)}}}{\Lambda_m^{(1)} e^{\Lambda_m^{(1)}} - \Lambda_m^{(2)} e^{\Lambda_m^{(2)}}} \tag{10.55a}$$

$$A_m^{(2)} \equiv \frac{\Lambda_m^{(1)} e^{\Lambda_m^{(1)}} - G_{D\nabla L}}{\Lambda_m^{(1)} e^{\Lambda_m^{(1)}} - \Lambda_m^{(2)} e^{\Lambda_m^{(2)}}} \tag{10.55b}$$

and the dimensionless eigenvalues as

$$\Lambda_m^{(n)} \equiv \frac{-\nu + (-1)^n \sqrt{\nu^2 + i4\eta_m^2}}{2} \quad (n = 1, 2). \tag{10.56}$$

Thus, by expressing the FRF in terms of dimensionless variables we have effectively reduced the total number of independent variables from six to four (X, ν, η_m, and $G_{D\nabla L}$).

To illustrate the results of this section, we constructed the dimensionless FRF solutions for the 12 cases listed in Table 10.2. Each case includes amplitude-versus-distance and phase-versus-distance plots for the three values $\nu = \ln 0.1$, $\nu = 0$, and $\nu = \ln 10$. Technically the cases corresponding to $\eta_m = 0$ (i.e., the zero-frequency case) are steady state rather than time periodic, but these are included because they illustrate the limiting behavior of the amplitude and phase functions. In all of these plots we define the dimensionless phase function as

$$\theta_D(X;\eta_m,\nu) \equiv \arg G_D(X;\eta_m,\nu).$$

Example 10.3 Dirichlet/Neuman Conditions: Zero-Frequency Limit—Case 1
We can use the dimensionless form of the FRF to derive the zero-frequency limit FRF. Consider the two cases $\nu \geq 0$ and $\nu < 0$ separately.

Case 1: $\nu \geq 0$. Substituting $\eta_m = 0$ in (10.56) gives

$$\Lambda_m^{(1)} = -\nu \tag{10.57a}$$

$$\Lambda_m^{(2)} = 0. \tag{10.57b}$$

Substituting the RHSs for the corresponding quantities on the RHSs of (10.55) gives

$$A_m^{(1)} = \frac{-G_{D\nabla L}}{\nu e^{-\nu}} \tag{10.58a}$$

$$A_m^{(2)} = \frac{G_{D\nabla L}}{\nu e^{-\nu}} + 1. \tag{10.58b}$$

Substituting the RHSs of the previous four equations for the corresponding coefficients on the RHS of (10.54) yields

$$G_D(X;0,\nu) = 1 + G_{D\nabla L}\left(\frac{1 - e^{-\nu X}}{\nu e^{-\nu}}\right), \quad \nu \geq 0. \tag{10.59}$$

◊

Example 10.4 Dirichlet/Neuman Conditions: Zero-Frequency Limit—Case 2
Case 2: $\nu < 0$. Substituting $\eta_m = 0$ in (10.56) gives

$$\Lambda_m^{(1)} = 0 \tag{10.60a}$$

$$\Lambda_m^{(2)} = -\nu. \tag{10.60b}$$

Substituting the RHSs for the corresponding quantities on the RHSs of (10.55) gives

$$A_m^{(1)} = \frac{G_{D\nabla L}}{\nu e^{-\nu}} + 1 \tag{10.61a}$$

$$A_m^{(2)} = \frac{-G_{D\nabla L}}{\nu e^{-\nu}}. \tag{10.61b}$$

Substituting the RHSs of the previous four equations for the corresponding coefficients on the RHS of (10.54) yields

$$G_D(X;0,\nu) = 1 + G_{D\nabla L}\left(\frac{1 - e^{-\nu X}}{\nu e^{-\nu}}\right), \quad \nu < 0. \tag{10.62}$$

Combining (10.59) and (10.62) leads to the final result:

$$G_D(X;0,\nu) = 1 + G_{D\nabla L}\left(\frac{1 - e^{-\nu X}}{\nu e^{-\nu}}\right). \tag{10.63}$$

◊

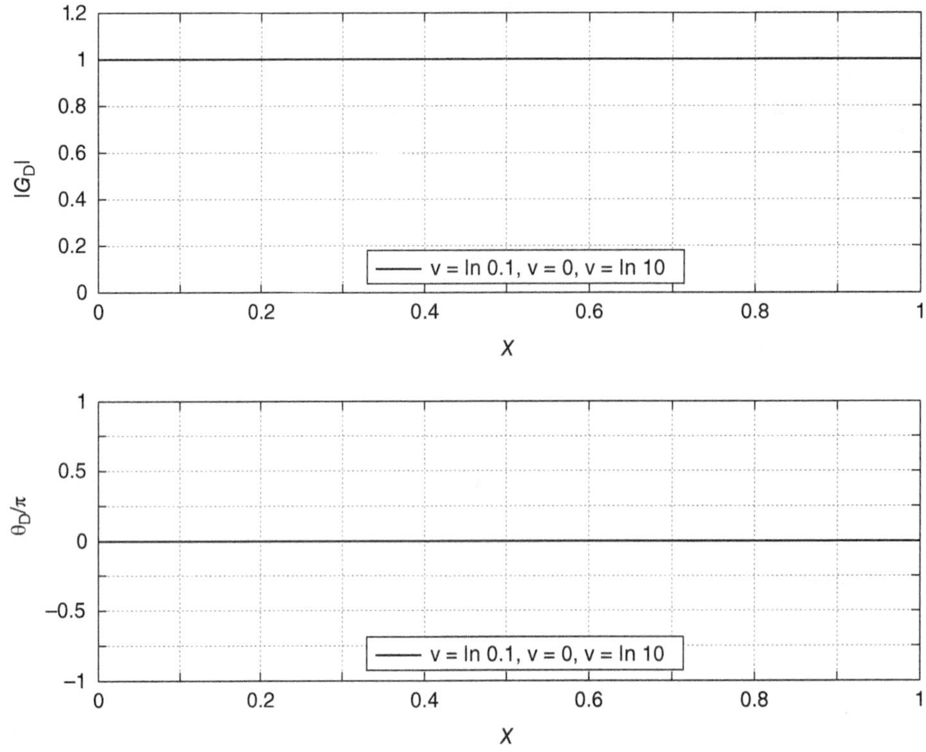

Figure 10.13 Case m-1 FRF (context: see Table 10.2).

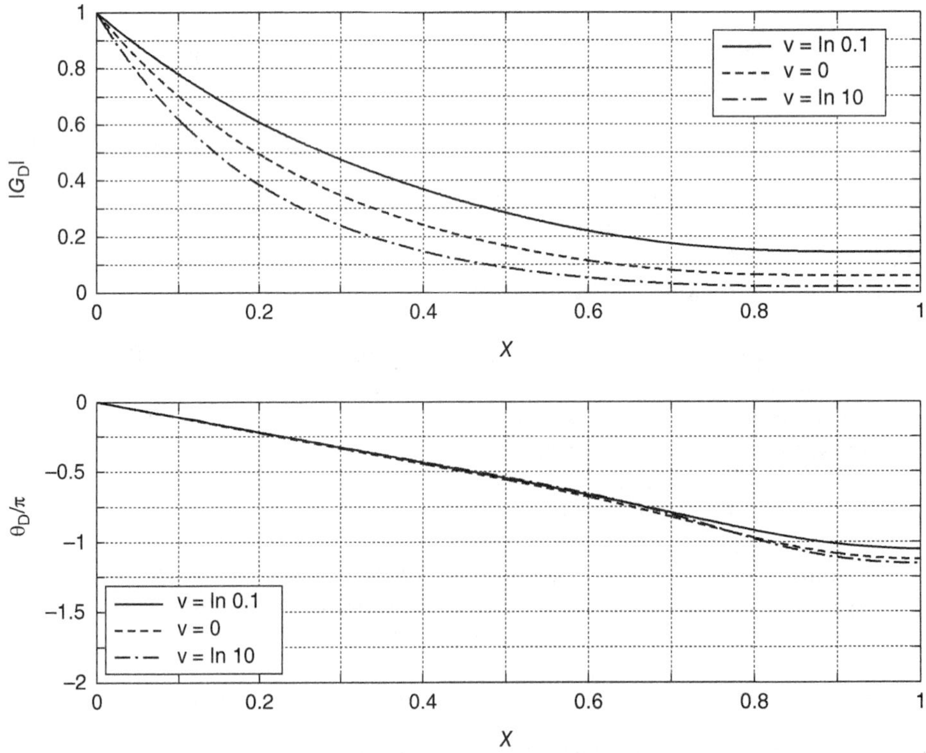

Figure 10.14 Case m-2 FRF (context: see Table 10.2).

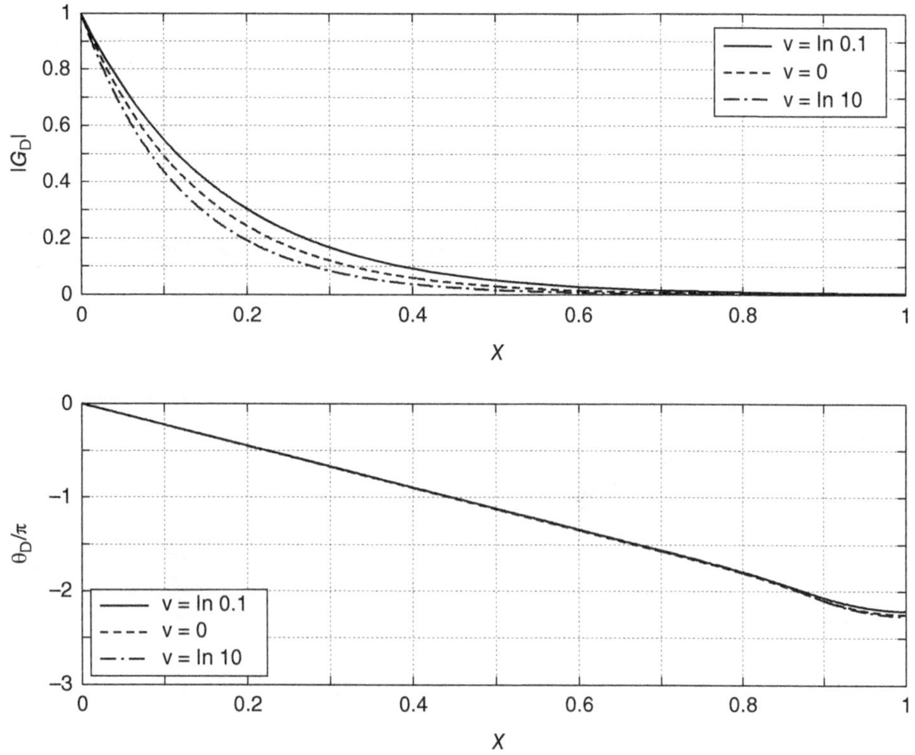

Figure 10.15 Case m-3 FRF (context: see Table 10.2).

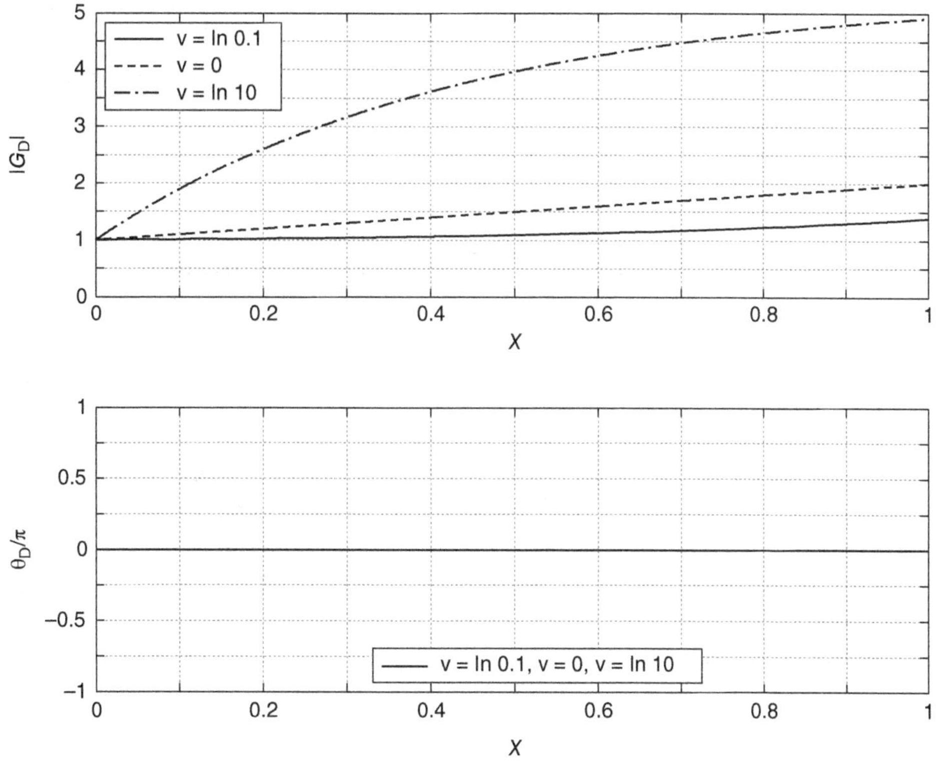

Figure 10.16 Case n-1 FRF (context: see Table 10.2).

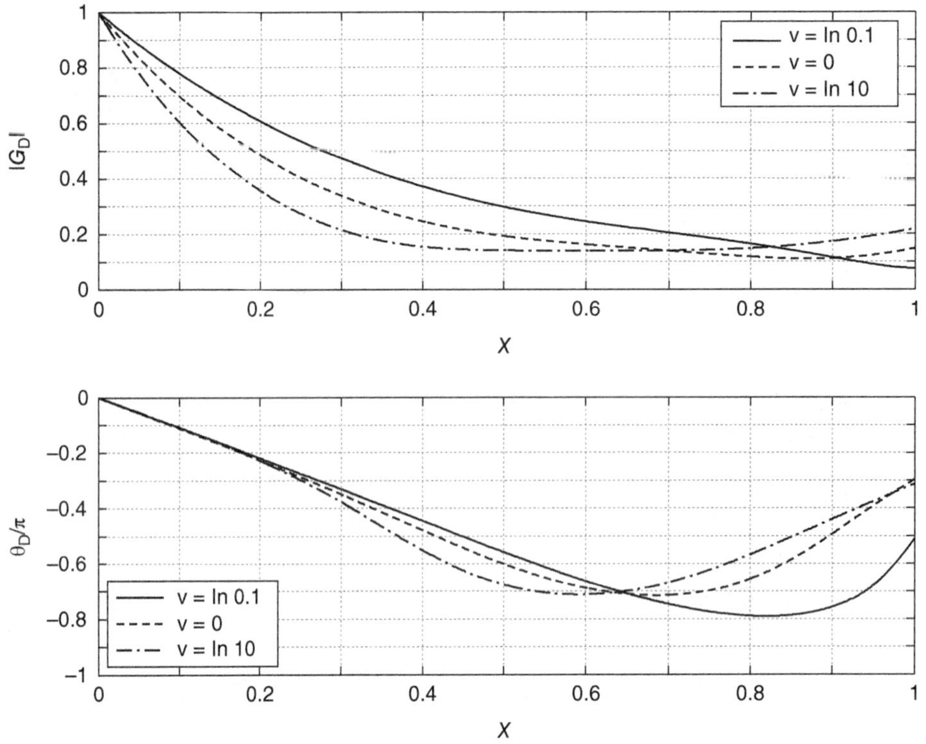

Figure 10.17 Case n-2 FRF (context: see Table 10.2).

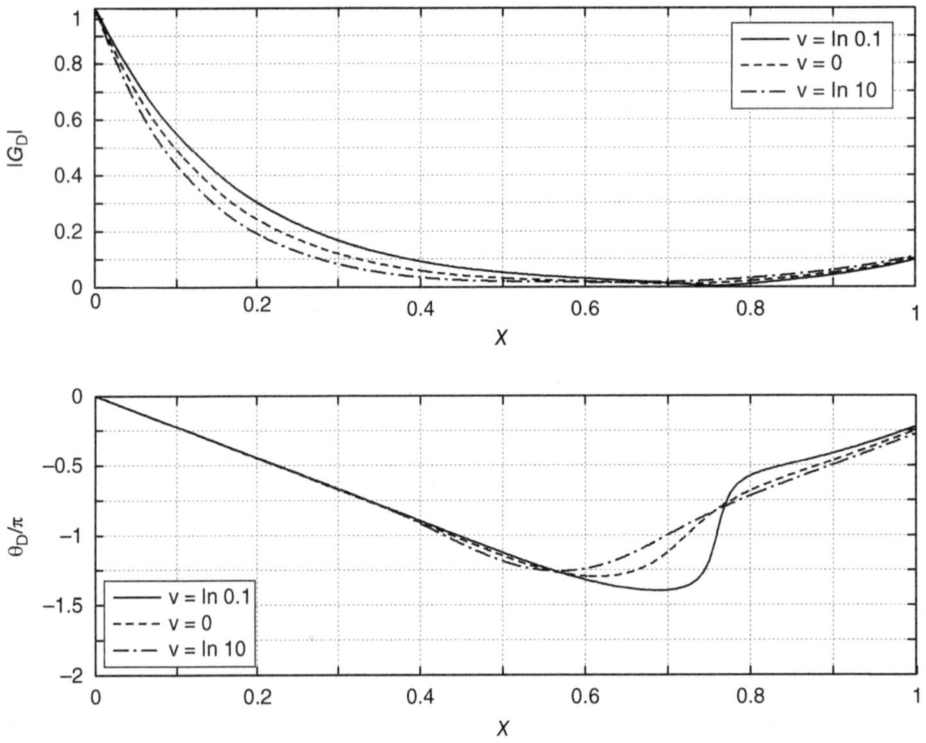

Figure 10.18 Case n-3 FRF (context: see Table 10.2).

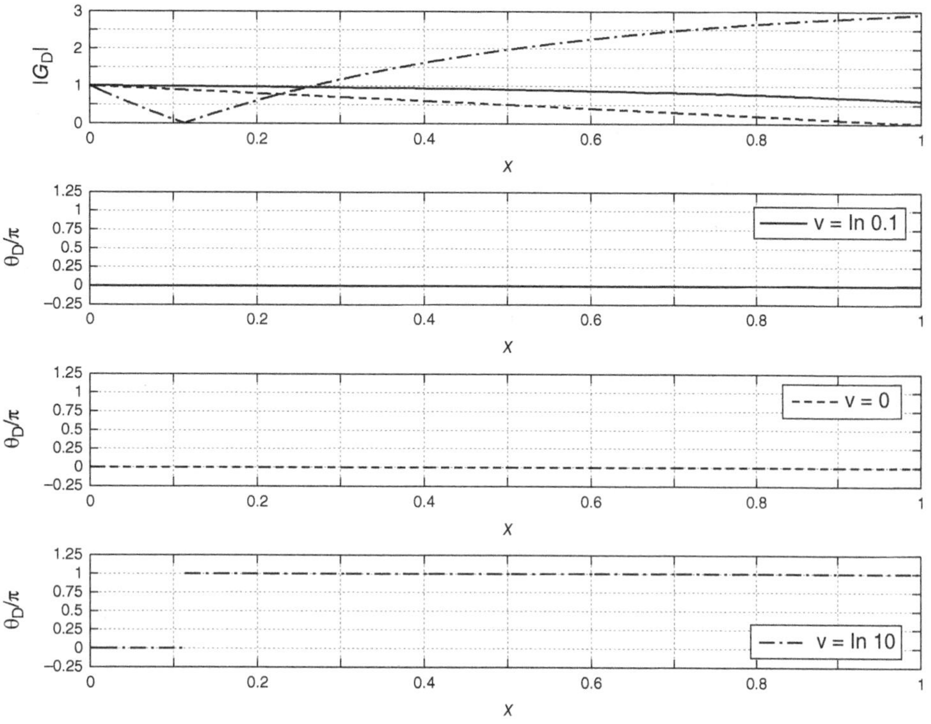

Figure 10.19 Case p-1 FRF (context: see Table 10.2).

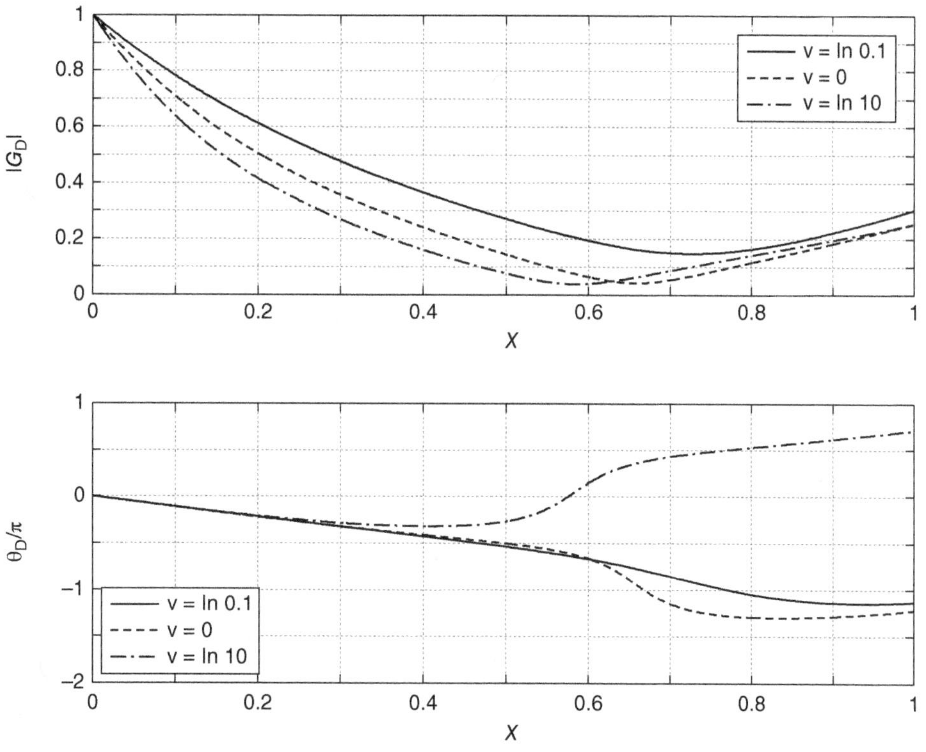

Figure 10.20 Case p-2 FRF (context: see Table 10.2).

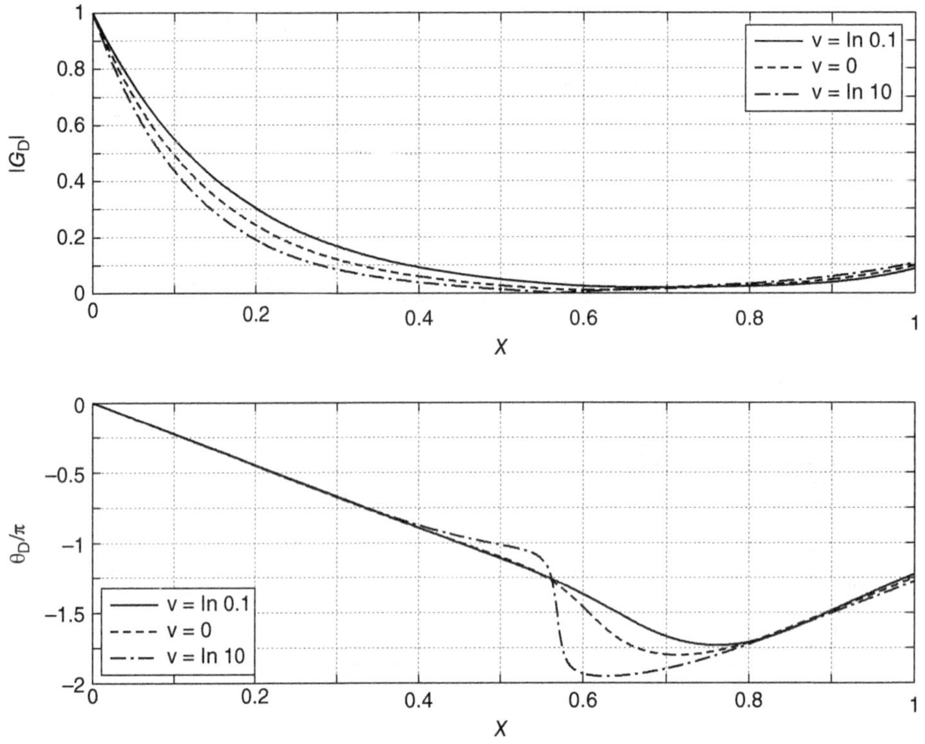

Figure 10.21 Case p-3 FRF (context: see Table 10.2).

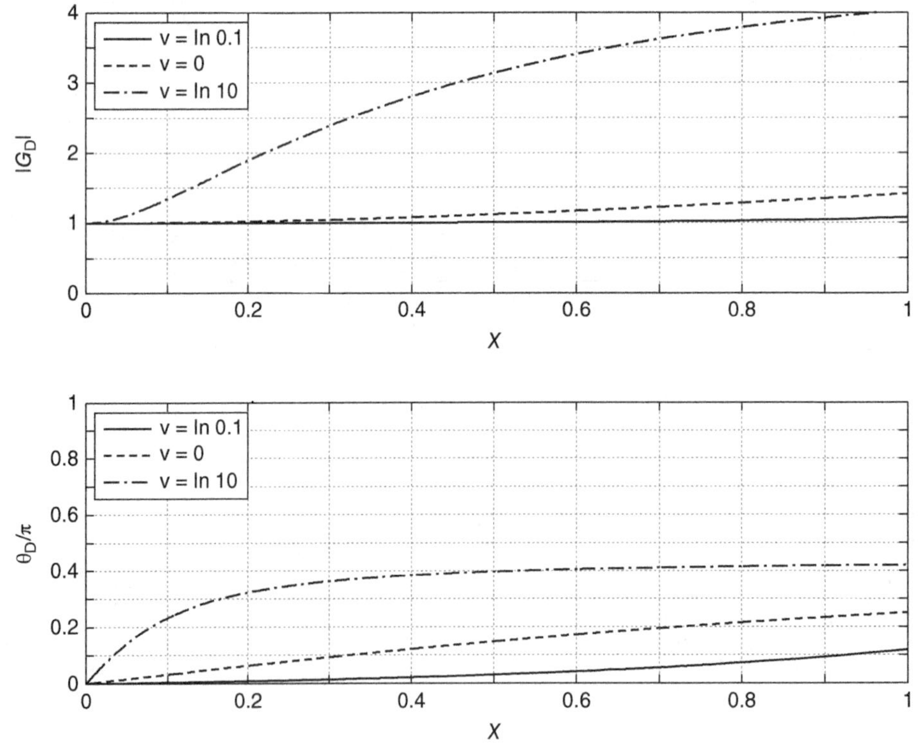

Figure 10.22 Case q-1 FRF (context: see Table 10.2).

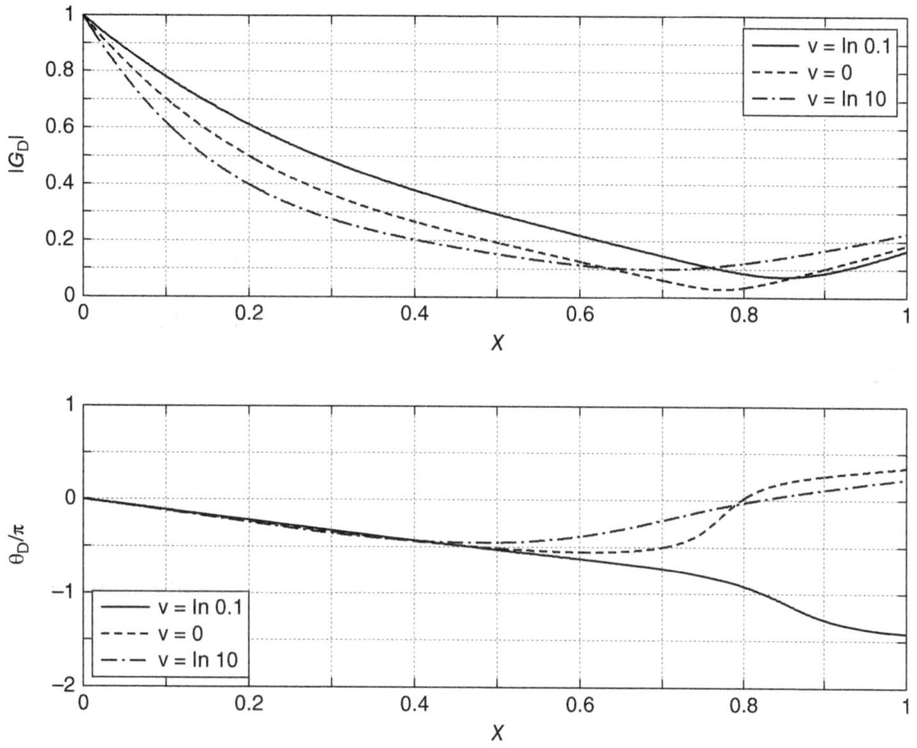

Figure 10.23 Case q-2 FRF (context: see Table 10.2).

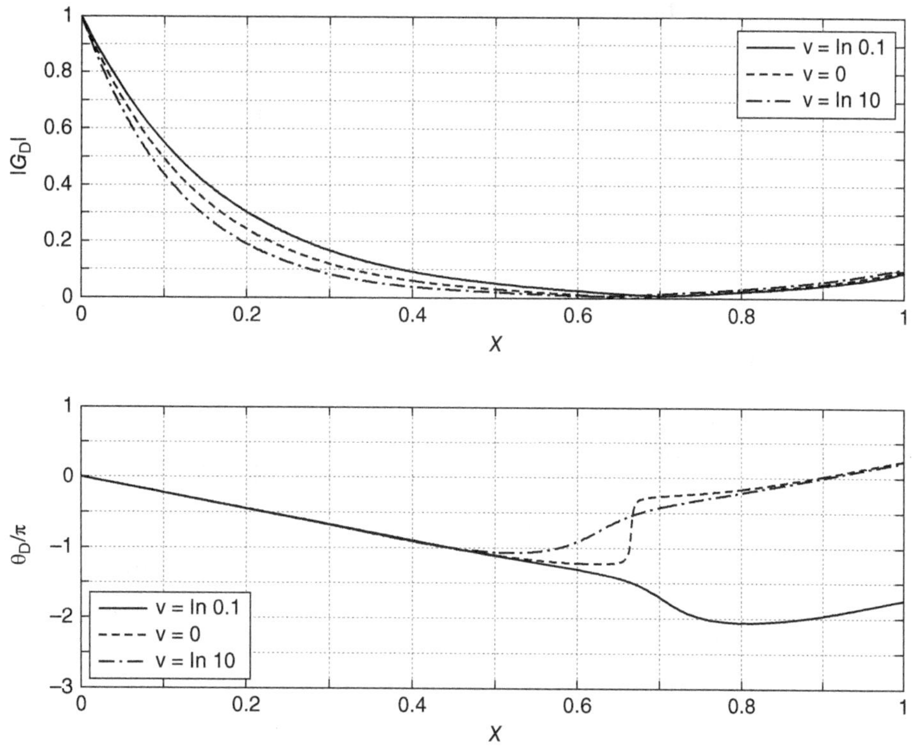

Figure 10.24 Case q-3 FRF (context: see Table 10.2).

Exercise 10.14 Zero Point of Dimensionless FRF

Problem: Suppose $G_{D\nabla L} = -1$ in the zero-frequency case (i.e., $\eta_m = 0$) with $\nu = \ln 10$. Figure 10.19 shows the amplitude function. Notice that it apparently vanishes near $X = 0.11$. Let X_{zp} denote the X coordinate of the amplitude function's zero point. Express X_{zp} as a function of ν, and then numerically evaluate the result specifically for $\nu = \ln 10$. Is your result consistent with the amplitude graph?

Solution: Substituting $G_{D\nabla L} = -1$ and $X = X_{zp}$ in (10.63) and then solving for X_{zp} give

$$X_{zp} = \frac{-1}{\nu} \ln\left[1 - \nu e^{-\nu}\right].$$

Evaluating the RHS for $\nu = \ln 10$ gives $X_{zp} \approx 0.1137$, which appears to be consistent with the amplitude graph. □

Exercise 10.15 Space Derivative of Zero-Frequency FRF

Problem: Consider the zero-frequency case (i.e., $\eta_m = 0$). Show that if $|G_{D\nabla L}| > 0$, then $|dG_D/dX| > 0$ within the domain $0 < X < 1$.

Solution: Differentiating (10.63) with respect to X yields

$$\frac{dG_D}{dX} = e^{\nu(1-X)} G_{D\nabla L}. \tag{10.64}$$

Both of the terms on the RHS are nonzero, so

$$\left|\frac{dG_D}{dX}\right| > 0.$$

□

10.3.2.1. No-Flow Boundary

A special type of Neumann boundary condition is the no-flow condition. Consider a time-periodic, 1D flow with a no-flow condition at the right boundary, i.e., $G_\nabla(l; \omega_m) = 0$. What values do the amplitude and phase gradients take at the no-flow boundary? We took up this question in Section 9.3.2, where we obtained

$$\left[\frac{d}{dx} M(x; \omega_m)\right]\Bigg|_{x=l} = 0 \tag{10.65a}$$

$$M(l; \omega_m)\left[\frac{d}{dx} \theta(x; \omega_m)\right]\Bigg|_{x=l} = 0. \tag{10.65b}$$

We derived these without making any assumptions about whether the medium is homogeneous or nonhomogeneous, so they are generally valid. Equation (10.65a) requires that the amplitude gradient vanishes at the no-flow boundary. Assuming the amplitude is nonzero there, equation (10.65b) requires that the phase gradient also vanishes at the no-flow boundary.

Exercise 10.16 Expression for FRF

Problem: Suppose the hydraulic head FRF is subject to a Dirichlet condition at the left boundary and a no-flow condition at the right boundary. Express the FRF compactly in terms of $G(0; \omega_m)$, $\lambda_m^{(1)}$, $\lambda_m^{(2)}$, l, and x.

Solution: Substituting $G_\nabla(l; \omega_m) = 0$ in (10.51) gives

$$a_m^{(1)} = \frac{-G(0; \omega_m)\lambda_m^{(2)} e^{\lambda_m^{(2)} l}}{\lambda_m^{(1)} e^{\lambda_m^{(1)} l} - \lambda_m^{(2)} e^{\lambda_m^{(2)} l}} \tag{10.66a}$$

$$a_m^{(2)} = \frac{G(0; \omega_m)\lambda_m^{(1)} e^{\lambda_m^{(1)} l}}{\lambda_m^{(1)} e^{\lambda_m^{(1)} l} - \lambda_m^{(2)} e^{\lambda_m^{(2)} l}}. \tag{10.66b}$$

Substituting the RHSs of (10.66) for $a_m^{(1)}$ and $a_m^{(2)}$ in (10.9) and then simplifying the result yield

$$G(x; \omega_m) = \frac{G(0; \omega_m)}{\lambda_m^{(1)} e^{\lambda_m^{(1)} l} - \lambda_m^{(2)} e^{\lambda_m^{(2)} l}} \times \left(-\lambda_m^{(2)} e^{\lambda_m^{(2)} l + \lambda_m^{(1)} x} + \lambda_m^{(1)} e^{\lambda_m^{(1)} l + \lambda_m^{(2)} x}\right).$$

□

10.3.3. Cauchy Conditions at Left Boundary

In this section we briefly consider problems in which the FRF and its derivative with respect to the space coordinate, both evaluated at the left boundary, are known:

$$G(0; \omega_m) = \text{known (constant for fixed } \omega_m) \tag{10.67a}$$

$$G_\nabla(0; \omega_m) \equiv \left[\frac{d}{dx} G(x; \omega_m)\right]\Bigg|_{x=0} = \text{known (constant for fixed } \omega_m). \tag{10.67b}$$

Exercise 10.17 FRF Coefficients

Problem: Suppose the hydraulic head FRF is subject to the boundary conditions (10.67). Express the coefficients $a_m^{(1)}$ and $a_m^{(2)}$ in terms of the known quantities $G(0; \omega_m)$, $G_\nabla(0; \omega_m)$, $\lambda_m^{(1)}$, and $\lambda_m^{(2)}$.

Solution: Differentiating (10.9) with respect to x gives

$$\frac{d}{dx} G(x; \omega_m) = a_m^{(1)} \lambda_m^{(1)} e^{\lambda_m^{(1)} x} + a_m^{(2)} \lambda_m^{(2)} e^{\lambda_m^{(2)} x}.$$

Substituting $x = 0$ in this equation and in (10.9) and combining the results with the boundary conditions (10.67) lead to

$$G(0;\omega_m) = a_m^{(1)} + a_m^{(2)} \tag{10.68a}$$

$$G_\nabla(0;\omega_m) = \lambda_m^{(1)} a_m^{(1)} + \lambda_m^{(2)} a_m^{(2)}. \tag{10.68b}$$

This is a system of two linear, generally nonhomogeneous equations in the two unknowns $a_m^{(1)}$ and $a_m^{(2)}$. Its solution is

$$a_m^{(1)} = \frac{G_\nabla(0;\omega_m) - \lambda_m^{(2)} G(0;\omega_m)}{\lambda_m^{(1)} - \lambda_m^{(2)}} \tag{10.69a}$$

$$a_m^{(2)} = \frac{\lambda_m^{(1)} G(0;\omega_m) - G_\nabla(0;\omega_m)}{\lambda_m^{(1)} - \lambda_m^{(2)}}. \tag{10.69b}$$

□

Exercise 10.18 Cauchy Boundary Conditions

Problem: Suppose the hydraulic head FRF is subject to the Cauchy boundary conditions (10.67), with an impermeable boundary at $x = 0$. Express the FRF compactly in terms of $G(0;\omega_m)$, $\lambda_m^{(1)}$, $\lambda_m^{(2)}$, and x.

Solution: Substituting $G_\nabla(0;\omega_m) = 0$ in (10.69) gives

$$a_m^{(1)} = \frac{-\lambda_m^{(2)} G(0;\omega_m)}{\lambda_m^{(1)} - \lambda_m^{(2)}} \tag{10.70a}$$

$$a_m^{(2)} = \frac{\lambda_m^{(1)} G(0;\omega_m)}{\lambda_m^{(1)} - \lambda_m^{(2)}}. \tag{10.70b}$$

Substituting these results for $a_m^{(1)}$ and $a_m^{(2)}$ in (10.9) yields

$$G(x;\omega_m) = \frac{G(0;\omega_m)}{\lambda_m^{(1)} - \lambda_m^{(2)}} \left(-\lambda_m^{(2)} e^{\lambda_m^{(1)} x} + \lambda_m^{(1)} e^{\lambda_m^{(2)} x} \right).$$

□

Exercise 10.19 Cauchy Boundary Conditions

Problem: Suppose the hydraulic head FRF is subject to the Cauchy boundary conditions (10.67) with $G(0;\omega_m) = 0$. Express the FRF compactly in terms of $G_\nabla(0;\omega_m)$, $\lambda_m^{(1)}$, $\lambda_m^{(2)}$, and x.

Solution: Substituting $G(0;\omega_m) = 0$ in (10.69) yields

$$a_m^{(1)} = \frac{G_\nabla(0;\omega_m)}{\lambda_m^{(1)} - \lambda_m^{(2)}} \tag{10.71a}$$

$$a_m^{(2)} = \frac{-G_\nabla(0;\omega_m)}{\lambda_m^{(1)} - \lambda_m^{(2)}}. \tag{10.71b}$$

Substituting the RHSs of these equations for $a_m^{(1)}$ and $a_m^{(2)}$ in (10.9) and then simplifying the result give

$$G(x;\omega_m) = \frac{G_\nabla(0;\omega_m)}{\lambda_m^{(1)} - \lambda_m^{(2)}} \left(e^{\lambda_m^{(1)} x} - e^{\lambda_m^{(2)} x} \right).$$

□

Exercise 10.20 Cauchy Boundary Conditions

Problem: Suppose the hydraulic head FRF is subject to the Cauchy boundary conditions (10.67). Does a solution exist for which both the FRF and its derivative vanish at the left boundary? Explain.

Solution: Substituting $G(0;\omega_m) = G_\nabla(0;\omega_m) = 0$ in (10.69) yields $a_m^{(1)} = a_m^{(2)} = 0$. Alternatively, substituting $G(0;\omega_m) = 0$ in (10.70) or $G_\nabla(0;\omega_m) = 0$ in (10.71) yields the same result. In any case the solution is the trivial one: $G(x;\omega_m) = 0$ for $0 < x < l$. Therefore, for these particular boundary conditions a nontrivial solution does not exist. □

10.4. SEMI-INFINITE DOMAIN

In this section we examine problems of time-periodic, 1D flow in a source-free, semi-infinite domain filled with an exponential medium.

10.4.1. BVP and Solution

Suppose the left boundary is periodically forced with a single constituent (i.e., ω_m fixed) and the fluctuations vanish as the distance from the left end increases without bound. The boundary conditions are

$$G(0;\omega_m) = \text{known (constant for fixed } \omega_m) \tag{10.72a}$$

$$\lim_{x\to\infty} G(x;\omega_m) = 0. \tag{10.72b}$$

Combining (10.72) with (10.9) and using (10.11) give

$$a_m^{(1)} = G(0;\omega_m) \tag{10.73a}$$

$$a_m^{(2)} = 0. \tag{10.73b}$$

Substituting these results for $a_m^{(1)}$ and $a_m^{(2)}$ in (10.9) yields the solution:

$$G(x;\omega_m) = G(0;\omega_m) e^{\lambda_m^{(1)} x} \tag{10.74}$$

where $\lambda_m^{(1)}$ is given by (10.11).

The case $\mu = 0$ corresponds to the Jacob-Ferris solution (see Section 9.4.1), so the solution (10.74) is a generalization of that result. The FRF for the case $\mu > 0$ is nonphysical for great distances X because in the limit as $X \to \infty$ such a medium's hydraulic conductivity and specific storage become infinite.

Example 10.5 Semi-Infinite, Nonuniform Medium—Dimensionless FRF
We can express the solution (10.74) in terms of dimensionless variables as

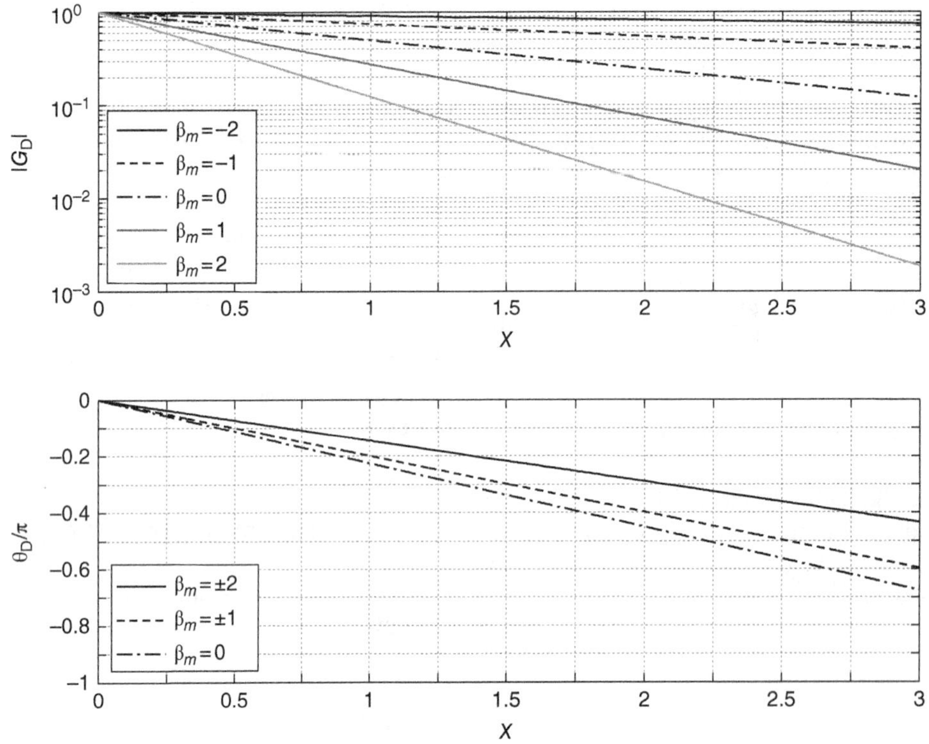

Figure 10.25 FRF for Example 10.5.

$$G_D(X;\beta_m) = e^{\Gamma_m^{(1)} X} \tag{10.75}$$

where

$$G_D \equiv \frac{G(x;\omega_m)}{G(0;\omega_m)} \tag{10.76a}$$

$$X \equiv \sqrt{\alpha_m} x \tag{10.76b}$$

$$\beta_m \equiv \frac{\mu}{\sqrt{\alpha_m}} \tag{10.76c}$$

and we define the dimensionless eigenvalue as

$$\Gamma_m^{(1)} \equiv \frac{-\beta_m - \sqrt{\beta_m^2 + i4}}{2}. \tag{10.77}$$

Figure 10.25 shows the corresponding amplitude and phase functions. We define the dimensionless phase function as

$$\theta_D(X;\beta_m) \equiv \arg G_D(X;\beta_m).$$

As X increases, the amplitude decreases exponentially and the phase decreases linearly. ◊

Exercise 10.21 Dimensionless Eigenvalue

Problem: Write the dimensionless eigenvalue (10.77) in rectangular form.

Solution: Begin by writing the principal square root $\sqrt{\beta_m^2 + i4}$ in rectangular form $x + iy$, where x and y are real valued. Define

$$z \equiv \beta_m^2 + i4.$$

Then $x = \beta_m^2$ and $y = 4$. Also, $r = \sqrt{\beta_m^4 + 16}$. Substituting these results for r, x, y, and z in (D.6) yields

$$\sqrt{\beta_m^2 + i4} = \sqrt{\frac{\sqrt{\beta_m^4 + 16} + \beta_m^2}{2}} + i\sqrt{\frac{\sqrt{\beta_m^4 + 16} - \beta_m^2}{2}}.$$

Substituting the RHS for $\sqrt{\beta_m^2 + i4}$ in (10.77) gives

$$\Gamma_m^{(1)} = \frac{-\beta_m}{2} - \sqrt{\frac{\sqrt{\beta_m^4 + 16} + \beta_m^2}{8}} - i\sqrt{\frac{\sqrt{\beta_m^4 + 16} - \beta_m^2}{8}}. \tag{10.78}$$

□

Exercise 10.22 Dimensionless Amplitude and Phase Functions

Problem: Consider the dimensionless FRF (10.75) in Example 10.5. Derive expressions for the dimensionless amplitude and phase functions. What numerical values do your expressions give for the slopes of the

Table 10.3 Numerical values of slopes for Exercise 10.22.

β_m	Slopes	
	b_1	b_2/π
−2	−0.04286	−0.1449
−1	−0.1304	−0.1989
0	−0.3071	−0.2251
1	−0.5647	−0.1989
2	−0.9114	−0.1449

$\log_{10}|G_D|$-versus-X and θ_D-versus-X curves? Are your results consistent with the graphs in Figure 10.25?

Solution: Taking the modulus and argument, respectively, of (10.75) yields

$$|G_D(X;\beta_m)| = \exp\left[\left(\mathrm{Re}\Gamma_m^{(1)}\right)X\right] \tag{10.79a}$$

$$\theta_D(X;\beta_m) = \left(\mathrm{Im}\Gamma_m^{(1)}\right)X. \tag{10.79b}$$

Substituting the real and imaginary parts of the RHS of (10.78) for the corresponding terms in the two equations above gives

$$|G_D(X;\beta_m)| = \exp\left\{-\left[\frac{\beta_m}{2}+\sqrt{\frac{\sqrt{\beta_m^4+16}+\beta_m^2}{8}}\right]X\right\} \tag{10.80a}$$

$$\theta_D(X;\beta_m) = -\sqrt{\frac{\sqrt{\beta_m^4+16}-\beta_m^2}{8}}\,X. \tag{10.80b}$$

Taking the base-10 logarithm of (10.80a) and then writing both of (10.80) compactly give

$$\log_{10}|G_D(X;\beta_m)| = b_1 X$$
$$\theta_D(X;\beta_m) = b_2 X$$

where the slopes (dimensionless) are

$$b_1 \equiv -(\log_{10} e)\left[\frac{\beta_m}{2}+\sqrt{\frac{\sqrt{\beta_m^4+16}+\beta_m^2}{8}}\right]$$

$$b_2 \equiv -\sqrt{\frac{\sqrt{\beta_m^4+16}-\beta_m^2}{8}}.$$

Table 10.3 lists numerical values of the slopes for the same values of β_m used to compile the graphs shown in Figure 10.25. The slope values appear to be consistent with those of the graphs. □

Exercise 10.23 Symmetry of Dimensionless Phase Function

Problem: As function of the dimensionless parameter β_m, is the dimensionless phase function θ_D even, odd, or neither?

Solution: Review Exercise 10.22. The RHS of (10.80b) is even in β_m, so θ_D is even in β_m. □

Exercise 10.24 High-Frequency Approximation

Problem: Suppose $\mu^2 \ll 4\alpha_m$. Derive an approximate form of the hydraulic head FRF (10.74) suitable for this case. What does your result imply about the behavior of the FRF at high frequencies?

Solution: Evaluating the RHS of (10.7) with $\mu \approx 0$ gives

$$\lambda_m^{(1)} \approx -\sqrt{i\alpha_m}.$$

Substituting the RHS for $\lambda_m^{(1)}$ in (10.74) yields the approximation

$$G(x;\omega_m) \approx G(0;\omega_m)\exp(-\sqrt{i\alpha_m}x). \tag{10.81}$$

The RHS is the same result one obtains when substituting $\mu = 0$ in (10.74). This implies that at high frequencies (i.e., for $\alpha_m \gg \mu^2/4$) the FRF approaches that of a homogeneous medium for which $K = K(0)$ and $S_s = S_s(0)$. □

Exercise 10.25 Low-Frequency Approximation

Problem: Suppose $\mu^2 \gg 4\alpha_m$. Derive an approximate form of the hydraulic head FRF (10.74) suitable for this case.

Solution: Evaluating the RHS of (10.7) with $\alpha_m \approx 0$ gives

$$\lambda_m^{(1)} \approx \frac{-\mu-|\mu|}{2} = 0.$$

Substituting the RHS for $\lambda_m^{(2)}$ in (10.74) yields the approximation

$$G(x;\omega_m) \approx G(0;\omega_m). \tag{10.82}$$

□

10.4.2. Parameter Estimation

Suppose the hydraulic head fluctuations have been measured and recorded at two known, distinct locations with space coordinates x_a and x_b for a sufficient period of time and with sufficient accuracy to reliably estimate the amplitude and phase at both locations. That is, the corresponding FRF values $G(x_a;\omega_m)$ and $G(x_b;\omega_m)$ are known, as is the constituent frequency ω_m. How can we use this information to extract information about the material hydrologic properties and what information can we extract?

At both monitoring locations the measured values of the FRF satisfy (10.74):

$$G(x_a;\omega_m) = G(0;\omega_m)e^{\lambda_m^{(1)} x_a}$$

$$G(x_b;\omega_m) = G(0;\omega_m)e^{\lambda_m^{(1)} x_b}.$$

Eliminating $G(0;\omega_m)$, and then solving for $\lambda_m^{(1)}$ give

$$\lambda_m^{(1)} = \frac{1}{x_a - x_b}\ln\frac{G(x_a;\omega_m)}{G(x_b;\omega_m)}. \qquad (10.83)$$

Successively taking the real and imaginary parts and using (10.11) to evaluate the LHS lead to

$$\frac{-\mu}{2} - \frac{1}{2}\sqrt{\frac{\sqrt{\mu^4 + 16\alpha_m^2} + \mu^2}{2}} = c_{\text{Re}} \qquad (10.84\text{a})$$

$$-\frac{1}{2}\sqrt{\frac{\sqrt{\mu^4 + 16\alpha_m^2} - \mu^2}{2}} = c_{\text{Im}} \qquad (10.84\text{b})$$

where we define the known, real constants c_{Re} and c_{Im} as

$$c_{\text{Re}} \equiv \frac{1}{x_a - x_b}\ln\frac{|G(x_a;\omega_m)|}{|G(x_b;\omega_m)|} \qquad (10.85\text{a})$$

$$c_{\text{Im}} \equiv \frac{\arg G(x_a;\omega_m) - \arg G(x_b;\omega_m)}{x_a - x_b} \qquad (10.85\text{b})$$

(dimensions: L^{-1}). Note that $c_{\text{Re}} < 0$ and $c_{\text{Im}} < 0$. Using algebra to solve the nonlinear system (10.84) yields

$$\mu = \frac{c_{\text{Im}}^2 - c_{\text{Re}}^2}{c_{\text{Re}}} \qquad (10.86\text{a})$$

$$\alpha_m = \frac{c_{\text{Im}}}{c_{\text{Re}}}\left(c_{\text{Im}}^2 + c_{\text{Re}}^2\right). \qquad (10.86\text{b})$$

Under ideal conditions (e.g., zero measurement error and complete satisfaction of all assumptions), the knowns c_{Re} and c_{Im} equal the real and imaginary parts, respectively, of the complex eigenvalue $\lambda_m^{(2)}$. Under experimental conditions, however, c_{Re} and c_{Im} should be considered estimates of the real and imaginary parts of the eigenvalue.

Exercise 10.26 Solving for the Coefficients

Problem: Derive the solution in (10.86) by solving the system of (10.84). *Hint*: Begin by eliminating the term $\sqrt{\mu^4 + 16\alpha_m^2}$.

Solution: After eliminating the term $\sqrt{\mu^4 + 16\alpha_m^2}$, solve for μ. Then back substitute to solve for α_m. □

Exercise 10.27 Solving for the Coefficients

Problem: Express $\sqrt{\mu^4 + 16\alpha_m^2}$ in terms of c_{Re} and c_{Im}.

Solution: Substituting the RHSs of (10.86) for μ and α_m in the expression above and then simplifying give

$$\sqrt{\mu^4 + 16\alpha_m^2} = \frac{c_{\text{Im}}^4 + 6c_{\text{Im}}^2 c_{\text{Re}}^2 + c_{\text{Re}}^4}{c_{\text{Re}}^2}. \qquad (10.87)$$

□

Exercise 10.28 Solving for the Coefficients

Problem: Express the quantity

$$\sqrt{\frac{\sqrt{\mu^4 + 16\alpha_m^2} \pm \mu^2}{2}}$$

in terms of c_{Re} and c_{Im}.

Solution: Substituting the RHSs of (10.87) and (10.86a), respectively, for $\sqrt{\mu^4 + 16\alpha_m^2}$ and μ in the expression above and then simplifying give

$$\sqrt{\frac{\sqrt{\mu^4 + 16\alpha_m^2} + \mu^2}{2}} = \frac{c_{\text{Re}}^2 + c_{\text{Im}}^2}{|c_{\text{Re}}|} \qquad (10.88\text{a})$$

$$\sqrt{\frac{\sqrt{\mu^4 + 16\alpha_m^2} - \mu^2}{2}} = 2|c_{\text{Im}}|. \qquad (10.88\text{b})$$

Recalling that $c_{\text{Re}} < 0$ and $c_{\text{Im}} < 0$, these yield

$$\sqrt{\frac{\sqrt{\mu^4 + 16\alpha_m^2} + \mu^2}{2}} = \frac{-(c_{\text{Re}}^2 + c_{\text{Im}}^2)}{c_{\text{Re}}} \qquad (10.89\text{a})$$

$$\sqrt{\frac{\sqrt{\mu^4 + 16\alpha_m^2} - \mu^2}{2}} = -2c_{\text{Im}}. \qquad (10.89\text{b})$$

□

Exercise 10.29 Solving for the Coefficients

Problem: Use the results obtained in Exercise 10.28 to verify that equations (10.86) are the solution of the system (10.84).

Solution: Combining (10.89) with (10.86a), we can write

$$\frac{-\mu}{2} - \frac{1}{2}\sqrt{\frac{\sqrt{\mu^4 + 16\alpha_m^2} + \mu^2}{2}} = \frac{c_{\text{Re}}^2 - c_{\text{Im}}^2}{2c_{\text{Re}}} - \left[\frac{-(c_{\text{Re}}^2 + c_{\text{Im}}^2)}{2c_{\text{Re}}}\right] \qquad (10.90\text{a})$$

$$-\frac{1}{2}\sqrt{\frac{\sqrt{\mu^4 + 16\alpha_m^2} - \mu^2}{2}} = c_{\text{Im}}. \qquad (10.90\text{b})$$

After simplifying the RHS of the first equation, these yield (10.84). □

10.5. ANALOGY: TIME-PERIODIC TEMPERATURE VARIATION

Stallman [1965] presented an analytical solution for time-periodic temperature variation in an infinitely deep porous medium through which groundwater flows vertically downward at a steady rate, while the surface temperature varies sinusoidally in time. Stallman's analysis has been used as the basis for methods to quantify surface water–groundwater interactions using thermal data [e.g., *Hatch et al*, 2006].

In this section we show that the problem of time-periodic, 1D, hydraulic head variation in a source-free, semi-infinite domain filled with an exponential medium, as described in Section 10.4, is mathematically equivalent to the problem of time-periodic, 1D, temperature variation, in a source-free, semi-infinite domain filled with a homogeneous medium, as described by *Stallman* [1965].

Define the following variables:

$$V \equiv \frac{-\mu}{2} \qquad (10.91a)$$

$$W \equiv \frac{\alpha_m}{2} \qquad (10.91b)$$

$$a \equiv -\mathrm{Re}\lambda_m^{(1)} \qquad (10.91c)$$

$$b \equiv -\mathrm{Im}\lambda_m^{(1)} \qquad (10.91d)$$

$$z \equiv x \qquad (10.91e)$$

$$T - T_{AZ} \equiv h' \qquad (10.91f)$$

$$\tau \equiv \frac{2\pi}{\omega_m}. \qquad (10.91g)$$

Stallman [1965] assumes the temperature difference at the upper boundary ($z = 0$) varies as

$$T(0;\omega_m) - T_{AZ} = \Delta T \sin(\omega_m t)$$

where ΔT is a real-valued constant. This requires $G(0;\omega_m) = -i\Delta T$, and (10.74) then yields the corresponding FRF as

$$G(z;\omega_m) = -i\Delta T e^{-az-ibz}.$$

Combining this result with (7.2a) and (10.3) then gives

$$T - T_{AZ} = \Delta T e^{-az} \sin\left(\frac{2\pi t}{\tau} - bz\right). \qquad (10.92)$$

Equation (10.92) is identical to equation (4) of *Stallman* [1965].

Exercise 10.30 Solving for the Coefficients

Problem: Use (10.10) for the principal square root $\sqrt{\mu^2 + i4\alpha_m}$ to show the following:

$$\mathrm{Re}\lambda_m^{(2)} = V - \sqrt{\sqrt{W^2 + V^4/4} + V^2/2} \qquad (10.93a)$$

$$\mathrm{Im}\lambda_m^{(2)} = -\sqrt{\sqrt{W^2 + V^4/4} - V^2/2}. \qquad (10.93b)$$

The expressions on the RHSs of these equations appear in the solution presented by *Stallman* [1965], except that Stallman uses his variable K in place of our variable W.

Solution: Taking the real and imaginary parts of (10.10) gives

$$\mathrm{Re}\sqrt{\mu^2 + i4\alpha_m} = \sqrt{\frac{\sqrt{\mu^4 + 16\alpha_m^2} + \mu^2}{2}} \qquad (10.94a)$$

$$\mathrm{Im}\sqrt{\mu^2 + i4\alpha_m} = \sqrt{\frac{\sqrt{\mu^4 + 16\alpha_m^2} - \mu^2}{2}}. \qquad (10.94b)$$

Solving (10.91a) and (10.91b) for μ and α_m, respectively, yields

$$\mu = -2V$$

$$\alpha_m = 2W.$$

Substituting the RHSs for μ and α_m, respectively, in (10.94) and simplifying lead to

$$\mathrm{Re}\sqrt{\mu^2 + i4\alpha_m} = \sqrt{\sqrt{W^2 + V^4/4} + V^2/2}$$

$$\mathrm{Im}\sqrt{\mu^2 + i4\alpha_m} = \sqrt{\sqrt{W^2 + V^4/4} - V^2/2}.$$

Combining these results with the definition of $\lambda_m^{(2)}$ (10.7) yields (10.93). □

Although the results obtained, in this section apply directly to semi-infinite 1D domains, extending the heat flow analogy to finite 1D domains is straightforward.

10.6. DISCUSSION

The following observations regard the FRFs presented in this chapter, exclusive of zero-frequency limit FRFs (i.e., those for which $\eta_m = 0$):

- One or two amplitude maxima occurred in every example. These always occurred at one or both of the boundaries; none occurred in the space domain. The same is true of phase maxima.
- One amplitude minimum occurred in every example, either in the space domain or at one of the boundaries (e.g., cases a-2, a-3, m-2, and m-3). The same is true of phase minima.
- In cases b-2, b-3, c-2, c-3, d-2, and d-3, both an amplitude minimum and a phase minimum occurred in the space domain. Generally the amplitude and phase minima did not occur at the same location. However, in

cases b-2, b-3, c-2, and c-3 they occurred at the same location when the medium was homogeneous ($\nu = 0$).

In cases c-2 and c-3, when the medium was homogeneous ($\nu = 0$), a zero-amplitude minimum occurred in the space domain at $X = 0.5$; the phase is discontinuous there. If we view each of these cases as two separate flow problems, one for the subregion $0 \leq X < 0.5$ and one for the subregion $0.5 < X \leq 1$, then the first two observations listed above apply to each of the two subregions in these examples as well.

10.7. CHAPTER SUMMARY

In this chapter we defined a particular type of continuously nonhomogeneous, isotropic media that we call *exponential media*. In such media the hydraulic conductivity and specific storage vary exponentially with one space coordinate and at the same relative rate (μ) so that the hydraulic diffusivity is constant. We derived elementary solutions of the complex-variable form of the space BVP for time-periodic, 1D flow in finite and semi-infinite domains filled with exponential media. For some of the solutions we also derived limiting forms and symmetry relations, both of which are useful for verifying numerical algorithms used to calculate the FRFs. Under these conditions the hydraulic head FRF depends not only on the hydraulic diffusivity and the angular frequency but on the reciprocal scale of spatial variation (μ) as well.

11
Examples: 1D Flow in Power Law Media

In this chapter we briefly explore time-periodic, 1D flow in yet another type of continuously nonhomogeneous media, which we call *power law media*.

11.1. ASSUMPTIONS

In addition to the assumptions listed in the introduction to Part III, throughout this chapter we assume that the material hydrologic properties vary continuously in space as follows:

- The scalar hydraulic conductivity varies as

$$K(\mathbf{x}) = K(x) = K(x_c)\left(\frac{x}{x_c}\right)^b \tag{11.1}$$

where $x > 0$, $K(x_c) > 0$, x_c is a positive constant (dimensions: L), b is a dimensionless constant, and all quantities are real valued.

- The specific storage varies as

$$S_s(\mathbf{x}) = S_s(x) = S_s(x_c)\left(\frac{x}{x_c}\right)^{b-2} \tag{11.2}$$

where, additionally, $S_s(x_c)$ is real valued and positive.

In this book, we call media that satisfy these assumptions *power law media*.

We also assume that the boundary conditions are such that the flow is purely 1D, so that we can represent the hydraulic head FRF as a function of a single Cartesian space coordinate (x):

$$F_h(\mathbf{x}; \omega_m) = G(x; \omega_m). \tag{11.3}$$

We can express the hydraulic conductivity and specific storage distributions of a power law medium in dimensionless form as

$$K_D(X) \equiv \frac{K(x)}{K(x_c)} = X^b$$

$$S_{sD}(X) \equiv \frac{S_s(x)}{S_s(x_c)} = X^{b-2}$$

where we define the dimensionless space coordinate as

$$X \equiv \frac{x}{x_c}. \tag{11.4}$$

Figure 11.1 shows the dimensionless distributions for a range of values of b.

Exercise 11.1 Spatial Distribution of Hydraulic Diffusivity

Problem: Consider a power law medium. Express the hydraulic diffusivity, $\kappa(x)$, compactly in terms of x and $\kappa(x_c)$, where

$$\kappa(x_c) \equiv \frac{K(x_c)}{S_s(x_c)}.$$

Solution: Substituting the RHSs of (11.1) and (11.2), respectively, for K and S_s in (2.18) leads to

$$\kappa(x) = \kappa(x_c)\left(\frac{x}{x_c}\right)^2.$$

□

In Section 11.3 we generalize the concept of power law medium to include the case where $x < 0$.

11.2. GOVERNING EQUATION AND GENERAL SOLUTION

Define the dimensionless frequency as

$$f_m \equiv \frac{\omega_m S_s(x_c)}{K(x_c)} x_c^2. \tag{11.5}$$

Consequently f_m is real valued and positive.

Hydrodynamics of Time-Periodic Groundwater Flow: Diffusion Waves in Porous Media, Geophysical Monograph 224, First Edition. Joe S. Depner and Todd C. Rasmussen.

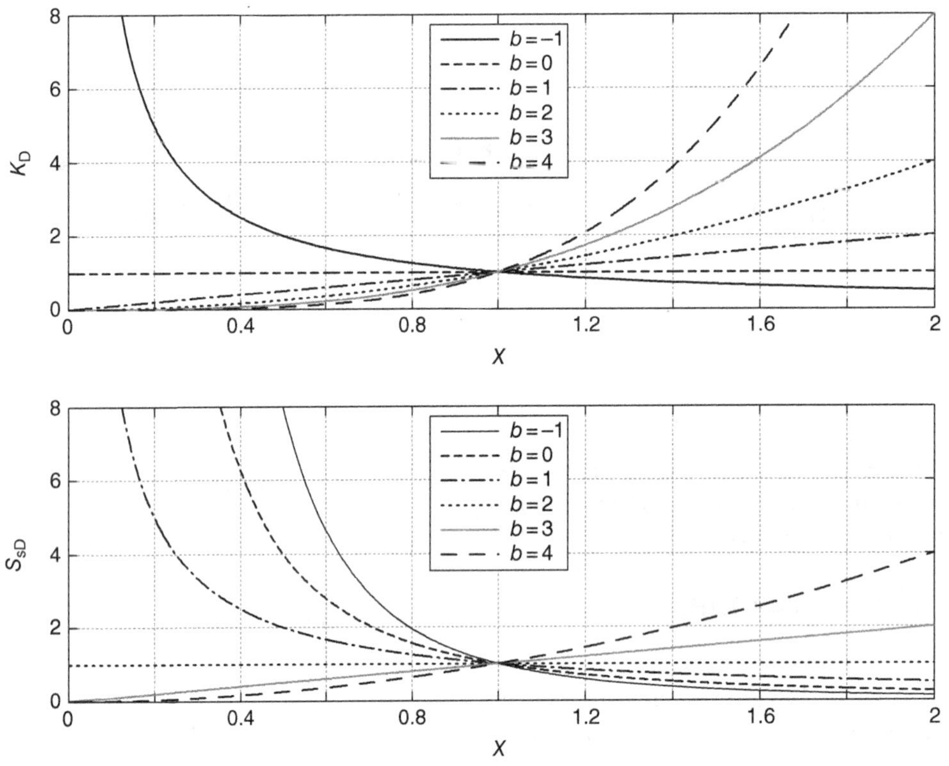

Figure 11.1 Dimensionless K and S_s distributions for power law medium.

Exercise 11.2 Governing Equation for 1D Flow in Power Law Medium

Problem: Consider time-periodic, 1D flow in a source-free domain filled with a power law medium. Write the simplified governing equation for the hydraulic head FRF for this particular case using the dimensionless parameters b and f_m.

Solution: Substituting the RHSs of (11.1) and (11.2), respectively, for K and S_s, zero for $F_u(\mathbf{x};\omega_m)$, and $G(x;\omega_m)$ for $F_h(\mathbf{x};\omega_m)$ in the governing equation (7.17a) and then simplifying yield

$$x^2\frac{d^2G}{dx^2} + bx\frac{dG}{dx} - if_mG = 0. \tag{11.6}$$

□

The governing equation (11.6) is a linear, second-order, homogeneous ODE with variable coefficients. Equation (11.6) is, in fact, a form of the *Euler differential equation* [*Weisstein*, 2016a] with complex coefficients.

Exercise 11.3 Trial Solution for 1D Flow in Power Law Medium

Problem: Regarding the governing equation (11.6), consider the trial solution

$$G(x;\omega_m) = \left(\frac{x}{x_c}\right)^{p_m} \tag{11.7}$$

where p_m is an undetermined constant (eigenvalue). Show that the trial function satisfies the governing equation by deriving the characteristic equation and then solving it to find the eigenvalue(s). Express p_m in terms of b and f_m. How many distinct eigenvalues are there?

Solution: Substituting the RHS of (11.7) for G in (11.6), carrying out the differentiation, and then dividing by G lead to the characteristic equation

$$p_m^2 + (b-1)p_m - if_m = 0$$

which is quadratic in p_m. Using the quadratic formula to solve it yields

$$p_m \equiv \frac{(1-b) \pm \sqrt{(1-b)^2 + i4f_m}}{2} \tag{11.8}$$

where $\sqrt{z}$ denotes the principal square root of z (see Section D.2.1). There are two distinct eigenvalues. □

We define the eigenvalues as

$$p_m^{(n)} \equiv \frac{(1-b) + (-1)^n\sqrt{(1-b)^2 + i4f_m}}{2} \quad (n = 1, 2). \tag{11.9}$$

where f_m is real and positive, so the eigenvalues are distinct. The corresponding FREs are

$$G^{(n)}(x;\omega_m) \equiv \left(\frac{x}{x_c}\right)^{p_m^{(n)}} \quad (n = 1, 2). \tag{11.10}$$

These are linearly independent, so we can construct a general solution to the homogeneous governing equation (11.6) by superposition as

$$G(x;\omega_m) = a_m^{(1)}\left(\frac{x}{x_c}\right)^{p_m^{(1)}} + a_m^{(2)}\left(\frac{x}{x_c}\right)^{p_m^{(2)}} \tag{11.11}$$

where the $a_m^{(n)}$ are complex constants (dimensions: L) to be determined by the boundary conditions.

Exercise 11.4 Eigenvalues—Rectangular Form

Problem: Express $\sqrt{(1-b)^2 + i4f_m}$ in rectangular form (i.e., in the form $x + iy$). Then use your result to express the eigenvalues $p_m^{(n)}$ $(n = 1, 2)$ in (11.9) in rectangular form.

Solution: Define

$$z = x + iy \equiv (1-b)^2 + i4f_m$$

so that

$$\sqrt{z} = \sqrt{(1-b)^2 + i4f_m}.$$

Using the notation in Section D.2.1, we have $x = (1-b)^2$, $y = 4f_m$, and $r = \sqrt{x^2+y^2} = \sqrt{(1-b)^4 + 16f_m^2}$. Equation (D.6) then gives, for the principal square root,

$$\sqrt{(1-b)^2 + i4f_m} = \sqrt{\frac{\sqrt{(1-b)^4 + 16f_m^2} + (1-b)^2}{2}} + i\sqrt{\frac{\sqrt{(1-b)^4 + 16f_m^2} - (1-b)^2}{2}}.$$

Substituting the RHS for the radical terms in (11.9) yields

$$p_m^{(n)} = \frac{(1-b)}{2} + \frac{(-1)^n}{2}\sqrt{\frac{\sqrt{(1-b)^4 + 16f_m^2} + (1-b)^2}{2}} + i\frac{(-1)^n}{2}\sqrt{\frac{\sqrt{(1-b)^4 + 16f_m^2} - (1-b)^2}{2}} \quad (n = 1, 2). \tag{11.12}$$

□

Exercise 11.5 Eigenvalues—Properties of Real and Imaginary Parts

Problem: What are the signs of the real and imaginary parts of $p_m^{(n)}$ $(n = 1, 2)$? Also, express $\mathrm{Re}\, p_m^{(1)} + \mathrm{Re}\, p_m^{(2)}$ and $\mathrm{Im}\, p_m^{(1)} + \mathrm{Im}\, p_m^{(2)}$ compactly in terms of b and f_m.

Solution: We deduce the following directly from (11.12):

$$\mathrm{Re}\, p_m^{(1)} < 0 \tag{11.13a}$$
$$\mathrm{Im}\, p_m^{(1)} < 0 \tag{11.13b}$$
$$\mathrm{Re}\, p_m^{(2)} > 0 \tag{11.13c}$$
$$\mathrm{Im}\, p_m^{(2)} > 0 \tag{11.13d}$$

and

$$\mathrm{Re}\, p_m^{(1)} + \mathrm{Re}\, p_m^{(2)} = 1 - b \tag{11.14a}$$
$$\mathrm{Im}\, p_m^{(1)} + \mathrm{Im}\, p_m^{(2)} = 0. \tag{11.14b}$$

□

Exercise 11.6 Algebraic Interpretation of FREs

Problem: Express the modulus and argument of the FREs (11.10) in terms of X [see (11.4)], Re $p_m^{(n)}$, and Im $p_m^{(n)}$.

Solution: Rewrite (11.10) as

$$\begin{aligned} G^{(n)}(x;\omega_m) &= X^{p_m^{(n)}} \\ &= X^{\mathrm{Re}\, p_m^{(n)} + i\mathrm{Im}\, p_m^{(n)}} \\ &= X^{\mathrm{Re}\, p_m^{(n)}} X^{i\mathrm{Im}\, p_m^{(n)}} \\ &= X^{\mathrm{Re}\, p_m^{(n)}} \left(e^{\ln X}\right)^{i\mathrm{Im}\, p_m^{(n)}} \\ &= X^{\mathrm{Re}\, p_m^{(n)}} e^{i(\mathrm{Im}\, p_m^{(n)})\ln X}. \end{aligned}$$

Taking the modulus and argument yields

$$\begin{aligned} \left|G^{(n)}(x;\omega_m)\right| &= X^{\mathrm{Re}\, p_m^{(n)}} \\ \arg G^{(n)}(x;\omega_m) &= \left(\mathrm{Im}\, p_m^{(n)}\right)\ln X \end{aligned} \tag{11.15}$$

$$n = 1, 2.$$

□

Exercise 11.7 Low-Frequency Limit FRF for Power Law Medium

Problem: Consider time-periodic, 1D flow in a finite, source-free domain filled with a power law medium. Investigate the low-frequency limit of the general solution (11.11) by considering the cases $b \leq 1$ and $b > 1$ separately.

Solution: In the limit as the constituent frequency approaches zero, (11.5) gives $f_m \to 0$ and (11.9) gives

$$\lim_{\omega_m \to 0} p_m^{(n)} \equiv \frac{(1-b) + (-1)^n|1-b|}{2}. \tag{11.16}$$

Case 1: $b \leq 1$. Then $|1 - b| = 1 - b$ and (11.16) produces

$$p_m^{(1)} \to 0$$
$$p_m^{(2)} \to 1 - b.$$

Substituting these results for the corresponding exponents in (11.11) then yields

$$\lim_{\omega_m \to 0} G(x; \omega_m) = a_m^{(1)} + a_m^{(2)} \left(\frac{x}{x_c} \right)^{1-b}. \qquad (11.17)$$

Case 2: $b > 1$. Then $|1 - b| = b - 1$ and (11.16) produces

$$p_m^{(1)} \to 1 - b$$
$$p_m^{(2)} \to 0.$$

Substituting these results for the corresponding exponents in (11.11) then yields

$$\lim_{\omega_m \to 0} G(x; \omega_m) = a_m^{(1)} \left(\frac{x}{x_c} \right)^{1-b} + a_m^{(2)}. \qquad (11.18)$$

□

Exercise 11.8 High-Frequency Form of FRF for Power Law Medium

Problem: Consider time-periodic, 1D flow in a finite, source-free domain filled with a power law medium. Investigate the high-frequency form of the FRF (11.11) by considering the special case where

$$f_m \gg \left(\frac{1 - b}{2} \right)^2. \qquad (11.19)$$

Express the eigenvalues $p_m^{(n)}$ $(n = 1, 2)$ in terms of b and f_m. Compare your results to those obtained for the special case where (11.19) is not necessarily satisfied but $b = 1$.

Solution: Condition (11.19) implies that (11.12) takes the following form:

$$p_m^{(n)} = (-1)^n \sqrt{\frac{f_m}{2}} (1 + i) \quad (n = 1, 2). \qquad (11.20)$$

For comparison, substituting $b = 1$ in (11.12) without incorporating (11.19) gives the same result. □

11.3. GENERALIZATION

In this section we generalize the concept of time-periodic, 1D flow in a power law medium to include the case where $x < 0$.

Example 11.1 Generalized Power Law Medium

We can generalize our definition of a power law medium in (11.1) and (11.2) as

$$K(x) = K(x_c) \left| \frac{x}{x_c} \right|^b \qquad (11.21a)$$

$$S_s(x) = S_s(x_c) \left| \frac{x}{x_c} \right|^{b-2} \qquad (11.21b)$$

where $|x| > 0$, $K(x_c) > 0$, and $S_s(x_c) > 0$; b is a real, dimensionless constant; and x_c is a real, positive constant (dimensions: L). In this case x can be negative. If we plot $K(x)$ and $S_s(x)$ on rectangular graphs, with x on the horizontal axis, then both graphs are symmetric about their respective vertical (K or S_s) axes. ◊

Exercise 11.9 FRF for Generalized Power Law Medium—Part 1 of 2

Problem: Consider time-periodic, 1D flow in a source-free domain filled with a generalized power law medium [see equations (11.21)]. Define the following hydraulic head FRF:

$$G(x; \omega_m) = a_m \left| \frac{x}{x_c} \right|^{p_m} \qquad (11.22)$$

where a_m and p_m are complex constants, p_m is dimensionless, $x_c > 0$, and $|x| > 0$. Evaluate dG/dx and d^2G/dx^2.

Solution: Noting that x is real valued and p_m generally is complex valued,

$$|x|^{p_m} = \begin{cases} (-1)^{p_m} x^{p_m} & x < 0 \\ x^{p_m} & x > 0. \end{cases}$$

Differentiating with respect to x gives

$$\frac{d}{dx}|x|^{p_m} = \begin{cases} (-1)^{p_m} p_m x^{p_m - 1} & x < 0 \\ p_m x^{p_m - 1} & x > 0. \end{cases}$$

Using this result to differentiate (11.22) with respect to x leads to

$$\frac{d}{dx} G(x; \omega_m) = \begin{cases} (-1)^{p_m} a_m x_c^{-p_m} p_m x^{p_m - 1} & x < 0 \\ a_m x_c^{-p_m} p_m x^{p_m - 1} & x > 0. \end{cases} \qquad (11.23)$$

Differentiating again with respect to x produces

$$\frac{d^2}{dx^2} G(x; \omega_m) = \begin{cases} (-1)^{p_m} a_m x_c^{-p_m} p_m (p_m - 1) x^{p_m - 2} & x < 0 \\ a_m x_c^{-p_m} p_m (p_m - 1) x^{p_m - 2} & x > 0. \end{cases} \qquad (11.24)$$

□

Exercise 11.10 FRF for Generalized Power Law Medium—Part 2 of 2

Problem: Consider time-periodic, 1D flow in a source-free domain filled with a generalized power law medium [see equations (11.21)]. Show that for a suitable choice of the dimensionless, complex constant p_m the hydraulic head FRF defined by (11.22) satisfies the governing equation (11.6). Express p_m in terms of b and f_m.

Solution: Consider the two cases $x < 0$ and $x > 0$ separately.

Case 1: $x < 0$. Substitute the RHSs of (11.22) and the forms of (11.23) and (11.24) that pertain to negative x values for the corresponding terms in the governing equation (11.6). After grouping terms this gives the characteristic equation

$$p_m^2 + (b-1)p_m - if_m = 0.$$

Case 2: $x > 0$. Substitute the RHSs of (11.22) and the forms of (11.23) and (11.24) that pertain to positive x values for the corresponding terms in the governing equation (11.6). Grouping terms then leads to the characteristic equation

$$p_m^2 + (b-1)p_m - if_m = 0.$$

Summary: The two cases produce the same characteristic equation. The governing equation is satisfied if the characteristic equation is satisfied. The characteristic equation is quadratic in p_m; using the quadratic formula to solve it produces

$$p_m \equiv \frac{(1-b) \pm \sqrt{(1-b)^2 + \mathrm{i}4f_m}}{2}$$

which is identical to (11.8). □

11.4. CHAPTER SUMMARY

In this chapter we defined a particular type of continuously nonhomogeneous, isotropic media that we call *power law media*. In such media the hydraulic conductivity and specific storage vary as power law functions of one space coordinate. When normalized, each power law function is characterized by two parameters—a length scale (x_c) and an exponent (b). Unlike the types of media considered in Chapters 9 and 10, in power law media the hydraulic diffusivity varies (quadratically) with the space coordinate. We derived elementary solutions of the complex-variable form of the space BVP for time-periodic, 1D flow in finite and semi-infinite domains filled with power law media.

12
Examples: 2D and 3D Flow in Ideal Media

In this chapter we explore elementary examples of FRFs for time-periodic, multidimensional flow in ideal media.

12.1. ASSUMPTIONS

In addition to the assumptions listed in the introduction to Part III, throughout this chapter we assume the porous medium is ideal (see Section 2.6.1).

12.2. GOVERNING AND BC EQUATIONS

Substituting $F_u = 0$ and the RHSs of (2.16) for the corresponding quantities in (7.17) and simplifying yield

$$\nabla^2 F_h(\mathbf{x}; \omega_m) - i\alpha_m F_h(\mathbf{x}; \omega_m) = 0 \quad \forall \mathbf{x} \in D \qquad (12.1a)$$

$$W\big[F_h(\mathbf{x}; \omega_m)\big] = F_\psi(\mathbf{x}; \omega_m) \quad \forall \mathbf{x} \in \Gamma \qquad (12.1b)$$

$$m = 1, 2, 3, \ldots$$

where (5.31) defines the real-valued constants α_m. The governing equations (12.1a) are valid only in the source-free portion of the domain.

12.3. 3D RADIAL SPHERICAL FLOW

This section explores elementary examples of time-periodic flow in three dimensions with "radial" geometry (i.e., purely convergent/divergent spherical flows). Then the hydraulic head FRF takes the form

$$F_h(\mathbf{x}; \omega_m) = G(r; \omega_m) \qquad (12.2)$$

where we define r, the radial space coordinate, as $r \equiv \sqrt{x_1^2 + x_2^2 + x_3^2}$.

12.3.1. Governing Equation and General Solution

Exercise 12.1 Derivation: Simplified Form of Governing Equation

Problem: Consider time-periodic, radially converging/diverging flow in three dimensions. What simplified form does the FRF governing equation take in this case?

Solution: Substituting the RHS of (12.2) for the FRF in the governing equation (12.1a) and evaluating the Laplacian (in spherical coordinates) yield

$$r^2 \frac{d^2 G}{dr^2} + 2r \frac{dG}{dr} - i\alpha_m r^2 G = 0. \qquad (12.3)$$

□

Exercise 12.2 Derivation: Modified Spherical Bessel Equation

Problem: Consider the coordinate transformation $R' \equiv \lambda_m r$, with

$$\lambda_m \equiv \sqrt{i\alpha_m} \qquad (12.4)$$

where $\sqrt{z}$ denotes the principal square root of z (see Section D.2.1). Show that one can use this to transform (12.3) into a modified spherical Bessel equation.

Solution: Define $g(R'; \omega_m) \equiv G(r; \omega_m)$. Then

$$\frac{dG}{dr} = \lambda_m \frac{dg}{dR'} \qquad (12.5a)$$

$$\frac{d^2 G}{dr^2} = \lambda_m^2 \frac{d^2 g}{dR'^2}. \qquad (12.5b)$$

Hydrodynamics of Time-Periodic Groundwater Flow: Diffusion Waves in Porous Media, Geophysical Monograph 224, First Edition. Joe S. Depner and Todd C. Rasmussen.

Substituting the RHSs of these equations for the corresponding terms of (12.3) and simplifying yield

$$R'^2 \frac{d^2 g}{dR'^2} + 2R' \frac{dg}{dR'} - R'^2 g = 0 \tag{12.6}$$

which is a spherical Bessel equation of order zero [see Section 10.47(i) of *Olver and Maximon*, 2016; also see *Weisstein*, 2016b]. □

The hydraulic head FRF is then the general solution of (12.6)

$$G(r; \omega_m) = g(R'; \omega_m) = a_m^{(1)} k_0(\lambda_m r) + a_m^{(2)} i_0(\lambda_m r) \tag{12.7}$$

where i_0 and k_0 are order-zero, modified, spherical Bessel functions of the first and second kinds, respectively [see Section 10.47 of *Olver and Maximon*, 2016], and $a_m^{(1)}$ and $a_m^{(2)}$ are complex constants (dimensions: L) to be determined by the boundary conditions.

Note 12.1 Order-Zero, Spherical Bessel Functions

According to *Weisstein* [2016c, d], we can represent the order-zero, modified, spherical Bessel functions of the first and second kinds, respectively, as

$$i_0(z) = \frac{\sinh z}{z} \tag{12.8a}$$

$$k_0(z) = \frac{e^{-z}}{z} \tag{12.8b}$$

where z generally is complex valued. △

12.3.2. Spherical Flow Due to a Point Source

Example 12.1 Spherical Flow Due to Point Source

Consider a point source located at the origin in a 3D, infinite domain filled with an ideal medium. The boundary conditions are [*Black and Kipp*, 1981]

$$\lim_{r \to 0} \left[-4\pi K r^2 \frac{d}{dr} G(r; \omega_m) \right] = Q_0(\omega_m) \tag{12.9a}$$

$$\lim_{r \to \infty} G(r; \omega_m) = 0 \tag{12.9b}$$

where $Q_0(\omega_m)$ denotes the complex amplitude of the point source [i.e., $|Q_0(\omega_m)|$ gives the maximum (peak) volumetric rate of recharge] (dimensions: $L^3 t^{-1}$). Applying the boundary conditions to the general solution (12.7) gives

$$G(r; \omega_m) = \frac{Q_0(\omega_m) \lambda_m}{4\pi K} k_0(\lambda_m r), \quad r > 0.$$

Using (12.8b), we can express this as

$$G(r; \omega_m) = \frac{Q_0(\omega_m)}{4\pi K} \frac{e^{-\lambda_m r}}{r}, \quad r > 0. \tag{12.10}$$

The FRF is undefined at $r = 0$. *Black and Kipp* [1981] reported this solution. ◊

Exercise 12.3 FRF Solution Satisfies Boundary Conditions

Problem: Consider Example 12.1. Verify that the FRF solution (12.10) satisfies the boundary conditions (12.9).□

Example 12.2 Spherical Flow Due to Point Source—Dimensionless FRF

We can express the FRF solution (12.10) in terms of dimensionless variables as

$$G_D(R) = \frac{e^{-\sqrt{i}R}}{R}, \quad R > 0 \tag{12.11}$$

where we define the dimensionless variables as

$$G_D \equiv \frac{4\pi K G(r; \omega_m)}{\sqrt{\alpha_m} Q_0(\omega_m)}$$

$$R \equiv \sqrt{\alpha_m} r.$$

The dimensionless FRF is undefined at $R = 0$. Figure 12.1 shows the corresponding amplitude and phase functions. We define the dimensionless phase function as

$$\theta_D(R) \equiv \arg G_D(R). \tag{12.12}$$

Notice that both functions decrease as the distance from the point source increases. ◊

Exercise 12.4 Dimensionless Amplitude and Phase Functions

Problem: Consider the dimensionless FRF for 3D spherical flow due to a point source (12.11). Find the corresponding dimensionless amplitude and phase functions.

Solution: Substituting the RHS of (D.8) for $\sqrt{i}$ in (12.11) and then rewriting the exponential term give

$$G_D(R) = \frac{e^{-R/\sqrt{2}}}{R} e^{-iR/\sqrt{2}}. \tag{12.13}$$

Taking the modulus of (12.13) gives the dimensionless amplitude function:

$$|G_D(R)| = \frac{e^{-R/\sqrt{2}}}{R}, \quad R > 0. \tag{12.14}$$

Dividing (12.13) by the modulus, taking the logarithm, and then dividing by i give the dimensionless phase function:

$$\theta_D(R) = \frac{-R}{\sqrt{2}}, \quad R > 0. \tag{12.15}$$

□

The point source solution of Example 12.1 is easily generalized to the case where the source is located at the point with coordinates $\mathbf{x} = \mathbf{x}_0$ by making the coordinate

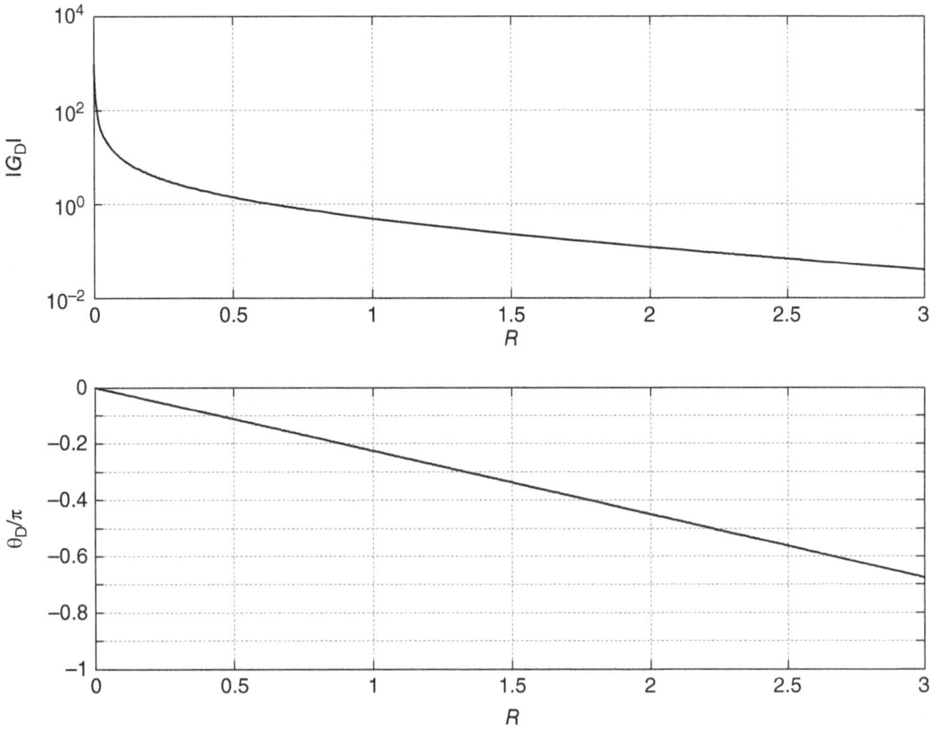

Figure 12.1 FRF for spherical flow due to a point source.

transformations $\mathbf{x}' \equiv \mathbf{x} - \mathbf{x}_0$ and $r' \equiv |\mathbf{x}'|$. Then $F_h(\mathbf{x}; \omega_m) = G(r'; \omega_m)$, where (12.10) defines the function G:

$$F_h(\mathbf{x}; \omega_m) = \frac{Q_0(\omega_m)}{4\pi K} \frac{e^{-\lambda_m |\mathbf{x}-\mathbf{x}_0|}}{|\mathbf{x} - \mathbf{x}_0|}, \quad |\mathbf{x} - \mathbf{x}_0| > 0. \quad (12.16)$$

12.3.3. Spherical Flow Due to Forced Outer Boundary

Example 12.3 Spherical Flow Due to Forced Outer Boundary
Consider a porous medium with a spherical outer boundary subject to time-periodic forcing with a single harmonic constituent. The boundary conditions are

$$\left[\frac{d}{dr} G(r; \omega_m)\right]\Bigg|_{r=0} = 0$$
$$G(r_{\max}; \omega_m) = \text{known(constant for fixed } \omega_m)$$

where $r_{\max} > 0$. Applying the boundary conditions to the general solution (12.7) gives

$$G(r; \omega_m) = \frac{G(r_{\max}; \omega_m)}{i_0(\lambda_m r_{\max})} i_0(\lambda_m r). \quad (12.17)$$

Using (12.8a), we can express this as

$$G(r; \omega_m) = G(r_{\max}; \omega_m) \frac{r_{\max}}{\sinh(\lambda_m r_{\max})} \frac{\sinh(\lambda_m r)}{r}. \quad (12.18)$$

◊

Exercise 12.5 FRF Values

Problem: Consider Example 12.3. What is the value of the FRF at the center ($r = 0$) of the spherical domain? What value must the outer radius ($r_{\max}$) take if the hydraulic head fluctuations are to decay to zero at the center?

Solution: Taking the limit of (12.18) as r approaches zero gives

$$G(0; \omega_m) = G(r_{\max}; \omega_m) \frac{\lambda_m r_{\max}}{\sinh(\lambda_m r_{\max})} \lim_{r \to 0} \frac{\sinh(\lambda_m r)}{\lambda_m r}.$$

The last term on the RHS is of the form 0/0; using l'Hôpital's rule to evaluate it yields

$$G(0; \omega_m) = G(r_{\max}; \omega_m) \frac{\lambda_m r_{\max}}{\sinh(\lambda_m r_{\max})}.$$

Although the modulus of the RHS approaches zero as $r_{\max}$ grows without bound, it does not vanish. Therefore, $|G(0; \omega_m)| > 0$ for *all* $r_{\max} > 0$. □

Example 12.4 Forced Outer Spherical Boundary—Dimensionless FRF
Consider the spherical flow described in Example 12.3, with $|G(r_{\max}; \omega_m)| > 0$. We can express the FRF in dimensionless form as

$$G_D(R;\eta_m) = \frac{\sinh(\sqrt{i}\eta_m R)}{R\sinh(\sqrt{i}\eta_m)} \quad (12.19)$$

where we define the dimensionless variables as

$$G_D \equiv \frac{G(r;\omega_m)}{G(r_{max};\omega_m)} \quad (12.20a)$$

$$R \equiv \frac{r}{r_{max}} \quad (12.20b)$$

$$\eta_m \equiv \sqrt{\alpha_m} r_{max}. \quad (12.20c)$$

Note the following:

- R and η_m are real valued;
- $0 \le R \le 1$, with $R = 1$ at the outer boundary; and
- $0 \le |G_D| \le 1$, with $G_D = 1$ at the outer boundary.

Figure 12.2 shows the corresponding amplitude and phase functions for various values of η_m. We define the dimensionless phase function as

$$\theta_D(R;\eta_m) \equiv \arg G_D(R;\eta_m).$$

◊

Note 12.2 Modulus and Argument of Hyperbolic Sine

Zucker [1972] (see paragraphs 4.5.54 and 4.5.55, respectively) gives the identities

$$|\sinh z| = \sqrt{\frac{1}{2}\Big[\cosh(2\text{Re } z) - \cos(2\text{Im } z)\Big]} \quad (12.21a)$$

$$\arg\sinh z = \arctan[\coth(\text{Re } z)\cdot\tan(\text{Im } z)]. \quad (12.21b)$$

△

Exercise 12.6 Dimensionless Amplitude and Phase Functions

Problem: Consider the dimensionless FRF for 3D spherical flow due to a forced outer boundary (12.19). Express the corresponding dimensionless amplitude and phase functions in terms of purely real variables. *Hint:* See Note 12.2.

Solution: Substituting $z = \sqrt{i}u$, where u is real valued, in identities (12.21) gives, respectively,

$$|\sinh(\sqrt{i}u)| = \sqrt{\frac{1}{2}\Big[\cosh(\sqrt{2}u) - \cos(\sqrt{2}u)\Big]} \quad (12.22a)$$

$$\arg\sinh(\sqrt{i}u) = \arctan\Big[\coth(u/\sqrt{2})\cdot\tan(u/\sqrt{2})\Big] \quad (12.22b)$$

because $\sqrt{i} = (1+i)/\sqrt{2}$. Taking the modulus and phase, respectively, of (12.19) leads to

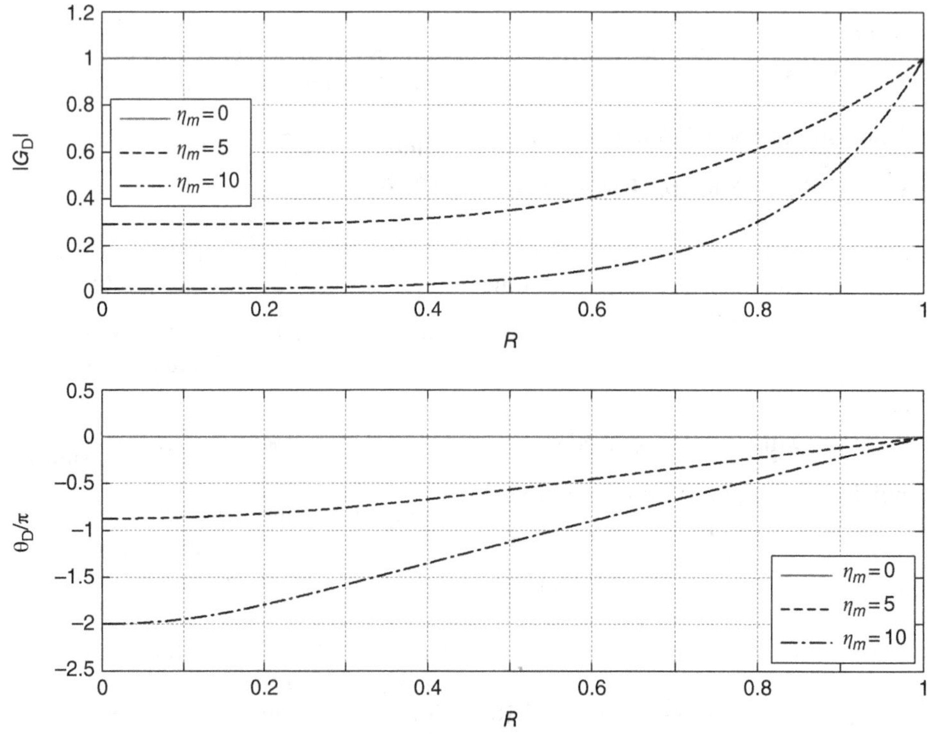

Figure 12.2 FRF for spherical flow due to forced outer boundary.

$$|G_D(R;\eta_m)| = \frac{|\sinh(\sqrt{i}\eta_m R)|}{R|\sinh(\sqrt{i}\eta_m)|} \quad (12.23a)$$

$$\theta_D(R;\eta_m) = \arg\sinh(\sqrt{i}\eta_m R) - \arg\sinh(\sqrt{i}\eta_m). \quad (12.23b)$$

Applying (12.22) to these, once with $u \equiv \eta_m R$ and once with $u \equiv \eta_m$, yields

$$|G_D(R;\eta_m)| = \frac{\sqrt{\cosh(\sqrt{2}\eta_m R) - \cos(\sqrt{2}\eta_m R)}}{R\sqrt{\cosh(\sqrt{2}\eta_m) - \cos(\sqrt{2}\eta_m)}} \quad (12.24a)$$

$$\theta_D(R;\eta_m) = \arctan\left[\coth(\eta_m R/\sqrt{2}) \cdot \tan(\eta_m R/\sqrt{2})\right] - \arctan\left[\coth(\eta_m/\sqrt{2}) \cdot \tan(\eta_m/\sqrt{2})\right]. \quad (12.24b)$$

□

Exercise 12.7 Generalizing the FRF Solution

Problem: Generalize the FRF solution obtained in Example 12.3 to the case where the spherical domain (of radius r_{max}) is centered at the point with coordinates $\mathbf{x} = \mathbf{x}_0$.

Solution: Define the transformed coordinates $\mathbf{x}' \equiv \mathbf{x} - \mathbf{x}_0$ and $r' \equiv |\mathbf{x}'|$. Then $F_h(\mathbf{x};\omega_m) = G(r';\omega_m)$, where (12.18) defines the function G:

$$F_h(\mathbf{x};\omega_m) = G(r_{max};\omega_m)\frac{r_{max}}{\sinh(\lambda_m r_{max})}\frac{\sinh(\lambda_m|\mathbf{x}-\mathbf{x}_0|)}{|\mathbf{x}-\mathbf{x}_0|}.$$

□

12.3.4. Radial Flow in a Spherical Shell

Exercise 12.8 Radial Flow in a Spherical Shell

Problem: Consider the domain that lies between two concentric spheres with different, nonzero diameters $r = r_{min}$ and $r = r_{max}$, where $r_{max} > r_{min} > 0$. Suppose the boundaries are periodically forced with a single harmonic constituent whose frequency is ω_m (m fixed). The boundary conditions are

$$G(r_{min};\omega_m) = \text{known (constant for fixed } \omega_m) \quad (12.25a)$$

$$G(r_{max};\omega_m) = \text{known (constant for fixed } \omega_m). \quad (12.25b)$$

Express the hydraulic head FRF, $G(r;\omega_m)$, in terms of $G(r_{min};\omega_m)$ and $G(r_{max};\omega_m)$.

Solution: Substituting the RHS of (12.7) for G in the boundary conditions (12.25) yields the linear system of two equations in two unknowns ($a_m^{(1)}$ and $a_m^{(2)}$):

$$a_m^{(1)} k_0(\lambda_m r_{min}) + a_m^{(2)} i_0(\lambda_m r_{min}) = G(r_{min};\omega_m) \quad (12.26a)$$

$$a_m^{(1)} k_0(\lambda_m r_{max}) + a_m^{(2)} i_0(\lambda_m r_{max}) = G(r_{max};\omega_m). \quad (12.26b)$$

The solution is (e.g., by Cramer's rule)

$$a_m^{(1)} = \frac{G(r_{min};\omega_m) i_0(\lambda_m r_{max}) - G(r_{max};\omega_m) i_0(\lambda_m r_{min})}{i_0(\lambda_m r_{max}) k_0(\lambda_m r_{min}) - i_0(\lambda_m r_{min}) k_0(\lambda_m r_{max})} \quad (12.27a)$$

$$a_m^{(2)} = \frac{G(r_{max};\omega_m) k_0(\lambda_m r_{min}) - G(r_{min};\omega_m) k_0(\lambda_m r_{max})}{i_0(\lambda_m r_{max}) k_0(\lambda_m r_{min}) - i_0(\lambda_m r_{min}) k_0(\lambda_m r_{max})} \quad (12.27b)$$

assuming that

$$\left| i_0(\lambda_m r_{max}) k_0(\lambda_m r_{min}) - i_0(\lambda_m r_{min}) k_0(\lambda_m r_{max}) \right| > 0.$$

Therefore, the BVP solution is given by (12.7), with $a_m^{(1)}$ and $a_m^{(2)}$ given by (12.27). □

Example 12.5 Radial Flow in a Spherical Shell

Consider radial flow in a spherical shell as described in Exercise 12.8, with $r_{min} = 0.2 r_{max}$ and $G(r_{min};\omega_m) = G(r_{max};\omega_m)$. Define the dimensionless variables:

$$G_D \equiv \frac{G(r;\omega_m)}{G(r_{min};\omega_m)} \quad (12.28a)$$

$$R \equiv \frac{r}{r_{max}} \quad (12.28b)$$

$$\eta_m \equiv \sqrt{\alpha_m} r_{max} \quad (12.28c)$$

$$\theta_D \equiv \arg G_D. \quad (12.28d)$$

Note the following:

- R, η_m, and θ_D are real valued,
- $0.2 \le R \le 1$,
- $G_D = 1$ at both of the boundaries, and
- $\theta_D = 0$ at both of the boundaries.

Figure 12.3 shows the corresponding amplitude and phase functions for various values of η_m. ◊

Exercise 12.9 Zero Sets of Amplitude and Phase Gradients

Problem: Consider time-periodic, radial flow in a spherical shell as described in Example 12.5 (see Figure 12.3). For a fixed value of the dimensionless frequency (say, $\eta_m = 0.5$), describe the geometry of the amplitude gradient's zero set. How is the zero set related to the shell's inner and outer boundaries? Repeat for the zero set of the phase gradient. How are the zero sets of the amplitude and phase gradients related to one another?

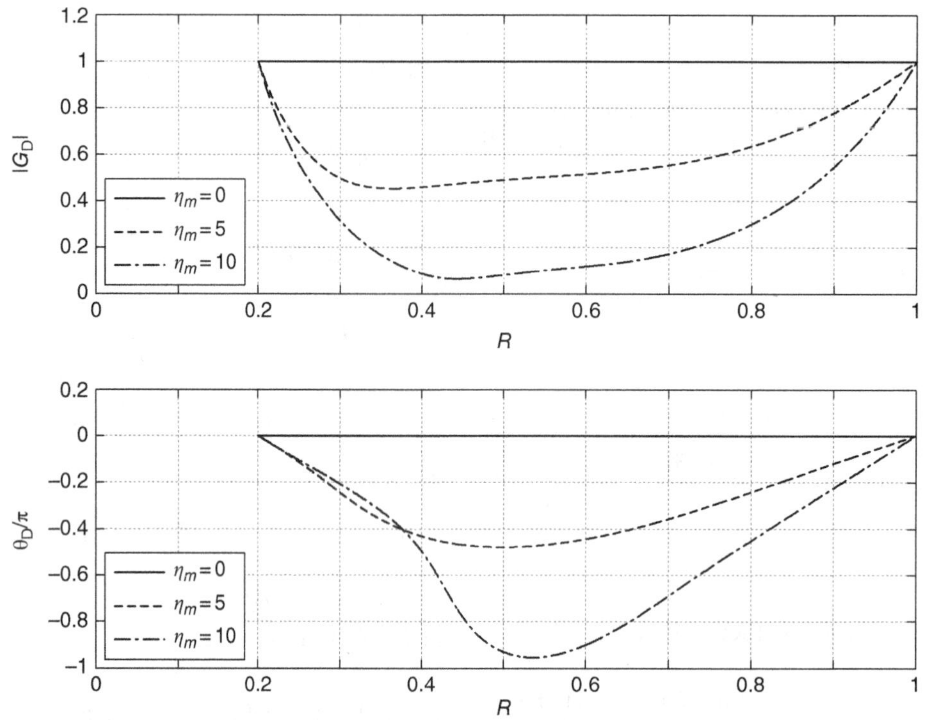

Figure 12.3 FRF for flow in spherical shell—Example 12.5.

Solution: The graph of the amplitude function for $\eta_m = 0.5$ in Figure 12.3 shows that the spherical shell contains a single amplitude minimum. This corresponds to a sphere that is concentric with, and whose radius is intermediate between those of, the shell's inner and outer boundaries. The same is true of the zero set of the phase gradient. Finally, the zero set of the amplitude gradient is a sphere whose radius is less than that of the zero set of the phase gradient. □

Exercise 12.10 Expression for FRF

Problem: Consider an infinite medium with a spherical cavity at its center. The cavity has diameter $r = r_{\min}$ where $r_{\min} > 0$. The inner boundary (i.e., $r = r_{\min}$) is periodically forced with a single harmonic constituent whose frequency is ω_m (m fixed). The boundary conditions are

$$G(r_{\min}; \omega_m) = \text{known (constant for fixed } \omega_m) \quad (12.29a)$$

$$\lim_{r\to\infty} G(r; \omega_m) = 0. \quad (12.29b)$$

Express the hydraulic head FRF, $G(r; \omega_m)$, in terms of $G(r_{\min}; \omega_m)$.

Solution: Substituting the RHS of (12.7) for G in the boundary condition (12.29b) yields $a_m^{(2)} = 0$. Substituting the RHS of (12.7) for G (with $a_m^{(2)} = 0$) in the boundary condition (12.29a) and solving for $a_m^{(1)}$ give

$$a_m^{(1)} = \frac{G(r_{\min}; \omega_m)}{k_0(\lambda_m r_{\min})}.$$

Substituting these results for $a_m^{(1)}$ and $a_m^{(2)}$ in (12.7) leads to

$$G(r; \omega_m) = \frac{G(r_{\min}; \omega_m)}{k_0(\lambda_m r_{\min})} k_0(\lambda_m r).$$

Noting that $k_0(z) = e^{-z}/z$, we can express this as

$$G(r; \omega_m) = G(r_{\min}; \omega_m) \frac{r_{\min}}{r} \exp\left[- \lambda_m (r - r_{\min}) \right].$$

□

12.4. 2D AXISYMMETRIC FLOW

This section explores elementary examples of time-periodic flow in two dimensions with azimuthal symmetry (i.e., purely convergent/divergent cylindrical flows). We will assume the flow has azimuthal symmetry in the $x_1 x_2$ plane. Then the hydraulic head FRF is uniform in the x_3 direction and takes the form

$$F_h(\mathbf{x}; \omega_m) = G(r; \omega_m) \quad (12.30)$$

where we define r, the radial space coordinate, as $r \equiv \sqrt{x_1^2 + x_2^2}$.

12.4.1. Governing Equation and General Solution

Exercise 12.11 Simplified Form of FRF Governing Equation

Problem: Consider time-periodic flow with azimuthal symmetry in the x_1x_2 plane. What simplified form does the FRF governing equation take in this case?

Solution: Substituting the RHS of (12.30) for the FRF in the governing equation (12.1a) and evaluating the Laplacian (in cylindrical coordinates) yield

$$r^2\frac{d^2G}{dr^2} + r\frac{dG}{dr} - i\alpha_m r^2 G = 0. \tag{12.31}$$

□

Exercise 12.12 Derivation: Modified Bessel Equation

Problem: Consider the coordinate transformation $R' \equiv \lambda_m r$, with

$$\lambda_m \equiv \sqrt{i\alpha_m} \tag{12.32}$$

where $\sqrt{z}$ denotes the principal square root of z (see Section D.2.1). Show that one can use this to transform (12.31) into a modified Bessel equation.

Solution: Define $g(R'; \omega_m) \equiv G(r; \omega_m)$. Then

$$\frac{dG}{dr} = \lambda_m \frac{dg}{dR'} \tag{12.33a}$$

$$\frac{d^2G}{dr^2} = \lambda_m^2 \frac{d^2g}{dR'^2}. \tag{12.33b}$$

Substituting the RHSs of these equations for the corresponding terms of (12.31) and simplifying yield

$$R'^2\frac{d^2g}{dR'^2} + R'\frac{dg}{dR'} - R'^2 g = 0 \tag{12.34}$$

which is a modifed Bessel, equation of order zero [see Section 10.25(i) of *Olver and Maximon*, 2016; also see *Weisstein*, 2016e]. □

The hydraulic head FRF is then the general solution of (12.34),

$$G(r; \omega_m) = g(R'; \omega_m) = a_m^{(1)} K_0(\lambda_m r) + a_m^{(2)} I_0(\lambda_m r) \tag{12.35}$$

where I_0 and K_0 are order-zero, modified Bessel functions of the first and second kinds, respectively [see Section 10.25 of *Olver and Maximon*, 2016], and $a_m^{(1)}$ and $a_m^{(2)}$ are complex constants (dimensions: L) to be determined by the boundary conditions.

12.4.2. Cylindrical Flow Due to a Line Source

In this section we examine time-periodic flow due to an infinitely long, single-constituent (ω_m, m fixed) line source in a 3D, infinite domain filled with an ideal medium. We will assume the line source is vertically oriented and located at $x_1 = x_2 = 0$. We will also assume that the source is uniform; that is, its complex amplitude (i.e., amplitude and phase) do not vary with position on the line.

Readers should review Appendix E before proceeding.

Example 12.6 Cylindrical Flow Due to Line Source
Consider the line source described at the beginning of this section. The boundary conditions are [*Black and Kipp*, 1981]

$$\lim_{r\to 0}\left[2\pi K r \frac{d}{dr} G(r; \omega_m)\right] = -\Delta Q_0(\omega_m) \tag{12.36a}$$

$$\lim_{r\to\infty} G(r; \omega_m) = 0 \tag{12.36b}$$

where $\Delta Q_0(\omega_m)$ denotes the complex amplitude of the line source [i.e., $|\Delta Q_0(\omega_m)|$ gives the maximum (peak) volumetric rate of recharge per unit length of source] (dimensions: L^2t^{-1}). Applying the boundary conditions to the general solution (12.35) yields [*Black and Kipp*, 1981]

$$G(r; \omega_m) = \frac{\Delta Q_0(\omega_m)}{2\pi K} K_0(\lambda_m r), \quad r > 0. \tag{12.37}$$

The FRF is undefined at $r = 0$. ◊

Exercise 12.13 FRF Solution Satisfies Boundary Conditions

Problem: Consider Example 12.6. Verify that the FRF solution (12.37) satisfies the boundary conditions (12.36). □

Example 12.7 Cylindrical Flow Due to Line Source—Dimensionless FRF
We can express the FRF solution (12.37) in terms of dimensionless variables as

$$G_D(R) = K_0(\sqrt{i}R), \quad R > 0 \tag{12.38}$$

where we define the dimensionless variables as

$$G_D \equiv \frac{2\pi K G(r; \omega_m)}{\Delta Q_0(\omega_m)}$$

$$R \equiv \sqrt{\alpha_m} r.$$

The dimensionless FRF is undefined at $R = 0$. Figure 12.4 shows the corresponding amplitude and phase functions. We define the dimensionless phase function as

$$\theta_D(R; \eta_m) \equiv \arg G_D(R; \eta_m). \tag{12.39}$$

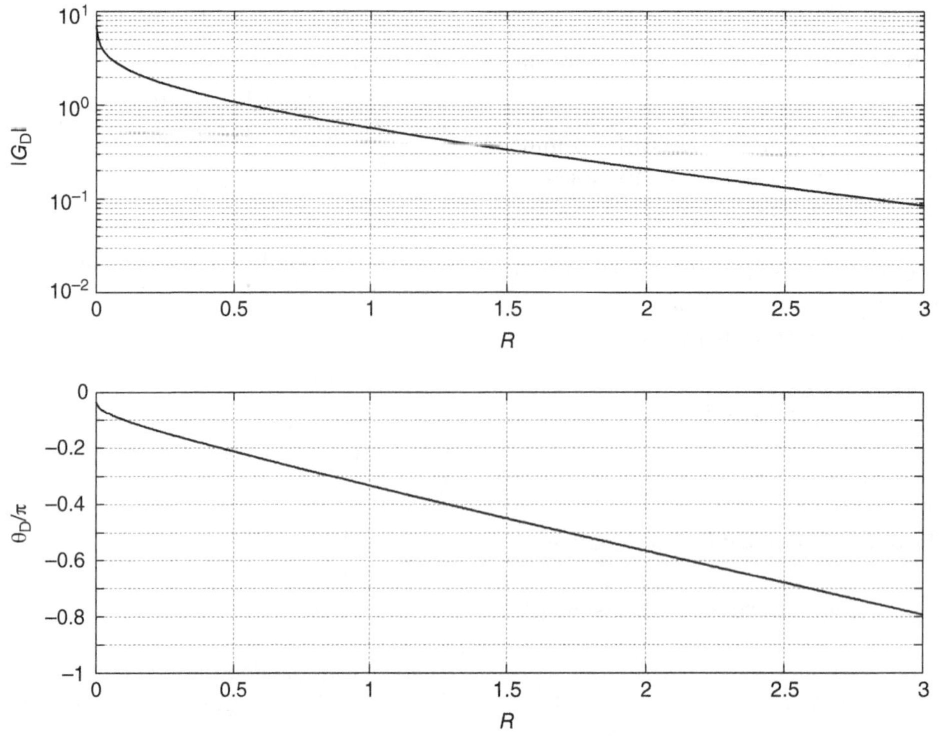

Figure 12.4 FRF for cylindrical flow due to a line source.

Notice that the phase decreases nonlinearly with increasing R near $R = 0$ and approximately linearly with increasing R as $R \to \infty$. ◊

Example 12.8 Line Source—Dimensionless Amplitude/Phase Functions

Consider the dimensionless FRF for 2D cylindrical flow due to a line source (12.38). We can express the corresponding dimensionless amplitude and phase functions in terms of Kelvin functions as follows. Substituting the RHS of (D.8) for $\sqrt{i}$ in (12.38) gives

$$G_D(R) = K_0\big(e^{i\pi/4}R\big), \quad R > 0.$$

Using (E.12) with $\nu = 0$ to evaluate the modulus and argument yields [*Black and Kipp*, 1981]

$$\big|G_D(R)\big| = N_0(R) \tag{12.40a}$$

$$\theta_D(R) = \phi_0(R) + n\pi \text{ rad}, \quad n \in \mathbb{Z} \tag{12.40b}$$

$$R > 0 \tag{12.40c}$$

where the particular value of n is determined by the spatial continuity requirement on $\theta_D(R)$. ◊

Exercise 12.14 Far-Field Behavior of Phase Function

Problem: Consider Example 12.8 and examine the phase function graph in Figure 12.4. Show that in the far field (i.e., for $R \gg 0$) the phase, as a function of R, asymptotically approaches a line and determine its equation. *Hint:* Look at the asymptotic expansions in Section 10.68 of *Olver and Maximon* [2016].

Solution: Substituting $\mu = \nu = 0$ in the asymptotic expansion in Section 10.68(iii) of *Olver and Maximon* [2016] gives

$$\phi_0(R) = \frac{-R}{\sqrt{2}} - \frac{\pi}{8} + O(R^{-1}), \quad R \to \infty.$$

Combining this result with (12.40b) yields the far-field approximation:

$$\theta_D(R) \approx \frac{-R}{\sqrt{2}} - \frac{\pi}{8}, \quad R \to \infty. \tag{12.41}$$

□

Exercise 12.15 Generalize FRF Solution for Line Source

Problem: Generalize the FRF solution obtained in Example 12.6 to the case where the line source is centered at the point with coordinates $\mathbf{x} = \mathbf{x}_0$.

Solution: Define the transformed coordinates $\mathbf{x}' \equiv \mathbf{x} - \mathbf{x}_0$ and $r' \equiv |\mathbf{x}'|$. Then $F_h(\mathbf{x}; \omega_m) = G(r'; \omega_m)$, where (12.37) defines the function G:

$$F_{\mathrm{h}}(\mathbf{x};\omega_m) = \frac{\Delta Q_0(\omega_m)}{2\pi K} K_0(\lambda_m|\mathbf{x}-\mathbf{x}_0|), \quad |\mathbf{x}-\mathbf{x}_0| > 0.$$

□

Example 12.9 Line Source FRF Derived from Point Source FRF

We can also derive the result obtained in Example 12.6 for the line source FRF (12.37) by superposition of the point source FRF. Imagine the line source as an infinite set of identical point sources distributed continuously and uniformly along the x_3 axis. Substitute the line source amplitude $\Delta Q_0(\omega_m)$ for the point source amplitude $Q_0(\omega_m)$ (because the line source is distributed continuously over the line) and substitute

$$\mathbf{x}_0 = \begin{bmatrix} 0 \\ 0 \\ v \end{bmatrix}$$

in (12.16). Then integrate the resulting expression with respect to v:

$$G(r;\omega_m) = \frac{\Delta Q_0(\omega_m)}{4\pi K} \times \int_{-\infty}^{\infty} \frac{\exp\left[-\lambda_m\sqrt{x_1^2+x_2^2+(x_3-v)^2}\right]}{\sqrt{x_1^2+x_2^2+(x_3-v)^2}}\,dv.$$

Substituting $r = \sqrt{x_1^2+x_2^2}$ and $u = v - x_3$ leads to

$$G(r;\omega_m) = \frac{\Delta Q_0(\omega_m)}{2\pi K}\int_0^{\infty}\frac{\exp\left(-\lambda_m\sqrt{r^2+u^2}\right)}{\sqrt{r^2+u^2}}\,du$$

where we have used the fact that the integrand is even in u to change the lower limit of integration. Finally, using equation 8.432 9 of *Gradshteyn and Ryzhik* [1980] to express the integral as the modified Bessel function of the second kind yields (12.37). ◊

12.4.3. Cylindrical Flow Due to Forced Outer Boundary

Example 12.10 Cylindrical Flow Due to Forced Outer Boundary

Consider a circular, horizontal, confined aquifer whose outer boundary ($r = r_{\max}$) is periodically forced. The boundary conditions are

$$\left[\frac{d}{dr}G(r;\omega_m)\right]\bigg|_{r=0} = 0 \tag{12.42a}$$

$$G(r_{\max};\omega_m) = \text{known(constant for fixed } \omega_m) \tag{12.42b}$$

where $r_{\max} > 0$. The solution is

$$G(r;\omega_m) = \frac{G(r_{\max};\omega_m)}{I_0(\lambda_m r_{\max})} I_0(\lambda_m r). \tag{12.43}$$

◊

Exercise 12.16 FRF Satisfies Boundary Conditions

Problem: Consider Example 12.10. Verify that the FRF solution (12.43) satisfies the boundary conditions (12.42). □

Example 12.11 Forced Outer Cylindrical Boundary—Dimensionless FRF

Consider the cylindrical flow described in Example 12.10 with $|G(r_{\max};\omega_m)| > 0$. We can express the FRF in dimensionless form as

$$G_{\mathrm{D}}(R;\eta_m) = \frac{I_0(\sqrt{i}\eta_m R)}{I_0(\sqrt{i}\eta_m)} \tag{12.44}$$

where we define the dimensionless variables as

$$G_{\mathrm{D}} \equiv \frac{G(r;\omega_m)}{G(r_{\max};\omega_m)}$$

$$R \equiv \frac{r}{r_{\max}}$$

$$\eta_m \equiv \sqrt{\alpha_m} r_{\max}.$$

Note the following:

- R and η_m are real valued;
- $0 \le R \le 1$, with $R = 1$ at the outer boundary; and
- $0 \le |G_{\mathrm{D}}| \le 1$, with $G_{\mathrm{D}} = 1$ at the outer boundary.

Figure 12.5 shows the corresponding amplitude and phase functions for various values of η_m. We define the dimensionless phase function as

$$\theta_{\mathrm{D}}(R;\eta_m) \equiv \arg G_{\mathrm{D}}(R;\eta_m).$$

◊

Exercise 12.17 Kelvin Modulus and Phase Functions

Problem: Consider the dimensionless FRF for 2D cylindrical flow due to a forced outer boundary (12.44). Following a procedure similar to that outlined in Example 12.8, express the corresponding dimensionless amplitude and phase functions in terms of Kelvin modulus and phase functions.

Solution: Substituting the RHS of (D.8) for $\sqrt{i}$ in (12.44) gives

$$G_{\mathrm{D}}(R;\eta_m) = \frac{I_0(e^{i\pi/4}\eta_m R)}{I_0(e^{i\pi/4}\eta_m)}. \tag{12.45}$$

Using (E.11) (with $\nu = 0$) to evaluate the modulus and argument yields

$$\left|G_{\mathrm{D}}(R;\eta_m)\right| = \frac{M_0(\eta_m R)}{M_0(\eta_m)} \tag{12.46a}$$

$$\theta_{\mathrm{D}}(R;\eta_m) = \theta_0(\eta_m R) - \theta_0(\eta_m) + n\pi \text{ rad}, \quad n \in \mathbb{Z} \tag{12.46b}$$

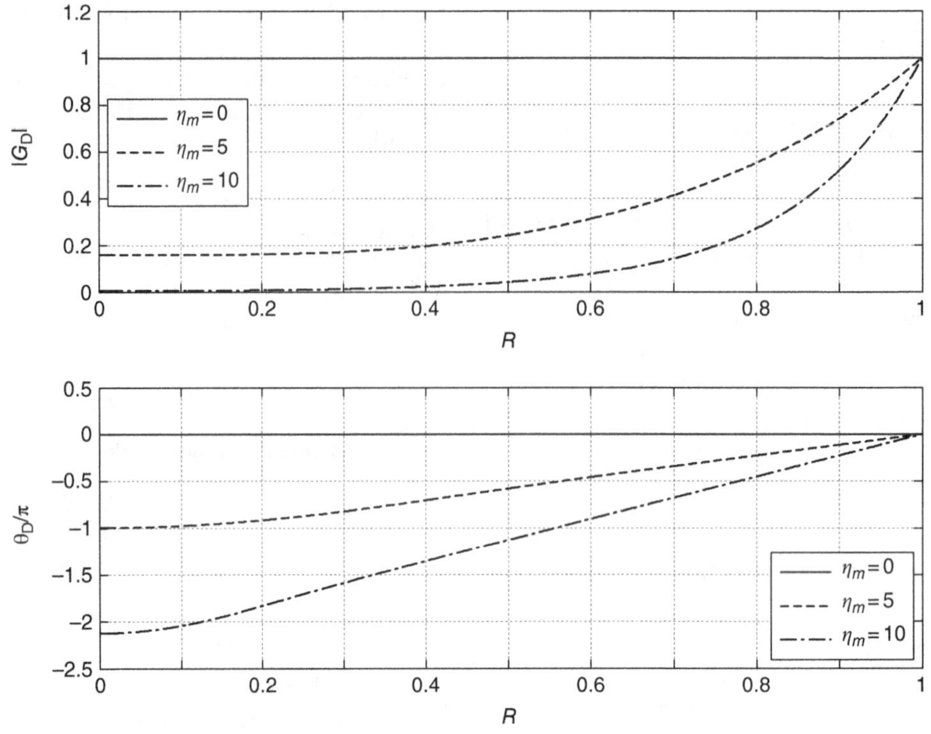

Figure 12.5 FRF for cylindrical flow due to forced outer boundary.

where the particular value of n is determined by the spatial continuity requirement on $\theta_D(R;\eta_m)$. □

12.4.4. Radial Flow in an Annulus

Exercise 12.18 Flow in an Annulus

Problem: Consider a circular annulus-shaped, horizontal, confined aquifer with inner and outer diameters $r = r_{\min}$ and $r = r_{\max}$, respectively, where $r_{\max} > r_{\min} > 0$. The boundaries are periodically forced with a single harmonic constituent whose frequency is ω_m (m fixed). The boundary conditions are

$$G(r_{\min};\omega_m) = \text{known (constant for fixed } \omega_m) \quad (12.47a)$$

$$G(r_{\max};\omega_m) = \text{known (constant for fixed } \omega_m). \quad (12.47b)$$

Express the hydraulic head FRF, $G(r;\omega_m)$, in terms of $G(r_{\min};\omega_m)$ and $G(r_{\max};\omega_m)$.

Solution: Substituting the RHS of (12.35) for G in the BC equations (12.47) yields the linear system of two equations in two unknowns ($a_m^{(1)}$ and $a_m^{(2)}$):

$$a_m^{(1)} K_0(\lambda_m r_{\min}) + a_m^{(2)} I_0(\lambda_m r_{\min}) = G(r_{\min};\omega_m) \quad (12.48a)$$

$$a_m^{(1)} K_0(\lambda_m r_{\max}) + a_m^{(2)} I_0(\lambda_m r_{\max}) = G(r_{\max};\omega_m). \quad (12.48b)$$

The solution is (e.g., by Cramer's rule)

$$a_m^{(1)} = \frac{G(r_{\max};\omega_m) I_0(\lambda_m r_{\min}) - G(r_{\min};\omega_m) I_0(\lambda_m r_{\max})}{I_0(\lambda_m r_{\min}) K_0(\lambda_m r_{\max}) - I_0(\lambda_m r_{\max}) K_0(\lambda_m r_{\min})} \quad (12.49a)$$

$$a_m^{(2)} = \frac{G(r_{\min};\omega_m) K_0(\lambda_m r_{\max}) - G(r_{\max};\omega_m) K_0(\lambda_m r_{\min})}{I_0(\lambda_m r_{\min}) K_0(\lambda_m r_{\max}) - I_0(\lambda_m r_{\max}) K_0(\lambda_m r_{\min})} \quad (12.49b)$$

assuming that

$$\left| I_0(\lambda_m r_{\min}) K_0(\lambda_m r_{\max}) - I_0(\lambda_m r_{\max}) K_0(\lambda_m r_{\min}) \right| > 0.$$

Therefore, (12.35) gives the BVP solution with $a_m^{(1)}$ and $a_m^{(2)}$ given above. □

Exercise 12.19 Expression for FRF

Problem: Consider a horizontally infinite, confined aquifer of constant thickness with a circular "hole" at its center. The hole has diameter $r = r_{\min}$ where $r_{\min} > 0$. The inner boundary (i.e., $r = r_{\min}$) is periodically forced with a single harmonic constituent whose frequency is ω_m (m fixed). The boundary conditions are

$$G(r_{\min};\omega_m) = \text{known (constant for fixed } \omega_m) \quad (12.50a)$$

$$\lim_{r\to\infty} G(r;\omega_m) = 0. \quad (12.50b)$$

Express the hydraulic head FRF, $G(r;\omega_m)$, in terms of $G(r_{\min};\omega_m)$.

Solution: Substituting the RHS of (12.35) for G in the boundary condition (12.50b) yields $a_m^{(2)} = 0$. Substituting the RHS of (12.35) for G (with $a_m^{(2)} = 0$) in the boundary condition (12.50a) and solving for $a_m^{(2)}$ give

$$a_m^{(1)} = \frac{G(r_{\min}; \omega_m)}{K_0(\lambda_m r_{\min})}.$$

Substituting these results for $a_m^{(1)}$ and $a_m^{(2)}$ in (12.35) yields the final solution:

$$G(r; \omega_m) = \frac{G(r_{\min}; \omega_m)}{K_0(\lambda_m r_{\min})} K_0(\lambda_m r).$$

□

Example 12.12 Radial Flow in Annulus

Consider radial flow in an annulus as described in Exercise 12.18 with $r_{\min} = 0.2 r_{\max}$ and $G(r_{\min}; \omega_m) = G(r_{\max}; \omega_m)$. Define the dimensionless variables

$$G_{\text{D}} \equiv \frac{G(r; \omega_m)}{G(r_{\min}; \omega_m)} \tag{12.51a}$$

$$R \equiv \frac{r}{r_{\max}} \tag{12.51b}$$

$$\eta_{\text{m}} \equiv \sqrt{\alpha_{\text{m}}} r_{\max} \tag{12.51c}$$

$$\theta_{\text{D}} \equiv \arg G_{\text{D}}. \tag{12.51d}$$

Note the following:

- R, η_m, and θ_{D} are real valued;
- $0.2 \leq R \leq 1$;
- $G_{\text{D}} = 1$ at both of the boundaries; and
- $\theta_{\text{D}} = 0$ at both of the boundaries.

Figure 12.6 shows the corresponding amplitude and phase functions for various values of η_m. ◊

12.4.5. Radial Flow to/from a Finite-Radius Well

Example 12.13 Radial Flow Due to/from Finite-Radius Well

Consider a circular, horizontal, confined aquifer with a periodically forced outer boundary ($r = r_{\max} > 0$). Suppose a well of radius r_{w} ($0 < r_{\text{w}} < r_{\max}$) is centered horizontally at $r = 0$, fully penetrates the aquifer, and is screened over the entire thickness (b) of the aquifer. Further suppose that the inner casing of the well has diameter r_{c} ($0 < r_{\text{c}} < r_{\max}$) within the depth interval over which the water level is changing. Assuming the well is fully efficient, a mass balance for the water in the well yields

$$r_{\text{c}}^2 \left[\frac{\partial}{\partial t} h'(r, t; \omega_m) \right] \Bigg|_{r=r_{\text{w}}} - 2 r_{\text{w}} T \left[\frac{\partial}{\partial r} h'(r, t; \omega_m) \right] \Bigg|_{r=r_{\text{w}}} = 0 \tag{12.52}$$

where $T \equiv bK$. This formulation neglects the compressibility of the water within the well and assumes that

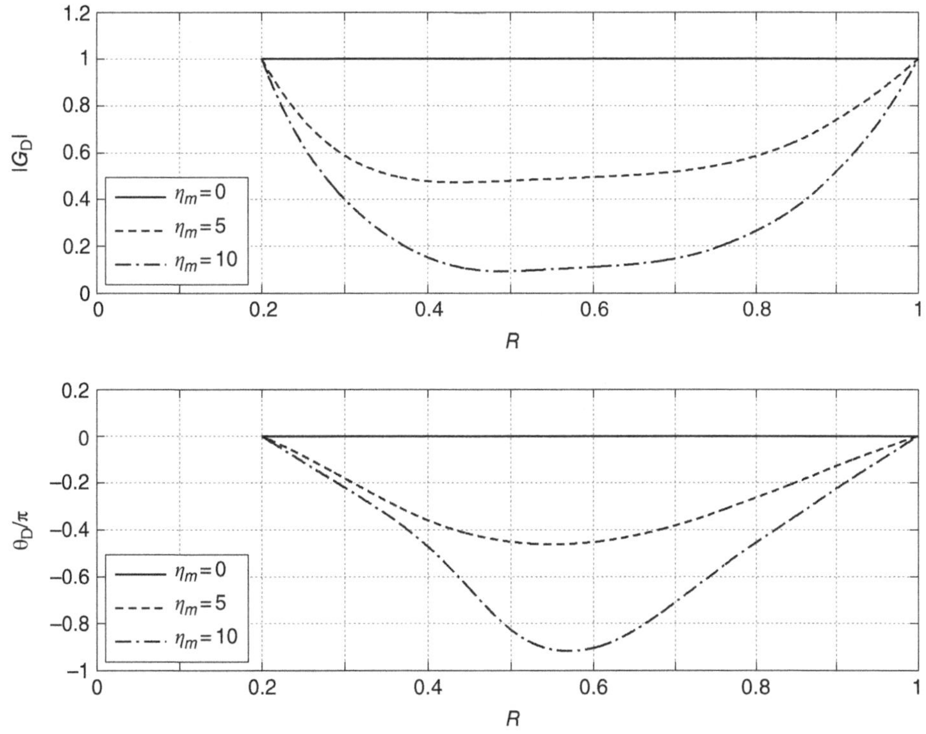

Figure 12.6 FRF for flow in annulus—Example 12.12.

water storage within the well changes only by horizontal inflow/outflow at the well face. It also neglects inertia and capillary forces within the well casing. ◊

Exercise 12.20 Boundary Conditions

Problem: Consider radial flow to/from a finite-radius well due to a forced outer boundary, as in Example 12.13. Write the corresponding boundary conditions in terms of the FRF $G(r; \omega_m)$.

Solution: In this case we write the harmonic constituent as

$$h'(r, t; \omega_m) = \mathrm{Re} H(r, t; \omega_m) = \mathrm{Re}\left[G(r; \omega_m) e^{i\omega_m t}\right].$$

Substituting the RHS for h' in (12.52) yields the inner boundary condition:

$$i\omega_m r_c^2 G(r_w; \omega_m) - 2r_w T\left[\frac{d}{dr} G(r; \omega_m)\right]\Bigg|_{r=r_w} = 0. \quad (12.53)$$

At the outer boundary we have the Dirichlet condition:

$$G(r_{\max}; \omega_m) = \text{known (constant for fixed } \omega_m). \quad (12.54)$$

□

Exercise 12.21 System of Equations for Coefficients

Problem: Consider radial flow to/from a finite-radius well due to a forced outer boundary, as in Example 12.13. Use the boundary conditions (Exercise 12.20) to write the linear system of two equations in the two complex-valued unknowns, $a_m^{(1)}$ and $a_m^{(2)}$ of (12.35).

Solution: Differentiating (12.35) with respect to r gives

$$\frac{d}{dr} G(r; \omega_m) = a_m^{(1)} \lambda_m \left[\frac{d}{du} K_0(u)\right]\Bigg|_{u=\lambda_m r} + a_m^{(2)} \lambda_m \left[\frac{d}{du} I_0(u)\right]\Bigg|_{u=\lambda_m r}.$$

Using the differentiation equations in Section 10.29(ii) of *Olver and Maximon* [2016], we can write this as

$$\frac{d}{dr} G(r; \omega_m) = a_m^{(1)} \lambda_m K_1(\lambda_m r) - a_m^{(2)} \lambda_m I_1(\lambda_m r)$$

where I_1 and K_1 are order-one, modified Bessel functions of the first and second kinds, respectively. Using this and (12.35) to write the inner boundary condition (12.53) in terms of Bessel functions leads to

$$i\omega_m r_c^2 \left[a_m^{(1)} K_0(\lambda_m r_w) + a_m^{(2)} I_0(\lambda_m r_w)\right] - 2r_w T\left[a_m^{(1)} \lambda_m K_1(\lambda_m r_w) - a_m^{(2)} \lambda_m I_1(\lambda_m r_w)\right] = 0.$$

Grouping coefficients for $a_m^{(1)}$ and $a_m^{(2)}$ in the previous equation and using (12.35) to write the outer boundary condition (12.54) in terms of Bessel functions yield

$$\left[i\omega_m r_c^2 K_0(\lambda_m r_w) - 2r_w T \lambda_m K_1(\lambda_m r_w)\right] a_m^{(1)} + \left[i\omega_m r_c^2 I_0(\lambda_m r_w) + 2r_w T \lambda_m I_1(\lambda_m r_w)\right] a_m^{(2)} = 0 \quad (12.55a)$$

$$K_0(\lambda_m r_{\max}) a_m^{(1)} + I_0(\lambda_m r_{\max}) a_m^{(2)} = G(r_{\max}; \omega_m) \quad (12.55b)$$

where $G(r_{\max}; \omega_m)$ is understood to be known. Equations (12.55) form a nonhomogeneous linear system of two equations in the two complex-valued unknowns, $a_m^{(1)}$ and $a_m^{(2)}$. □

12.5. CHAPTER SUMMARY

In this chapter we briefly derived some elementary solutions of the complex-variable form of the space BVP for time-periodic, multidimensional flow in domains filled with ideal media. We considered radial spherical flow in domains having simple spherical geometry and axisymmetric flow in domains having simple cylindrical geometry.

13
Examples: Uniform-Gradient Flow

In this chapter we present a particular family of FRFs for time-periodic, multidimensional flow in a source-free domain filled with an exponential medium. The corresponding FREs are exponential functions of the space coordinates. The logarithm of each FRE has a gradient that is spatially uniform, so we call them *uniform-gradient FREs*. These probably are some of the simplest possible nontrivial solutions for time-periodic, multidimensional flow in continuously nonhomogeneous media.

Note 13.1 Collinear and Noncollinear Vectors
Suppose $\mathbf{a}$ and $\mathbf{b}$ are nonzero, real-valued vectors in three dimensions. We say that they are *collinear* vectors if they point either in the same direction or in opposite directions. Conversely, we say that they are *noncollinear* vectors if they point in different but not opposite directions.

Mathematically, the nonzero, real-valued vectors $\mathbf{a}$ and $\mathbf{b}$ are collinear if and only if one of the following conditions is satisfied:

1. $|\mathbf{a} \cdot \mathbf{b}| = |\mathbf{a}||\mathbf{b}|$;
2. $\mathbf{a} = d\mathbf{b}$, where d is a nonzero, real-valued scalar; and
3. $\mathbf{a} \times \mathbf{b} = \mathbf{0}$.

It does not matter which condition is used to test for or to prove collinearity; the three conditions are equivalent. △

The family of solutions (FRFs) presented in this chapter differ from those presented in Chapter 10 in that generally no two of the hydraulic head amplitude gradient, hydraulic head phase gradient, and log conductivity gradient need be collinear. In this respect the solutions presented in this chapter constitute a generalization of those FRFs presented in Chapter 10.

13.1. ASSUMPTIONS

In addition to the assumptions listed in the introduction to Part III, throughout this chapter we assume the following:

• The material hydrologic properties of the medium vary continuously in space as follows. First, the scalar hydraulic conductivity varies as

$$K(\mathbf{x}) = K(\mathbf{0})e^{\boldsymbol{\mu}\cdot\mathbf{x}} \tag{13.1}$$

where $K(\mathbf{0})$ (scalar) and $\boldsymbol{\mu}$ (vector) are real constants and $K(\mathbf{0}) > 0$. Second, the specific storage varies as

$$S_s(\mathbf{x}) = S_s(\mathbf{0})e^{\boldsymbol{\mu}\cdot\mathbf{x}} \tag{13.2}$$

where $S_s(\mathbf{0})$ is a real, positive constant. Recall from Section 10.1 that in this book we call media that satisfy these assumptions *exponential media*.

• The boundary conditions are such that we can represent the FRF as the superposition of two FREs as

$$F_h(\mathbf{x};\omega_m) = a_m^{(1)}F^{(1)}(\mathbf{x};\omega_m) + a_m^{(2)}F^{(2)}(\mathbf{x};\omega_m) \tag{13.3}$$

where the $a_m^{(n)}$ are complex-valued constants, the $F^{(n)}$ are FREs (i.e., linearly independent solutions of the governing equation for the hydraulic head FRF), and

$$F^{(n)}(\mathbf{x};\omega_m) = \exp\left[\mathbf{w}^{(n)}(\omega_m)\cdot\mathbf{x}\right] \quad (n = 1, 2) \tag{13.4}$$

where the $\mathbf{w}^{(n)}(\omega_m)$ are complex-valued *eigenvectors* that generally depend on the constituent frequency.

Hydrodynamics of Time-Periodic Groundwater Flow: Diffusion Waves in Porous Media, Geophysical Monograph 224,
First Edition. Joe S. Depner and Todd C. Rasmussen.

Exercise 13.1 Hydraulic Diffusivity of Exponential Medium

Problem: Consider an exponential medium as defined by (13.1) and (13.2). How does the hydraulic diffusivity (2.18) depend on position? How do the gradients of the logarithms of K and S_s depend on position? What is the significance of the special case where $\boldsymbol{\mu} = 0$?

Solution: Substituting the RHSs of (13.1) and (13.2) for $K(\mathbf{x})$ and $S_s(\mathbf{x})$, respectively, in (2.18) and then simplifying the result give

$$\kappa \equiv \frac{K(\mathbf{0})}{S_s(\mathbf{0})}$$

from which it follows that the hydraulic diffusivity is a real, positive constant. Therefore, its gradient vanishes:

$$\nabla\kappa = \mathbf{0}.$$

Taking the gradients of (13.1) and (13.2) gives

$$\nabla \ln K(\mathbf{x}) = \nabla \ln S_s(\mathbf{x}) = \boldsymbol{\mu}. \tag{13.5}$$

That is, the gradients of the log hydraulic conductivity and log specific storage are spatially uniform and identical. The special case $\boldsymbol{\mu} = 0$ corresponds to an ideal medium. □

Define component amplitude and phase functions as

$$a_m^{(n)} F^{(n)}(\mathbf{x};\omega_m) = M^{(n)}(\mathbf{x};\omega_m) \exp\left[i\theta^{(n)}(\mathbf{x};\omega_m)\right] \quad (n = 1, 2) \tag{13.6}$$

where $M^{(n)}(\mathbf{x};\omega_m)$ and $\theta^{(n)}(\mathbf{x};\omega_m)$ are real valued. Define the logarithm of the component amplitude as

$$Y^{(n)}(\mathbf{x};\omega_m) \equiv \ln M^{(n)}(\mathbf{x};\omega_m) \quad (n = 1, 2). \tag{13.7}$$

Exercise 13.2 Space Variation of Component Gradients

Problem: Express $\nabla Y^{(n)}$ and $\nabla\theta^{(n)}$ in terms of $\mathbf{x}$, ω_m, and the eigenvectors $\mathbf{w}^{(n)}$. What do your results imply about the space variation of $\nabla Y^{(n)}$ and $\nabla\theta^{(n)}$?

Solution: Taking the logarithm of (13.6) and using (13.7) give

$$\begin{aligned} \ln a_m^{(n)} + \ln F^{(n)}(\mathbf{x};\omega_m) &= Y^{(n)}(\mathbf{x};\omega_m) \\ &\quad + i\theta^{(n)}(\mathbf{x};\omega_m) \quad (n = 1, 2). \end{aligned}$$

Taking the gradient, substituting the RHS of (13.4) for $F^{(n)}(\mathbf{x};\omega_m)$, and simplifying give

$$\mathbf{w}^{(n)}(\omega_m) = \nabla Y^{(n)}(\mathbf{x};\omega_m) + i\nabla\theta^{(n)}(\mathbf{x};\omega_m) \quad (n = 1, 2). \tag{13.8}$$

Taking the real and imaginary parts leads to

$$\begin{aligned} \nabla Y^{(n)}(\mathbf{x};\omega_m) &= \operatorname{Re} \mathbf{w}^{(n)}(\omega_m) = \nabla Y^{(n)}(\omega_m) \\ \nabla\theta^{(n)}(\mathbf{x};\omega_m) &= \operatorname{Im} \mathbf{w}^{(n)}(\omega_m) = \nabla\theta^{(n)}(\omega_m) \\ n &= 1, 2. \end{aligned}$$

Both $\nabla Y^{(n)}$ and $\nabla\theta^{(n)}$ are space invariant. □

As Exercise 13.2 shows, in this case $\nabla \ln F^{(n)}$, $\nabla Y^{(n)}$, and $\nabla\theta^{(n)}$ are spatially uniform. Thus, in this book we sometimes refer to FRFs defined by (13.4) as *uniform-gradient FRFs*. Taking the gradient of (13.3) and using (13.4) yield the gradient of the log FRF:

$$\frac{\nabla F(\mathbf{x};\omega_m)}{F(\mathbf{x};\omega_m)} = \frac{\left\{ \begin{array}{l} a_m^{(1)} \exp\left[\mathbf{w}^{(1)}(\omega_m)\cdot\mathbf{x}\right]\mathbf{w}^{(1)}(\omega_m) \\ +a_m^{(2)} \exp\left[\mathbf{w}^{(2)}(\omega_m)\cdot\mathbf{x}\right]\mathbf{w}^{(2)}(\omega_m) \end{array} \right\}}{\left\{ \begin{array}{l} a_m^{(1)} \exp\left[\mathbf{w}^{(1)}(\omega_m)\cdot\mathbf{x}\right] \\ +a_m^{(2)} \exp\left[\mathbf{w}^{(2)}(\omega_m)\cdot\mathbf{x}\right] \end{array} \right\}}. \tag{13.9}$$

Notice that the RHS generally is spatially nonuniform even though the gradients of the log FREs (13.4) are spatially uniform.

Because for fixed n and ω_m generally the real and imaginary parts of $\mathbf{w}^{(n)}(\omega_m)$ may differ, $\nabla Y^{(n)}$ and $\nabla\theta^{(n)}$ need not point in the same direction. This is an important difference between the family of uniform-gradient FREs and the FRE solutions presented in Chapters 11 and 12.

13.2. GOVERNING EQUATIONS

Exercise 13.3 Governing Equation for Flow in Exponential Medium

Problem: Consider time-periodic, multidimensional flow in a source-free domain filled with an exponential medium. Write the simplified governing equation for the hydraulic head FRF for this particular case using the parameters α_m and $\boldsymbol{\mu}$.

Solution: Substituting $F_u = 0$ and the RHSs of (13.1) and (13.2) for $\mathbf{K}$ and S_s, respectively, in (7.17a) leads to

$$\nabla^2 F_h + \boldsymbol{\mu}\cdot\nabla F_h - i\alpha_m F_h = 0 \tag{13.10}$$

where (5.31) defines the real-valued constants α_m. □

Equation (13.10) is a linear, second-order, homogeneous PDE with constant coefficients. Also note the similarity between (13.10) and (10.4) for 1D flow in an exponential medium.

Exercise 13.4 Characteristic Equation

Problem: Regarding (13.10), consider the assumed FRE (13.4), where $\mathbf{w}^{(n)}(\omega_m)$ is an undetermined eigenvector. Show that $F^{(n)}(\mathbf{x};\omega_m)$ satisfies the governing equation by deriving a characteristic equation for $\mathbf{w}^{(n)}(\omega_m)$. Then express the characteristic equation in terms of $\nabla Y^{(n)}$ and $\nabla\theta^{(n)}$.

Solution: Substituting the RHS of (13.4) for F_h in (13.10) and then simplifying produce the characteristic equation

$$\mathbf{w}^{(n)}(\omega_m)\cdot\mathbf{w}^{(n)}(\omega_m)+\boldsymbol{\mu}\cdot\mathbf{w}^{(n)}(\omega_m)-i\alpha_m=0 \quad (n=1,2). \tag{13.11}$$

Substituting the RHS of (13.8) for $\mathbf{w}^{(n)}(\omega_m)$ and then taking the real and imaginary parts successively yield

$$\left|\nabla Y^{(n)}(\omega_m)\right|^2+\boldsymbol{\mu}\cdot\nabla Y^{(n)}(\omega_m)-\left|\nabla\theta^{(n)}(\omega_m)\right|^2=0 \tag{13.12a}$$

$$2\nabla Y^{(n)}(\omega_m)\cdot\nabla\theta^{(n)}(\omega_m)+\boldsymbol{\mu}\cdot\nabla\theta^{(n)}(\omega_m)=\alpha_m \tag{13.12b}$$

$$n=1,2.$$

□

Exercise 13.5 Purely Real and Purely Imaginary Eigenvectors

Problem: Does the characteristic equation (13.11) admit solutions that are purely real valued? Does it admit solutions that are purely imaginary valued?

Solution: Inspecting (13.11) and noting that $\operatorname{Im}\boldsymbol{\mu}=\mathbf{0}$, it is evident that purely real valued and purely imaginary valued solutions imply $\alpha_m=0$. This violates our basic assumption that $\alpha_m>0$. Therefore, purely real valued and purely imaginary valued solutions for the eigenvectors $\mathbf{w}^{(n)}(\omega_m)$ are prohibited. Mathematically,

$$\left|\operatorname{Re}\mathbf{w}^{(n)}(\omega_m)\right|=\left|\nabla Y^{(n)}(\omega_m)\right|>0$$

$$\left|\operatorname{Im}\mathbf{w}^{(n)}(\omega_m)\right|=\left|\nabla\theta^{(n)}(\omega_m)\right|>0$$

$$n=1,2.$$

□

Exercise 13.6 Oppositely Directed Component Gradients

Problem: Prove that (13.12) prohibits solutions for which the log amplitude gradient $\nabla Y^{(n)}(\omega_m)$ and the corresponding phase gradient $\nabla\theta^{(n)}(\omega_m)$ are oppositely directed.

Solution: We will use the method of contradiction. Suppose that, for some fixed m and n, $\nabla Y^{(n)}(\omega_m)$ and $\nabla\theta^{(n)}(\omega_m)$ are oppositely directed. Then

$$\nabla Y^{(n)}(\omega_m)=d_m^{(n)}\nabla\theta^{(n)}(\omega_m) \tag{13.13}$$

where $d_m^{(n)}$ is real valued, negative, and dimensionless. Substituting the RHS for $\nabla Y^{(n)}$ in (13.12) gives

$$\left(d_m^{(n)}\right)^2\left|\nabla\theta^{(n)}\right|^2+d_m^{(n)}\boldsymbol{\mu}\cdot\nabla\theta^{(n)}-\left|\nabla\theta^{(n)}\right|^2=0 \tag{13.14a}$$

$$2d_m^{(n)}\left|\nabla\theta^{(n)}\right|^2+\boldsymbol{\mu}\cdot\nabla\theta^{(n)}=\alpha_m. \tag{13.14b}$$

Multiplying (13.14b) by $d_m^{(n)}$, subtracting (13.14a), and then solving for α_m lead to

$$\alpha_m=\frac{\left[\left(d_m^{(n)}\right)^2+1\right]}{d_m^{(n)}}\left|\nabla\theta^{(n)}\right|^2 \tag{13.15}$$

which implies that $\alpha_m<0$. This violates our general assumption that $\alpha_m>0$, so $d_m^{(n)}\geq 0$. Consequently, $\nabla Y^{(n)}$ and $\nabla\theta^{(n)}$ are not oppositely directed. □

Exercise 13.7 Mutually Orthogonal Gradients

Problem: Do (13.12) allow solutions for which the three vectors $\nabla Y^{(n)}(\omega_m)$, $\nabla\theta^{(n)}(\omega_m)$, and $\boldsymbol{\mu}$ are mutually orthogonal? If so, under what conditions?

Solution: If the three vectors are mutually orthogonal, then the inner-product terms on the LHS of (13.12b) vanish, yielding $\alpha_m=0$. This violates our general assumption that $\alpha_m>0$, so such solutions are prohibited. □

Exercise 13.8 Orthogonal Component Gradients

Problem: Do the gradient conditions (13.12) allow solutions for which $\nabla Y^{(n)}(\omega_m)$ is orthogonal to $\nabla\theta^{(n)}(\omega_m)$? If so, under what conditions?

Solution: If $\nabla Y^{(n)}$ and $\nabla\theta^{(n)}$ are orthogonal, then the first inner-product term on the LHS of (13.12b) vanishes, yielding $\boldsymbol{\mu}\cdot\nabla\theta^{(n)}=\alpha_m$. Generally this is allowed. However, if $\boldsymbol{\mu}\cdot\nabla\theta^{(n)}=0$ (e.g., if the medium is homogeneous or if $\boldsymbol{\mu}$ is nonzero but perpendicular to $\nabla\theta^{(n)}$), this gives $\alpha_m=0$, which violates our general assumption that $\alpha_m>0$. Therefore, solutions for which the gradients of the log amplitude and the phase are orthogonal generally are allowed subject to the condition $\boldsymbol{\mu}\cdot\nabla\theta^{(n)}>0$. □

Exercise 13.9 Orthogonal Component Gradients

Problem: Consider uniform-gradient solutions for time-periodic flow in a source-free domain filled with a nonhomogeneous, exponential medium ($|\boldsymbol{\mu}|>0$). Suppose $\nabla\theta^{(n)}$ and $\boldsymbol{\mu}$ point in the same direction and the gradients of the amplitude and phase are perpendicular. Express $|\nabla Y^{(n)}|$ and $|\nabla\theta^{(n)}|$ in terms of $|\boldsymbol{\mu}|$ and α_m.

Solution: By assumption

$$\boldsymbol{\mu}\cdot\nabla\theta^{(n)}(\omega_m)=|\boldsymbol{\mu}|\left|\nabla\theta^{(n)}(\omega_m)\right|$$

$$\nabla Y^{(n)}(\omega_m)\cdot\nabla\theta^{(n)}(\omega_m)=0.$$

Collectively these imply

$$\nabla Y^{(n)}(\omega_m) \cdot \boldsymbol{\mu} = 0.$$

Substituting these in (13.12) gives

$$\left|\nabla Y^{(n)}(\omega_m)\right| = \left|\nabla\theta^{(n)}(\omega_m)\right|$$

$$\left|\boldsymbol{\mu}\right|\left|\nabla\theta^{(n)}(\omega_m)\right| = \alpha_m.$$

Dividing the second equation by $|\boldsymbol{\mu}|$ and combining the result with the first equation yield

$$\left|\nabla Y^{(n)}(\omega_m)\right| = \left|\nabla\theta^{(n)}(\omega_m)\right| = \frac{\alpha_m}{|\boldsymbol{\mu}|}, \qquad |\boldsymbol{\mu}| > 0.$$

□

13.3. COMPLIMENTARY EIGENVECTORS AND LINEAR INDEPENDENCE

In this book we define *complimentary eigenvectors* as any pair of eigenvectors $\mathbf{w}^{(1)}(\omega_m)$ and $\mathbf{w}^{(2)}(\omega_m)$ for which the corresponding FRFs (13.4) are linearly independent eigenfunctions. If the two FREs are to be linearly independent, then their ratio must not be space invariant. Using (13.4) we obtain

$$\frac{F^{(1)}(\mathbf{x};\omega_m)}{F^{(2)}(\mathbf{x};\omega_m)} = \exp\left\{\left[\mathbf{w}^{(1)}(\omega_m) - \mathbf{w}^{(2)}(\omega_m)\right]\cdot\mathbf{x}\right\} \quad (13.16)$$

which is space invariant for all $\mathbf{x}$ if and only if $\mathbf{w}^{(1)}(\omega_m) = \mathbf{w}^{(2)}(\omega_m)$. Therefore, if two distinct eigenvectors satisfy the characteristic equations (13.11), then they are complimentary. Equivalently, if two distinct eigenvectors satisfy the characteristic equations (13.11), then the two corresponding FREs are linearly independent.

Exercise 13.10 Particular Eigenvector Relation

Problem: In the case of time-periodic, 1D flow in a source-free domain filled with an exponential medium, (10.13) indicate the eigenvalues satisfy $\lambda_m^{(1)} + \lambda_m^{(2)} = -\mu$. By analogy this suggests that for uniform-gradient flow through an exponential medium we explore the possible eigenvector relation

$$\mathbf{w}^{(1)}(\omega_m) + \mathbf{w}^{(2)}(\omega_m) = -\boldsymbol{\mu}. \quad (13.17)$$

Show that if $\mathbf{w}^{(1)}(\omega_m)$ satisfies (13.11), then $\mathbf{w}^{(2)}(\omega_m)$ also satisfies (13.11), where

$$\mathbf{w}^{(2)}(\omega_m) = -\mathbf{w}^{(1)}(\omega_m) - \boldsymbol{\mu} \quad (13.18)$$

so that (13.17) is satisfied.

Solution: Substituting the RHS of (13.17) for $\mathbf{w}^{(2)}(\omega_m)$ gives

$$\begin{aligned}
&\mathbf{w}^{(2)}(\omega_m)\cdot\mathbf{w}^{(2)}(\omega_m) + \boldsymbol{\mu}\cdot\mathbf{w}^{(2)}(\omega_m) - i\alpha_m \\
&= \left[-\mathbf{w}^{(1)}(\omega_m) - \boldsymbol{\mu}\right]\cdot\left[-\mathbf{w}^{(1)}(\omega_m) - \boldsymbol{\mu}\right] \\
&\quad + \boldsymbol{\mu}\cdot\left[-\mathbf{w}^{(1)}(\omega_m) - \boldsymbol{\mu}\right] - i\alpha_m \\
&= \mathbf{w}^{(1)}(\omega_m)\cdot\mathbf{w}^{(1)}(\omega_m) + \boldsymbol{\mu}\cdot\mathbf{w}^{(1)}(\omega_m) - i\alpha_m.
\end{aligned}$$

Equation (13.11) with $n = 1$ implies that the RHS vanishes, leaving

$$\mathbf{w}^{(2)}(\omega_m)\cdot\mathbf{w}^{(2)}(\omega_m) + \boldsymbol{\mu}\cdot\mathbf{w}^{(2)}(\omega_m) - i\alpha_m = 0$$

which is (13.11) with $n = 2$. □

In Exercise 13.10, we showed the complimentary eigenvector $\mathbf{w}^{(2)}(\omega_m)$ defined by (13.18) satisfies the characteristic equation (13.11) with $n = 2$. However, we have *not* shown that this is the only possible solution of (13.11) with $n = 2$. In fact, additional solutions (i.e., additional complimentary eigenvectors) exist that do not satisfy (13.18). Exercise 13.11 gives an example.

Exercise 13.11 Particular Eigenvector Relation

Problem: Consider time-periodic, uniform-gradient flow in a source-free domain filled with a nonhomogeneous, exponential medium. Suppose $\mathbf{w}^{(1)}(\omega_m)$ satisfies (13.11). Show that $\mathbf{w}^{(2)}(\omega_m)$ also satisfies (13.11), where

$$\mathbf{w}^{(2)}(\omega_m) \equiv \frac{\alpha_m}{\alpha_m + i\boldsymbol{\mu}\cdot\mathbf{w}^{(1)}(\omega_m)}\mathbf{w}^{(1)}(\omega_m) - \boldsymbol{\mu}. \quad (13.19)$$

How can one be sure that the denominator on the RHS does not vanish?

Solution: Substituting the RHS of (13.19) for $\mathbf{w}^{(2)}(\omega_m)$ and then rearranging eventually lead to

$$\begin{aligned}
&\mathbf{w}^{(2)}(\omega_m)\cdot\mathbf{w}^{(2)}(\omega_m) + \boldsymbol{\mu}\cdot\mathbf{w}^{(2)}(\omega_m) - i\alpha_m \\
&= \frac{\alpha_m^2\left[\mathbf{w}^{(1)}(\omega_m)\cdot\mathbf{w}^{(1)}(\omega_m) + \boldsymbol{\mu}\cdot\mathbf{w}^{(1)}(\omega_m) - i\alpha_m\right]}{\left[\alpha_m + i\boldsymbol{\mu}\cdot\mathbf{w}^{(1)}(\omega_m)\right]^2}.
\end{aligned}$$

Equation (13.11) with $n = 1$ implies that the numerator on the RHS vanishes, leaving

$$\mathbf{w}^{(2)}(\omega_m)\cdot\mathbf{w}^{(2)}(\omega_m) + \boldsymbol{\mu}\cdot\mathbf{w}^{(2)}(\omega_m) - i\alpha_m = 0$$

which is (13.11) with $n = 2$.

The result obtained in Exercise 13.5 ensures that $\left|\mathrm{Re}\,\mathbf{w}^{(1)}(\omega_m)\right| > 0$. Also, by assumption, $\alpha_m > 0$ and $\mathrm{Im}\,\boldsymbol{\mu} = 0$. Together these imply that the denominator on the RHS of (13.19) is nonzero. □

Exercise 13.12 Linear Independence of FREs

Problem: Consider the particular case where the eigenvector relation (13.17) is satisfied. Show that the corresponding FREs are linearly independent.

Solution: We will use the method of contradiction. Suppose the two FREs are linearly dependent. Then

$$\mathbf{w}^{(1)}(\omega_m) = \mathbf{w}^{(2)}(\omega_m).$$

Substituting the RHS of (13.18) for $\mathbf{w}^{(2)}(\omega_m)$ gives

$$\mathbf{w}^{(1)}(\omega_m) = -\mathbf{w}^{(1)}(\omega_m) - \boldsymbol{\mu}.$$

Solving for $\mathbf{w}^{(1)}(\omega_m)$ yields

$$\mathbf{w}^{(1)}(\omega_m) = \frac{-1}{2}\boldsymbol{\mu}$$

which implies that $\mathbf{w}^{(1)}(\omega_m)$ is purely real valued. As shown in Exercise 13.5, this is prohibited. Therefore, the two FREs are linearly independent. □

Exercise 13.13 Linear Independence of FREs

Problem: Consider the particular case where the eigenvector relation (13.19) is satisfied. Determine the conditions under which the corresponding FREs are linearly independent.

Solution: If the two FREs are linearly dependent, then

$$\mathbf{w}^{(1)}(\omega_m) = \mathbf{w}^{(2)}(\omega_m).$$

Substituting the RHS of (13.19) for $\mathbf{w}^{(2)}(\omega_m)$ gives

$$\mathbf{w}^{(1)}(\omega_m) = \frac{\alpha_m}{\alpha_m + i\boldsymbol{\mu}\cdot\mathbf{w}^{(1)}(\omega_m)}\mathbf{w}^{(1)}(\omega_m) - \boldsymbol{\mu}.$$

Solving for $\mathbf{w}^{(1)}(\omega_m)$ produces

$$\mathbf{w}^{(1)}(\omega_m) = \left[\frac{\alpha_m + i\boldsymbol{\mu}\cdot\mathbf{w}^{(1)}(\omega_m)}{\alpha_m}\right]\boldsymbol{\mu} \tag{13.20}$$

which is of the form

$$\mathbf{w}^{(1)}(\omega_m) = c\boldsymbol{\mu} \tag{13.21}$$

where c is a complex-valued scalar. Substituting the RHS for $\mathbf{w}^{(1)}(\omega_m)$ in (13.20) and then rearranging yield

$$|\boldsymbol{\mu}|^2 c^2 + |\boldsymbol{\mu}|^2 c - i\alpha_m = 0$$

which is quadratic in c. Using the quadratic formula to solve it, we obtain

$$c = \frac{-|\boldsymbol{\mu}| \pm \sqrt{|\boldsymbol{\mu}|^2 + i4\alpha_m}}{2|\boldsymbol{\mu}|}.$$

In conclusion, by combining this result with (13.21), we deduce that the two FREs are linearly independent unless

$$\mathbf{w}^{(1)}(\omega_m) = \left[\frac{-|\boldsymbol{\mu}| \pm \sqrt{|\boldsymbol{\mu}|^2 + i4\alpha_m}}{2|\boldsymbol{\mu}|}\right]\boldsymbol{\mu}.$$

□

Exercise 13.14 Complimentary Eigenvectors in Terms of Gradients

Problem: Suppose the particular eigenvector relation (13.17) is satisfied. Express (13.17) in terms of the gradients of the component log amplitude and phase.

Solution: Substituting the RHS of (13.8) for $\mathbf{w}^{(n)}(\omega_m)$ in (13.17) and then taking the real and imaginary parts successively yield

$$\nabla Y^{(1)}(\omega_m) + \nabla Y^{(2)}(\omega_m) = -\boldsymbol{\mu} \tag{13.22a}$$

$$\nabla\theta^{(1)}(\omega_m) + \nabla\theta^{(2)}(\omega_m) = 0. \tag{13.22b}$$

□

Exercise 13.15 Particular Eigenvector Relation—Algebraic Result

Problem: Consider the particular eigenvector relation (13.19). Express the denominator on the RHS of (13.19) in the form $u + iv$, where u and v are real valued. Then express its reciprocal in the form $a + ib$, where a and b are real valued.

Solution: The denominator on the RHS of (13.19) is

$$\alpha_m + i\boldsymbol{\mu}\cdot\mathbf{w}^{(1)}(\omega_m).$$

Substituting the RHS of (13.8), with $n = 1$, for $\mathbf{w}^{(1)}(\omega_m)$ leads to

$$\alpha_m + i\boldsymbol{\mu}\cdot\mathbf{w}^{(1)}(\omega_m) = \left[\alpha_m - \boldsymbol{\mu}\cdot\nabla\theta^{(1)}(\omega_m)\right] + i\boldsymbol{\mu}\cdot\nabla Y^{(1)}(\omega_m).$$

Taking the reciprocal and then simplifying yield

$$\frac{1}{\alpha_m + i\boldsymbol{\mu}\cdot\mathbf{w}^{(1)}(\omega_m)} = \frac{\left[\alpha_m - \boldsymbol{\mu}\cdot\nabla\theta^{(1)}(\omega_m)\right] - i\boldsymbol{\mu}\cdot\nabla Y^{(1)}(\omega_m)}{\left[\alpha_m - \boldsymbol{\mu}\cdot\nabla\theta^{(1)}(\omega_m)\right]^2 + \left[\boldsymbol{\mu}\cdot\nabla Y^{(1)}(\omega_m)\right]^2}. \tag{13.23}$$

□

Exercise 13.16 Complimentary Eigenvectors in Terms of Gradients

Problem: Suppose the particular eigenvector relation (13.19) is satisfied. Using the result (13.23), express $\nabla Y^{(2)}(\omega_m)$ in terms of $\nabla Y^{(1)}(\omega_m)$, $\nabla\theta^{(1)}(\omega_m)$, α_m, and $\boldsymbol{\mu}$. Then do the same for $\nabla\theta^{(2)}(\omega_m)$.

Solution: Substituting the RHSs of (13.8) for $\mathbf{w}^{(n)}(\omega_m)$ ($n = 1, 2$) in (13.19), taking the real and imaginary parts, and then simplifying lead to

$$\nabla Y^{(2)}(\omega_m) = c(\omega_m)\Big\{\big[\alpha_m - \boldsymbol{\mu}\cdot\nabla\theta^{(1)}(\omega_m)\big]\nabla Y^{(1)}(\omega_m) + \big[\boldsymbol{\mu}\cdot\nabla Y^{(1)}(\omega_m)\big]\nabla\theta^{(1)}(\omega_m)\Big\} - \boldsymbol{\mu} \tag{13.24a}$$

$$\nabla\theta^{(2)}(\omega_m) = c(\omega_m)\Big\{\big[\alpha_m - \boldsymbol{\mu}\cdot\nabla\theta^{(1)}(\omega_m)\big]\nabla\theta^{(1)}(\omega_m) - \big[\boldsymbol{\mu}\cdot\nabla Y^{(1)}(\omega_m)\big]\nabla Y^{(1)}(\omega_m)\Big\} \tag{13.24b}$$

where we define the real-valued parameter $c(\omega_m)$ as

$$c(\omega_m) \equiv \frac{\alpha_m}{\big[\alpha_m - \boldsymbol{\mu}\cdot\nabla\theta^{(1)}(\omega_m)\big]^2 + \big[\boldsymbol{\mu}\cdot\nabla Y^{(1)}(\omega_m)\big]^2}.$$

□

Refer to the general equations (4.25) and (4.26). For any single-component FRF (i.e., $a_m^{(n)}F^{(n)}$, n fixed), the gradients $\nabla Y^{(n)}$ and $\nabla\theta^{(n)}$ are necessarily coplanar (see Note 13.2) but generally noncollinear, so the corresponding flow component is effectively 2D. In the special case where $\nabla Y^{(n)}$ and $\nabla\theta^{(n)}$ are collinear, the corresponding flow component is only 1D.

Note 13.2 Coplanar and Noncoplanar Vectors

Suppose **a**, **b**, and **c** are nonzero, real-valued vectors in three dimensions. We say that they are *coplanar* vectors if each vector is parallel to the plane defined by the other two. Conversely, we say that they are *noncoplanar* vectors if any one of the three is not parallel to the plane defined by the other two.

Mathematically, the nonzero, real-valued vectors **a**, **b**, and **c** are coplanar if and only if

$$(\mathbf{a} \times \mathbf{b}) \cdot \mathbf{c} = \mathbf{0}.$$

△

Exercise 13.17 Dimensionality of Single-Component Flows

Problem: Consider time-periodic, uniform-gradient flow in a source-free domain filled with an exponential medium with $|a_m^{(1)}| > 0$ and $a_m^{(2)} = 0$. Can the flow be 2D in this case? Can it be three dimensional? Give the rationale for your response in terms of $\nabla Y^{(1)}$ and $\nabla\theta^{(1)}$.

Solution: The flow cannot be 3D in this case. If $\nabla Y^{(1)}$ and $\nabla\theta^{(1)}$ point in different directions, the flow is two dimensional; if they point in the same direction, the flow is 1D. □

13.4. UNIT GRADIENT VECTORS

When working with uniform-gradient FREs, it is convenient to define the unit gradient vectors for the component FRFs as

$$\hat{\mathbf{M}}^{(n)}(\omega_m) \equiv \frac{\nabla M^{(n)}(\mathbf{x};\omega_m)}{\big|\nabla M^{(n)}(\mathbf{x};\omega_m)\big|} \tag{13.25a}$$

$$\hat{\boldsymbol{\theta}}^{(n)}(\omega_m) \equiv \frac{\nabla\theta^{(n)}(\omega_m)}{\big|\nabla\theta^{(n)}(\omega_m)\big|} \tag{13.25b}$$

$$n = 1, 2.$$

These give

$$\nabla Y^{(n)}(\omega_m) = \big|\nabla Y^{(n)}(\omega_m)\big|\hat{\mathbf{M}}^{(n)}(\omega_m) \tag{13.26a}$$

$$\nabla\theta^{(n)}(\omega_m) = \big|\nabla\theta^{(n)}(\omega_m)\big|\hat{\boldsymbol{\theta}}^{(n)}(\omega_m) \tag{13.26b}$$

$$n = 1, 2.$$

Exercise 13.18 Space Invariance of Unit Gradient Vectors

Problem: Consider uniform-gradient flow in a source-free domain filled with an exponential medium. Our notation in (13.25a) indicates that the unit gradients $\hat{\mathbf{M}}^{(n)}$ ($n = 1, 2$) are space invariant. Verify that this is the case.

Solution: We can express the component FRF amplitude gradient on the RHS of (13.25a) in terms of the log amplitude gradient as follows:

$$\hat{\mathbf{M}}^{(n)} = \frac{M^{(n)}(\mathbf{x};\omega_m)\nabla Y^{(n)}(\mathbf{x};\omega_m)}{\big|M^{(n)}(\mathbf{x};\omega_m)\nabla Y^{(n)}(\mathbf{x};\omega_m)\big|} \quad (n = 1, 2).$$

Canceling the amplitude terms and using the results of Exercise 13.2, we obtain

$$\hat{\mathbf{M}}^{(n)}(\omega_m) = \frac{\nabla Y^{(n)}(\omega_m)}{\big|\nabla Y^{(n)}(\omega_m)\big|} \quad (n = 1, 2).$$

The RHS is space invariant, so the $\hat{\mathbf{M}}^{(n)}$ ($n = 1, 2$) are space invariant. □

The FRE $F^{(n)}(\mathbf{x};\omega_m)$ corresponds to 1D flow if and only if

$$\big|\hat{\mathbf{M}}^{(n)}(\omega_m) \times \hat{\boldsymbol{\theta}}^{(n)}(\omega_m)\big| = 0. \tag{13.27}$$

Conversely, $F^{(n)}(\mathbf{x};\omega_m)$ corresponds to 2D flow if and only if

$$\big|\hat{\mathbf{M}}^{(n)}(\omega_m) \times \hat{\boldsymbol{\theta}}^{(n)}(\omega_m)\big| > 0. \tag{13.28}$$

13.5. 1D FLOW

In this section we briefly explore 1D flows with uniform-gradient FREs. We examine two cases:

Case 1: One FRF component.
Case 2: Two FRF components.
Chapter 10 examined flows parallel to the log conductivity gradient ($\boldsymbol{\mu}$), so in this section we consider only those 1D flows that are not parallel to $\boldsymbol{\mu}$.

13.5.1. 1D Flow with One FRF Component

Here $|a_m^{(1)}| > 0$ and $a_m^{(2)} = 0$. For 1D flow we require

$$\nabla Y^{(1)}(\omega_m) = d_m^{(1)} \nabla\theta^{(1)}(\omega_m) \tag{13.29}$$

where for any given harmonic constituent (i.e., m fixed) $d_m^{(1)}$ is a real-valued, dimensionless constant. Based on the result obtained in Exercise 13.6, $\nabla Y^{(1)}(\omega_m)$ and $\nabla\theta^{(1)}(\omega_m)$ point in the same direction so $d_m^{(1)} > 0$.

Substituting $n = 1$ and the RHS of (13.29) for $\nabla Y^{(1)}$ in (13.12) and then eliminating $\boldsymbol{\mu}\cdot\nabla\theta^{(1)}$ from the resulting pair of equations lead to

$$|\nabla\theta^{(1)}|^2 (d_m^{(1)})^2 - \alpha_m d_m^{(1)} + |\nabla\theta^{(1)}|^2 = 0$$

which is quadratic in $d_m^{(1)}$. Using the quadratic formula to solve it yields

$$d_m^{(1)} = \frac{\alpha_m \pm \sqrt{\alpha_m^2 - 4|\nabla\theta^{(1)}(\omega_m)|^4}}{2|\nabla\theta^{(1)}(\omega_m)|^2}. \tag{13.30}$$

Exercise 13.19 Log Amplitude Gradient in Terms of Phase Gradient

Problem: Consider 1D, uniform-gradient flow in a source-free domain filled with an exponential medium and suppose there is only one FRF component. Express $|\nabla Y^{(1)}(\omega_m)|$ in terms of $\nabla\theta^{(1)}(\omega_m)$ and α_m.

Solution: Substituting the RHS of (13.30) for $d_m^{(1)}$ in (13.29), we obtain

$$\nabla Y^{(1)}(\omega_m) = \frac{\alpha_m \pm \sqrt{\alpha_m^2 - 4|\nabla\theta^{(1)}(\omega_m)|^4}}{2|\nabla\theta^{(1)}(\omega_m)|}\,\hat{\boldsymbol{\theta}}^{(1)}(\omega_m).$$

Therefore,

$$|\nabla Y^{(1)}(\omega_m)| = \frac{\alpha_m \pm \sqrt{\alpha_m^2 - 4|\nabla\theta^{(1)}(\omega_m)|^4}}{2|\nabla\theta^{(1)}(\omega_m)|}.$$

□

13.5.1.1. Bounds on Phase Gradient

Because $d_m^{(1)}$ is real valued, the radical term in the numerator on the RHS of (13.30) must be real valued. Thus,

$$|\nabla\theta^{(1)}(\omega_m)| \le \sqrt{\frac{\alpha_m}{2}}. \tag{13.31}$$

For flow not parallel to the log conductivity gradient, the absolute value of the cosine of the angle between $\boldsymbol{\mu}$ and $\nabla\theta^{(1)}$ must be less than unity, so

$$|\boldsymbol{\mu}\cdot\nabla\theta^{(1)}| < |\boldsymbol{\mu}||\nabla\theta^{(1)}|. \tag{13.32}$$

Substituting the RHS of (13.30) for $d_m^{(1)}$ in (13.29) and then substituting the resulting expression for $\nabla Y^{(1)}$ in (13.12b) produce

$$\boldsymbol{\mu}\cdot\nabla\theta^{(1)} = \mp\sqrt{\alpha_m^2 - 4|\nabla\theta^{(1)}|^4}.$$

Substituting the RHS for $\boldsymbol{\mu}\cdot\nabla\theta^{(1)}$ in (13.32) gives

$$\sqrt{\alpha_m^2 - 4|\nabla\theta^{(1)}|^4} < |\boldsymbol{\mu}||\nabla\theta^{(1)}|.$$

Squaring and then grouping terms result in

$$4|\nabla\theta^{(1)}|^4 + |\boldsymbol{\mu}|^2|\nabla\theta^{(1)}|^2 - \alpha_m^2 > 0$$

which is a quadratic inequality in $|\nabla\theta^{(1)}|^2$. Using the quadratic formula to solve it gives

$$|\nabla\theta^{(1)}| > \frac{\sqrt{\sqrt{|\boldsymbol{\mu}|^4 + 16\alpha_m^2} - |\boldsymbol{\mu}|^2}}{2\sqrt{2}}.$$

Combining this with (13.31) yields the final result:

$$\frac{\sqrt{\sqrt{|\boldsymbol{\mu}|^4 + 16\alpha_m^2} - |\boldsymbol{\mu}|^2}}{2\sqrt{2}} < |\nabla\theta^{(1)}(\omega_m)| \le \sqrt{\frac{\alpha_m}{2}}. \tag{13.33}$$

Exercise 13.20 1D Flow Perpendicular to Log Conductivity Gradient

Problem: Consider time-periodic, 1D flow in a source-free domain filled with an exponential medium. Suppose (13.3) and (13.4) give the hydraulic head FRF with $|a_m^{(1)}| > 0$ and $a_m^{(2)} = 0$. Under these conditions is it possible for the flow to be perpendicular to the log conductivity gradient? If so, find simple expressions for $d_m^{(1)}$ and $|\nabla\theta^{(1)}(\omega_m)|$ for this special case. Is (13.33) satisfied? Also, express $\nabla Y^{(1)}(\omega_m)$ in terms of $\nabla\theta^{(1)}(\omega_m)$.

Solution: For 1D flow perpendicular to the log conductivity gradient, we have $\boldsymbol{\mu}\cdot\nabla Y^{(1)} = 0$ and $\boldsymbol{\mu}\cdot\nabla\theta^{(1)} = 0$. Substituting these and $n = 1$ in (13.12) gives

$$|\nabla Y^{(1)}(\omega_m)| = |\nabla\theta^{(1)}(\omega_m)| \tag{13.34a}$$

$$\nabla Y^{(1)}(\omega_m)\cdot\nabla\theta^{(1)}(\omega_m) = \frac{\alpha_m}{2}. \tag{13.34b}$$

Together these imply

$$\nabla Y^{(1)}(\omega_m) = \nabla\theta^{(1)}(\omega_m) \tag{13.35a}$$

$$|\nabla\theta^{(1)}(\omega_m)| = \sqrt{\frac{\alpha_m}{2}} \tag{13.35b}$$

so that $d_m^{(1)} = 1$ and (13.33) is satisfied. Yes, under the stated conditions it is possible for the flow to be perpendicular to the log conductivity gradient. □

13.5.2. 1D Flow with Two FRF Components

Here $|a_m^{(1)} a_m^{(2)}| > 0$. For 1D flow we require

$$\nabla Y^{(n)}(\omega_m) = d_m^{(n)} \nabla \theta^{(n)}(\omega_m) \quad (n = 1, 2). \tag{13.36}$$

where the $d_m^{(n)}$ $(n = 1, 2)$ are real valued, positive, and dimensionless (see Exercise 13.6).

For 1D flow, in addition to the requirements derived for case 1, the gradients of the two component log amplitudes and phases must be collinear, i.e.,

$$\nabla Y^{(1)}(\omega_m) = b_m \nabla Y^{(2)}(\omega_m) \tag{13.37a}$$

$$\nabla \theta^{(1)}(\omega_m) = g_m \nabla \theta^{(2)}(\omega_m) \tag{13.37b}$$

where b_m and g_m are real valued and dimensionless.

Exercise 13.21 Signs of Coefficients

Problem: Consider time-periodic, 1D flow in a source-free domain filled with an exponential medium. Suppose that for some fixed m we have $|a_m^{(1)} a_m^{(2)}| > 0$. Prove that the signs of b_m and g_m must be identical.

Solution: From Exercise 13.6 we have

$$\nabla Y^{(1)}(\omega_m) \cdot \nabla \theta^{(1)}(\omega_m) > 0 \tag{13.38a}$$

$$\nabla Y^{(2)}(\omega_m) \cdot \nabla \theta^{(2)}(\omega_m) > 0. \tag{13.38b}$$

Substituting the RHSs of (13.37) for the corresponding terms on the LHS of (13.38a) yields

$$\nabla Y^{(2)}(\omega_m) \cdot \nabla \theta^{(2)}(\omega_m) b_m g_m > 0.$$

Combining this with (13.38b) then leads to

$$b_m g_m > 0$$

which requires that for fixed m the signs of b_m and g_m are identical. □

Exercise 13.22 Special Case: Flow Parallel to Conductivity Gradient

Problem: Consider time-periodic, 1D flow in a source-free domain filled with an exponential medium. Suppose that for some fixed m we have $|a_m^{(1)} a_m^{(2)}| > 0$ and the eigenvector relation (13.18) is satisfied. Show that flow in a direction not parallel to the log conductivity gradient is prohibited in this case.

Solution: Using (13.8) we can rewrite (13.18) as

$$\nabla Y^{(2)}(\omega_m) = -\nabla Y^{(1)}(\omega_m) - \boldsymbol{\mu} \tag{13.39a}$$

$$\nabla \theta^{(2)}(\omega_m) = -\nabla \theta^{(1)}(\omega_m). \tag{13.39b}$$

Substituting the RHSs for the corresponding terms on the RHSs of (13.37) leads to

$$\nabla Y^{(1)}(\omega_m) = \left(\frac{-b_m}{1 + b_m} \right) \boldsymbol{\mu}$$

$$g_m = -1.$$

These and (13.37) together imply that each of the two gradients $\nabla Y^{(n)}$ $(n = 1, 2)$ must point either in the same direction as $\boldsymbol{\mu}$ or opposite $\boldsymbol{\mu}$. And because the flow is 1D, this implies that each of the two gradients $\nabla \theta^{(n)}$ $(n = 1, 2)$ also must point either in the same direction as $\boldsymbol{\mu}$ or opposite $\boldsymbol{\mu}$. That is, the flow must parallel $\boldsymbol{\mu}$. □

13.5.3. Summary

In summary, for 1D flow not parallel to the log conductivity gradient, we must have $|a_m^{(1)}| > 0$ and $a_m^{(2)} = 0$, or vice versa. Section 13.5.1 treats this case.

13.6. 2D FLOW

In this section we briefly explore 2D flows with uniform-gradient FREs. We examine two cases:
Case 1: One FRF component.
Case 2: Two FRF components.

13.6.1. 2D Flow with One FRF Component

Here $|a_m^{(1)}| > 0$ and $a_m^{(2)} = 0$. As discussed in Exercise 13.17, if the flow is to be 2D, then $\nabla Y^{(1)}$ and $\nabla \theta^{(1)}$ must point in different but not opposite directions:

$$\left| \nabla Y^{(1)}(\omega_m) \cdot \nabla \theta^{(1)}(\omega_m) \right| < \left| \nabla Y^{(1)}(\omega_m) \right| \left| \nabla \theta^{(1)}(\omega_m) \right|. \tag{13.40}$$

Otherwise, if $\nabla Y^{(1)}$ and $\nabla \theta^{(1)}$ point in the same direction, then the flow is 1D. Recall that Exercise 13.6 also showed that $\nabla Y^{(1)}$ and $\nabla \theta^{(1)}$ cannot be oppositely directed.

Exercise 13.23 2D Flow with Perpendicular Component Gradients

Problem: Consider time-periodic, uniform-gradient flow in a source-free domain filled with an exponential medium. Suppose $|a_m^{(1)}| > 0$, $a_m^{(2)} = 0$, $\hat{\mathbf{M}}^{(1)}$ and $\hat{\boldsymbol{\theta}}^{(1)}$ are perpendicular, and $\boldsymbol{\mu} \cdot \hat{\boldsymbol{\theta}}^{(1)} > 0$. Express $|\nabla Y^{(1)}|$ and $|\nabla \theta^{(1)}|$ in terms of $\boldsymbol{\mu}$, α_m, $\hat{\mathbf{M}}^{(1)}$, and $\hat{\boldsymbol{\theta}}^{(1)}$.

Solution: Substituting (13.26) with $n = 1$ in (13.12) and noting that $\hat{\mathbf{M}}^{(1)} \cdot \hat{\boldsymbol{\theta}}^{(1)} = 0$ lead to

$$\left|\nabla Y^{(1)}\right|^2 + \left(\boldsymbol{\mu} \cdot \hat{\mathbf{M}}^{(1)}\right)\left|\nabla Y^{(1)}\right| - \left|\nabla \theta^{(1)}\right|^2 = 0 \quad (13.41a)$$

$$\left(\boldsymbol{\mu} \cdot \hat{\boldsymbol{\theta}}^{(1)}\right)\left|\nabla \theta^{(1)}\right| = \alpha_m. \quad (13.41b)$$

Solving (13.41b) for the magnitude of the phase gradient yields

$$\left|\nabla \theta^{(1)}(\omega_m)\right| = \frac{\alpha_m}{\boldsymbol{\mu} \cdot \hat{\boldsymbol{\theta}}^{(1)}}. \quad (13.42)$$

Substituting the RHS for $\left|\nabla \theta^{(1)}\right|$ in (13.41a) produces

$$\left|\nabla Y^{(1)}\right|^2 + \left(\boldsymbol{\mu} \cdot \hat{\mathbf{M}}^{(1)}\right)\left|\nabla Y^{(1)}\right| - \left(\frac{\alpha_m}{\boldsymbol{\mu} \cdot \hat{\boldsymbol{\theta}}^{(1)}}\right)^2 = 0$$

which is a quadratic equation in $\left|\nabla Y^{(1)}\right|$. Using the quadratic formula to solve it gives

$$\left|\nabla Y^{(1)}(\omega_m)\right| = \frac{1}{2}\left[-\boldsymbol{\mu} \cdot \hat{\mathbf{M}}^{(1)} + \sqrt{\left(\boldsymbol{\mu} \cdot \hat{\mathbf{M}}^{(1)}\right)^2 + 4\left(\frac{\alpha_m}{\boldsymbol{\mu} \cdot \hat{\boldsymbol{\theta}}^{(1)}}\right)^2}\,\right]. \quad (13.43)$$

□

In Exercise 13.23 the flow is 2D, being confined to the plane defined by the gradients of the amplitude and phase. Under the particular conditions of this exercise, the log conductivity gradient generally need not lie in the plane of the flow. The following example illustrates.

Example 13.1 2D Flow with Perpendicular Component Gradients
Consider time-periodic, uniform-gradient flow in a source-free domain filled with an exponential medium with $\left|a_m^{(1)}\right| > 0$ and $a_m^{(2)} = 0$. Suppose the component FRF's unit gradient vectors are

$$\hat{\mathbf{M}}^{(1)} = \frac{\sqrt{2}}{2}\begin{bmatrix} 1 \\ -1 \\ 0 \end{bmatrix}$$

$$\hat{\boldsymbol{\theta}}^{(1)} = \frac{\sqrt{2}}{2}\begin{bmatrix} 1 \\ 1 \\ 0 \end{bmatrix}$$

and the log conductivity gradient is

$$\boldsymbol{\mu} = |\boldsymbol{\mu}|\frac{\sqrt{5}}{5}\begin{bmatrix} 2 \\ 0 \\ 1 \end{bmatrix}. \quad (13.44)$$

Then

$$\boldsymbol{\mu} \cdot \hat{\mathbf{M}}^{(1)} = |\boldsymbol{\mu}|\frac{\sqrt{10}}{5} \quad (13.45a)$$

$$\boldsymbol{\mu} \cdot \hat{\boldsymbol{\theta}}^{(1)} = |\boldsymbol{\mu}|\frac{\sqrt{10}}{5}. \quad (13.45b)$$

As in Exercise 13.23, $\hat{\mathbf{M}}^{(1)} \cdot \hat{\boldsymbol{\theta}}^{(1)} = 0$ and $\boldsymbol{\mu} \cdot \hat{\boldsymbol{\theta}}^{(1)} > 0$. Substituting these results in (13.42) and (13.43) gives, respectively,

$$\left|\nabla \theta^{(1)}(\omega_m)\right| = \frac{\sqrt{10}\alpha_m}{2|\boldsymbol{\mu}|} \quad (13.46a)$$

$$\left|\nabla Y^{(1)}(\omega_m)\right| = \frac{1}{2}\left[\sqrt{\frac{2}{5}|\boldsymbol{\mu}|^2 + \frac{10\alpha_m^2}{|\boldsymbol{\mu}|^2}} - \frac{\sqrt{10}}{5}|\boldsymbol{\mu}|\right]. \quad (13.46b)$$

◊

Exercise 13.24 2D Flow with ∇Y Perpendicular to $\boldsymbol{\mu}$

Problem: Consider time-periodic, uniform-gradient flow in a source-free domain filled with an exponential medium with $\left|a_m^{(1)}\right| > 0$ and $a_m^{(2)} = 0$. Suppose $\hat{\mathbf{M}}^{(1)}$ and $\boldsymbol{\mu}$ are perpendicular and $\hat{\mathbf{M}}^{(1)}$ and $\hat{\boldsymbol{\theta}}^{(1)}$ are nonperpendicular. Express $\left|\nabla Y^{(1)}\right|$ and $\left|\nabla \theta^{(1)}\right|$ in terms of $\boldsymbol{\mu}$, α_m, $\hat{\mathbf{M}}^{(1)}$, and $\hat{\boldsymbol{\theta}}^{(1)}$.

Solution: Substituting the RHS of (13.26) with $n = 1$ for the corresponding terms in (13.12) and noting that $\boldsymbol{\mu} \cdot \hat{\mathbf{M}}^{(1)} = 0$ lead to

$$\left|\nabla Y^{(1)}\right|^2 - \left|\nabla \theta^{(1)}\right|^2 = 0 \quad (13.47a)$$

$$2\left(\hat{\mathbf{M}}^{(1)} \cdot \hat{\boldsymbol{\theta}}^{(1)}\right)\left|\nabla Y^{(1)}\right|^2 + \left(\boldsymbol{\mu} \cdot \hat{\boldsymbol{\theta}}^{(1)}\right)\left|\nabla Y^{(1)}\right| - \alpha_m = 0. \quad (13.47b)$$

The second of these is a quadratic equation in $\left|\nabla Y^{(1)}\right|$. Using the quadratic formula to solve this system yields

$$\left|\nabla Y^{(1)}(\omega_m)\right| = \left|\nabla \theta^{(1)}(\omega_m)\right| = \frac{-\boldsymbol{\mu} \cdot \hat{\boldsymbol{\theta}}^{(1)} + \sqrt{\left(\boldsymbol{\mu} \cdot \hat{\boldsymbol{\theta}}^{(1)}\right)^2 + 8\left(\hat{\mathbf{M}}^{(1)} \cdot \hat{\boldsymbol{\theta}}^{(1)}\right)\alpha_m}}{4\hat{\mathbf{M}}^{(1)} \cdot \hat{\boldsymbol{\theta}}^{(1)}}. \quad (13.48)$$

□

Example 13.2 2D Flow with ∇Y Perpendicular to $\boldsymbol{\mu}$
Consider time-periodic, uniform-gradient flow in a source-free domain filled with an exponential medium with $\left|a_m^{(1)}\right| > 0$ and $a_m^{(2)} = 0$. Suppose the component FRF's unit gradient vectors are

$$\hat{\mathbf{M}}^{(1)} = \frac{\sqrt{2}}{2}\begin{bmatrix}1\\1\\0\end{bmatrix}$$

$$\hat{\boldsymbol{\theta}}^{(1)} = \frac{\sqrt{5}}{5}\begin{bmatrix}2\\1\\0\end{bmatrix}$$

and the log conductivity gradient is

$$\boldsymbol{\mu} = |\boldsymbol{\mu}|\frac{\sqrt{2}}{2}\begin{bmatrix}1\\-1\\1\end{bmatrix}.$$

Then

$$\boldsymbol{\mu}\cdot\hat{\mathbf{M}}^{(1)} = 0$$

$$\boldsymbol{\mu}\cdot\hat{\boldsymbol{\theta}}^{(1)} = |\boldsymbol{\mu}|\frac{\sqrt{10}}{10}.$$

As in Exercise 13.24, $\hat{\mathbf{M}}^{(1)}\cdot\hat{\boldsymbol{\theta}}^{(1)} = 3\sqrt{10}/10$, so $|\hat{\mathbf{M}}^{(1)}\cdot\hat{\boldsymbol{\theta}}^{(1)}| > 0$. Substituting these results in (13.48) gives

$$\left|\nabla Y^{(1)}(\omega_m)\right| = \left|\nabla\theta^{(1)}(\omega_m)\right| = \frac{-|\boldsymbol{\mu}| + \sqrt{|\boldsymbol{\mu}|^2 + 24\sqrt{10}\alpha_m}}{12}.$$

◊

Exercise 13.25 2D Flow in Plane Oblique to $\boldsymbol{\mu}$

Problem: Consider time-periodic, uniform-gradient flow in a source-free domain filled with an exponential medium for which $\boldsymbol{\mu} = (1.5\times10^{-2}\text{m}^{-1})\mathbf{e}_3$. Furthermore, suppose $|a_m^{(1)}| > 0$, $a_m^{(2)} = 0$, and

$$\nabla Y^{(1)}(\omega_m) = \begin{bmatrix}5\times10^{-3}\\0\\\sqrt{3}/2\times10^{-2}\end{bmatrix}\text{m}^{-1} \qquad (13.49a)$$

$$\hat{\boldsymbol{\theta}}^{(1)}(\omega_m) = (\sqrt{2}/2)\begin{bmatrix}0\\1\\1\end{bmatrix}. \qquad (13.49b)$$

Find the numerical values of α_m and $\left|\nabla\theta^{(1)}(\omega_m)\right|$.

Solution: Substituting $n = 1$ and the numerical values given above for $\boldsymbol{\mu}$ and $\nabla Y^{(1)}(\omega_m)$ for the corresponding variables in (13.12a) and then solving for $\left|\nabla\theta^{(1)}(\omega_m)\right|$ give

$$\left|\nabla\theta^{(1)}(\omega_m)\right| \approx 1.516\times10^{-2}\text{m}^{-1}. \qquad (13.50)$$

Substituting the numerical values given above for the appropriate terms in (13.12b) yields

$$\alpha_m \approx 3.465\times10^{-4}\text{m}^{-2}. \qquad (13.51)$$

□

13.6.2. 2D Flow with Two FRF Components

Here $|a_m^{(1)}a_m^{(2)}| > 0$. If the flow is to be 2D, then one of the following must hold:

Case 1: Each of the two FRF components corresponds to a 1D flow, but the two component flows are non-collinear.

Case 2: One or both of the two FRF components corresponds to a 2D flow (i.e., at least one of the FRF components satisfies the conditions discussed in Section 13.6.1), and the gradients $\nabla\theta^{(1)}(\omega_m)$, $\nabla Y^{(1)}(\omega_m)$, $\nabla\theta^{(2)}(\omega_m)$, and $\nabla Y^{(2)}(\omega_m)$ are coplanar.

13.6.2.1. Case 1: Two 1D Component FRFs

In this case each of the component FRFs satisfies the equivalent of (13.29):

$$\nabla Y^{(n)}(\omega_m) = d_m^{(n)}\nabla\theta^{(n)}(\omega_m) \quad (n = 1,2) \qquad (13.52)$$

where for any given harmonic constituent (i.e., fixed m), the $d_m^{(n)}$ are real-valued, positive, dimensionless constants that satisfy the equivalent of (13.30):

$$d_m^{(n)} = \frac{\alpha_m \pm \sqrt{\alpha_m^2 - 4\left|\nabla\theta^{(n)}(\omega_m)\right|^4}}{2\left|\nabla\theta^{(n)}(\omega_m)\right|^2} \quad (n = 1,2). \qquad (13.53)$$

Consequently, for 2D flow,

$$\left|\hat{\mathbf{M}}^{(1)}(\omega_m)\cdot\hat{\mathbf{M}}^{(2)}(\omega_m)\right| < 1 \qquad (13.54a)$$

$$\left|\hat{\boldsymbol{\theta}}^{(1)}(\omega_m)\cdot\hat{\boldsymbol{\theta}}^{(2)}(\omega_m)\right| < 1. \qquad (13.54b)$$

Exercise 13.26 Admissibility of Particular Eigenvector Relation

Problem: Consider 2D, uniform-gradient flow in a source-free domain filled with an ideal medium. Suppose that for some fixed m we have $|a_m^{(1)}a_m^{(2)}| > 0$ and (13.52) and (13.53) are satisfied. Is it possible for the particular eigenvector relation (13.17) to be satisfied in this case?

Solution: The eigenvector relation (13.17) is equivalent to (13.22). Substituting $\boldsymbol{\mu} = \mathbf{0}$ in (13.22) leads to

$$\hat{\mathbf{M}}^{(1)}(\omega_m)\cdot\hat{\mathbf{M}}^{(2)}(\omega_m) = -1$$

$$\hat{\boldsymbol{\theta}}^{(1)}(\omega_m)\cdot\hat{\boldsymbol{\theta}}^{(2)}(\omega_m) = -1.$$

These contradict (13.54) and with (13.52) and (13.53) imply that the flow is 1D rather than 2D. Therefore, it is not possible for the particular eigenvector relation (13.17) to be satisfied in this case. □

Exercise 13.27 Admissibility of Particular Eigenvector Relation

Problem: Consider 2D, uniform-gradient flow in a source-free domain filled with a nonhomogeneous, exponential medium. Suppose that for some fixed m we have $|a_m^{(1)} a_m^{(2)}| > 0$, (13.52) and (13.53) are satisfied, and $\boldsymbol{\mu} \cdot \nabla Y^{(1)}(\omega_m) = 0$. Show that the particular eigenvector relation (13.19) cannot be satisfied under these conditions.

Solution: We will use the method of contradiction. Suppose the particular eigenvector relation (13.19) is satisfied. Combining (13.52) with (13.8) gives

$$\mathbf{w}^{(n)}(\omega_m) = \left(d_m^{(n)} + i\right)\nabla\theta^{(n)}(\omega_m) \quad (n = 1, 2).$$

If $\boldsymbol{\mu} \cdot \nabla Y^{(1)}(\omega_m) = 0$, then $\boldsymbol{\mu} \cdot \nabla\theta^{(1)}(\omega_m) = 0$ and $\boldsymbol{\mu} \cdot \mathbf{w}^{(1)}(\omega_m) = 0$. Using these results in (13.19) gives

$$\mathbf{w}^{(2)}(\omega_m) = \mathbf{w}^{(1)}(\omega_m) - \boldsymbol{\mu}.$$

Taking the imaginary part yields $\nabla\theta^{(2)}(\omega_m) = \nabla\theta^{(1)}(\omega_m)$. This result and (13.52) together imply that the flow is 1D, thus violating our assumption that the flow is 2D. Therefore, the particular eigenvector relation (13.19) cannot be satisfied under these conditions. □

Example 13.3 2D Flow in Plane Perpendicular to $\boldsymbol{\mu}$

Consider time-periodic, uniform-gradient flow in a source-free domain filled with an exponential, nonhomogeneous medium for which

$$\boldsymbol{\mu} = |\boldsymbol{\mu}| \begin{bmatrix} 0 \\ 0 \\ 1 \end{bmatrix}.$$

Suppose that for some fixed m we have two component flows (i.e., $|a_m^{(1)} a_m^{(2)}| > 0$) and each component flow is one dimensional [i.e., (13.52) and (13.53) are satisfied]. Furthermore, suppose the unit-gradient vectors are

$$\hat{\mathbf{M}}^{(1)}(\omega_m) = \hat{\boldsymbol{\theta}}^{(1)}(\omega_m) = \begin{bmatrix} 1 \\ 0 \\ 0 \end{bmatrix}$$

$$\hat{\mathbf{M}}^{(2)}(\omega_m) = \hat{\boldsymbol{\theta}}^{(2)}(\omega_m) = \begin{bmatrix} 0 \\ 1 \\ 0 \end{bmatrix}$$

and the gradient magnitudes are

$$|\nabla Y^{(n)}(\omega_m)| = |\nabla\theta^{(n)}(\omega_m)| = \sqrt{\frac{\alpha_m}{2}} \quad (n = 1, 2).$$

Readers can easily verify that the gradients $\nabla Y^{(1)}(\omega_m)$ and $\nabla\theta^{(1)}(\omega_m)$ jointly satisfy (approximately) the characteristic equations (13.12), as do the gradients $\nabla Y^{(2)}(\omega_m)$ and $\nabla\theta^{(2)}(\omega_m)$. ◊

Example 13.4 2D Flow in Plane Oblique to $\boldsymbol{\mu}$

Consider time-periodic, uniform-gradient flow in a source-free domain filled with an exponential, nonhomogeneous medium for which

$$\boldsymbol{\mu} = \begin{bmatrix} 0 \\ 0 \\ 2 \times 10^{-2} \end{bmatrix} \mathrm{m}^{-1}. \tag{13.55}$$

Suppose that for some fixed m we have $\alpha_m = 3 \times 10^{-4}\mathrm{m}^{-2}$ and two component flows (i.e., $|a_m^{(1)} a_m^{(2)}| > 0$). Furthermore, suppose the unit-gradient vectors are

$$\hat{\mathbf{M}}^{(1)}(\omega_m) = \hat{\boldsymbol{\theta}}^{(1)}(\omega_m) = \begin{bmatrix} \cos(5\pi/8) \\ 0 \\ \sin(5\pi/8) \end{bmatrix} \tag{13.56a}$$

$$\hat{\mathbf{M}}^{(2)}(\omega_m) = \hat{\boldsymbol{\theta}}^{(2)}(\omega_m) = \begin{bmatrix} 0 \\ \cos(\pi/4) \\ \sin(\pi/4) \end{bmatrix} \tag{13.56b}$$

and the gradient magnitudes are

$$\begin{aligned} \left|\nabla Y^{(1)}(\omega_m)\right| = \left|\nabla Y^{(2)}(\omega_m)\right| &\approx 4.855 \times 10^{-3}\mathrm{m}^{-1} \\ \left|\nabla\theta^{(1)}(\omega_m)\right| &\approx 1.064 \times 10^{-2}\mathrm{m}^{-1} \\ \left|\nabla\theta^{(2)}(\omega_m)\right| &\approx 1.258 \times 10^{-2}\mathrm{m}^{-1}. \end{aligned} \tag{13.57}$$

Readers can easily verify that the gradients $\nabla Y^{(1)}(\omega_m)$ and $\nabla\theta^{(1)}(\omega_m)$ jointly satisfy (approximately) the characteristic equations (13.12), as do the gradients $\nabla Y^{(2)}(\omega_m)$ and $\nabla\theta^{(2)}(\omega_m)$. ◊

Exercise 13.28 2D Flow in Plane Oblique to $\boldsymbol{\mu}$

Problem: Consider time-periodic, uniform-gradient flow in a source-free domain filled with an exponential, nonhomogeneous medium for which

$$\boldsymbol{\mu} = \begin{bmatrix} 0 \\ 0 \\ 2 \times 10^{-2} \end{bmatrix} \mathrm{m}^{-1}. \tag{13.58}$$

Suppose that for some fixed m we have $\alpha_m = 3 \times 10^{-4}\mathrm{m}^{-2}$ and two component flows (i.e., $|a_m^{(1)} a_m^{(2)}| > 0$). Furthermore, suppose the unit-gradient vectors are

$$\hat{\mathbf{M}}^{(1)}(\omega_m) = \hat{\boldsymbol{\theta}}^{(1)}(\omega_m) = \begin{bmatrix} \cos(5\pi/8) \\ 0 \\ \sin(5\pi/8) \end{bmatrix} \tag{13.59a}$$

$$\hat{\mathbf{M}}^{(2)}(\omega_m) = \hat{\boldsymbol{\theta}}^{(2)}(\omega_m) = \begin{bmatrix} 0 \\ \cos(5\pi/8) \\ \sin(5\pi/8) \end{bmatrix}. \tag{13.59b}$$

Calculations indicate that the gradient magnitudes are approximated by

$$\left|\nabla Y^{(1)}(\omega_m)\right| = \left|\nabla Y^{(2)}(\omega_m)\right| \approx 4.855 \times 10^{-3}\mathrm{m}^{-1} \tag{13.60a}$$

$$\left|\nabla\theta^{(1)}(\omega_m)\right| = \left|\nabla\theta^{(2)}(\omega_m)\right| \approx 1.064 \times 10^{-2}\mathrm{m}^{-1}. \tag{13.60b}$$

Do the gradients $\nabla Y^{(1)}(\omega_m)$ and $\nabla\theta^{(1)}(\omega_m)$ jointly satisfy (approximately) the characteristic equations (13.12)? What about the gradients $\nabla Y^{(2)}(\omega_m)$ and $\nabla\theta^{(2)}(\omega_m)$?

Solution: Substituting the values given above for the corresponding variables in the characteristic equations (13.12), first with $n = 1$ and then with $n = 2$, leads approximately to identities of the form $0 = 0$. Thus, the gradients do approximately satisfy the characteristic equations. □

13.6.2.2. Case 2: At Least One 2D Component FRF

In this case at least one of the two component FRFs corresponds to a 2D flow; the other component FRF corresponds to either 1D or 2D flow.

Example 13.5 Particular Eigenvector Relation, Nonhomogeneous Medium

Consider time-periodic, uniform-gradient flow in a source-free domain filled with an exponential, nonhomogeneous medium. Suppose that for some fixed m we have $|a_m^{(1)} a_m^{(2)}| > 0$, (13.29) and (13.30) are satisfied, and

$$\boldsymbol{\mu}\cdot\nabla Y^{(1)}(\omega_m) = 0 \qquad (13.61a)$$

$$\boldsymbol{\mu}\cdot\nabla\theta^{(1)}(\omega_m) = 0. \qquad (13.61b)$$

That is, the flow corresponding to the $n = 1$ component FRF is 1D and perpendicular to the log conductivity gradient (i.e., the flow of Exercise 13.20). Further suppose the particular eigenvector relation (13.19) is satisfied. Substituting the RHSs of (13.61) for the corresponding terms of (13.24) yields

$$\nabla Y^{(2)}(\omega_m) = \nabla Y^{(1)}(\omega_m) - \boldsymbol{\mu} \qquad (13.62a)$$

$$\nabla\theta^{(2)}(\omega_m) = \nabla\theta^{(1)}(\omega_m). \qquad (13.62b)$$

It follows that

- $\nabla\theta^{(2)}(\omega_m)$ is perpendicular to $\boldsymbol{\mu}$;
- $\nabla Y^{(2)}(\omega_m)$ and $\nabla\theta^{(2)}(\omega_m)$ are noncollinear, so the flow corresponding to the $n = 2$ FRF component is 2D;
- $\nabla Y^{(2)}(\omega_m)$ is neither parallel nor perpendicular to $\boldsymbol{\mu}$;

and

- $\nabla\theta^{(1)}(\omega_m)$, $\nabla Y^{(1)}(\omega_m)$, $\nabla\theta^{(2)}(\omega_m)$, and $\nabla Y^{(2)}(\omega_m)$ are coplanar, so the flows corresponding to the two FRF components are coplanar.

Consequently, the net flow is 2D. The time-varying hydraulic gradient lies in a fixed plane that contains $\boldsymbol{\mu}$. ◊

Example 13.6 2D Flow in Plane Oblique to $\boldsymbol{\mu}$

Consider time-periodic, uniform-gradient flow in a source-free domain filled with an exponential, nonhomogeneous medium for which

$$\boldsymbol{\mu} = \left(1.5\times10^{-2}\mathrm{m}^{-1}\right)\begin{bmatrix}0\\ \cos(3\pi/8)\\ \sin(3\pi/8)\end{bmatrix} \approx \begin{bmatrix}0\\ 5.7403\times10^{-3}\\ 1.3858\times10^{-2}\end{bmatrix}\mathrm{m}^{-1}. \qquad (13.63)$$

Suppose that for some fixed m we have $\alpha_m = 4.3844\times 10^{-4}\mathrm{m}^{-2}$ and two component flows (i.e., $|a_m^{(1)} a_m^{(2)}| > 0$). Furthermore, suppose

$$\hat{\mathbf{M}}^{(1)}(\omega_m) = \hat{\boldsymbol{\theta}}^{(1)}(\omega_m) = \begin{bmatrix}\cos(3\pi/8)\\ 0\\ \sin(3\pi/8)\end{bmatrix}$$

$$\left|\nabla Y^{(1)}(\omega_m)\right| \approx 9.1120\times10^{-3}\mathrm{m}^{-1}$$

$$\left|\nabla\theta^{(1)}(\omega_m)\right| \approx 1.4131\times10^{-2}\mathrm{m}^{-1}$$

and

$$\nabla Y^{(2)}(\omega_m) = \left(1\times10^{-2}\mathrm{m}^{-1}\right)\begin{bmatrix}\cos(5\pi/12)\\ 0\\ \sin(5\pi/12)\end{bmatrix} \approx \begin{bmatrix}2.5882\times10^{-3}\\ 0\\ 9.6593\times10^{-3}\end{bmatrix}\mathrm{m}^{-1} \qquad (13.64a)$$

$$\nabla\theta^{(2)}(\omega_m) \approx \left(1.4131\times10^{-2}\mathrm{m}^{-1}\right)\begin{bmatrix}\cos(5\pi/8)\\ 0\\ \sin(5\pi/8)\end{bmatrix} \approx \begin{bmatrix}-5.8522\times10^{-3}\\ 0\\ 1.4128\times10^{-2}\end{bmatrix}\mathrm{m}^{-1}. \qquad (13.64b)$$

Notice that while $\nabla Y^{(1)}(\omega_m)$ and $\nabla\theta^{(1)}(\omega_m)$ are collinear, $\nabla Y^{(2)}(\omega_m)$ and $\nabla\theta^{(2)}(\omega_m)$ are noncollinear. Also, while $\nabla Y^{(1)}$, $\nabla\theta^{(1)}$, $\nabla Y^{(2)}$, and $\nabla\theta^{(2)}$ lie in the x_1x_3 plane, $\boldsymbol{\mu}$ lies in the x_2x_3 plane. Therefore, the plane of the flow is oblique to the vector $\boldsymbol{\mu}$. ◊

13.7. 3D FLOW

In this section we explore 3D flows with uniform-gradient FREs. For 3D flow there must be two linearly independent FRF components, so $|a_m^{(1)} a_m^{(2)}| > 0$. Additionally, because the flow corresponding to each component FRF is at most 2D, the two component flows must be noncoplanar; that is, $\nabla\theta^{(1)}(\omega_m)$, $\nabla Y^{(1)}(\omega_m)$, $\nabla\theta^{(2)}(\omega_m)$, and $\nabla Y^{(2)}(\omega_m)$ must be noncoplanar. This implies that at least one of the component flows is 2D. Therefore, one of the following must hold:

- Case 1: One of the FRF components corresponds to a 1D flow and the other corresponds to a 2D flow.
- Case 2: Both of the FRF components correspond to 2D flows.

13.7.1. Case 1: 1D and 2D Component FRFs

In this case one of the FRF components (say, $F^{(1)}$) corresponds to a 1D flow; it satisfies (13.29) where $d_m^{(1)}$ is a real-valued, positive, dimensionless constant that satisfies (13.30). The other FRF component, $F^{(2)}$, corresponds to a 2D flow; it satisfies (13.28) with $n = 2$. Then if the net flow is to be 3D, the gradients must satisfy

$$\left|\nabla Y^{(1)}(\omega_m) \cdot \left[\nabla Y^{(2)}(\omega_m) \times \nabla\theta^{(2)}(\omega_m)\right]\right| > 0. \quad (13.65)$$

Example 13.7 3D Flow, with 1D and 2D Component FRFs
Consider time-periodic, uniform-gradient flow in a source-free domain filled with an exponential, nonhomogeneous medium for which

$$\boldsymbol{\mu} = \left(1.5 \times 10^{-2}\text{m}^{-1}\right)\begin{bmatrix} 0 \\ \cos(\pi/3) \\ \sin(\pi/3) \end{bmatrix} \approx \begin{bmatrix} 0 \\ 7.5000 \times 10^{-3} \\ 1.2990 \times 10^{-2} \end{bmatrix} \text{m}^{-1}.$$

Suppose that for some fixed m we have $\alpha_m = 4.4182 \times 10^{-4}\text{m}^{-2}$ and two component flows (i.e., $|a_m^{(1)} a_m^{(2)}| > 0$). Furthermore, suppose

$$\hat{\mathbf{M}}^{(1)}(\omega_m) = \hat{\boldsymbol{\theta}}^{(1)}(\omega_m) = \begin{bmatrix} \cos(3\pi/8) \\ 0 \\ \sin(3\pi/8) \end{bmatrix}$$

$$\left|\nabla Y^{(1)}(\omega_m)\right| \approx 9.4797 \times 10^{-3}\text{m}^{-1} \quad (13.66)$$
$$\left|\nabla\theta^{(1)}(\omega_m)\right| \approx 1.4270 \times 10^{-2}\text{m}^{-1}$$

and

$$\nabla Y^{(2)}(\omega_m) \approx \begin{bmatrix} 2.5882 \times 10^{-3} \\ 1.5110 \times 10^{-3} \\ 9.5403 \times 10^{-3} \end{bmatrix} \text{m}^{-1} \quad (13.67\text{a})$$

$$\nabla\theta^{(2)}(\omega_m) \approx \begin{bmatrix} -5.8697 \times 10^{-3} \\ 2.2168 \times 10^{-3} \\ 1.3996 \times 10^{-2} \end{bmatrix} \text{m}^{-1}. \quad (13.67\text{b})$$

Notice that $\nabla Y^{(1)}(\omega_m)$ and $\nabla\theta^{(1)}(\omega_m)$ are collinear and $\nabla Y^{(2)}(\omega_m)$ and $\nabla\theta^{(2)}(\omega_m)$ are noncollinear. Also notice that $\nabla Y^{(1)}$ and $\nabla\theta^{(1)}$ do not lie in the same plane as $\nabla Y^{(2)}$ and $\nabla\theta^{(2)}$. Therefore, the flow is 3D. ◊

13.7.2. Case 2: 2D Component FRFs

In this case both of the FRF components correspond to 2D flows. Then if the net flow is to be 3D, the gradients must satisfy

$$\left|\left[\nabla Y^{(1)}(\omega_m) \times \nabla\theta^{(1)}(\omega_m)\right] \times \left[\nabla Y^{(2)}(\omega_m) \times \nabla\theta^{(2)}(\omega_m)\right]\right| > 0. \quad (13.68)$$

Example 13.8 3D Flow with Only 2D Component FRFs
Consider time-periodic, uniform-gradient flow in a source-free domain filled with an exponential, nonhomogeneous medium for which

$$\boldsymbol{\mu} = \left(1.5 \times 10^{-2}\text{m}^{-1}\right)\begin{bmatrix} 0 \\ \cos(\pi/5) \\ \sin(\pi/5) \end{bmatrix} \approx \begin{bmatrix} 0 \\ 8.0902 \times 10^{-2} \\ 5.8779 \times 10^{-2} \end{bmatrix} \text{m}^{-1}.$$

Suppose that for some fixed m we have $\alpha_m = 5.9588 \times 10^{-4}\text{m}^{-2}$ and two component flows (i.e., $|a_m^{(1)} a_m^{(2)}| > 0$). Furthermore, suppose

$$\nabla Y^{(1)}(\omega_m) \approx \begin{bmatrix} -9.3524 \times 10^{-4} \\ 1.1583 \times 10^{-3} \\ 1.8346 \times 10^{-4} \end{bmatrix} \text{m}^{-1} \quad (13.69\text{a})$$

$$\nabla\theta^{(1)}(\omega_m) \approx \begin{bmatrix} 7.9781 \times 10^{-3} \\ 6.4417 \times 10^{-3} \\ 1.2636 \times 10^{-3} \end{bmatrix} \text{m}^{-1} \quad (13.69\text{b})$$

and

$$\nabla Y^{(2)}(\omega_m) \approx \begin{bmatrix} 2.4497 \times 10^{-4} \\ 9.0297 \times 10^{-4} \\ 1.4302 \times 10^{-4} \end{bmatrix} \text{m}^{-1} \quad (13.70\text{a})$$

$$\nabla\theta^{(2)}(\omega_m) \approx \begin{bmatrix} -3.4728 \times 10^{-3} \\ 1.3116 \times 10^{-3} \\ 8.2809 \times 10^{-3} \end{bmatrix} \text{m}^{-1}. \quad (13.70\text{b})$$

Notice that $\nabla Y^{(1)}(\omega_m)$ and $\nabla\theta^{(1)}(\omega_m)$ are noncollinear and $\nabla Y^{(2)}(\omega_m)$ and $\nabla\theta^{(2)}(\omega_m)$ are noncollinear. Also notice that $\nabla Y^{(1)}$ and $\nabla\theta^{(1)}$ do not lie in the same plane as $\nabla Y^{(2)}$ and $\nabla\theta^{(2)}$. Therefore, the flow is 3D. ◊

13.8. IDEAL MEDIA

For ideal media, (13.5) gives $\boldsymbol{\mu} = \mathbf{0}$. Substituting $\boldsymbol{\mu} = \mathbf{0}$ in (13.12) gives

$$\left|\nabla Y^{(n)}(\omega_m)\right| = \left|\nabla\theta^{(n)}(\omega_m)\right| \quad (13.71\text{a})$$

$$\nabla Y^{(n)}(\omega_m) \cdot \nabla\theta^{(n)}(\omega_m) = \frac{\alpha_m}{2} \quad (13.71\text{b})$$

$$n = 1, 2.$$

Let $\epsilon_m^{(n)}$ denote the nonnegative angle between $\nabla Y^{(n)}(\omega_m)$ and $\nabla\theta^{(n)}(\omega_m)$. Then equations (13.71) give

$$\cos\epsilon_m^{(n)} = \frac{\alpha_m}{2\left|\nabla\theta^{(n)}(\omega_m)\right|^2}. \quad (13.72)$$

By assumption, $\alpha_m > 0$, so this implies that $0 \leq \epsilon_m^{(n)} < \pi/2$ rad. Therefore,

$$\left|\nabla\theta^{(n)}(\omega_m)\right| \geq \sqrt{\frac{\alpha_m}{2}}.$$

Example 13.9 Particular Eigenvector Relation, Ideal Media

Consider time-periodic, 2D, uniform-gradient flow in a source-free domain filled with an ideal medium. Suppose $\left|a_m^{(1)} a_m^{(2)}\right| > 0$ and the particular eigenvector relation (13.17) is satisfied. Substituting $\boldsymbol{\mu} = \mathbf{0}$ in (13.22) gives

$$\nabla Y^{(1)}(\omega_m) + \nabla Y^{(2)}(\omega_m) = \mathbf{0} \qquad (13.73a)$$

$$\nabla\theta^{(1)}(\omega_m) + \nabla\theta^{(2)}(\omega_m) = \mathbf{0}. \qquad (13.73b)$$

Combining (13.73) with (13.71) and (13.72) yields $\epsilon_m^{(1)} = \epsilon_m^{(2)}$. Regarding the dimensionality of the flow, there are two possible cases:

- Case 1: Each component's amplitude and phase gradients point in the same direction. Then because of (13.73) the other component's amplitude and phase gradients both point in the opposite direction. Consequently, the flow is 1D.
- Case 2: Each component's amplitude and phase gradients point in different directions. Condition (13.73) requires that the two amplitude gradients and the two phase gradients are coplanar. Consequently, the flow is 2D. ◊

13.9. CHAPTER SUMMARY

In this chapter we presented uniform-gradient FRF solutions for time-periodic, multidimensional flow in exponential media. Under suitable boundary conditions the hydraulic head FRF depends not only on the hydraulic diffusivity and the angular frequency but also on the gradient of the log hydraulic conductivity ($\boldsymbol{\mu}$) as well.

The FREs described in this chapter exhibit the following properties:

$$\theta^{(n)}(\mathbf{x} + \mathbf{s}; \omega_m) - \theta^{(n)}(\mathbf{x}; \omega_m) = \nabla\theta^{(n)}(\omega_m) \cdot \mathbf{s}$$

$$\frac{M^{(n)}(\mathbf{x} + \mathbf{s}; \omega_m)}{M^{(n)}(\mathbf{x}; \omega_m)} = e^{\nabla Y^{(n)}(\omega_m)\cdot\mathbf{s}}$$

$$n = 1, 2$$

where the gradients $\nabla\theta^{(n)}(\omega_m)$ and $\nabla Y^{(n)}(\omega_m)$ are space invariant. That is, the phase difference and amplitude ratio for any two points depend only on the vector difference (i.e., $\mathbf{s}$) between the two points, rather than on their absolute locations. Under these conditions, no extrema of $M^{(n)}(\mathbf{x}; \omega_m)$ or $\theta^{(n)}(\mathbf{x}; \omega_m)$ occur in the space domain. Also, for a given harmonic constituent the corresponding amplitude and phase gradients are either globally collinear or globally noncollinear.

Part IV
Essential Concepts

In this part we introduce important basic concepts in time-periodic groundwater flow. In Chapter 14 we explain attenuation and delay, and then we discuss collinearity of hydraulic head amplitude and phase gradients. Then in Chapter 15 we explore local time variation of the specific-discharge vector harmonic constituent.

Hydrodynamics of Time-Periodic Groundwater Flow: Diffusion Waves in Porous Media, Geophysical Monograph 224,
First Edition. Joe S. Depner and Todd C. Rasmussen.

14
Attenuation, Delay, and Gradient Collinearity

In this chapter we explain some fundamental concepts that are essential for general analytical work with time-periodic groundwater flow—attenuation and delay—and define basic parameters for their characterization. Then we introduce another important concept—gradient collinearity.

14.1. ATTENUATION

In general, *attenuation* (also *damping*) is the decrease of amplitude from some reference value. This section discusses attenuation of hydraulic head harmonic constituents. Section 20.5 discusses spatial attenuation in the context of hydraulic head wave propagation.

In the context of hydraulic head harmonic constituents, we generally associate attenuation with a change in position, frequency, or some other parameter such as a boundary value or source distribution.

14.1.1. Attenuation Measures

We can define both scalar and vector attenuation measures.

A basic, unitless, scalar measure of relative attenuation suitable for time-periodic groundwater flow is

$$\begin{aligned}\nu_{\mathrm{h}} &\equiv \frac{-\Delta M_{\mathrm{h}}}{M_{\mathrm{h}0}(\mathbf{x}_0;\omega_{\mathrm{p}0})}, \quad M_{\mathrm{h}0}(\mathbf{x}_0;\omega_{\mathrm{p}0}) > 0 \\ &\equiv \frac{M_{\mathrm{h}0}(\mathbf{x}_0;\omega_{\mathrm{p}0}) - M_{\mathrm{h}}(\mathbf{x};\omega_m)}{M_{\mathrm{h}0}(\mathbf{x}_0;\omega_{\mathrm{p}0})} \\ &= 1 - \frac{M_{\mathrm{h}}(\mathbf{x};\omega_m)}{M_{\mathrm{h}0}(\mathbf{x}_0;\omega_{\mathrm{p}0})} \end{aligned} \tag{14.1}$$

where

$M_{\mathrm{h}0}(\mathbf{x};\omega_m)$ denotes a reference hydraulic head amplitude field;

$\mathbf{x}_0$ is a reference location; and

$\omega_{\mathrm{p}0}$ is a reference frequency.

To characterize the relative attenuation due solely to a change in position at a fixed frequency ω_m (i.e., the *spatial attenuation*), set $\omega_{\mathrm{p}0} = \omega_m$ and $M_{\mathrm{h}0} \equiv M_{\mathrm{h}}$. To characterize the relative attenuation due solely to a change in frequency at a fixed position $\mathbf{x}$ (i.e., the *frequency attenuation*), set $\mathbf{x}_0 = \mathbf{x}$ and $M_{\mathrm{h}0} \equiv M_{\mathrm{h}}$. Finally, to characterize the relative attenuation due to a change in some other parameter at a fixed frequency ω_m and a fixed position $\mathbf{x}$, set $\mathbf{x}_0 = \mathbf{x}$, $\omega_{\mathrm{p}0} = \omega_m$, and $M_{\mathrm{h}0}$ equal to the amplitude field corresponding to the other parameter value.

We can also express the relative attenuation in terms of dimensionless units, such as percent (%) or decibels (dB). We define *percent attenuation* as

$$\nu_{\mathrm{h}}^{(\%)} \equiv \left[1 - \frac{M_{\mathrm{h}}(\mathbf{x};\omega_m)}{M_{\mathrm{h}0}(\mathbf{x}_0;\omega_{\mathrm{p}0})}\right] \cdot 100\%. \tag{14.2}$$

We define *decibel attenuation* as

$$\nu_{\mathrm{h}}^{(\mathrm{dB})} \equiv -20\,\mathrm{dB} \cdot \log_{10}\left[\frac{M_{\mathrm{h}}(\mathbf{x};\omega_m)}{M_{\mathrm{h}0}(\mathbf{x}_0;\omega_{\mathrm{p}0})}\right]. \tag{14.3}$$

For example, if the amplitude ratio ($M_{\mathrm{h}}/M_{\mathrm{h}0}$) is 0.25, then the relative attenuation is 0.75 (75%), or 12 dB.

Exercise 14.1 Numerical Ranges for Attenuation Measures

Problem: Give ranges for the relative attenuation, percent attenuation, and decibel attenuation for the following two cases:

Hydrodynamics of Time-Periodic Groundwater Flow: Diffusion Waves in Porous Media, Geophysical Monograph 224,
First Edition. Joe S. Depner and Todd C. Rasmussen.

1. The amplitude may be less than, equal to, or greater than the reference amplitude.
2. The amplitude is less than or equal to the reference amplitude.

Solution:

Case 1:

$$-\infty \le \nu_{\mathrm{h}} \le 1$$
$$-\infty \le \nu_{\mathrm{h}}^{(\%)} \le 100\%$$
$$-\infty \le \nu_{\mathrm{h}}^{(\mathrm{dB})} \le \infty.$$

Case 2:

$$0 \le \nu_{\mathrm{h}} \le 1$$
$$0 \le \nu_{\mathrm{h}}^{(\%)} \le 100\%$$
$$0 \le \nu_{\mathrm{h}}^{(\mathrm{dB})} \le \infty.$$

□

As (14.1) shows, the attenuation is a monotone decreasing function of the amplitude M_{h}. Therefore, for a given harmonic constituent (i.e., $\omega_{\mathrm{p}0} = \omega_m =$ constant), if the reference amplitude is independent of position, then amplitude maxima and minima correspond to attenuation minima and maxima, respectively.

We define the following measures of the local spatial attenuation rate:

Dimensionless attenuation rate: $-\nabla M_{\mathrm{h}}$ (dimensionless).

Relative attenuation rate: $-\nabla Y_{\mathrm{h}} = \frac{-\nabla M_{\mathrm{h}}}{M_{\mathrm{h}}}$, $M_{\mathrm{h}} > 0$ (dimensions: L^{-1}).

Spatial attenuation scale: $\delta_{\mathrm{h}} \equiv \frac{1}{|\nabla Y_{\mathrm{h}}|}$, $|\nabla Y_{\mathrm{h}}| > 0$ (dimensions: L).

14.2. DELAY

Consider a single hydraulic head harmonic constituent, as per (4.3a):

$$h'(\mathbf{x}, t; \omega_m) = M_{\mathrm{h}}(\mathbf{x}; \omega_m) \cos\left[\omega_m t + \theta_{\mathrm{h}}(\mathbf{x}; \omega_m)\right]. \quad (14.4)$$

For a fixed position $\mathbf{x}$, the unit-amplitude constituent (i.e., the cosine term) is a function of time that we can think of as having been shifted from $\omega_m t$ to $\omega_m\left[t + \tau_{\mathrm{h}}(\mathbf{x}; \omega_m)\right]$, where

$$\tau_{\mathrm{h}}(\mathbf{x}; \omega_m) = \frac{\theta_{\mathrm{h}}(\mathbf{x}; \omega_m)}{\omega_m}. \quad (14.5)$$

That is, the *time shift* τ_{h} of the time-periodic variation, relative to the time origin, is directly proportional to the phase. When $\tau_{\mathrm{h}} > 0$, we say the periodic variation is *advanced*; when $\tau_{\mathrm{h}} < 0$, we say it is *delayed* or *retarded*. Equivalently, we define the *time delay* (also *delay*) as $-\tau_{\mathrm{h}}$. Consequently, a maximum in the phase corresponds to a minimum in the delay and vice versa.

14.3. PHASE SHIFT

Phase shift is a change in the phase from a more or less arbitrarily defined reference value. We generally associate phase shift with a change in position, frequency, or some other parameter such as a boundary value or source distribution. We can define both scalar and vector phase shift measures, which typically are dimensionless.

A very basic, general definition of phase shift is

$$\Delta\theta_{\mathrm{h}} \equiv \theta_{\mathrm{h}}(\mathbf{x}; \omega_m) - \theta_{\mathrm{h}0}(\mathbf{x}_0; \omega_{\mathrm{p}0}) \quad (14.6)$$

where

$\theta_{\mathrm{h}0}(\mathbf{x}; \omega_m)$ is a reference phase field;
$\mathbf{x}_0$ is a reference location; and
$\omega_{\mathrm{p}0}$ is a reference frequency.

To characterize the phase shift due solely to a change in position at a fixed frequency ω_m, set $\omega_{\mathrm{p}0} = \omega_m$ and $\theta_{\mathrm{h}0} \equiv \theta_{\mathrm{h}}$. Alternatively, to characterize the phase shift due solely to a change in frequency at a fixed position $\mathbf{x}$, set $\mathbf{x}_0 = \mathbf{x}$ and $\theta_{\mathrm{h}0} \equiv \theta_{\mathrm{h}}$. Finally, to characterize the attenuation due to a change in some other parameter at a fixed position $\mathbf{x}$ and a fixed frequency ω_m, set $\mathbf{x}_0 = \mathbf{x}$, $\omega_{\mathrm{p}0} = \omega_m$, and $\theta_{\mathrm{h}0}$ equal to the phase field corresponding to the other parameter value.

Exercise 14.2 Gradient of Logarithm of FRF

Problem: Consider the natural logarithm of the hydraulic head FRF. Express the real and imaginary parts of its gradient in terms of the relative attenuation vector and the phase gradient.

Solution: Using calculus to expand the gradient of the logarithm,

$$\begin{aligned}\nabla \ln F_{\mathrm{h}}(\mathbf{x}; \omega_m) &= \frac{\nabla F_{\mathrm{h}}}{F_{\mathrm{h}}} = \frac{\nabla\left[M_{\mathrm{h}} e^{i\theta_{\mathrm{h}}}\right]}{M_{\mathrm{h}} e^{i\theta_{\mathrm{h}}}} \\ &= \frac{\nabla M_{\mathrm{h}} e^{i\theta_{\mathrm{h}}} + \mathrm{i} M_{\mathrm{h}} e^{i\theta_{\mathrm{h}}} \nabla\theta_{\mathrm{h}}}{M_{\mathrm{h}} e^{i\theta_{\mathrm{h}}}} \\ &= \frac{\nabla M_{\mathrm{h}}}{M_{\mathrm{h}}} + i\nabla\theta_{\mathrm{h}} \\ &= \nabla Y_{\mathrm{h}}(\mathbf{x}; \omega_m) + i\nabla\theta_{\mathrm{h}}(\mathbf{x}; \omega_m). \end{aligned} \quad (14.7)$$

where Y_{h} and θ_{h} are real valued, so ∇Y_{h} and $\nabla\theta_{\mathrm{h}}$ are the real and imaginary parts, respectively, of $\nabla \ln F_{\mathrm{h}}$. □

Exercise 14.3 Gradients for 1D Flow in Semi-Infinite Domain

Problem: Consider time-periodic, 1D flow in a semi-infinite, source-free domain filled with an exponential medium. Suppose the flow parallels the x coordinate axis. Use the results of Exercise 14.2 to express the relative attenuation vector and the phase gradient in terms of the real and imaginary parts of the eigenvalue $\lambda_m^{(2)}$.

Solution: Taking the gradient of (10.74) and rearranging give

$$\begin{aligned}\nabla \ln F_{\mathrm{h}}(\mathbf{x};\omega_m) &= \nabla \ln G(x_1;\omega_m)\\ &= \lambda_m^{(2)}\hat{\mathbf{x}} \qquad (14.8)\\ &= \mathrm{Re}\,\lambda_m^{(2)}\hat{\mathbf{x}} + i\mathrm{Im}\,\lambda_m^{(2)}\hat{\mathbf{x}}.\end{aligned}$$

Comparing the RHS to that of (14.7) leads to

$$\begin{aligned}\nabla Y_{\mathrm{h}}(\mathbf{x};\omega_m) &= \mathrm{Re}\,\lambda_m^{(2)}\hat{\mathbf{x}}\\ \nabla\theta_{\mathrm{h}}(\mathbf{x};\omega_m) &= \mathrm{Im}\,\lambda_m^{(2)}\hat{\mathbf{x}}.\end{aligned}$$

□

14.4. IN-PHASE AND QUADRATURE COMPONENTS

Consider a general homogeneous linear operation (L) on the hydraulic head harmonic constituent. The result of such an operation is a harmonic constituent of the same frequency as the operand but whose amplitude and phase generally differ from those of the operand:

$$L\left[h'(\mathbf{x},t;\omega_m)\right] = M_{\mathrm{L}}(\mathbf{x};\omega_m)\cos\left[\omega_m t + \theta_{\mathrm{L}}(\mathbf{x};\omega_m)\right].$$

Rewrite the RHS as a phase-shifted version of the operand's sinusoid:

$$L\left[h'(\mathbf{x},t;\omega_m)\right] = M_{\mathrm{L}}(\mathbf{x};\omega_m)\cos\left[\omega_m t + \theta_{\mathrm{h}}(\mathbf{x};\omega_m) + \Delta\theta(\mathbf{x};\omega_m)\right] \qquad (14.9)$$

where the phase shift is

$$\Delta\theta(\mathbf{x};\omega_m) \equiv \theta_{\mathrm{L}}(\mathbf{x};\omega_m) - \theta_{\mathrm{h}}(\mathbf{x};\omega_m).$$

Next we show how one can resolve $L\left[h'(\mathbf{x},t;\omega_m)\right]$ into components that are in phase and out of phase with respect to the operand. Using trigonometric identity (B.2a) to rewrite the cosine term on the RHS of (14.9) yields

$$\begin{aligned}L\left[h'(\mathbf{x},t;\omega_m)\right] &= A_{\mathrm{L}}(\mathbf{x};\omega_m)\cos\left[\omega_m t + \theta_{\mathrm{h}}(\mathbf{x};\omega_m)\right]\\ &\quad - B_{\mathrm{L}}(\mathbf{x};\omega_m)\sin\left[\omega_m t + \theta_{\mathrm{h}}(\mathbf{x};\omega_m)\right] \qquad (14.10)\end{aligned}$$

where we define the amplitude coefficients as

$$A_{\mathrm{L}}(\mathbf{x};\omega_m) \equiv M_{\mathrm{L}}(\mathbf{x};\omega_m)\cos\Delta\theta(\mathbf{x};\omega_m) \qquad (14.11a)$$

$$B_{\mathrm{L}}(\mathbf{x};\omega_m) \equiv M_{\mathrm{L}}(\mathbf{x};\omega_m)\sin\Delta\theta(\mathbf{x};\omega_m). \qquad (14.11b)$$

Solving these for the amplitude (M_{L}) and phase (θ_{L}) leads to

$$M_{\mathrm{L}}(\mathbf{x};\omega_m) = \sqrt{A_{\mathrm{L}}^2(\mathbf{x};\omega_m) + B_{\mathrm{L}}^2(\mathbf{x};\omega_m)} \qquad (14.12a)$$

$$\theta_{\mathrm{L}}(\mathbf{x};\omega_m) = \arctan\frac{B_{\mathrm{L}}(\mathbf{x};\omega_m)}{A_{\mathrm{L}}(\mathbf{x};\omega_m)} + \theta_{\mathrm{h}}(\mathbf{x};\omega_m). \qquad (14.12b)$$

The cosine term on the RHS of (14.10) has the same phase as the operand, so we call it the *in-phase* component:

$$\left\{L\left[h'(\mathbf{x},t;\omega_m)\right]\right\}_{\mathrm{Ip}} \equiv A_{\mathrm{L}}(\mathbf{x};\omega_m)\cos\left[\omega_m t + \theta_{\mathrm{h}}(\mathbf{x};\omega_m)\right]. \qquad (14.13)$$

The sine term on the RHS of (14.10) leads the operand by $\pi/2$ rad, so we call it the *quadrature* (or *out-of-phase*) component:

$$\begin{aligned}\left\{L[h'(\mathbf{x},t;\omega_m)]\right\}_{\mathrm{Qu}} &\equiv -B_{\mathrm{L}}(\mathbf{x};\omega_m)\sin\left[\omega_m t + \theta_{\mathrm{h}}(\mathbf{x};\omega_m)\right]\\ &= B_{\mathrm{L}}(\mathbf{x};\omega_m)\cos\left[\omega_m t + \theta_{\mathrm{h}}(\mathbf{x};\omega_m) + \pi/2\right]. \qquad (14.14)\end{aligned}$$

Similar results hold when the result of the linear operation is a vector rather than a scalar; Exercises 14.6 and 14.7 provide relevant examples.

Exercise 14.4 Space Differentiation: In-Phase and Quadrature Components

Problem: Consider the linear operation

$$L\left[h'(\mathbf{x},t;\omega_m)\right] = \frac{\partial}{\partial x_p}h'(\mathbf{x},t;\omega_m). \qquad (14.15)$$

Resolve the result of the linear operation into components that are in phase and in quadrature with respect to the phase of the operand and identify the *in-phase* and quadrature components. Express the coefficients (A_{L}, B_{L}), amplitude (M_{L}), and phase (θ_{L}) in terms of the amplitude and phase of the operand (M_{h}, θ_{h}).

Solution: Differentiating (4.3a) with respect to x_p leads to

$$\begin{aligned}\frac{\partial}{\partial x_p}h'(\mathbf{x},t;\omega_m) &= \frac{\partial M_{\mathrm{h}}}{\partial x_p}\cos\left[\omega_m t + \theta_{\mathrm{h}}\right]\\ &\quad - M_{\mathrm{h}}\frac{\partial\theta_{\mathrm{h}}}{\partial x_p}\sin\left[\omega_m t + \theta_{\mathrm{h}}\right]\end{aligned}$$

so the in-phase and quadrature components, respectively, are

$$\left[\frac{\partial}{\partial x_p}h'(\mathbf{x},t;\omega_m)\right]_{\mathrm{Ip}} = \frac{\partial M_{\mathrm{h}}}{\partial x_p}\cos\left[\omega_m t + \theta_{\mathrm{h}}\right]$$

$$\left[\frac{\partial}{\partial x_p}h'(\mathbf{x},t;\omega_m)\right]_{\mathrm{Qu}} = -M_{\mathrm{h}}\frac{\partial\theta_{\mathrm{h}}}{\partial x_p}\sin\left[\omega_m t + \theta_{\mathrm{h}}\right].$$

Therefore

$$A_{\mathrm{L}}(\mathbf{x};\omega_m) = \frac{\partial M_{\mathrm{h}}}{\partial x_p} \qquad (14.16a)$$

$$B_{\mathrm{L}}(\mathbf{x};\omega_m) = M_{\mathrm{h}}\frac{\partial\theta_{\mathrm{h}}}{\partial x_p}. \qquad (14.16b)$$

Substituting the RHSs of these equations for the corresponding terms of (14.12) yields

$$M_L(\mathbf{x};\omega_m) = \sqrt{\left(\frac{\partial M_h}{\partial x_p}\right)^2 + M_h^2\left(\frac{\partial \theta_h}{\partial x_p}\right)^2} \quad (14.17a)$$

$$\theta_L(\mathbf{x};\omega_m) = \arctan\frac{M_h \partial\theta_h/\partial x_p}{\partial M_h/\partial x_p} + \theta_h. \quad (14.17b)$$

□

Exercise 14.5 Time Differentiation: In-Phase and Quadrature Components

Problem: Consider the linear operation

$$L\left[h'(\mathbf{x},t;\omega_m)\right] = \frac{\partial}{\partial t}h'(\mathbf{x},t;\omega_m).$$

Resolve the result of the linear operation into components that are in phase and in quadrature with respect to the phase of the operand and identify the in-phase and quadrature components. Express the coefficients (A_L, B_L), amplitude (M_L), and phase (θ_L) in terms of the amplitude and phase of the operand (M_h, θ_h).

Solution: Differentiating the hydraulic head harmonic constituent (4.3a) with respect to time gives

$$\frac{\partial h'}{\partial t} = -\omega_m M_h(\mathbf{x};\omega_m)\sin\left[\omega_m t + \theta_h\right]$$

so the in-phase and quadrature components, respectively, are

$$\left(\frac{\partial h'}{\partial t}\right)_{\mathrm{Ip}} = 0$$

$$\left(\frac{\partial h'}{\partial t}\right)_{\mathrm{Qu}} = -\omega_m M_h\sin\left[\omega_m t + \theta_h\right].$$

Therefore

$$A_L(\mathbf{x};\omega_m) = 0$$

$$B_L(\mathbf{x};\omega_m) = \omega_m M_h(\mathbf{x};\omega_m).$$

Substituting the RHSs of these equations for the corresponding terms of (14.12) yields

$$M_L(\mathbf{x};\omega_m) = \omega_m M_h(\mathbf{x};\omega_m) \quad (14.18a)$$

$$\theta_L(\mathbf{x};\omega_m) = \theta_h(\mathbf{x};\omega_m) + \pi/2 \text{ rad}. \quad (14.18b)$$

□

Exercise 14.6 In-Phase/Quadrature Components of Hydraulic Gradient

Problem: Show that one can resolve each hydraulic gradient vector harmonic constituent into component vectors that are in phase and in quadrature with respect to the corresponding hydraulic head harmonic constituent and give expressions for the component vectors.

Solution: Taking the gradient of the hydraulic head harmonic constituent (4.3a) gives the hydraulic gradient harmonic constituent,

$$(\nabla h')(\mathbf{x},t;\omega_m) = (\nabla h')_{\mathrm{Ip}}(\mathbf{x},t;\omega_m) + (\nabla h')_{\mathrm{Qu}}(\mathbf{x},t;\omega_m) \quad \forall\, \mathbf{x} \in D_+(\omega_m) \quad (14.19)$$

where we define the in-phase and quadrature component vectors as

$$(\nabla h')_{\mathrm{Ip}}(\mathbf{x},t;\omega_m) \equiv \nabla M_h(\mathbf{x};\omega_m)\cos\left[\omega_m t + \theta_h(\mathbf{x};\omega_m)\right] \quad (14.20a)$$

$$(\nabla h')_{\mathrm{Qu}}(\mathbf{x},t;\omega_m) \equiv -M_h(\mathbf{x};\omega_m)\nabla\theta_h(\mathbf{x};\omega_m)\sin\left[\omega_m t + \theta_h(\mathbf{x};\omega_m)\right] \quad (14.20b)$$

$$\forall\, \mathbf{x} \in D_+(\omega_m).$$

Recall that these same results were obtained earlier [see (5.27)] in a different context. □

Exercise 14.7 In-Phase/Quadrature Components of Specific Discharge

Problem: Show that one can resolve each specific-discharge vector harmonic constituent into component vectors that are in phase and in quadrature with respect to the corresponding hydraulic head harmonic constituent and give expressions for the component vectors.

Solution: Using the results of Exercise 14.6 and Darcy's law, we can write the specific-discharge vector harmonic constituent as

$$\mathbf{q}'(\mathbf{x},t;\omega_m) = \mathbf{q}'_{\mathrm{Ip}}(\mathbf{x},t;\omega_m) + \mathbf{q}'_{\mathrm{Qu}}(\mathbf{x},t;\omega_m) \quad \forall\, \mathbf{x} \in D_+(\omega_m) \quad (14.21)$$

where we define the in-phase and quadrature component vectors as

$$\mathbf{q}'_{\mathrm{Ip}}(\mathbf{x},t;\omega_m) \equiv -\mathbf{K}(\mathbf{x})\nabla M_h(\mathbf{x};\omega_m)\cos\left[\omega_m t + \theta_h(\mathbf{x};\omega_m)\right] \quad (14.22a)$$

$$\mathbf{q}'_{\mathrm{Qu}}(\mathbf{x},t;\omega_m) \equiv M_h(\mathbf{x};\omega_m)\mathbf{K}(\mathbf{x})\nabla\theta_h(\mathbf{x};\omega_m)\sin\left[\omega_m t + \theta_h(\mathbf{x};\omega_m)\right] \quad (14.22b)$$

$$\forall\, \mathbf{x} \in D_+(\omega_m).$$

□

Exercise 14.8 No-Flow Condition: In-Phase and Quadrature Components

Problem: Suppose that for some harmonic constituent (i.e., ω_m fixed) the point $\mathbf{x} = \mathbf{x}_{\mathrm{nf}}$ lies on a no-flow boundary:

$$\mathbf{q}'(\mathbf{x}_{\mathrm{nf}}, t; \omega_m) \cdot \hat{\mathbf{n}}(\mathbf{x}_{\mathrm{nf}}) = 0 \quad \forall\, t \tag{14.23}$$

where $\mathbf{x}_{\mathrm{nf}} \in \Gamma_+(\omega_m)$ and $\hat{\mathbf{n}}(\mathbf{x})$ denotes the unit vector that is outward normal to the boundary at the point $\mathbf{x}$. What can you infer about the vectors $\mathbf{K}\nabla\theta_{\mathrm{h}}$, $\mathbf{K}\nabla M_{\mathrm{h}}$, $\mathbf{q}'_{\mathrm{Ip}}$, and $\mathbf{q}'_{\mathrm{Qu}}$ at the point $\mathbf{x} = \mathbf{x}_{\mathrm{nf}}$?

Solution: Combining (14.21) and (14.22) with (14.23) gives

$$\begin{aligned}&-\left[\mathbf{K}(\mathbf{x}_{\mathrm{nf}})\nabla M_{\mathrm{h}}(\mathbf{x}_{\mathrm{nf}}; \omega_m)\right] \cdot \hat{\mathbf{n}}(\mathbf{x}_{\mathrm{nf}}) \cos\left[\omega_m t + \theta_{\mathrm{h}}(\mathbf{x}_{\mathrm{nf}}; \omega_m)\right] \\ &+ M_{\mathrm{h}}(\mathbf{x}_{\mathrm{nf}}; \omega_m)\left[\mathbf{K}(\mathbf{x}_{\mathrm{nf}})\nabla\theta_{\mathrm{h}}(\mathbf{x}_{\mathrm{nf}}; \omega_m)\right] \\ &\cdot \hat{\mathbf{n}}(\mathbf{x}_{\mathrm{nf}}) \sin\left[\omega_m t + \theta_{\mathrm{h}}(\mathbf{x}_{\mathrm{nf}}; \omega_m)\right] = 0 \quad \forall\, t.\end{aligned}$$

The LHS is a trigonometric series. Using the results derived in Section B.2 (see, e.g., Section B.2.2) to evaluate the series coefficients leads to

$$\left[\mathbf{K}(\mathbf{x}_{\mathrm{nf}})\nabla M_{\mathrm{h}}(\mathbf{x}_{\mathrm{nf}}; \omega_m)\right] \cdot \hat{\mathbf{n}}(\mathbf{x}_{\mathrm{nf}}) = 0 \tag{14.24a}$$

$$\left[\mathbf{K}(\mathbf{x}_{\mathrm{nf}})\nabla\theta_{\mathrm{h}}(\mathbf{x}_{\mathrm{nf}}; \omega_m)\right] \cdot \hat{\mathbf{n}}(\mathbf{x}_{\mathrm{nf}}) = 0. \tag{14.24b}$$

Combining these results with (14.22) yields

$$\mathbf{q}'_{\mathrm{Ip}}(\mathbf{x}_{\mathrm{nf}}, t; \omega_m) \cdot \hat{\mathbf{n}}(\mathbf{x}_{\mathrm{nf}}) = 0 \tag{14.25a}$$

$$\mathbf{q}'_{\mathrm{Qu}}(\mathbf{x}_{\mathrm{nf}}, t; \omega_m) \cdot \hat{\mathbf{n}}(\mathbf{x}_{\mathrm{nf}}) = 0 \tag{14.25b}$$

$$\forall\, t.$$

Thus, every no-flow boundary for the vector harmonic constituent $\mathbf{q}'$ also is a no-flow boundary for the corresponding in-phase and quadrature component vectors $\mathbf{q}'_{\mathrm{Ip}}$ and $\mathbf{q}'_{\mathrm{Qu}}$. □

14.5. COLLINEARITY OF AMPLITUDE AND PHASE GRADIENTS

Suppose that for some harmonic constituent (i.e., ω_m fixed) and some fixed location $\mathbf{x}$ the corresponding amplitude and phase gradients are collinear (see Note 13.1):

$$\left|\nabla M_{\mathrm{h}}(\mathbf{x}; \omega_m) \cdot \nabla\theta_{\mathrm{h}}(\mathbf{x}; \omega_m)\right| = \left|\nabla M_{\mathrm{h}}(\mathbf{x}; \omega_m)\right|\left|\nabla\theta_{\mathrm{h}}(\mathbf{x}; \omega_m)\right|.$$

Then we say that the amplitude and phase gradients for that constituent are *locally collinear* at the point $\mathbf{x}$. If the gradients are collinear throughout the space domain [i.e., $\forall\, \mathbf{x} \in D_+(\omega_m)$], then we say that the gradients are *globally collinear*.

Collinearity of a constituent's amplitude and phase gradients is relevant to both the classification of FRF solutions and the local time variation of the specific-discharge vector harmonic constituent (see Chapter 15).

In all of the examples presented in Chapters 9–12, and some of the examples presented in Chapter 13, for each constituent the corresponding amplitude and phase gradients are globally collinear. In contrast, in some of the examples presented in Chapter 13, for each constituent the corresponding amplitude and phase gradients are globally noncollinear. Under other conditions (e.g., see Example 14.1), a constituent's amplitude and phase gradients may be neither globally collinear nor globally noncollinear.

The examples of uniform-gradient FRF solutions with globally noncollinear gradients presented in Chapter 13 suggest that noncollinear gradients can result in part from particular configurations of the boundary conditions. And as Example 14.1 illustrates, particular source configurations can result in flows with locally noncollinear gradients.

Example 14.1 Noncollinear Amplitude/Phase Gradients: Point Sources
Consider time-periodic, 3D flow in an ideal medium of infinite extent due to three single-constituent point sources. The sources, which are individually identified by an integer index j ($j = 1, 2, 3$), have the same constituent frequency (ω_m; m fixed).

Define the dimensionless space coordinate and dimensionless FRF, respectively, as

$$\mathbf{X} \equiv \sqrt{\alpha_m}\mathbf{x} \tag{14.26a}$$

$$F_{\mathrm{D}}(\mathbf{X}) \equiv \sqrt{\alpha_m} F_{\mathrm{h}}(\mathbf{x}; \omega_m). \tag{14.26b}$$

Let $\mathbf{x}^{(j)}$, $F_{\mathrm{h}}^{(j)}(\mathbf{x}; \omega_m)$, and $Q^{(j)}(\omega_m)$ denote the space coordinates, hydraulic head FRF, and complex amplitude (strength), respectively, of the jth point source. Define corresponding dimensionless variables as follows:

$$\mathbf{X}^{(j)} \equiv \sqrt{\alpha_m}\mathbf{x}^{(j)} \tag{14.27a}$$

$$F_{\mathrm{D}}^{(j)}(\mathbf{X}) \equiv \sqrt{\alpha_m} F_{\mathrm{h}}^{(j)}(\mathbf{x}; \omega_m) \tag{14.27b}$$

$$Q_{\mathrm{D}}^{(j)} \equiv \frac{\alpha_m Q^{(j)}(\omega_m)}{K}. \tag{14.27c}$$

Using these we can express the dimensionless FRF for the jth point source (12.16) in terms of dimensionless quantities as follows:

$$F_{\mathrm{D}}^{(j)}(\mathbf{X}) \equiv \frac{Q_{\mathrm{D}}^{(j)}}{4\pi} \frac{e^{-\sqrt{i}|\mathbf{X}-\mathbf{X}^{(j)}|}}{|\mathbf{X}-\mathbf{X}^{(j)}|}, \quad |\mathbf{X}-\mathbf{X}^{(j)}| > 0; \quad j = 1, 2, 3. \tag{14.28}$$

By superposition,

$$F_{\mathrm{D}}(\mathbf{X}) = \sum_{j=1}^{3} F_{\mathrm{D}}^{(j)}(\mathbf{X})$$

where (14.28) gives the point FRFs $F_{\mathrm{D}}^{(j)}$.

Suppose the three point sources are situated at the vertices of an equilateral triangle in the horizontal ($x_3 = 0$) plane, with the dimensionless coordinates and

Table 14.1 Point-source data (dimensionless) for Example 14.1.

j	$x_1^{(j)}$	$x_2^{(j)}$	$x_3^{(j)}$	$Q_D^{(j)}$
1	$\frac{-1}{2}$	$\frac{-\sqrt{3}}{4}$	0	1
2	$\frac{1}{2}$	$\frac{-\sqrt{3}}{4}$	0	$e^{i\pi/2}$
3	0	$\frac{\sqrt{3}}{4}$	0	$\frac{1}{2}$

complex amplitudes given by Table 14.1. Because of symmetry, at all points in this plane the FRF amplitude and phase gradients are horizontally oriented vectors. For at least some of the points in this plane (e.g., the origin), the FRF amplitude and phase gradients are noncollinear. ◊

In Note 14.1 we relate the collinearity of the specific-discharge constituent's in-phase and quadrature components to the collinearity of the corresponding amplitude and phase gradients.

Note 14.1 A Vector Product Equivalence

In this note we show that the question

Are the two vectors $\mathbf{K}\nabla M_h$ and $\mathbf{K}\nabla\theta_h$ collinear or noncollinear?

is equivalent to

Are the two vectors ∇M_h and $\nabla\theta_h$ collinear or noncollinear?

That is, we establish the following equivalence:

$$(\mathbf{K}\nabla M_h) \times (\mathbf{K}\nabla\theta_h) = \mathbf{0} \quad \text{if and only if} \quad \nabla M_h \times \nabla\theta_h = \mathbf{0}. \tag{14.29}$$

Recall that $\mathbf{K}$ is a real-valued, 3×3, nonsingular, positive-definite matrix and ∇M_h and $\nabla\theta_h$ are nonzero, real-valued, column 3-vectors. Substituting $\mathbf{R} \equiv \mathbf{K}$, $\mathbf{a} \equiv \nabla M_h$, and $\mathbf{b} \equiv \nabla\theta_h$ in identity (c.17) and noting that $\mathbf{K}$ is symmetric, we obtain

$$(\mathbf{K}\nabla M_h) \times (\mathbf{K}\nabla\theta_h) = (\det \mathbf{K})\mathbf{K}^{-1}(\nabla M_h \times \nabla\theta_h). \tag{14.30}$$

Note that $|\det \mathbf{K}| > 0$ because $\mathbf{K}$ is nonsingular.

First we show that the first equation of (14.29) implies the second. If $(\mathbf{K}\nabla M_h) \times (\mathbf{K}\nabla\theta_h) = \mathbf{0}$, then (14.30) gives $\mathbf{0} = (\det \mathbf{K})\mathbf{K}^{-1}(\nabla M_h \times \nabla\theta_h)$. Premultiplying by $\mathbf{K}$ then yields $\mathbf{0} = \nabla M_h \times \nabla\theta_h$. Thus, the first equation of (14.29) implies the second.

Next we show that the second equation of (14.29) implies the first. If $\nabla M_h \times \nabla\theta_h = \mathbf{0}$, then (14.30) gives $(\mathbf{K}\nabla M_h) \times (\mathbf{K}\nabla\theta_h) = \mathbf{0}$. Thus, the second equation of (14.29) implies the first. Therefore, the equivalence (14.29) is established. △

Exercise 14.9 Globally Collinear Gradients in Ideal Media

Problem: Consider time-periodic flow with a single harmonic constituent (i.e., ω_m fixed) in a source-free domain filled with an ideal medium. Show that if the amplitude and phase gradients are globally collinear, then we can express the polar form governing equations (6.11) as

$$|\nabla\theta_h(\mathbf{x};\omega_m)| = \sqrt{\nabla^2 Y_h(\mathbf{x};\omega_m) + |\nabla Y_h(\mathbf{x};\omega_m)|^2} \tag{14.31a}$$

$$|\nabla Y_h(\mathbf{x};\omega_m)| = \frac{|\alpha_m - \nabla^2\theta_h(\mathbf{x};\omega_m)|}{2|\nabla\theta_h(\mathbf{x};\omega_m)|} \tag{14.31b}$$

$$\forall\, \mathbf{x} \in D_+(\omega_m); \quad m = 1, 2, 3, \ldots$$

where $Y_h \equiv \ln M_h$.

Solution: Substitute $\mathbf{K}(\mathbf{x}) = K\mathbf{I}$, $K = \text{constant}$, $S_s = \text{constant}$, and $M_u = 0$ in (6.11) and use the definition of α_m (5.31). Using algebra and calculus we obtain

$$\nabla^2 M_h(\mathbf{x};\omega_m) - M_h(\mathbf{x};\omega_m)|\nabla\theta_h(\mathbf{x};\omega_m)|^2 = 0 \tag{14.32a}$$

$$\nabla \cdot [M_h^2(\mathbf{x};\omega_m)\nabla\theta_h(\mathbf{x};\omega_m)] - \alpha_m M_h^2(\mathbf{x};\omega_m) = 0. \tag{14.32b}$$

Solving (14.32a) for $|\nabla\theta_h|$ gives

$$|\nabla\theta_h(\mathbf{x};\omega_m)| = \sqrt{\frac{\nabla^2 M_h}{M_h}}.$$

Substituting $M_h = \exp Y_h$ and then simplifying the result yield (14.31a). Expanding the LHS of (14.32b) leads to

$$2M_h\nabla M_h \cdot \nabla\theta_h + M_h^2\nabla^2\theta_h - \alpha_m M_h^2 = 0. \tag{14.33}$$

The gradients are collinear, so $\nabla M_h \cdot \nabla\theta_h = \pm|\nabla M_h||\nabla\theta_h|$. Substituting this result in (14.33) and then dividing by M_h^2 produce

$$\pm 2|\nabla Y_h||\nabla\theta_h| + \nabla^2\theta_h - \alpha_m = 0.$$

Solving for $|\nabla Y_h|$ then gives (14.31b). □

The flows described in Exercise 14.9 are noteworthy in that if one knows either the log amplitude or the phase throughout the space domain, then in principle one can integrate one of equations (14.31) to infer the other to within an unknown additive constant.

14.6. CHAPTER SUMMARY

In this chapter we explained the concepts of attenuation and delay in the context of time-periodic flow. We defined scalar and vector measures of attenuation and related the concept of delay to those of in-phase versus quadrature components of the harmonic constituents. We also discussed collinearity of the harmonic constituent's amplitude and phase gradients. We defined local and global collinearity and listed examples of elementary FRF solutions in which the constituent amplitude and phase gradients are globally collinear, globally noncollinear, or neither. Finally, we showed that the question of whether the in-phase and quadrature components of a particular specific-discharge constituent are collinear is mathematically equivalent to the question of whether the corresponding hydraulic head constituent's amplitude and phase gradients are collinear, even in a nonuniformly anisotropic medium.

15

Time Variation of Specific-Discharge Constituent

In this chapter we explore the time variation of the specific-discharge harmonic constituent, $\mathbf{q}'(\mathbf{x}, t; \omega_m)$.

15.1. PERIODICITY OF SPECIFIC-DISCHARGE CONSTITUENT

Consider the specific-discharge constituent (14.21) with the in-phase and quadrature components given by (14.22). Combining these gives

$$\begin{aligned} \mathbf{q}'(\mathbf{x}, t; \omega_m) = & -\mathbf{K}(\mathbf{x})\nabla M_\mathrm{h}(\mathbf{x}; \omega_m) \cos \xi_\mathrm{h}(\mathbf{x}, t; \omega_m) \\ & + M_\mathrm{h}(\mathbf{x}; \omega_m)\mathbf{K}(\mathbf{x})\nabla\theta_\mathrm{h}(\mathbf{x}; \omega_m) \\ & \times \sin \xi_\mathrm{h}(\mathbf{x}, t; \omega_m) \quad \forall\, \mathbf{x} \in D_+(\omega_m) \end{aligned} \tag{15.1}$$

where

$$\xi_\mathrm{h}(\mathbf{x}, t; \omega_m) \equiv \omega_m t + \theta_\mathrm{h}(\mathbf{x}; \omega_m). \tag{15.2}$$

Equations (15.1) and (15.2) show that for a given frequency ω_m and a fixed location $\mathbf{x}$ the in-phase component of the specific-discharge harmonic constituent varies continuously and sinuoidally in time between $\pm\mathbf{K}\nabla M_\mathrm{h}$ and at all times points in either the same or the opposite direction as $\mathbf{K}\nabla M_\mathrm{h}$ or it vanishes. Similarly, the quadrature component oscillates continuously and sinusoidally in time between $\pm M_\mathrm{h}\mathbf{K}\nabla\theta_\mathrm{h}$ and at all times points in either the same or the opposite direction as $\mathbf{K}\nabla\theta_\mathrm{h}$ or it vanishes. Both components have period $T_m = 2\pi \text{ rad}/\omega_m$.

Consequently, the specific-discharge constituent, $\mathbf{q}'(\mathbf{x}, t; \omega_m)$, is periodic in time with period $T_m = 2\pi \text{ rad}/\omega_m$. Also, $\mathbf{q}'(\mathbf{x}, t; \omega_m)$ is a periodic function of the angle $\xi_\mathrm{h}(\mathbf{x}, t; \omega_m)$ with period 2π rad. We say that $\mathbf{q}'(\mathbf{x}, t; \omega_m)$ *oscillates* in time because its direction and its magnitude generally vary periodically with time.

Because $\cos\left[\xi_\mathrm{h} + (2k+1)\pi\right] = -\cos\xi_\mathrm{h}$ and $\sin\left[\xi_\mathrm{h} + (2k+1)\pi\right] = -\sin\xi_\mathrm{h}$, where k is any integer, we have

$$\mathbf{q}'\left(\mathbf{x}, t + (2k+1)\pi/\omega_m; \omega_m\right) = -\mathbf{q}'(\mathbf{x}, t; \omega_m) \quad \forall\, k \in \mathbb{Z}. \tag{15.3}$$

One implication of (15.3) is that for every time interval of duration $2\pi/\omega_m$, if $\mathbf{q}'(\mathbf{x}, t; \omega_m)$ takes some value, then it also takes the opposite value at some other time in that interval.

Exercise 15.1 Period of Specific-Discharge Magnitude

Problem: Is the magnitude $\left|\mathbf{q}'(\mathbf{x}, t; \omega_m)\right|$ time periodic? If so, what is its period? How does this compare to the period of $\mathbf{q}'(\mathbf{x}, t; \omega_m)$?

Solution: Taking the magnitude of (15.3) gives

$$\left|\mathbf{q}'\left(\mathbf{x}, t + (2k+1)\pi/\omega_m; \omega_m\right)\right| = \left|\mathbf{q}'(\mathbf{x}, t; \omega_m)\right| \quad \forall\, k \in \mathbb{Z}. \tag{15.4}$$

Therefore, the magnitude $\left|\mathbf{q}'(\mathbf{x}, t; \omega_m)\right|$ is time periodic with period $\pi \text{ rad}/\omega_m$. The period of $\left|\mathbf{q}'(\mathbf{x}, t; \omega_m)\right|$ is one half that of $\mathbf{q}'(\mathbf{x}, t; \omega_m)$. □

15.2. TIME EXTREMA OF SPECIFIC-DISCHARGE MAGNITUDE

15.2.1. Critical Times

Consider the magnitude of the specific-discharge constituent vector $\left|\mathbf{q}'(\mathbf{x}, t; \omega_m)\right|$. For fixed frequency and

Hydrodynamics of Time-Periodic Groundwater Flow: Diffusion Waves in Porous Media, Geophysical Monograph 224, First Edition. Joe S. Depner and Todd C. Rasmussen.

position, $|\mathbf{q}'(\mathbf{x}, t; \omega_m)|$ is a function of time only. If extrema (with respect to time) of $|\mathbf{q}'(\mathbf{x}, t; \omega_m)|$ exist, then they occur at critical times t_c, which we define by

$$\left[\frac{\partial}{\partial t}\left|\mathbf{q}'(\mathbf{x}, t; \omega_m)\right|\right]\Bigg|_{t=t_c} = 0. \tag{15.5}$$

Using (15.1) we can express the magnitude as

$$|\mathbf{q}'| = M_h \sqrt{\left|\mathbf{K}\nabla Y_h\right|^2 \cos^2 \xi_h - 2\left(\mathbf{K}\nabla Y_h\right) \cdot \left(\mathbf{K}\nabla\theta_h\right) \cos \xi_h \sin \xi_h + \left|\mathbf{K}\nabla\theta_h\right|^2 \sin^2 \xi_h}. \tag{15.6}$$

Substituting the RHS of (15.6) for the magnitude of the specific discharge in (15.5), differentiating with respect to time, and then substituting $t = t_c$ lead to

$$\left(\left|\mathbf{K}\nabla\theta_h\right|^2 - \left|\mathbf{K}\nabla Y_h\right|^2\right) \cos \xi_c \sin \xi_c + \left(\mathbf{K}\nabla Y_h\right) \cdot \left(\mathbf{K}\nabla\theta_h\right)\left(1 - 2\cos^2 \xi_c\right) = 0 \tag{15.7}$$

where

$$\xi_c \equiv \omega_m t_c + \theta_h(\mathbf{x}; \omega_m). \tag{15.8}$$

We can use equations 4.21.27 and 4.21.28 of *Roy and Olver* [2016] to express (15.7) as

$$\left(\left|\mathbf{K}\nabla\theta_h\right|^2 - \left|\mathbf{K}\nabla Y_h\right|^2\right) \sin\left(2\xi_c\right) - 2\left(\mathbf{K}\nabla Y_h\right) \cdot \left(\mathbf{K}\nabla\theta_h\right) \cos\left(2\xi_c\right) = 0. \tag{15.9}$$

The following two mutually exclusive cases collectively exhaust the universe of possibilities:

- Case 1: Both $\left|\mathbf{K}\nabla\theta_h\right| = \left|\mathbf{K}\nabla Y_h\right|$ and $\left(\mathbf{K}\nabla Y_h\right) \cdot \left(\mathbf{K}\nabla\theta_h\right) = 0$.
- Case 2: Either $\left|\mathbf{K}\nabla\theta_h\right| \neq \left|\mathbf{K}\nabla Y_h\right|$ or $\left(\mathbf{K}\nabla Y_h\right) \cdot \left(\mathbf{K}\nabla\theta_h\right) \neq 0$ or both.

In case 1, (15.9) is satisfied identically, so every time is a critical time. Then $\left|\mathbf{q}'(\mathbf{x}, t; \omega_m)\right|$ is time invariant and equal to $M_h(\mathbf{x}; \omega_m)\left|\mathbf{K}(\mathbf{x})\nabla\theta_h(\mathbf{x}; \omega_m)\right|$, so there are no magnitude extrema with respect to time. Also see Example 15.5.

In case 2, (15.9) leads to

$$\tan\left(2\xi_c\right) = \frac{2\left(\mathbf{K}\nabla Y_h\right) \cdot \left(\mathbf{K}\nabla\theta_h\right)}{\left|\mathbf{K}\nabla\theta_h\right|^2 - \left|\mathbf{K}\nabla Y_h\right|^2}. \tag{15.10}$$

The function $\tan(k\xi)$, where k is a positive integer, has period π rad$/k$. Therefore, every interval of length 2π rad contains four critical values (ξ_c), and numerically consecutive critical values differ by $\pi/2$ rad. Equivalently, every period of duration 2π rad$/\omega_m$ contains four critical times (t_c), and numerically consecutive critical times differ by π rad$/(2\omega_m)$. Combining (15.10) with (15.8) yields the following, which the critical times must satisfy:

$$t_c = \frac{1}{\omega_m}\left[\frac{1}{2}\arctan\frac{2\left(\mathbf{K}\nabla Y_h\right) \cdot \left(\mathbf{K}\nabla\theta_h\right)}{\left|\mathbf{K}\nabla\theta_h\right|^2 - \left|\mathbf{K}\nabla Y_h\right|^2} - \theta_h\right].$$

Maxima at numerically consecutive critical times are prohibited, as are minima at numerically consecutive critical times. Consequently every period of duration 2π rad$/\omega_m$ contains two maxima and two minima, and the maxima and minima alternate in time.

Exercise 15.2 Derivation of Intermediate Results

Problem: Suppose that either $\left|\mathbf{K}\nabla\theta_h\right| \neq \left|\mathbf{K}\nabla Y_h\right|$ or $\left(\mathbf{K}\nabla Y_h\right) \cdot \left(\mathbf{K}\nabla\theta_h\right) \neq 0$ or both. Express $\cos\left(2\xi_c\right)$ and $\sin\left(2\xi_c\right)$ in terms of $\mathbf{K}\nabla Y_h$ and $\mathbf{K}\nabla\theta_h$. Then do the same for $\cos \xi_c$ and $\sin \xi_c$.

Solution: At critical times (15.10) is satisfied, which implies

$$\cos\left(2\xi_c\right) = \frac{\pm\left(\left|\mathbf{K}\nabla\theta_h\right|^2 - \left|\mathbf{K}\nabla Y_h\right|^2\right)}{\sqrt{\left(\left|\mathbf{K}\nabla\theta_h\right|^2 - \left|\mathbf{K}\nabla Y_h\right|^2\right)^2 + 4\left[\left(\mathbf{K}\nabla Y_h\right) \cdot \left(\mathbf{K}\nabla\theta_h\right)\right]^2}} \tag{15.11a}$$

$$\sin\left(2\xi_c\right) = \frac{\pm 2\left(\mathbf{K}\nabla Y_h\right) \cdot \left(\mathbf{K}\nabla\theta_h\right)}{\sqrt{\left(\left|\mathbf{K}\nabla\theta_h\right|^2 - \left|\mathbf{K}\nabla Y_h\right|^2\right)^2 + 4\left[\left(\mathbf{K}\nabla Y_h\right) \cdot \left(\mathbf{K}\nabla\theta_h\right)\right]^2}} \tag{15.11b}$$

where either both of the plus-or-minus signs are positive or both are negative (i.e., the signs must match). Using the half-angle relations (B.4), these give

$$\cos \xi_c = \pm\frac{\sqrt{2}}{2}\sqrt{1 \pm \frac{\left(\left|\mathbf{K}\nabla\theta_h\right|^2 - \left|\mathbf{K}\nabla Y_h\right|^2\right)}{\sqrt{\left(\left|\mathbf{K}\nabla\theta_h\right|^2 - \left|\mathbf{K}\nabla Y_h\right|^2\right)^2 + 4\left[\left(\mathbf{K}\nabla Y_h\right) \cdot \left(\mathbf{K}\nabla\theta_h\right)\right]^2}}} \tag{15.12a}$$

$$\sin \xi_c = \pm\frac{\sqrt{2}}{2}\sqrt{1 \pm \frac{\left(\left|\mathbf{K}\nabla Y_h\right|^2 - \left|\mathbf{K}\nabla\theta_h\right|^2\right)}{\sqrt{\left(\left|\mathbf{K}\nabla\theta_h\right|^2 - \left|\mathbf{K}\nabla Y_h\right|^2\right)^2 + 4\left[\left(\mathbf{K}\nabla Y_h\right) \cdot \left(\mathbf{K}\nabla\theta_h\right)\right]^2}}} \tag{15.12b}$$

where the plus-or-minus signs under the radicals on the RHSs must match. □

In case 2, substituting $\xi_h = \xi_c$ in (15.1) yields

$$\mathbf{q}'(\mathbf{x}, t_c; \omega_m) = -\mathbf{K}\nabla M_h \cos \xi_c + M_h \mathbf{K}\nabla\theta_h \sin \xi_c$$

where $\cos \xi_c$ and $\sin \xi_c$ are given by (15.12). This generally gives four distinct values for $\mathbf{q}'(\mathbf{x}, t_c; \omega_m)$, depending on the signs of $\cos \xi_c$ and $\sin \xi_c$.

Exercise 15.3 Time Extrema of Specific-Discharge Constituent

Problem: Suppose that either $|\mathbf{K}\nabla\theta_\mathrm{h}| \neq |\mathbf{K}\nabla Y_\mathrm{h}|$ or $|(\mathbf{K}\nabla Y_\mathrm{h}) \cdot (\mathbf{K}\nabla\theta_\mathrm{h}| > 0$ or both. Express $|\mathbf{q}'(\mathbf{x}, t_\mathrm{c}; \omega_m)|$ compactly in terms of $\mathbf{K}(\mathbf{x})\nabla Y_\mathrm{h}(\mathbf{x}; \omega_m)$ and $\mathbf{K}(\mathbf{x})\nabla\theta_\mathrm{h}(\mathbf{x}; \omega_m)$. How many distinct values of $|\mathbf{q}'(\mathbf{x}, t_\mathrm{c}; \omega_m)|$ exist in every time interval of duration 2π rad$/\omega_m$?

Solution: Substituting $\xi_\mathrm{h} = \xi_\mathrm{c}$ in (15.6) gives

$$|\mathbf{q}'(\mathbf{x}, t_\mathrm{c}; \omega_m)| = M_\mathrm{h}\sqrt{\begin{aligned}&|\mathbf{K}\nabla Y_\mathrm{h}|^2 \cos^2\xi_\mathrm{c} - 2(\mathbf{K}\nabla Y_\mathrm{h})\\ &\cdot(\mathbf{K}\nabla\theta_\mathrm{h})\cos\xi_\mathrm{c}\sin\xi_\mathrm{c} + |\mathbf{K}\nabla\theta_\mathrm{h}|^2\sin^2\xi_\mathrm{c}.\end{aligned}}$$

Using equations 4.21.27 and 4.21.28 of *Roy and Olver* [2016], we can write this as

$$\frac{|\mathbf{q}'(\mathbf{x}, t_\mathrm{c}; \omega_m)|}{M_\mathrm{h}} = \sqrt{\begin{aligned}&|\mathbf{K}\nabla Y_\mathrm{h}|^2\left[\frac{1+\cos(2\xi_\mathrm{c})}{2}\right] - (\mathbf{K}\nabla Y_\mathrm{h})\\ &\cdot(\mathbf{K}\nabla\theta_\mathrm{h})\sin(2\xi_\mathrm{c}) + |\mathbf{K}\nabla\theta_\mathrm{h}|^2\left[\frac{1-\cos(2\xi_\mathrm{c})}{2}\right].\end{aligned}}$$

Squaring and then grouping terms lead to

$$\begin{aligned}\frac{|\mathbf{q}'(\mathbf{x}, t_\mathrm{c}; \omega_m)|^2}{M_\mathrm{h}^2} &= \left(\frac{|\mathbf{K}\nabla\theta_\mathrm{h}|^2 + |\mathbf{K}\nabla Y_\mathrm{h}|^2}{2}\right)\\ &+ \left(\frac{|\mathbf{K}\nabla Y_\mathrm{h}|^2 - |\mathbf{K}\nabla\theta_\mathrm{h}|^2}{2}\right)\cos(2\xi_\mathrm{c})\\ &- (\mathbf{K}\nabla Y_\mathrm{h})\cdot(\mathbf{K}\nabla\theta_\mathrm{h})\sin(2\xi_\mathrm{c}).\end{aligned}$$

Substituting the RHSs of (15.11) for $\cos(2\xi_\mathrm{c})$ and $\sin(2\xi_\mathrm{c})$ and then simplifying produce

$$|\mathbf{q}'(\mathbf{x}, t_\mathrm{c}; \omega_m)|^2 = \frac{M_\mathrm{h}^2}{2} \times \left[\left(|\mathbf{K}\nabla\theta_\mathrm{h}|^2 + |\mathbf{K}\nabla Y_\mathrm{h}|^2\right)^2 \pm\sqrt{\left(|\mathbf{K}\nabla\theta_\mathrm{h}|^2 - |\mathbf{K}\nabla Y_\mathrm{h}|^2\right)^2 + 4\left[(\mathbf{K}\nabla Y_\mathrm{h})\cdot(\mathbf{K}\nabla\theta_\mathrm{h})\right]^2}\right].$$

Taking the square root yields

$$|\mathbf{q}'(\mathbf{x}, t_\mathrm{c}; \omega_m)| = \frac{M_\mathrm{h}}{\sqrt{2}} \times \sqrt{\left(|\mathbf{K}\nabla\theta_\mathrm{h}|^2 + |\mathbf{K}\nabla Y_\mathrm{h}|^2\right)^2 \pm\sqrt{\begin{aligned}&\left(|\mathbf{K}\nabla\theta_\mathrm{h}|^2 - |\mathbf{K}\nabla Y_\mathrm{h}|^2\right)^2\\ &+4\left[(\mathbf{K}\nabla Y_\mathrm{h})\cdot(\mathbf{K}\nabla\theta_\mathrm{h})\right]^2\end{aligned}}}. \tag{15.13}$$

This gives two distinct values of $|\mathbf{q}'(\mathbf{x}, t_\mathrm{c}; \omega_m)|$ in every interval of duration 2π rad$/\omega_m$. □

Exercise 15.4 Time Extrema of Specific-Discharge Constituent—Special Case

Problem: Suppose the in-phase and quadrature components of the specific discharge at the point $\mathbf{x}$ are collinear:

$$\begin{aligned}&[\mathbf{K}(\mathbf{x})\nabla Y_\mathrm{h}(\mathbf{x}; \omega_m)]\cdot[\mathbf{K}(\mathbf{x})\nabla\theta_\mathrm{h}(\mathbf{x}; \omega_m)]\\ &\quad = \pm|\mathbf{K}(\mathbf{x})\nabla Y_\mathrm{h}(\mathbf{x}; \omega_m)||\mathbf{K}(\mathbf{x})\nabla\theta_\mathrm{h}(\mathbf{x}; \omega_m)|.\end{aligned} \tag{15.14}$$

What are the critical values $|\mathbf{q}'(\mathbf{x}, t_\mathrm{c}; \omega_m)|$? How many distinct critical values $|\mathbf{q}'(\mathbf{x}, t_\mathrm{c}; \omega_m)|$ exist in every time interval of duration 2π rad$/\omega_m$?

Solution: Substituting the RHS of (15.14) for $(\mathbf{K}\nabla Y_\mathrm{h}) \cdot (\mathbf{K}\nabla\theta_\mathrm{h})$ in (15.13) and then simplifying give

$$|\mathbf{q}'(\mathbf{x}, t_\mathrm{c}; \omega_m)| = \frac{M_\mathrm{h}}{\sqrt{2}}\sqrt{|\mathbf{K}\nabla\theta_\mathrm{h}|^2 + |\mathbf{K}\nabla Y_\mathrm{h}|^2 \pm \left(|\mathbf{K}\nabla\theta_\mathrm{h}|^2 + |\mathbf{K}\nabla Y_\mathrm{h}|^2\right)}.$$

Evaluating the RHS and reverting to explicit notation lead to

$$|\mathbf{q}'(\mathbf{x}, t_\mathrm{c}; \omega_m)| = \begin{cases}\sqrt{\begin{aligned}&M_\mathrm{h}^2(\mathbf{x}; \omega_m)|\mathbf{K}(\mathbf{x})\nabla\theta_\mathrm{h}(\mathbf{x}; \omega_m)|^2\\ &+|\mathbf{K}(\mathbf{x})\nabla M_\mathrm{h}(\mathbf{x}; \omega_m)|^2\end{aligned}}\\ \text{or}\\ 0.\end{cases} \tag{15.15}$$

Therefore, every interval of duration 2π rad$/\omega_m$ contains two distinct values of $|\mathbf{q}'(\mathbf{x}, t_\mathrm{c}; \omega_m)|$. □

15.2.2. Perpendicularity of Extrema

In this section we show that if the magnitude of the specific-discharge constituent has extrema with respect to time (see case 2 of Section 15.2.1), then the specific-discharge constituent evaluated at any two numerically consecutive critical times yields two perpendicular vectors.

As shown in Section 15.2.1, for every numerically consecutive pair of critical times $t_{\mathrm{c}1}$ and $t_{\mathrm{c}2}$ we have the corresponding numerically consecutive critical values $\xi_{\mathrm{c}1}$ and $\xi_{\mathrm{c}2}$, where

$$\xi_{\mathrm{c}2} \equiv \xi_{\mathrm{c}1} + \frac{\pi}{2}\ \text{rad}. \tag{15.16}$$

Substituting $\xi_h = \xi_{c1}$ and $\xi_h = \xi_{c2}$ in (15.1), respectively, yields

$$\mathbf{q}'(\mathbf{x}, t_{c1}; \omega_m) = -\mathbf{K}\nabla M_h \cos\xi_{c1} + M_h\mathbf{K}\nabla\theta_h \sin\xi_{c1} \quad (15.17a)$$

$$\mathbf{q}'(\mathbf{x}, t_{c2}; \omega_m) = -\mathbf{K}\nabla M_h \cos\xi_{c2} + M_h\mathbf{K}\nabla\theta_h \sin\xi_{c2}. \quad (15.17b)$$

Using trigonometric identities (B.2a) and (B.2c) with (15.16) leads to

$$\cos\xi_{c2} = -\sin\xi_{c1}$$

$$\sin\xi_{c2} = \cos\xi_{c1}.$$

Substituting these results for $\cos\xi_{c2}$ and $\sin\xi_{c2}$ in (15.17b) gives

$$\mathbf{q}'(\mathbf{x}, t_{c2}; \omega_m) = \mathbf{K}\nabla M_h \sin\xi_{c1} + M_h\mathbf{K}\nabla\theta_h \cos\xi_{c1}.$$

Using this result with (15.17a) to evaluate the scalar product of the specific-discharge vectors at the two critical times produces, after grouping terms,

$$\begin{aligned}\mathbf{q}'(\mathbf{x}, t_{c1}; \omega_m) \cdot \mathbf{q}'(\mathbf{x}, t_{c2}; \omega_m) &= M_h^2\Big[\Big(|\mathbf{K}\nabla\theta_h|^2 - |\mathbf{K}\nabla Y_h|^2\Big)\cos\xi_{c1}\sin\xi_{c1} \quad (15.18)\\ &+ (\mathbf{K}\nabla Y_h)\cdot(\mathbf{K}\nabla\theta_h)\big(1 - 2\cos^2\xi_{c1}\big)\Big].\end{aligned}$$

Equation (15.7) implies that the RHS vanishes, leaving

$$\mathbf{q}'(\mathbf{x}, t_{c1}; \omega_m) \cdot \mathbf{q}'(\mathbf{x}, t_{c2}; \omega_m) = 0. \quad (15.19)$$

Thus, the specific-discharge constituent evaluated at any two numerically consecutive critical times yields two perpendicular vectors.

15.3. DISCHARGE SPACE AND DISCHARGE ENVELOPE

To describe and visualize the time oscillation of the specific-discharge constituent, we use the concepts of discharge space and discharge envelope.

15.3.1. Discharge Space

Discharge space is an abstract, 3D, right-handed Cartesian space in which the coordinates are q_1, q_2, and q_3, corresponding to the Cartesian components of the specific-discharge vector, rather than the usual space coordinates x_1, x_2, and x_3.

A single value of a discharge vector at a fixed location and fixed time is plotted as a single point in discharge space. If one imagines the vector's tail located at the origin (i.e., the point $\mathbf{q} = \mathbf{0}$), then the point is plotted at the location of the vector's head. Thus, the plot point conveys both the direction of the vector (i.e., the orientation of the radial line that contains the point) and its relative magnitude (i.e., the distance from the origin to the point).

Table 15.1 Measures in regular space and discharge space.

	Dimensions	
Measure	Regular space	Discharge space
Length	L	Lt^{-1}
Area	L^2	L^2t^{-2}
Volume	L^3	L^3t^{-3}

Because of the particular way we have defined the discharge space, measures of length, area, and volume take on slightly different meanings in this context. For perspective, Table 15.1 summarizes the differences of interpretation in regular space and discharge space.

15.3.2. Discharge Envelope

15.3.2.1. Description

We can visualize how the specific-discharge constituent $\mathbf{q}'$ varies with time for a given frequency ω_m and a fixed location $\mathbf{x}$ by plotting its values over an entire constituent period in discharge space. That is, for each time t in some period $t_0 < t \leq t_0 + 2\pi/\omega_m$ we plot a single point in discharge space corresponding to the head of the vector $\mathbf{q}'(\mathbf{x}, t; \omega_m)$. We call the resulting curve, which is a continuous trajectory, the *envelope* of the specific-discharge constituent at the point $\mathbf{x}$. Figure 15.1 shows an example of an envelope plot.

Although the discharge envelope is a useful concept for visualizing the time variation of the specific-discharge constituent, it has limitations. For instance, by itself the envelope does not convey any information about the phase of the time variation. Also, by itself the envelope does not convey any information about spatial variation.

Exercise 15.5 Identical Envelopes

Problem: Suppose $\mathbf{x}_1$ and $\mathbf{x}_2$ are two distinct (i.e., $\mathbf{x}_1 \neq \mathbf{x}_2$), hydraulically connected points in a time-periodic groundwater flow field. Further suppose the specific-discharge envelope for the point $\mathbf{x}_1$ is identical to that for the point $\mathbf{x}_2$. Can we then conclude that $\mathbf{q}'(\mathbf{x}_1, t; \omega_m) = \mathbf{q}'(\mathbf{x}_2, t; \omega_m)$? Explain.

Solution: No, we cannot conclude that $\mathbf{q}'(\mathbf{x}_1, t; \omega_m) = \mathbf{q}'(\mathbf{x}_2, t; \omega_m)$, because although the two envelopes are identical, $\mathbf{q}'(\mathbf{x}_1, t; \omega_m)$ may not be in phase with $\mathbf{q}'(\mathbf{x}_2, t; \omega_m)$. Stated another way, although $\mathbf{q}'(\mathbf{x}_1, t; \omega_m)$ and $\mathbf{q}'(\mathbf{x}_2, t; \omega_m)$ take on the same *set* of values over a

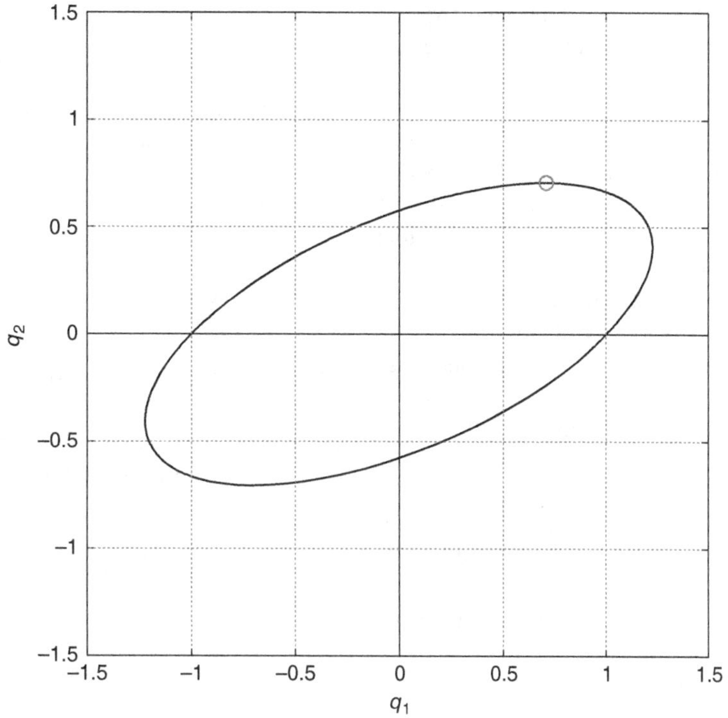

Figure 15.1 Example of discharge-envelope plot (black line is envelope; small circle marks $t = 0$; time increases anticlockwise).

constituent period, because of a possible phase difference the two might not take identical values simultaneously. □

By making some simple modifications to the envelope plot, it can be made to convey more information. For example, one can index the envelope to a standard reference time (e.g., $t = 0$) by distinctly marking the point on the envelope that corresponds to the reference time. Also, one can indicate the sense of time increase (clockwise or anticlockwise) at the reference time by either (a) using an explanatory note or (b) placing a small arrow on or near the envelope. The envelope plot in Figure 15.1 uses an index point and a note to indicate the sense of time increase.

15.3.2.2. Envelopes of In-Phase and Quadrature Components

Define the unit vectors

$$\hat{\mathbf{u}}(\mathbf{x};\omega_m) \equiv \frac{\mathbf{K}(\mathbf{x})\nabla M_{\mathrm{h}}(\mathbf{x};\omega_m)}{\left|\mathbf{K}(\mathbf{x})\nabla M_{\mathrm{h}}(\mathbf{x};\omega_m)\right|}, \qquad \left|\nabla M_{\mathrm{h}}(\mathbf{x};\omega_m)\right| > 0 \tag{15.20a}$$

$$\hat{\mathbf{v}}(\mathbf{x};\omega_m) \equiv \frac{\mathbf{K}(\mathbf{x})\nabla \theta_{\mathrm{h}}(\mathbf{x};\omega_m)}{\left|\mathbf{K}(\mathbf{x})\nabla \theta_{\mathrm{h}}(\mathbf{x};\omega_m)\right|}, \qquad \left|\nabla \theta_{\mathrm{h}}(\mathbf{x};\omega_m)\right| > 0. \tag{15.20b}$$

By combining these with (15.1), we can write the in-phase and quadrature components of the specific-discharge constituent as, respectively,

$$\mathbf{q}'_{\mathrm{Ip}}(\mathbf{x},t;\omega_m) = -\left|\mathbf{K}(\mathbf{x})\nabla M_{\mathrm{h}}(\mathbf{x};\omega_m)\right| \cos\xi_{\mathrm{h}}(\mathbf{x},t;\omega_m)\hat{\mathbf{u}}(\mathbf{x};\omega_m) \tag{15.21a}$$

$$\mathbf{q}'_{\mathrm{Qu}}(\mathbf{x},t;\omega_m) = M_{\mathrm{h}}(\mathbf{x};\omega_m)\left|\mathbf{K}(\mathbf{x})\nabla \theta_{\mathrm{h}}(\mathbf{x};\omega_m)\right| \sin\xi_{\mathrm{h}}(\mathbf{x},t;\omega_m)\hat{\mathbf{v}}(\mathbf{x};\omega_m). \tag{15.21b}$$

As the in-phase component of the specific-discharge constituent oscillates, it points in the direction of $\pm\hat{\mathbf{u}}(\mathbf{x};\omega_m)$ or it vanishes. Similarly, as the quadrature component of the specific-discharge constituent oscillates, it points in the direction of $\pm\hat{\mathbf{v}}(\mathbf{x};\omega_m)$ or it vanishes.

Example 15.1 Envelope of In-Phase Component

Imagine plotting the in-phase component in discharge space over an entire constituent period. Referring to (15.21a), the locus (i.e., the envelope of the in-phase component) is a segment of the radial line with unit direction vector $\hat{\mathbf{u}}(\mathbf{x};\omega_m)$. The segment has length $2\left|\mathbf{K}(\mathbf{x})\nabla M_{\mathrm{h}}(\mathbf{x};\omega_m)\right|$ and is centered at the origin. ◊

Exercise 15.6 Envelope of Quadrature Component

Problem: Describe the envelope of the quadrature component of the specific-discharge constituent.

Solution: Referring to (15.21b), the envelope of the quadrature component is a segment of the radial line with unit direction vector $\hat{\mathbf{v}}(\mathbf{x};\omega_m)$. The segment has length $2M_h(\mathbf{x};\omega_m)\left|\mathbf{K}(\mathbf{x})\nabla\theta_h(\mathbf{x};\omega_m)\right|$ and is centered at the origin. □

15.3.2.3. Plan for Envelope Analysis

In the following sections we analyze the specific-discharge constituent's envelope and time rotation (if any). Our analysis, which relies on the vector product equivalence (14.29), examines the following cases:

- The constituent's amplitude gradient or its phase gradient (or both) vanishes locally. Consequently at least one of $\hat{\mathbf{u}}(\mathbf{x};\omega_m)$ and $\hat{\mathbf{v}}(\mathbf{x};\omega_m)$ is undefined. See Section 15.4.
- The constituent's amplitude and phase gradients are locally nonzero, so that both $\hat{\mathbf{u}}(\mathbf{x};\omega_m)$ and $\hat{\mathbf{v}}(\mathbf{x};\omega_m)$ are well defined.
  - The constituent amplitude and phase gradients are locally collinear; i.e., $\hat{\mathbf{u}}(\mathbf{x};\omega_m)$ and $\hat{\mathbf{v}}(\mathbf{x};\omega_m)$ are locally collinear. See Section 15.5.
  - The constituent amplitude and phase gradients are locally noncollinear; i.e., $\hat{\mathbf{u}}(\mathbf{x};\omega_m)$ and $\hat{\mathbf{v}}(\mathbf{x};\omega_m)$ are locally noncollinear. See Section 15.6.

15.4. LOCALLY VANISHING GRADIENTS

Throughout this section we assume that for a particular harmonic constituent (i.e., ω_m fixed) and a fixed location $\mathbf{x}$ the corresponding amplitude gradient, phase gradient, or both vanish:

$$\left|\nabla M_h(\mathbf{x};\omega_m)\right|\left|\nabla\theta_h(\mathbf{x};\omega_m)\right| = 0.$$

Exercise 15.7 Amplitude and Phase Gradients Vanish Locally

Problem: Suppose that for some fixed constituent (i.e., ω_m fixed) and some fixed location $\mathbf{x}$ the corresponding amplitude and phase gradients vanish:

$$\nabla M_h(\mathbf{x};\omega_m) = \mathbf{0} \qquad (15.22a)$$

$$\nabla\theta_h(\mathbf{x};\omega_m) = \mathbf{0}. \qquad (15.22b)$$

Determine the envelope of the corresponding specific-discharge constituent.

Solution: Substituting the RHSs of (15.22) for the corresponding gradients in (15.1) produces $\mathbf{q}'(\mathbf{x},t;\omega_m) = \mathbf{0}$. Therefore, the envelope of the corresponding specific-discharge constituent at the location $\mathbf{x}$ is a single point at the origin of discharge space. □

Exercise 15.8 Amplitude Gradient Vanishes Locally

Problem: Suppose that for some constituent (i.e., ω_m fixed) and some fixed location $\mathbf{x}$ the amplitude gradient vanishes but the phase gradient does not:

$$\nabla M_h(\mathbf{x};\omega_m) = \mathbf{0} \qquad (15.23a)$$

$$\left|\nabla\theta_h(\mathbf{x};\omega_m)\right| > 0. \qquad (15.23b)$$

Describe the envelope of the corresponding specific-discharge constituent.

Solution: Substituting $\nabla M_h(\mathbf{x};\omega_m) = \mathbf{0}$ in (15.1) produces

$$\mathbf{q}'(\mathbf{x},t;\omega_m) = M_h(\mathbf{x};\omega_m)\mathbf{K}(\mathbf{x})\nabla\theta_h(\mathbf{x};\omega_m)\sin\left[\omega_m t + \theta_h(\mathbf{x};\omega_m)\right]. \qquad (15.24)$$

Therefore, the envelope of the corresponding specific-discharge constituent at the location $\mathbf{x}$ is a line segment centered at the origin of discharge space. The segment lies on the radial line whose unit direction vector is $\hat{\mathbf{v}}(\mathbf{x};\omega_m)$. Its length is

$$2\max_t\left|\mathbf{q}'(\mathbf{x},t;\omega_m)\right| = 2M_h(\mathbf{x};\omega_m)\left|\mathbf{K}(\mathbf{x})\nabla\theta_h(\mathbf{x};\omega_m)\right| \qquad (15.25)$$

where $\max_t$ denotes the maximum value with respect to time. □

Exercise 15.9 Phase Gradient Vanishes Locally

Problem: Suppose that for some constituent (i.e., ω_m fixed) and some fixed location $\mathbf{x}$ the phase gradient vanishes but the amplitude gradient does not:

$$\left|\nabla M_h(\mathbf{x};\omega_m)\right| > 0$$

$$\nabla\theta_h(\mathbf{x};\omega_m) = \mathbf{0}.$$

Describe the envelope of the corresponding specific-discharge constituent.

Solution: Substituting $\nabla\theta_h(\mathbf{x};\omega_m) = \mathbf{0}$ in (15.1) produces

$$\mathbf{q}'(\mathbf{x},t;\omega_m) = -\mathbf{K}(\mathbf{x})\nabla M_h(\mathbf{x};\omega_m)\cos\left[\omega_m t + \theta_h(\mathbf{x};\omega_m)\right]. \qquad (15.26)$$

Therefore, the envelope of the corresponding specific-discharge constituent at the location $\mathbf{x}$ is a line segment centered at the origin of discharge space. The segment lies on the radial line whose unit direction vector is $\hat{\mathbf{u}}(\mathbf{x};\omega_m)$. Its length is

$$2\max_t\left|\mathbf{q}'(\mathbf{x},t;\omega_m)\right| = 2\left|\mathbf{K}(\mathbf{x})\nabla M_h(\mathbf{x};\omega_m)\right|$$

where $\max_t$ denotes the maximum value with respect to time. □

15.5. COLLINEAR AMPLITUDE AND PHASE GRADIENTS

Throughout this section we assume that for a particular harmonic constituent (i.e., ω_m fixed) and a fixed location

$\mathbf{x}$ the corresponding amplitude and phase gradients are nonzero and $\hat{\mathbf{u}}(\mathbf{x};\omega_m)$ and $\hat{\mathbf{v}}(\mathbf{x};\omega_m)$ are locally collinear:

$$\left[\mathbf{K}(\mathbf{x})\nabla M_{\mathrm{h}}(\mathbf{x};\omega_m)\right] \times \left[\mathbf{K}(\mathbf{x})\nabla\theta_{\mathrm{h}}(\mathbf{x};\omega_m)\right] = \mathbf{0}.$$

Then

$$\hat{\mathbf{v}}(\mathbf{x};\omega_m) = \pm\hat{\mathbf{u}}(\mathbf{x};\omega_m). \tag{15.27}$$

By combining (15.21) and (15.27), we can write the specific-discharge constituent (15.1) as

$$\mathbf{q}'(\mathbf{x},t;\omega_m) = \Big[- \left|\mathbf{K}(\mathbf{x})\nabla M_{\mathrm{h}}(\mathbf{x};\omega_m)\right| \cos\xi_{\mathrm{h}}(\mathbf{x},t;\omega_m) \pm M_{\mathrm{h}}(\mathbf{x};\omega_m)\left|\mathbf{K}(\mathbf{x})\nabla\theta_{\mathrm{h}}(\mathbf{x};\omega_m)\right| \sin\xi_{\mathrm{h}}(\mathbf{x},t;\omega_m)\Big]\hat{\mathbf{u}}(\mathbf{x};\omega_m). \tag{15.28}$$

Then the specific-discharge constituent's envelope is a segment of the radial line with unit direction vector $\hat{\mathbf{u}}(\mathbf{x};\omega_m)$. The envelope has the origin as midpoint.

Using algebra we can write (15.28) as

$$\mathbf{q}'(\mathbf{x},t;\omega_m) = \sqrt{\left|\mathbf{K}(\mathbf{x})\nabla M_{\mathrm{h}}(\mathbf{x};\omega_m)\right|^2 + M_{\mathrm{h}}^2(\mathbf{x};\omega_m)\left|\mathbf{K}(\mathbf{x})\nabla\theta_{\mathrm{h}}(\mathbf{x};\omega_m)\right|^2} \times \Big[\cos\epsilon(\mathbf{x};\omega_m)\cos\xi_{\mathrm{h}}(\mathbf{x},t;\omega_m) \mp \sin\epsilon(\mathbf{x};\omega_m)\sin\xi_{\mathrm{h}}(\mathbf{x},t;\omega_m)\Big]\hat{\mathbf{u}}(\mathbf{x};\omega_m) \tag{15.29}$$

where

$$\epsilon(\mathbf{x};\omega_m) \equiv \arctan\frac{\pm M_{\mathrm{h}}(\mathbf{x};\omega_m)\left|\mathbf{K}(\mathbf{x})\nabla\theta_{\mathrm{h}}(\mathbf{x};\omega_m)\right|}{\left|\mathbf{K}(\mathbf{x})\nabla M_{\mathrm{h}}(\mathbf{x};\omega_m)\right|}.$$

Using trigonometric identities (B.2a) and (B.2b) to rewrite the RHS of (15.29) yields

$$\mathbf{q}'(\mathbf{x},t;\omega_m) = \sqrt{\left|\mathbf{K}(\mathbf{x})\nabla M_{\mathrm{h}}(\mathbf{x}_0;\omega_m)\right|^2 + M_{\mathrm{h}}^2(\mathbf{x};\omega_m)\left|\mathbf{K}(\mathbf{x})\nabla\theta_{\mathrm{h}}(\mathbf{x};\omega_m)\right|^2} \times \cos\left[\omega_m t + \theta_{\mathrm{h}}(\mathbf{x};\omega_m) \pm \epsilon(\mathbf{x};\omega_m)\right]\hat{\mathbf{u}}(\mathbf{x};\omega_m). \tag{15.30}$$

The length of the envelope in discharge space is then

$$2\max_t\left|\mathbf{q}'(\mathbf{x},t;\omega_m)\right| = 2\sqrt{\left|\mathbf{K}(\mathbf{x})\nabla M_{\mathrm{h}}(\mathbf{x};\omega_m)\right|^2 + M_{\mathrm{h}}^2(\mathbf{x};\omega_m)\left|\mathbf{K}(\mathbf{x})\nabla\theta_{\mathrm{h}}(\mathbf{x};\omega_m)\right|^2}$$

where $\max_t$ denotes the maximum value with respect to time. This result is consistent with the result (15.15) obtained earlier for the magnitude extrema.

Example 15.2 Envelope for 1D Flow
Consider time-periodic, 1D flow. In such a flow the amplitude and phase functions depend only on the space coordinate x and the constituent frequency, so at every point in the domain the amplitude and phase gradients are collinear. Consequently, at every point in the flow field the envelope of the specific-discharge constituent is a line segment in discharge space. ◊

Example 15.3 Envelope for 2D, Axisymmetric Flow
Consider time-periodic, 2D, axisymmetric flow in a source-free domain filled with an ideal medium as described in Section 12.4. In such a flow the amplitude and phase functions depend only on the radial space coordinate r $(=\sqrt{x_1^2+x_2^2})$ and the constituent frequency, so at every point in the space domain the amplitude and phase gradients are collinear. Consequently, at every point in the flow field the envelope of the specific-discharge constituent is a segment of a radial line in discharge space. ◊

Example 15.4 Envelope for 3D, Radial Spherical Flow
Consider time-periodic, 3D, radial spherical flow in a source-free domain filled with an ideal medium as described in Section 12.3. In such a flow the amplitude and phase functions depend only on the radial space coordinate r $(=\sqrt{x_1^2+x_2^2+x_3^2})$ and the constituent frequency, so at every point in the space domain the amplitude and phase gradients are collinear. Consequently, at every point in the flow field the envelope of the specific-discharge constituent is a segment of a radial line in discharge space. ◊

Exercise 15.10 Special Case: Envelope Intersects Origin

Problem: Suppose $M_{\mathrm{h}}(\mathbf{x};\omega_m)$, $\left|\mathbf{K}(\mathbf{x})\nabla M_{\mathrm{h}}(\mathbf{x};\omega_m)\right|$, and $\left|\mathbf{K}(\mathbf{x})\nabla\theta_{\mathrm{h}}(\mathbf{x};\omega_m)\right|$ are positive. Is it possible for the specific-discharge constituent's envelope to intersect the polar plot origin? If so, what does this imply about the vectors $\mathbf{K}(\mathbf{x})\nabla Y_{\mathrm{h}}(\mathbf{x};\omega_m)$ and $\mathbf{K}(\mathbf{x})\nabla\theta_{\mathrm{h}}(\mathbf{x};\omega_m)$? Assuming the intersection occurs at $t = t_0$, express t_0 in terms of ξ_0, where $\xi_0 \equiv \xi_{\mathrm{h}}(\mathbf{x},t_0;\omega_m)$. How many times does the intersection occur in every time period of duration 2π rad$/\omega_m$?

Solution: If the envelope intersects the polar plot origin, then $\left|\mathbf{q}'(\mathbf{x},t_0;\omega_m)\right| = 0$ for some time t_0, and vice versa. Substituting $\mathbf{q}'(\mathbf{x},t_0;\omega_m) = \mathbf{0}$ in (15.1) yields

$$\mathbf{0} = M_{\mathrm{h}}(\mathbf{x};\omega_m)\left[-\mathbf{K}(\mathbf{x})\nabla Y_{\mathrm{h}}(\mathbf{x};\omega_m)\cos\xi_0 + \mathbf{K}(\mathbf{x})\nabla\theta_{\mathrm{h}}(\mathbf{x};\omega_m)\sin\xi_0\right]. \tag{15.31}$$

This requires the vectors $\hat{\mathbf{u}}(\mathbf{x};\omega_m)$ and $\hat{\mathbf{v}}(\mathbf{x};\omega_m)$ to be collinear. Substituting the RHSs for the corresponding terms of (15.31) and then rearranging lead to

$$\tan\xi_0 = \pm\frac{\left|\mathbf{K}(\mathbf{x})\nabla Y_{\mathrm{h}}(\mathbf{x};\omega_m)\right|}{\left|\mathbf{K}(\mathbf{x})\nabla\theta_{\mathrm{h}}(\mathbf{x};\omega_m)\right|}.$$

Solving for t_0 gives

$$t_0 = \frac{1}{\omega_m}\left\{\arctan\left[\pm\frac{\left|\mathbf{K}(\mathbf{x})\nabla Y_{\mathrm{h}}(\mathbf{x};\omega_m)\right|}{\left|\mathbf{K}(\mathbf{x})\nabla\theta_{\mathrm{h}}(\mathbf{x};\omega_m)\right|}\right] - \theta_{\mathrm{h}}(\mathbf{x};\omega_m)\right\}. \tag{15.32}$$

The tangent function has period π rad, so this equation gives four possible times t_0 in every period of duration 2π rad$/\omega_m$. However, in any given situation the sign ambiguity represented by the plus-or-minus sign on the RHS of (15.32) is resolved; that is, the sign is one or the other, not both. Therefore, the intersection occurs at two times t_0 in every period of duration 2π rad$/\omega_m$. □

Exercise 15.11 Minimum Sweep Time for Segment

Problem: Suppose that, for a given harmonic constituent (i.e., ω_m fixed) and some fixed location $\mathbf{x}$, $\hat{\mathbf{u}}(\mathbf{x};\omega_m)$ and $\hat{\mathbf{v}}(\mathbf{x};\omega_m)$ are collinear. What is the minimum amount of time required for the specific-discharge constituent $\mathbf{q}'(\mathbf{x},t;\omega_m)$ to sweep its envelope? What constraints, if any, must the time interval of the sweep period satisfy?

Solution: Equation (15.30) indicates that the envelope (segment) is swept twice during every full constituent period. Therefore, the minimum amount of time required to sweep the segment is one half the constituent period, or π rad$/\omega_m$. However, not just any period of duration π rad$/\omega_m$ will result in a full sweep. The minimum sweep period must correspond to

$$\begin{aligned} k\pi \text{ rad} &\le \omega_m t + \theta_{\mathrm{h}}(\mathbf{x};\omega_m) \pm \epsilon(\mathbf{x};\omega_m) \\ &\le (k+1)\pi \text{ rad} \quad \forall\, k \in \mathbb{Z} \end{aligned}$$

which means the time interval for the minimum sweep period must satisfy

$$\frac{k\pi \text{ rad} - \theta_{\mathrm{h}}(\mathbf{x};\omega_m) \mp \epsilon(\mathbf{x};\omega_m)}{\omega_m} \le t \le \frac{(k+1)\pi \text{ rad} - \theta_{\mathrm{h}}(\mathbf{x};\omega_m) \mp \epsilon(\mathbf{x};\omega_m)}{\omega_m} \quad \forall\, k \in \mathbb{Z}.$$

□

15.6. NONCOLLINEAR AMPLITUDE AND PHASE GRADIENTS

Throughout this section we assume that for a particular harmonic constituent (i.e., ω_m fixed) and a fixed location $\mathbf{x}$ the corresponding amplitude and phase gradients are nonzero and $\hat{\mathbf{u}}(\mathbf{x};\omega_m)$ and $\hat{\mathbf{v}}(\mathbf{x};\omega_m)$ are locally noncollinear:

$$\left[\mathbf{K}(\mathbf{x})\nabla M_{\mathrm{h}}(\mathbf{x};\omega_m)\right] \times \left[\mathbf{K}(\mathbf{x})\nabla\theta_{\mathrm{h}}(\mathbf{x};\omega_m)\right] \neq \mathbf{0}.$$

(Example 14.1 essentially demonstrated that such conditions are in principle physically realizable.) Then the envelopes of the in-phase and quadrature components of the corresponding specific-discharge constituent lie on two distinct (nonparallel) radial lines in discharge space (see Section 15.3.2), so we can uniquely define the plane that contains both:

$$\left[\hat{\mathbf{v}}(\mathbf{x};\omega_m) \times \hat{\mathbf{u}}(\mathbf{x};\omega_m)\right] \cdot \mathbf{q} = 0. \tag{15.33}$$

We call this plane, which intersects the origin in discharge space, the *plane of oscillation* for the specific-discharge constituent. In this case the specific-discharge constituent vector, $\mathbf{q}'(\mathbf{x},t;\omega_m)$ rotates within this plane as it oscillates.

15.6.1. Time Rotation of Specific-Discharge Constituent

15.6.1.1. Azimuth Angle

For our analysis of the specific-discharge constituent rotation, it is useful to establish a convention for defining a reference coordinate system in discharge space. We do this in Note 15.1.

Note 15.1 Reference Coordinate System for Specific-Discharge Constituent

Define a *reference system* in discharge space as a right-handed, Cartesian coordinate system whose basis is the orthonormal set $\{\hat{\mathbf{q}}_1, \hat{\mathbf{q}}_2, \hat{\mathbf{q}}_3\}$, where $\hat{\mathbf{q}}_1$ and $\hat{\mathbf{q}}_2$ lie in the plane of oscillation. For such a system the basis vectors satisfy

$$\begin{aligned} \hat{\mathbf{q}}_1 &= \hat{\mathbf{q}}_2 \times \hat{\mathbf{q}}_3 \\ \hat{\mathbf{q}}_2 &= \hat{\mathbf{q}}_3 \times \hat{\mathbf{q}}_1 \\ \hat{\mathbf{q}}_3 &= \hat{\mathbf{q}}_1 \times \hat{\mathbf{q}}_2 \end{aligned}$$

and

$$\left|\hat{\mathbf{q}}_k\right| = 1 \quad (k = 1,2,3).$$

It follows that

$$\hat{\mathbf{q}}_3 = \pm\frac{\hat{\mathbf{u}}(\mathbf{x};\omega_m) \times \hat{\mathbf{v}}(\mathbf{x};\omega_m)}{\left|\hat{\mathbf{u}}(\mathbf{x};\omega_m) \times \hat{\mathbf{v}}(\mathbf{x};\omega_m)\right|}.$$

For our convention we arbitrarily choose the negative sign:

$$\hat{\mathbf{q}}_3 = -\frac{\hat{\mathbf{u}}(\mathbf{x};\omega_m) \times \hat{\mathbf{v}}(\mathbf{x};\omega_m)}{\left|\hat{\mathbf{u}}(\mathbf{x};\omega_m) \times \hat{\mathbf{v}}(\mathbf{x};\omega_m)\right|}.$$

Then the equation of the plane of oscillation (15.33) becomes

$$q_3 = 0.$$

That is, we orient our reference coordinate system in discharge space so that the horizontal ($q_1 q_2$) plane coincides with the plane of oscillation. △

We can express the vectors $\mathbf{K}\nabla M_h$ and $\mathbf{K}\nabla\theta_h$, both of which are parallel to the plane of oscillation, as follows:

$$\mathbf{K}(\mathbf{x})\nabla M_h(\mathbf{x};\omega_m) = M_h\left\{\left[\left(\mathbf{K}\nabla Y_h\right)\cdot\hat{\mathbf{q}}_1\right]\hat{\mathbf{q}}_1 + \left[\left(\mathbf{K}\nabla Y_h\right)\cdot\hat{\mathbf{q}}_2\right]\hat{\mathbf{q}}_2\right\} \qquad (15.34a)$$

$$\mathbf{K}(\mathbf{x})\nabla\theta_h(\mathbf{x};\omega_m) = \left[\left(\mathbf{K}\nabla\theta_h\right)\cdot\hat{\mathbf{q}}_1\right]\hat{\mathbf{q}}_1 + \left[\left(\mathbf{K}\nabla\theta_h\right)\cdot\hat{\mathbf{q}}_2)\right]\hat{\mathbf{q}}_2. \qquad (15.34b)$$

Combining these with (15.1) gives

$$\mathbf{q}'(\mathbf{x},t;\omega_m) = q_1'\hat{\mathbf{q}}_1 + q_2'\hat{\mathbf{q}}_2 \qquad (15.35)$$

where

$$q_k' = M_h\left[-\left(\mathbf{K}\nabla Y_h\right)\cdot\hat{\mathbf{q}}_k\cos\xi_h + \left(\mathbf{K}\nabla\theta_h\right)\cdot\hat{\mathbf{q}}_k\sin\xi_h\right] \quad (k = 1,2). \qquad (15.36)$$

Let $\Omega(\mathbf{x},t;\omega_m)$ denote the angle the specific-discharge vector makes in the plane of oscillation at time t and let positive angles Ω be measured anticlockwise from the positive q_1 axis when viewed from above the plane of oscillation. We call $\Omega(\mathbf{x},t;\omega_m)$ the *azimuth angle* for the specific-discharge constituent. Then

$$\Omega(\mathbf{x},t;\omega_m) = \arctan\frac{q_2'}{q_1'}. \qquad (15.37)$$

15.6.1.2. Rotation Rate

We define the *rotation rate* of the specific-discharge constituent as the time derivative

$$\dot{\Omega}(\mathbf{x},t;\omega_m) \equiv \frac{d}{dt}\Omega(\mathbf{x},t;\omega_m). \qquad (15.38)$$

The rotation rate has the dimensions of reciprocal time. A positive value for $\dot{\Omega}(\mathbf{x},t;\omega_m)$ indicates anticlockwise rotation when viewed from above; a negative value indicates clockwise rotation when viewed from above.

Combining (15.38) with (15.37), we obtain

$$\dot{\Omega}(\mathbf{x},t;\omega_m) = \frac{1}{|\mathbf{q}'|^2}\left[q_1'\frac{d}{dt}q_2' - q_2'\frac{d}{dt}q_1'\right].$$

Substituting the RHSs of (15.36) with $k = 1$ and $k = 2$, respectively, for q_1' and q_2' and then simplifying yield

$$\dot{\Omega}(\mathbf{x},t;\omega_m) = \frac{\omega_m M_h^2}{|\mathbf{q}'|^2}\left\{\left[\left(\mathbf{K}\nabla\theta_h\right)\cdot\hat{\mathbf{q}}_1\right]\left[\left(\mathbf{K}\nabla Y_h\right)\cdot\hat{\mathbf{q}}_2\right] - \left[\left(\mathbf{K}\nabla\theta_h\right)\cdot\hat{\mathbf{q}}_2\right]\left[\left(\mathbf{K}\nabla Y_h\right)\cdot\hat{\mathbf{q}}_1\right]\right\}. \qquad (15.39)$$

Using the definition of the vector (cross) product in Section 1.6 of *Roy et al* [2016] leads to

$$\left[\left(\mathbf{K}\nabla\theta_h\right)\cdot\hat{\mathbf{q}}_1\right]\left[\left(\mathbf{K}\nabla Y_h\right)\cdot\hat{\mathbf{q}}_2\right]-\left[\left(\mathbf{K}\nabla\theta_h\right)\cdot\hat{\mathbf{q}}_2\right]\left[\left(\mathbf{K}\nabla Y_h\right)\cdot\hat{\mathbf{q}}_1\right] = \left[\left(\mathbf{K}\nabla\theta_h\right)\times\left(\mathbf{K}\nabla Y_h\right)\right]\cdot\hat{\mathbf{q}}_3.$$

We defined our local coordinate system (see Note 15.1) in discharge space so that the vectors $\left(\mathbf{K}\nabla\theta_h\right)\times\left(\mathbf{K}\nabla Y_h\right)$ and $\hat{\mathbf{q}}_3$ point in the same direction, so

$$\left[\left(\mathbf{K}\nabla\theta_h\right)\cdot\hat{\mathbf{q}}_1\right]\left[\left(\mathbf{K}\nabla Y_h\right)\cdot\hat{\mathbf{q}}_2\right]-\left[\left(\mathbf{K}\nabla\theta_h\right)\cdot\hat{\mathbf{q}}_2\right]\left[\left(\mathbf{K}\nabla Y_h\right)\cdot\hat{\mathbf{q}}_1\right] = \left|\left(\mathbf{K}\nabla\theta_h\right)\times\left(\mathbf{K}\nabla Y_h\right)\right|.$$

Substituting the RHS for the corresponding term in the numerator on the RHS of (15.39) gives

$$\dot{\Omega}(\mathbf{x},t;\omega_m) = \frac{\omega_m M_h(\mathbf{x};\omega_m)\left|\left[\mathbf{K}(\mathbf{x})\nabla\theta_h(\mathbf{x};\omega_m)\right]\times\left[\mathbf{K}(\mathbf{x})\nabla M_h(\mathbf{x};\omega_m)\right]\right|}{\left|\mathbf{q}'(\mathbf{x},t;\omega_m)\right|^2} \qquad (15.40)$$

which we can also write as

$$\dot{\Omega}(\mathbf{x},t;\omega_m) = \omega_m M_h(\mathbf{x};\omega_m)\sin\gamma(\mathbf{x};\omega_m) \times \frac{\left|\mathbf{K}(\mathbf{x})\nabla\theta_h(\mathbf{x};\omega_m)\right|\left|\mathbf{K}(\mathbf{x})\nabla M_h(\mathbf{x};\omega_m)\right|}{\left|\mathbf{q}'(\mathbf{x},t;\omega_m)\right|^2} \qquad (15.41)$$

where $\gamma(\mathbf{x};\omega_m)$ is the smallest positive angle ($0 < \gamma < \pi$ rad) formed by the two vectors $\hat{\mathbf{u}}(\mathbf{x};\omega_m)$ and $\hat{\mathbf{v}}(\mathbf{x};\omega_m)$.

Regarding (15.40) and (15.41), we observe the following:

- The rotation rate is time periodic, because $|\mathbf{q}'|$ is time periodic.
- The rotation rate is positive for all times. By assumption $0 < \gamma < \pi$ rad, so $\hat{\mathbf{u}}(\mathbf{x};\omega_m)$ and $\hat{\mathbf{v}}(\mathbf{x};\omega_m)$ are noncollinear and the plane of oscillation is uniquely defined. In this case $\dot{\Omega} > 0$, so the specific-discharge constituent appears to rotate anticlockwise when viewed from above the plane of oscillation and clockwise when viewed from below.
- The numerator on the RHS is time invariant. The denominator, $|\mathbf{q}'|^2$, generally is not time invariant, so the rotation rate is time varying. Consequently, the azimuth angle, Ω, generally varies nonlinearly with time (and with ξ_h). In some special cases (e.g., see Example 15.5) $|\mathbf{q}'(\mathbf{x},t;\omega_m)|$ is time invariant; then the rotation rate is time invariant.
- The rotation rate is inversely proportional to the square of the specific-discharge constituent magnitude. Thus, the constituent rotates faster at those times in its cycle when its magnitude is small than at those times when its magnitude is large.

Equation (15.40) implies that if maxima with respect to time of the rotation rate $\dot{\Omega}(\mathbf{x},t;\omega_m)$ exist, then they correspond to minima with respect to time of the magnitude $\left|\mathbf{q}'(\mathbf{x},t;\omega_m)\right|$, and vice versa.

Exercise 15.12 Time Extrema of Rotation Rate for Specific Discharge

Problem: Suppose $\nabla M_h(\mathbf{x};\omega_m) \times \nabla\theta_h(\mathbf{x};\omega_m) \neq \mathbf{0}$. Formally show that if extrema with respect to time of the rotation rate $\dot{\Omega}(\mathbf{x},t;\omega_m)$ exist, then they coincide in time with extrema with respect to time of the magnitude $|\mathbf{q}'(\mathbf{x},t;\omega_m)|$.

Solution: If extrema of the rotation rate $\dot{\Omega}(\mathbf{x},t;\omega_m)$ exist, then they occur at those critical times t_r which we define as

$$\left[\frac{d}{dt}\dot{\Omega}(\mathbf{x},t;\omega_m)\right]\bigg|_{t=t_r} = 0.$$

Substituting the RHS of (15.40) for $\dot{\Omega}$ and then simplifying yield

$$\left[\frac{d}{dt}\left|\mathbf{q}'(\mathbf{x},t;\omega_m)\right|\right]\bigg|_{t=t_r} = 0 \tag{15.42}$$

which implies that the critical times t_r of the rotation rate also are critical times of the magnitude $|\mathbf{q}'(\mathbf{x},t;\omega_m)|$. □

Note 15.2 Rotation Vector for Specific-Discharge Constituent
We define the *rotation vector*, $\boldsymbol{\Gamma}(\mathbf{x},t;\omega_m)$, for the specific-discharge harmonic constituent as

$$\boldsymbol{\Gamma}(\mathbf{x},t;\omega_m) \equiv \frac{\omega_m M_h(\mathbf{x};\omega_m)}{\left|\mathbf{q}'(\mathbf{x},t;\omega_m)\right|^2}\left[\mathbf{K}(\mathbf{x})\nabla\theta_h(\mathbf{x};\omega_m)\right] \times \left[\mathbf{K}(\mathbf{x})\nabla M_h(\mathbf{x};\omega_m)\right].$$

The rotation vector has the following characteristics:

- $\boldsymbol{\Gamma}$ has the dimensions of reciprocal time.
- $\boldsymbol{\Gamma}$ is perpendicular to the plane of rotation (i.e., the plane of oscillation).
- Using the right-hand rule, the direction of $\boldsymbol{\Gamma}$ gives the sense of rotation, clockwise or anticlockwise, of the specific-discharge constituent.
- The magnitude of $\boldsymbol{\Gamma}$ gives the absolute rate of rotation:

$$\left|\boldsymbol{\Gamma}(\mathbf{x},t;\omega_m)\right| = \left|\dot{\Omega}(\mathbf{x},t;\omega_m)\right|.$$

The orientation of the plane of oscillation does not change with time, so the direction of the rotation vector is time invariant; the rotation appears anticlockwise when viewed from above the plane of oscillation. However, the rotation *rate* generally is not time invariant, so the magnitude of the rotation vector generally is not time invariant. △

15.6.1.3. Time Variation of Azimuth Angle

In this section we derive a closed-form expression for the azimuth angle $\Omega(\mathbf{x},t;\omega_m)$ as a function of time by integrating (15.40).

Substituting the RHS of (15.6) for $|\mathbf{q}'|$ in (15.40) gives

$$\dot{\Omega}(\mathbf{x},t;\omega_m) = \frac{\omega_m\left|(\mathbf{K}\nabla\theta_h)\times(\mathbf{K}\nabla Y_h)\right|}{\left\{\left(\left|\mathbf{K}\nabla Y_h\right|^2\cos^2\xi_h - 2(\mathbf{K}\nabla Y_h)\cdot(\mathbf{K}\nabla\theta_h)\cos\xi_h\sin\xi_h + \left|\mathbf{K}\nabla\theta_h\right|^2\sin^2\xi_h\right)\right\}}. \tag{15.43}$$

Using equations 4.21.27 and 4.21.28 of *Roy and Olver* [2016], we can write this as

$$\dot{\Omega}(\mathbf{x},t;\omega_m) = \frac{2\omega_m\left|(\mathbf{K}\nabla\theta_h)\times(\mathbf{K}\nabla Y_h)\right|}{a + b\cos(2\xi_h) + c\sin(2\xi_h)} \tag{15.44}$$

where

$$a \equiv \left|\mathbf{K}\nabla Y_h\right|^2 + \left|\mathbf{K}\nabla\theta_h\right|^2 \tag{15.45a}$$

$$b \equiv \left|\mathbf{K}\nabla Y_h\right|^2 - \left|\mathbf{K}\nabla\theta_h\right|^2 \tag{15.45b}$$

$$c \equiv -2(\mathbf{K}\nabla Y_h)\cdot(\mathbf{K}\nabla\theta_h). \tag{15.45c}$$

Substituting $t = u$ in (15.44) and then integrating with respect to u, from $u = 0$ to $u = t$, give

$$\Omega(\mathbf{x},t;\omega_m) - \Omega(\mathbf{x},0;\omega_m) = 2\omega_m\left|(\mathbf{K}\nabla\theta_h)\times(\mathbf{K}\nabla Y_h)\right| \times \int_0^t \frac{du}{a + b\cos(2\xi) + c\sin(2\xi)} \tag{15.46}$$

where $\xi \equiv \omega_m u + \theta_h(\mathbf{x};\omega_m)$. Define a new variable of integration as $v \equiv 2\xi$ so that $dv = 2\omega_m\,du$. Substituting these in (15.46) yields

$$\Omega(\mathbf{x},t;\omega_m) - \Omega(\mathbf{x},0;\omega_m) = 2\omega_m\left|(\mathbf{K}\nabla\theta_h)\times(\mathbf{K}\nabla Y_h)\right| \times \int_0^t \frac{du}{a + b\cos v + c\sin v}. \tag{15.47}$$

We use equations 2.558(4) of *Gradshteyn and Ryzhik* [1980] to carry out the integration. *Gradshteyn and Ryzhik* [1980] distinguish four cases, only one of which is relevant in the present context—the case where $a^2 > b^2 + c^2$. Equivalently, $\left|(\mathbf{K}\nabla Y_h)\cdot(\mathbf{K}\nabla\theta_h)\right| < \left|\mathbf{K}\nabla\theta_h\right|\left|\mathbf{K}\nabla Y_h\right|$. Then (15.47) evaluates to

$$\Omega(\mathbf{x},t;\omega_m) = \Omega(\mathbf{x},0;\omega_m) + \frac{\left|(\mathbf{K}\nabla\theta_h)\times(\mathbf{K}\nabla Y_h)\right|}{\sqrt{\left|\mathbf{K}\nabla Y_h\right|^2\left|\mathbf{K}\nabla\theta_h\right|^2 - \left[(\mathbf{K}\nabla Y_h)\cdot(\mathbf{K}\nabla\theta_h)\right]^2}} \times \left[\arctan\frac{\left|\mathbf{K}\nabla\theta_h\right|^2\tan(v/2) - (\mathbf{K}\nabla Y_h)\cdot(\mathbf{K}\nabla\theta_h)}{\sqrt{\left|\mathbf{K}\nabla Y_h\right|^2\left|\mathbf{K}\nabla\theta_h\right|^2 - \left[(\mathbf{K}\nabla Y_h)\cdot(\mathbf{K}\nabla\theta_h)\right]^2}}\right]\Bigg|_{2\theta_h}^{2\xi_h}. \tag{15.48}$$

Note 15.3 Vector Algebra: Lagrange's Identity
The vector identity

$$|\mathbf{a} \times \mathbf{b}|^2 = |\mathbf{a}|^2|\mathbf{b}|^2 - (\mathbf{a} \cdot \mathbf{b})^2 \qquad (15.49)$$

is a form of *Lagrange's identity* [*Weisstein*, 2016f]. △

Now we turn to Lagrange's identity (see Note 15.3). Letting $\mathbf{a} \equiv \mathbf{K}\nabla\theta_h$ and $\mathbf{b} \equiv \mathbf{K}\nabla Y_h$ in (15.49) and then taking the square root give

$$\left|(\mathbf{K}\nabla\theta_h) \times (\mathbf{K}\nabla Y_h)\right| = \sqrt{|\mathbf{K}\nabla\theta_h|^2|\mathbf{K}\nabla Y_h|^2 - \left[(\mathbf{K}\nabla Y_h)\cdot(\mathbf{K}\nabla\theta_h)\right]^2}.$$

Using this result we can write (15.48) as

$$\Omega(\mathbf{x}, t; \omega_m) = \Omega(\mathbf{x}, 0; \omega_m) + \left[\arctan \frac{|\mathbf{K}\nabla\theta_h|^2 \tan(v/2) - (\mathbf{K}\nabla Y_h)\cdot(\mathbf{K}\nabla\theta_h)}{\sqrt{|\mathbf{K}\nabla Y_h|^2|\mathbf{K}\nabla\theta_h|^2 - \left[(\mathbf{K}\nabla Y_h)\cdot(\mathbf{K}\nabla\theta_h)\right]^2}}\right]_{2\theta_h}^{2\xi_h}.$$

After evaluating the term within square brackets at the upper and lower limits, we obtain

$$\begin{aligned}\Omega(\mathbf{x}, t; \omega_m) &= \Omega(\mathbf{x}, 0; \omega_m) \\ &\quad + \arctan \frac{|\mathbf{K}\nabla\theta_h|^2 \tan\xi_h - (\mathbf{K}\nabla Y_h)\cdot(\mathbf{K}\nabla\theta_h)}{\left|(\mathbf{K}\nabla\theta_h) \times (\mathbf{K}\nabla Y_h)\right|} \\ &\quad - \arctan \frac{|\mathbf{K}\nabla\theta_h|^2 \tan\theta_h - (\mathbf{K}\nabla Y_h)\cdot(\mathbf{K}\nabla\theta_h)}{\left|(\mathbf{K}\nabla\theta_h) \times (\mathbf{K}\nabla Y_h)\right|}\end{aligned} \qquad (15.50)$$

where $\xi_h \equiv \omega_m t + \theta_h(\mathbf{x}; \omega_m)$.

Equation (15.50) shows that generally the azimuth angle Ω is nonlinear in ξ_h and thus nonlinear in t. However, special cases exist in which Ω is linear in ξ_h and thus linear in t. For example, if both,

$$\hat{\mathbf{u}}(\mathbf{x}; \omega_m) \cdot \hat{\mathbf{v}}(\mathbf{x}; \omega_m) = 0$$
$$\left|\mathbf{K}(\mathbf{x})\nabla Y_h(\mathbf{x}; \omega_m)\right| = \left|\mathbf{K}(\mathbf{x})\nabla\theta_h(\mathbf{x}; \omega_m)\right|$$

are satisfied, then (15.50) gives

$$\Omega(\mathbf{x}, t; \omega_m) = \Omega(\mathbf{x}, 0; \omega_m) + \omega_m t.$$

15.6.2. Envelope of Specific-Discharge Constituent

In this section we resume our investigation of the envelope of the specific-discharge constituent in the case where $\nabla\theta_h(\mathbf{x}; \omega_m) \times \nabla M_h(\mathbf{x}; \omega_m) \neq \mathbf{0}$.

What does the envelope look like? We know the following:

• It lies entirely within the plane of oscillation, because $\mathbf{q}'(\mathbf{x}, t; \omega_m)$ parallels the plane of oscillation.

• It is a closed curve, because $\mathbf{q}'(\mathbf{x}, t; \omega_m)$ is time periodic.

• It is symmetric about the origin in discharge space, based on (15.3).

• It contains the endpoints of the two vectors $\pm\mathbf{K}(\mathbf{x})\nabla M_h(\mathbf{x}; \omega_m)$, because when

$$t = \frac{k\pi \text{ rad} - \theta_h(\mathbf{x}; \omega_m)}{\omega_m}, \quad k \in \mathbb{Z}$$

the specific-discharge constituent takes these values.

• It contains the endpoints of the two vectors $\pm M_h(\mathbf{x}; \omega_m)\mathbf{K}(\mathbf{x})\nabla\theta_h(\mathbf{x}; \omega_m)$, because when

$$t = \frac{(k + 1/2)\pi \text{ rad} - \theta_h(\mathbf{x}; \omega_m)}{\omega_m}, \quad k \in \mathbb{Z}$$

the specific-discharge constituent takes these values.

Note 15.4 Specific-Discharge Envelope Plots
We can think of the discharge envelope as a curve described by $r(\Omega) = |\mathbf{q}'|$, where r is a radius that varies with the azimuth angle Ω. For an example, see the upper graph in Figure 15.2. In this and all other polar plots of this book, displayed azimuth (Ω) grid values are in degrees.

For convenience in plotting, we define the *dimensionless envelope* of the specific-discharge constituent as

$$r_D(\mathbf{x}, t; \omega_m) \equiv \frac{\left|\mathbf{q}'(\mathbf{x}, t; \omega_m)\right|}{\max_t \left|\mathbf{q}'(\mathbf{x}, t; \omega_m)\right|} \qquad (15.51)$$

where $\max_t$ denotes the maximum with respect to time (t). Then $0 \leq r_D \leq 1$. Below we present polar plots of the dimensionless envelope plotted in the plane of oscillation.

We also present rectangular (xy) plots of r_D versus Ω and r_D versus ξ_h. For examples, see the middle and lower graphs in Figure 15.2. The r_D-versus-Ω plots show the same information as the polar plots, but in a different format. The r_D-versus-ξ_h plots essentially show how the magnitude varies with time. △

Define the *time-scaled*, specific-discharge constituent as

$$\mathbf{q}'_{\text{scaled}}(\mathbf{x}, t; \omega_m) \equiv \frac{\mathbf{q}'(\mathbf{x}, t; \omega_m)}{\omega_m}. \qquad (15.52)$$

The scaled specific discharge has dimensions of length.

Over every time interval whose duration is equal to one constituent period ($T_m = 2\pi \text{ rad}/\omega_m$), the time-scaled specific-discharge constituent sweeps the same 2D region in its plane of oscillation—the region enclosed by the envelope.

Example 15.5 Special Case: Perpendicular Amplitude and Phase Gradients
Suppose that for some harmonic constituent (i.e., ω_m fixed) and some fixed location $\mathbf{x} = \mathrm{x}$, $\hat{\mathbf{u}}(\mathbf{x}; \omega_m)$ and

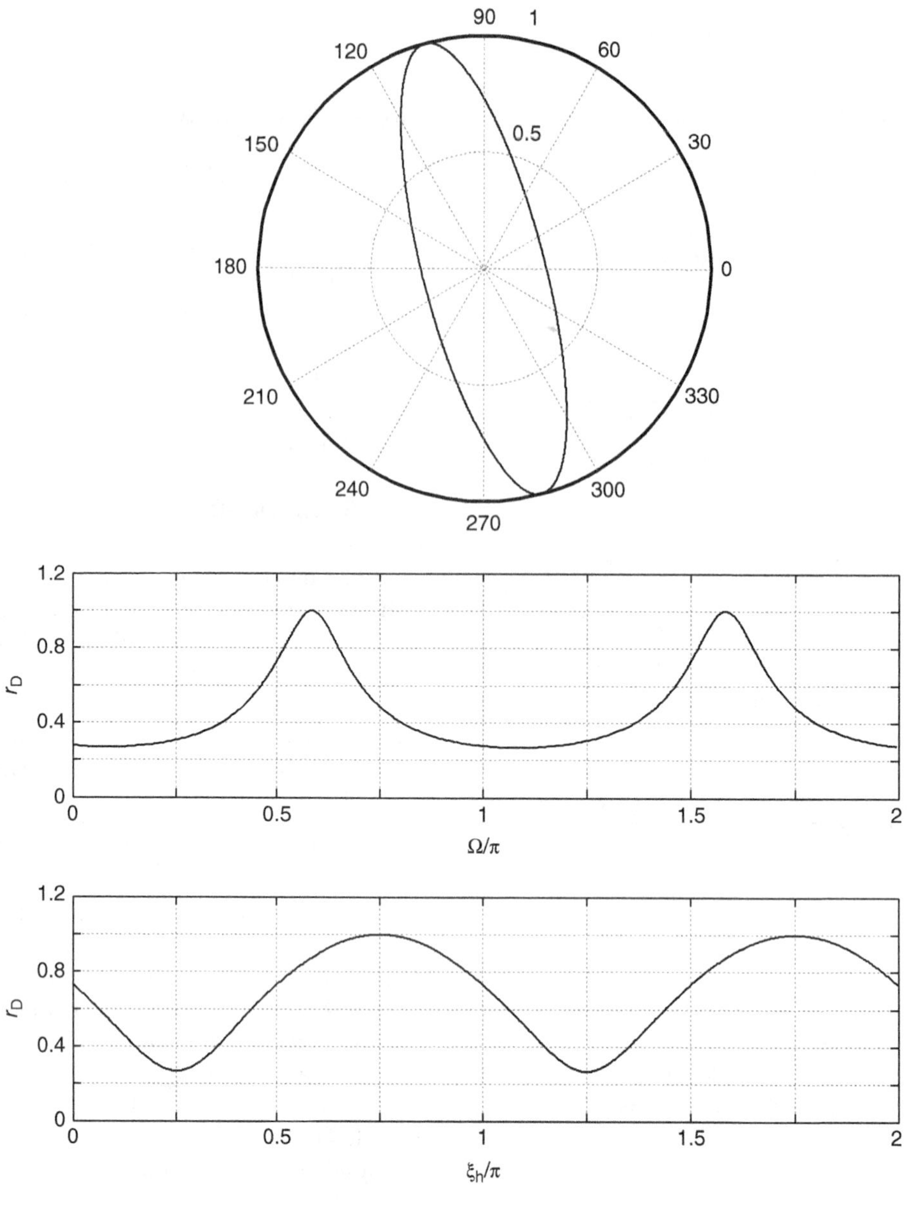

Figure 15.2 Example of dimensionless envelope for specific-discharge constituent.

$\hat{\mathbf{v}}(\mathbf{x};\omega_m)$ are perpendicular. In this case we can define our reference system in discharge space as follows:

$$\hat{\mathbf{q}}_1 \equiv \hat{\mathbf{v}}(\mathbf{x};\omega_m)$$
$$\hat{\mathbf{q}}_2 \equiv \hat{\mathbf{u}}(\mathbf{x};\omega_m)$$
$$\hat{\mathbf{q}}_3 \equiv \hat{\mathbf{q}}_1 \times \hat{\mathbf{q}}_2.$$

Combining these with (15.21) leads to

$$\begin{aligned}\mathbf{q}'(\mathbf{x},t;\omega_m) = &-\left|\mathbf{K}(\mathbf{x})\nabla M_\mathrm{h}(\mathbf{x};\omega_m)\right| \cos\xi_\mathrm{h}(\mathbf{x},t;\omega_m)\hat{\mathbf{q}}_2\\ &+ M_\mathrm{h}(\mathbf{x};\omega_m)\left|\mathbf{K}(\mathbf{x})\nabla\theta_\mathrm{h}(\mathbf{x};\omega_m)\right|\\ &\times \sin\xi_\mathrm{h}(\mathbf{x},t;\omega_m)\hat{\mathbf{q}}_1.\end{aligned}$$

Therefore, the plane of oscillation has the equation

$$q_3 = 0 \tag{15.53}$$

and the endpoints of the scaled specific-discharge harmonic constituent form an ellipse in the plane of oscillation. The ellipse is centered at the origin, and its major and minor axes are parallel to the $\hat{\mathbf{q}}_1$ and $\hat{\mathbf{q}}_2$ vectors. We have the following three possibilities:

- If $\left|\mathbf{K}(\mathbf{x})\nabla M_\mathrm{h}(\mathbf{x};\omega_m)\right| > M_\mathrm{h}(\mathbf{x};\omega_m)\left|\mathbf{K}(\mathbf{x})\nabla\theta_\mathrm{h}(\mathbf{x};\omega_m)\right|$, then the semimajor axis parallels the vector $\hat{\mathbf{q}}_2$ and has length $\omega_m^{-1}\left|\mathbf{K}(\mathbf{x})\nabla M_\mathrm{h}(\mathbf{x};\omega_m)\right|$. The semiminor axis parallels the vector $\hat{\mathbf{q}}_1$ and has length $\omega_m^{-1} M_\mathrm{h}(\mathbf{x};\omega_m)\left|\mathbf{K}(\mathbf{x})\nabla\theta_\mathrm{h}(\mathbf{x};\omega_m)\right|$.

- If $|\mathbf{K}(\mathbf{x})\nabla M_h(\mathbf{x};\omega_m)| = M_h(\mathbf{x};\omega_m)|\mathbf{K}(\mathbf{x})\nabla\theta_h(\mathbf{x};\omega_m)|$, then the ellipse is a circle of radius $\omega_m^{-1}|\mathbf{K}(\mathbf{x})\nabla M_h(\mathbf{x};\omega_m)|$.
- If $|\mathbf{K}(\mathbf{x})\nabla M_h(\mathbf{x};\omega_m)| < M_h(\mathbf{x};\omega_m)|\mathbf{K}(\mathbf{x})\nabla\theta_h(\mathbf{x};\omega_m)|$, then the semimajor axis parallels the vector $\hat{\mathbf{q}}_1$ and has length $\omega_m^{-1}M_h(\mathbf{x};\omega_m)|\mathbf{K}(\mathbf{x})\nabla\theta_h(\mathbf{x};\omega_m)|$. The semiminor axis parallels the vector $\hat{\mathbf{q}}_2$ and has length $\omega_m^{-1}|\mathbf{K}(\mathbf{x})\nabla M_h(\mathbf{x};\omega_m)|$.

◊

If $\nabla M_h(\mathbf{x};\omega_m) \times \nabla\theta_h(\mathbf{x};\omega_m) \neq \mathbf{0}$, then the area swept by the scaled, specific-discharge constituent between times 0 and t, where $0 \le t \le 2\pi/\omega_m$, is

$$A(\mathbf{x},t;\omega_m) = \int_{\Omega_0}^{\Omega} \frac{|\mathbf{q}'_{\text{scaled}}|^2}{2}\,d\beta \tag{15.54}$$

where

$$\Omega_0 \equiv \Omega(\mathbf{x},0;\omega_m)$$
$$\Omega \equiv \Omega(\mathbf{x},t;\omega_m).$$

Figure 15.3 shows an example of the plane of oscillation. In Figure 15.3, the elliptical curve (blue line) is the envelope, and the shaded region is the area swept by the scaled specific-discharge constituent between time $t = 0$ and a later time $t > 0$.

Exercise 15.13 Kepler-Type Law for Scaled, Specific-Discharge Constituent

Problem: Using calculus, show that the instantaneous, areal sweep rate of the scaled, specific-discharge constituent (i.e., dA/dt) is constant, independent of time. This characteristic of the specific-discharge oscillation is analogous to Kepler's second law of planetary motion, which states that a line segment joining a planet and the Sun sweeps out equal areas during equal intevals of time [*Goldstein*, 1980]. Also, express the total area swept during each constituent period in terms of M_h, $\mathbf{K}\nabla M_h$, $\mathbf{K}\nabla\theta_h$, and ω_m.

Solution: From (15.54), the differential of the area is

$$dA = \frac{|\mathbf{q}'_{\text{scaled}}|^2}{2}\,d\Omega.$$

The corresponding time derivative gives the instantaneous areal sweep rate:

$$\frac{dA}{dt} = \frac{|\mathbf{q}'_{\text{scaled}}|^2}{2}\dot{\Omega}.$$

Substituting the RHS of (15.52) for $\mathbf{q}'_{\text{scaled}}$ and the RHS of (15.40) for $\dot{\Omega}$ yields

$$\frac{dA}{dt} = \frac{M_h(\mathbf{x};\omega_m)}{2\omega_m}|\mathbf{K}(\mathbf{x})\nabla\theta_h(\mathbf{x};\omega_m) \times \mathbf{K}(\mathbf{x})\nabla M_h(\mathbf{x};\omega_m)|. \tag{15.55}$$

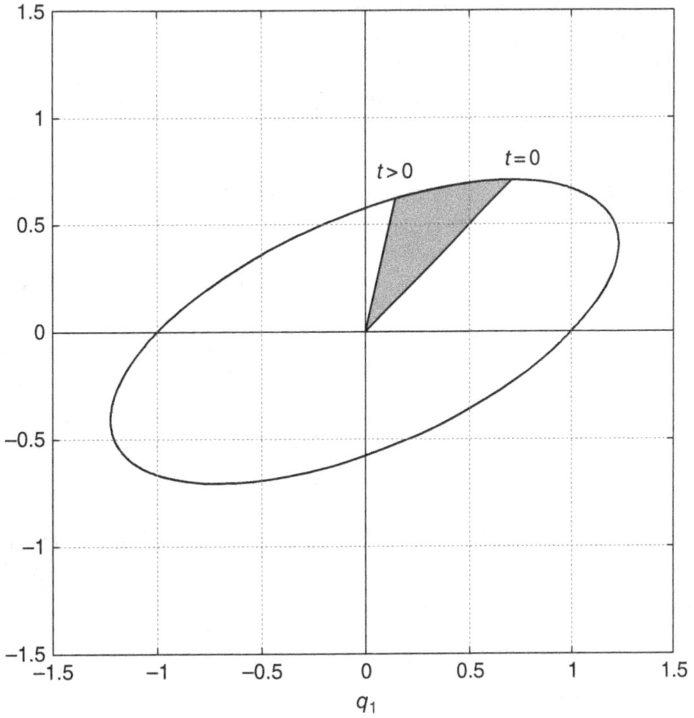

Figure 15.3 Example of plane of oscillation.

The RHS is time invariant. Therefore, the areal sweep rate is constant, independent of time. Integrating (15.55) with respect to time, from $t = 0$ to $t = 2\pi/\omega_m$, gives the total area swept during one constituent period as

$$A(\mathbf{x}, 2\pi/\omega_m; \omega_m) = \frac{\pi M_h(\mathbf{x}; \omega_m)}{\omega_m^2} \left| \mathbf{K}(\mathbf{x})\nabla\theta_h(\mathbf{x}; \omega_m) \times \mathbf{K}(\mathbf{x})\nabla M_h(\mathbf{x}; \omega_m) \right|.$$

□

15.6.2.1. Critical Angles of Envelope

Consider the envelope as a function of the azimuth angle Ω:

$$|\mathbf{q}'(\mathbf{x}, t; \omega_m)| = f(\Omega).$$

We define the critical azimuth angles (Ω_c) of the envelope, if any exist, as follows:

$$\left(\frac{d}{d\Omega} |\mathbf{q}'(\mathbf{x}, t; \omega_m)| \right) \bigg|_{\Omega=\Omega_c} = 0. \qquad (15.56)$$

If the magnitude of the specific-discharge constituent has a local extremum (maximum or minimum) with respect to the azimuth angle Ω, then it occurs at $\Omega = \Omega_c$.

15.6.2.2. Correspondence of Critical Angles and Critical Times

At any fixed location, the azimuth angle Ω is a function of time: $\Omega = \Omega(\mathbf{x}, t; \omega_m)$. In Section 15.6.1 we saw that the rotation rate $\dot{\Omega}$ is nonnegative. If we restrict our analysis to those situations in which $\hat{\mathbf{u}}(\mathbf{x}; \omega_m)$ and $\hat{\mathbf{v}}(\mathbf{x}; \omega_m)$ are noncollinear, then $\nabla M_h(\mathbf{x}; \omega_m) \times \nabla\theta_h(\mathbf{x}; \omega_m) \neq \mathbf{0}$ and the rotation rate is positive. Multiplying (15.56) by the rotation rate, we obtain

$$\left(\frac{d}{d\Omega} |\mathbf{q}'(\mathbf{x}, t; \omega_m)| \right) \bigg|_{\Omega=\Omega_c} \left(\frac{d\Omega}{dt} \right) \bigg|_{\Omega=\Omega_c} = 0$$

which we can write as

$$\left[\left(\frac{d}{d\Omega} |\mathbf{q}'(\mathbf{x}, t; \omega_m)| \right) \frac{d\Omega}{dt} \right] \bigg|_{\Omega=\Omega_c} = 0.$$

Using the chain rule to rewrite the LHS leads to

$$\left(\frac{d}{dt} |\mathbf{q}'(\mathbf{x}, t; \omega_m)| \right) \bigg|_{\Omega=\Omega_c} = 0.$$

Comparing this equation to (15.5), evidently the critical angles satisfy

$$\Omega_c = \Omega(\mathbf{x}, t_c; \omega_m) \qquad (15.57)$$

where the t_c are the critical times for the specific-discharge constituent magnitude [see (15.5)]. In words, the critical angles are those values that the azimuth angle takes at the critical times.

15.6.2.3. Azimuthal Distribution of Critical Angles

In Section 15.2 we saw that if every time interval of duration 2π rad$/\omega_m$ contains a finite number of critical angles, then that number is four. In the previous section we found that within every time interval of duration $2\pi/\omega_m$ there is a correspondence between the critical times t_c and the critical angles Ω_c. Therefore, in every interval of length 2π rad that has a finite number of critical angles, that number is four.

Equation (15.19) implies that every numerically consecutive pair of critical angles differ by $\pi/2$ rad. No two numerically consecutive critical values can be maxima, and no two consecutive values can be minima; the maxima and minima must occur alternately. Consequently, in every interval of length 2π rad with a finite number of critical angles, two of the critical angles correspond to magnitude maxima and two correspond to minima. Because the envelope is symmetric about the origin, the two maxima are equal and the two minima are equal.

In summary, if $\nabla M_h(\mathbf{x}; \omega_m) \times \nabla\theta_h(\mathbf{x}; \omega_m) \neq \mathbf{0}$, then one of the following is true:

- Case 1: Every azimuth angle $\Omega(\mathbf{x}, t; \omega_m)$ is a critical angle, because $|\mathbf{q}'(\mathbf{x}, t; \omega_m)|$ is independent of Ω. Then the envelope is a circle and $|\mathbf{q}'(\mathbf{x}, t; \omega_m)|$ has no extrema.
- Case 2: Every interval of length 2π rad contains four critical angles (Ω_c) and numerically consecutive critical angles differ by $\pi/2$ rad.

Exercise 15.14 Calculation and Plotting of Envelope

Problem: Suppose that for some harmonic constituent (i.e., ω_m fixed) and a fixed location $\mathbf{x}$ we have

$$\mathbf{K}(\mathbf{x})\nabla Y_h(\mathbf{x}; \omega_m) = \left| \mathbf{K}(\mathbf{x})\nabla Y_h(\mathbf{x}; \omega_m) \right| \frac{\sqrt{2}}{2} (\hat{\mathbf{x}}_1 + \hat{\mathbf{x}}_2) \qquad (15.58a)$$

$$\mathbf{K}(\mathbf{x})\nabla\theta_h(\mathbf{x}; \omega_m) = \left| \mathbf{K}(\mathbf{x})\nabla Y_h(\mathbf{x}; \omega_m) \right| \hat{\mathbf{x}}_1. \qquad (15.58b)$$

What are the critical values of ξ_h? Complete the entries of Table 15.2 shown by question marks (?). Using the data in your completed table and the fact that $|\mathbf{q}'|$ is symmetric about the origin, sketch the scaled envelope on a 2D polar plot. What are the critical angles Ω_c?

Solution: Combining (15.1) with (15.58) allows us, in this case, to write the specific-discharge constituent as

$$\frac{\mathbf{q}'(\mathbf{x}, t; \omega_m)}{\left| \mathbf{K}(\mathbf{x})\nabla M_h(\mathbf{x}; \omega_m) \right|} = -\hat{\mathbf{u}}(\mathbf{x}; \omega_m) \cos\xi_h(\mathbf{x}, t; \omega_m) + \hat{\mathbf{v}}(\mathbf{x}; \omega_m) \sin\xi_h(\mathbf{x}, t; \omega_m) \qquad (15.59)$$

Table 15.2 Problem of Exercise 15.14.

ξ_h (rad)	$\cos\xi_h$	$\sin\xi_h$	$\frac{q'_1}{\lvert\mathbf{K}\nabla M_h\rvert}$	$\frac{q'_2}{\lvert\mathbf{K}\nabla M_h\rvert}$	$\frac{\lvert\mathbf{q}'\rvert}{\lvert\mathbf{K}\nabla M_h\rvert}$	r_D	Ω (rad)
0	1	0	?	?	?	?	?
$\frac{\pi}{4}$	$\frac{\sqrt{2}}{2}$	$\frac{\sqrt{2}}{2}$	?	?	?	?	?
$\frac{\pi}{2}$	0	1	?	?	?	?	?
$\frac{3\pi}{4}$	$-\frac{\sqrt{2}}{2}$	$\frac{\sqrt{2}}{2}$	?	?	?	?	?

where (15.20) define $\hat{\mathbf{u}}$ and $\hat{\mathbf{v}}$ and

$$\hat{\mathbf{u}}(\mathbf{x};\omega_m) = \frac{\sqrt{2}}{2}(\hat{\mathbf{x}}_1 + \hat{\mathbf{x}}_2) \quad (15.60a)$$

$$\hat{\mathbf{v}}(\mathbf{x};\omega_m) = \hat{\mathbf{x}}_1. \quad (15.60b)$$

Equations (15.58) imply that $|\mathbf{K}\nabla\theta_h| = |\mathbf{K}\nabla Y_h|$. Substituting these in (15.10) gives the critical values of ξ_h as $\xi_c = (2k+1)\pi/4$ where $k \in \mathbb{Z}$. Using (15.36) with (15.59) and (15.60) yields

$$\frac{(\mathbf{q}')_1}{|\mathbf{K}\nabla M_h|} = -\frac{\sqrt{2}}{2}\cos\xi_h + \sin\xi_h \quad (15.61a)$$

$$\frac{(\mathbf{q}')_2}{|\mathbf{K}\nabla M_h|} = -\frac{\sqrt{2}}{2}\cos\xi_h. \quad (15.61b)$$

Use these to calculate the x_1 and x_2 components of $\mathbf{q}'/|\mathbf{K}\nabla M_h|$. Use (15.51) to calculate the dimensionless envelope value, r_D. Use (15.37) to calculate the azimuth angle. See Table 15.3 and Figure 15.4. □

Exercise 15.15 Special Case: Perpendicular Components

Problem: Suppose that for some harmonic constituent (i.e., ω_m fixed) and some fixed location $\mathbf{x}$ the vectors $\mathbf{K}(\mathbf{x})\nabla M_h(\mathbf{x};\omega_m)$ and $\mathbf{K}(\mathbf{x})\nabla\theta_h(\mathbf{x};\omega_m)$ are perpendicular. Use (15.50) to show that each numerically consecutive pair of critical angles Ω_c differ by $\pi/2$ rad.

Solution: If $(\mathbf{K}\nabla\theta_h)\cdot(\mathbf{K}\nabla Y_h) = 0$, then $|(\mathbf{K}\nabla\theta_h)\times(\mathbf{K}\nabla Y_h)| = |\mathbf{K}\nabla\theta_h||\mathbf{K}\nabla Y_h|$. Substituting these results in (15.50) gives

$$\Omega(\mathbf{x},t;\omega_m) = \Omega(\mathbf{x},0;\omega_m) + \arctan\frac{|\mathbf{K}\nabla\theta_h|\tan\xi_h}{|\mathbf{K}\nabla Y_h|} - \arctan\frac{|\mathbf{K}\nabla\theta_h|\tan\theta_h}{|\mathbf{K}\nabla Y_h|}. \quad (15.62)$$

Let Ω_{c1} and Ω_{c2} be two numerically consecutive critical angles corresponding to the numerically consecutive critical times t_{c1} and t_{c2}, respectively, i.e.,

$$\Omega_{c1} \equiv \Omega(\mathbf{x},t_{c1};\omega_m) \quad (15.63a)$$

$$\Omega_{c2} \equiv \Omega(\mathbf{x},t_{c2};\omega_m). \quad (15.63b)$$

Then using (15.62) we can express their difference as

$$\Omega_{c2}-\Omega_{c1} = \arctan\frac{|\mathbf{K}\nabla\theta_h|\tan\xi_{c2}}{|\mathbf{K}\nabla Y_h|} - \arctan\frac{|\mathbf{K}\nabla\theta_h|\tan\xi_{c1}}{|\mathbf{K}\nabla Y_h|}. \quad (15.64)$$

Substituting $(\mathbf{K}\nabla\theta_h)\cdot(\mathbf{K}\nabla Y_h) = 0$ in (15.10) and then solving for ξ_c indicate that the critical values are integer multiples of $\pi/2$ rad:

$$\xi_c = \frac{k\pi}{2}\text{rad} \quad \forall k \in \mathbb{Z}. \quad (15.65)$$

Then for any two numerically consecutive critical values (ξ_{c1},ξ_{c2}) we have either (1) $|\tan\xi_{c1}| \to \infty$ and $\tan\xi_{c2} = 0$ or (2) $\tan\xi_{c1} = 0$ and $|\tan\xi_{c2}| \to \infty$. Using these results to evaluate the RHS of (15.64) yields

$$|\Omega_{c2} - \Omega_{c1}| = \frac{\pi}{2}. \quad (15.66)$$

□

15.6.2.4. *Envelope as Ellipse*

In this section we show that when $\nabla\theta_h(\mathbf{x};\omega_m) \times \nabla M_h(\mathbf{x};\omega_m) \neq \mathbf{0}$ the envelope is an ellipse.

Define a reference coordinate system in accordance with Note 15.1. Additionally, for this section only, define the q_1 and q_2 coordinate axes so that they are parallel to the directions of maximum and minimum $|\mathbf{q}'|$, respectively.

We can express (15.1) as

$$\mathbf{q}'(\mathbf{x},t;\omega_m) = -M_h(\mathbf{x};\omega_m)\left[\mathbf{u}\cos\xi_h(\mathbf{x},t;\omega_m) - \mathbf{v}\sin\xi_h(\mathbf{x},t;\omega_m)\right]$$

where

$$\mathbf{u} \equiv \mathbf{K}(\mathbf{x})\nabla Y_h(\mathbf{x};\omega_m)$$

$$\mathbf{v} \equiv \mathbf{K}(\mathbf{x})\nabla\theta_h(\mathbf{x};\omega_m).$$

The q_1 and q_2 components of the specific-discharge constituent are then, respectively,

$$q'_1(\mathbf{x},t;\omega_m) = -M_h(\mathbf{x};\omega_m)\left[u_1\cos\xi_h(\mathbf{x},t;\omega_m) - v_1\sin\xi_h(\mathbf{x},t;\omega_m)\right]$$

$$q'_2(\mathbf{x},t;\omega_m) = -M_h(\mathbf{x};\omega_m)\left[u_2\cos\xi_h(\mathbf{x},t;\omega_m) - v_2\sin\xi_h(\mathbf{x},t;\omega_m)\right].$$

Using the cosine angle-sum relation (B.2a), we can rewrite these as

$$q'_1(\mathbf{x},t;\omega_m) = -R_1\cos\left[\xi_h(\mathbf{x},t;\omega_m) + \alpha_1\right] \quad (15.67a)$$

$$q'_2(\mathbf{x},t;\omega_m) = -R_2\cos\left[\xi_h(\mathbf{x},t;\omega_m) + \alpha_2\right] \quad (15.67b)$$

where

$$R_1 \equiv M_h(\mathbf{x};\omega_m)\sqrt{u_1^2 + v_1^2}$$

$$R_2 \equiv M_h(\mathbf{x};\omega_m)\sqrt{u_2^2 + v_2^2}$$

Table 15.3 Solution for Exercise 15.14.

ξ_h (rad)	$\cos \xi_h$	$\sin \xi_h$	$\frac{q_1'}{\lvert \mathbf{K}\nabla M_h \rvert}$	$\frac{q_2'}{\lvert \mathbf{K}\nabla M_h \rvert}$	$\frac{\lvert \mathbf{q}' \rvert}{\lvert \mathbf{K}\nabla M_h \rvert}$	r_D	Ω (rad)
0	1	0	$\frac{-\sqrt{2}}{2}$	$\frac{-\sqrt{2}}{2}$	1	$\frac{2}{\sqrt{4+2\sqrt{2}}} \approx 0.77$	$\frac{\pi}{4}$
$\frac{\pi}{4}$	$\frac{\sqrt{2}}{2}$	$\frac{\sqrt{2}}{2}$	$\frac{-1+\sqrt{2}}{2}$	$\frac{-1}{2}$	$\frac{\sqrt{4-2\sqrt{2}}}{2}$	$\frac{\sqrt{4-2\sqrt{2}}}{\sqrt{4+2\sqrt{2}}} \approx 0.41$	$\frac{5\pi}{8}$
$\frac{\pi}{2}$	0	1	1	0	1	$\frac{2}{\sqrt{4+2\sqrt{2}}}$	π
$\frac{3\pi}{4}$	$-\frac{\sqrt{2}}{2}$	$\frac{\sqrt{2}}{2}$	$\frac{1+\sqrt{2}}{2}$	$\frac{1}{2}$	$\frac{\sqrt{4+2\sqrt{2}}}{2}$	1	$\frac{9\pi}{8}$

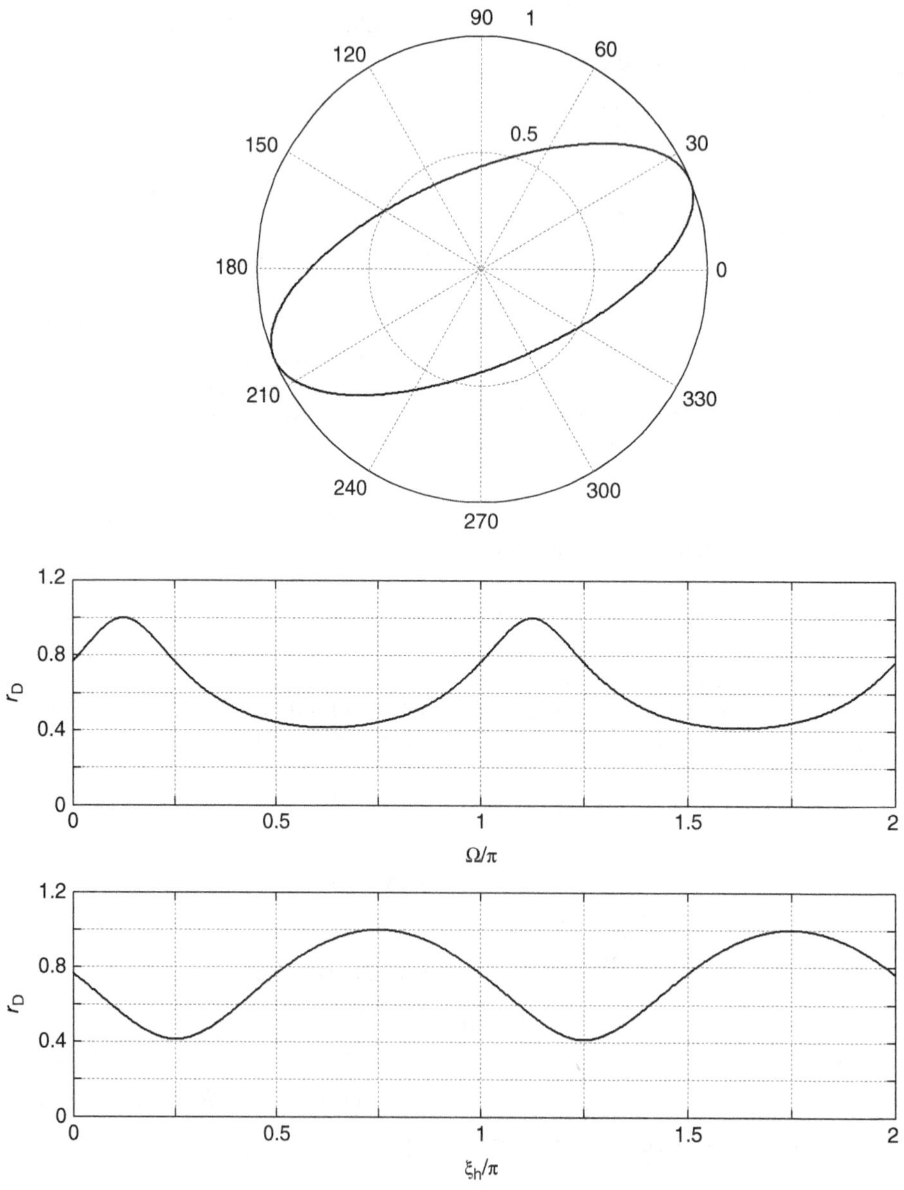

Figure 15.4 Dimensionless envelope for specific-discharge constituent in Exercise 15.14.

and

$$\tan\alpha_1 = \frac{u_1}{v_1}$$
$$\tan\alpha_2 = \frac{u_2}{v_2}.$$

When $|\mathbf{q}'|$ is a maximum (i.e., when $\Omega = n\pi$, where n is integer), we have $|\mathbf{q}'| = |q_1'|$ and $q_2' = 0$. Then (15.67) leads to

$$\cos\left[\xi_h(\mathbf{x}, t_c; \omega_m) + \alpha_1\right] = \pm 1 \quad (15.68a)$$
$$\cos\left[\xi_h(\mathbf{x}, t_c; \omega_m) + \alpha_2\right] = 0. \quad (15.68b)$$

These imply that $\alpha_2 - \alpha_1 = (k+1/2)\pi$ where k is an integer, so we can write (15.67) as

$$q_1'(\mathbf{x}, t; \omega_m) = -R_1 \cos\left[\xi_h(\mathbf{x}, t; \omega_m) + \alpha_1\right] \quad (15.69a)$$
$$q_2'(\mathbf{x}, t; \omega_m) = \pm R_2 \sin\left[\xi_h(\mathbf{x}, t; \omega_m) + \alpha_1\right]. \quad (15.69b)$$

These satisfy

$$\frac{(q_1')^2}{R_1^2} + \frac{(q_2')^2}{R_2^2} = 1. \quad (15.70)$$

This is the equation for an ellipse in canonical position (i.e., center at origin and major and minor semiaxes aligned with q_1 and q_2 coordinate axes, respectively) in discharge space. Here R_1 and R_2 are the major and minor semiaxes, respectively.

15.7. DISCUSSION

15.7.1. Rotation Condition

Exercise 15.16 Test for Time Rotation of Specific-Discharge Constituent

Problem: Suppose that for some constituent (i.e., ω_m fixed) and some fixed location $\mathbf{x}$ we wish to know if the corresponding specific-discharge vector constituent rotates with time. Can you think of a concise mathematical condition for making this determination?

Solution: Based on the results obtained in Sections 15.4, 15.5, and 15.6, the specific-discharge vector constituent rotates with time if and only if

$$\left|\left[\mathbf{K}(\mathbf{x})\nabla M_h(\mathbf{x}; \omega_m)\right] \times \left[\mathbf{K}(\mathbf{x})\nabla\theta_h(\mathbf{x}; \omega_m)\right]\right| > 0. \quad (15.71)$$

We can use the vector product equivalence (14.29) to express this more concisely as follows: The specific-discharge vector constituent rotates with time if and only if

$$\left|\nabla M_h(\mathbf{x}; \omega_m) \times \nabla\theta_h(\mathbf{x}; \omega_m)\right| > 0. \quad (15.72)$$

□

We call (15.72) the *rotation condition* for the specific-discharge vector harmonic constituent, $\mathbf{q}'(\mathbf{x}, t; \omega_m)$.

Exercise 15.17 Test for Time Rotation of Hydraulic Gradient Constituent

Problem: Suppose that for some constituent (i.e., ω_m fixed) and some fixed location $\mathbf{x}$ we wish to know if the corresponding hydraulic gradient constituent rotates with time. Derive a rotation condition suitable for this determination. How does your result compare to the rotation condition for the specific-discharge constituent (15.72)?

Solution: The hydraulic gradient constituent rotates with time if and only if

$$\left|\nabla M_h(\mathbf{x}; \omega_m) \times \nabla\theta_h(\mathbf{x}; \omega_m)\right| > 0. \quad (15.73)$$

The rotation condition is the same for the hydraulic gradient and specific-discharge constituents. □

15.7.2. Comparison with Other Work

As part of their investigation of flow visualization, *Smith et al.* [2005] analyzed the envelope of the mean pore water velocity harmonic constituent (they called it the *modal velocity ellipse*) in 2D flow. They concluded that the envelope is either a line segment or an ellipse, and they presented expressions [see their equations (4)–(10)] for the envelope. Their results for the envelope appear to be consistent with the more general results presented in this chapter.

15.8. CHAPTER SUMMARY

In this chapter we explored the time variation of the specific-discharge vector harmonic constituent, $\mathbf{q}'(\mathbf{x}, t; \omega_m)$. We introduced the concepts of discharge space and discharge envelope. Then we showed that under rather general conditions (i.e., 3D flow through nonhomogeneous, anisotropic media) one of the following cases describes the time variation:

1. $\nabla M_h(\mathbf{x}; \omega_m) = \mathbf{0}$ and $\nabla\theta_h(\mathbf{x}; \omega_m) = \mathbf{0}$. Then $\mathbf{q}'(\mathbf{x}, t; \omega_m)$ is time invariant because it vanishes for all times. The envelope is a point.
2. $\nabla M_h(\mathbf{x}; \omega_m)$ and $\nabla\theta_h(\mathbf{x}; \omega_m)$ are locally collinear or one of them (but not both) vanishes locally. Then $\mathbf{q}'(\mathbf{x}, t; \omega_m)$ alternates direction and magnitude with time but does not rotate. The envelope is a line segment.
3. $\nabla M_h(\mathbf{x}; \omega_m)$ and $\nabla\theta_h(\mathbf{x}; \omega_m)$ are locally nonzero and locally noncollinear. Then $\mathbf{q}'(\mathbf{x}, t; \omega_m)$ rotates within the plane of oscillation, and its magnitude also may vary periodically with time. The rotation rate generally varies with time, and the rotation satisfies a Kepler-type law (see Exercise 15.13). The envelope is an ellipse.

In all cases the envelope of the specific-discharge constituent is symmetric about the origin in discharge space.

Part V
Stationary Points

In this part we explore stationary points of the hydraulic head FRF.

In Part III we explored numerous examples of time-periodic flow through source-free domains. In all of the examples that we considered, relative maxima of the amplitude functions occurred only at domain boundaries—never in the space domain. In some cases relative minima occurred in the space domain. We observed the same for the phase functions. Are these merely coincidence, arising solely from the particular examples we chose to examine, or do these observations reflect underlying general rules or principles?

In this part we conduct theoretical analyses of the stationary points of the hydraulic head harmonic constituent coefficient functions in source-free domains to infer general principles governing their possible existence and nature. Such principles provide insights for flow modeling and a theoretical basis for understanding physical features such as stagnation zones. They also provide a baseline for the investigation of more complex problems involving nonzero source distributions.

Hydrodynamics of Time-Periodic Groundwater Flow: Diffusion Waves in Porous Media, Geophysical Monograph 224,
First Edition. Joe S. Depner and Todd C. Rasmussen.

16
Stationary Points: Basic Concepts

In this chapter we consider the stationary points of the rectangular form coefficient functions—i.e., $A_{\mathrm{h}}(\mathbf{x};\omega_m)$ and $B_{\mathrm{h}}(\mathbf{x};\omega_m)$—that we defined in Chapter 4. The analysis presented in this chapter relies heavily on the results derived in Chapter 5 and is based on the same assumptions.

Recall that throughout this book we assume that S_{s} is positive and $\mathbf{K}$ is positive definite, everywhere in the space domain.

Consider a general function of the space coordinates, $f(\mathbf{x})$. The stationary points of $f(\mathbf{x})$ are those points at which the gradient of the function vanishes. That is, the point with coordinates $\mathbf{x} = \mathbf{x}_{\mathrm{c}}$ is a stationary point of f if and only if $\nabla f(\mathbf{x}_{\mathrm{c}}) = \mathbf{0}$. If f varies smoothly in space, then each of its stationary points (if any exist) corresponds to a local (possibly global) extremum or to a saddle point.

16.1. ELLIPTIC OPERATORS

The development in this section is adapted from the discussion of elliptic operators presented by *Protter and Weinberger* [1999].

Consider the linear, second-order differential operator, L:

$$L[f] \equiv \sum_{m=1}^{N}\sum_{n=1}^{N} a_{mn}(\mathbf{x})\frac{\partial^2 f}{\partial x_m \partial x_n} + \sum_{n=1}^{N} b_n(\mathbf{x})\frac{\partial f}{\partial x_n}. \qquad (16.1)$$

We can write this in vector notation as

$$L[f] \equiv \mathbf{A}(\mathbf{x}):\mathbf{H}[f] + \mathbf{b}(\mathbf{x})\cdot\nabla f \qquad (16.2)$$

where

$\mathbf{A}$ is the real-valued, symmetric matrix whose (m,n)th element is

$$A_{mn} \equiv \frac{1}{2}\left[a_{mn}(\mathbf{x}) + a_{nm}(\mathbf{x})\right]$$

$\mathbf{b}$ is the real-valued vector whose nth element is $b_n(\mathbf{x})$;

$\mathbf{H}[f]$ is the Hessian matrix for the real-valued function $f(\mathbf{x})$, i.e.,

$$H_{mn} \equiv \frac{\partial^2 f}{\partial x_m \partial x_n}$$

and ":" denotes a double inner product of two (conformable) matrices, i.e.,

$$\mathbf{A}:\mathbf{H} = \sum_{m=1}^{M}\sum_{n=1}^{N} A_{mn}H_{mn}.$$

Exercise 16.1 Expressions for Operator Coefficients

Problem: Show that one can express the operator $L[f] \equiv \nabla\cdot\left[\mathbf{K}(\mathbf{x})\nabla f(\mathbf{x})\right]$ in the form (16.1) by expressing the corresponding $a_{np}(\mathbf{x})$ and $b_n(\mathbf{x})$ coefficients in terms of $\mathbf{K}$.

Solution: Expand L in terms of its first- and second-order derivative terms as

$$\begin{aligned} L[f] &= \nabla\cdot\left[\mathbf{K}(\mathbf{x})\nabla f(\mathbf{x})\right] \\ &= \mathbf{K}(\mathbf{x}):\mathbf{H}[f] + \left[\nabla\cdot\mathbf{K}(\mathbf{x})\right]\cdot\nabla f. \end{aligned} \qquad (16.3)$$

Hydrodynamics of Time-Periodic Groundwater Flow: Diffusion Waves in Porous Media, Geophysical Monograph 224,
First Edition. Joe S. Depner and Todd C. Rasmussen.

This latter align is equivalent to (16.2) with the substitutions

$$\mathbf{A}(\mathbf{x}) = \mathbf{K}(\mathbf{x})$$
$$\mathbf{b}(\mathbf{x}) = \nabla \cdot \mathbf{K}(\mathbf{x}).$$

Equivalently, in scalar notation these are

$$a_{np}(\mathbf{x}) = K_{np}(\mathbf{x}); \quad n = 1, 2, \ldots, N; \; p = 1, 2, \ldots, N \tag{16.4a}$$

$$b_n(\mathbf{x}) = \sum_{p=1}^{N} \frac{\partial K_{pn}(\mathbf{x})}{\partial x_p}; \quad n = 1, 2, \ldots, N. \tag{16.4b}$$

□

The operator L is said to be *elliptic* at a point $\mathbf{x}$ if and only if a positive quantity $C(\mathbf{x})$ exists such that

$$\sum_{n=1}^{N} \sum_{p=1}^{N} a_{np}(\mathbf{x})\, \xi_n\, \xi_p \geq C(\mathbf{x}) \sum_{n=1}^{N} \xi_n^2 \tag{16.5}$$

for all real-valued N-vectors $\boldsymbol{\xi}$. We can express (16.5) in vector notation as

$$\boldsymbol{\xi}^{\mathrm{T}} \mathbf{A}(\mathbf{x}) \boldsymbol{\xi} \geq C(\mathbf{x})\, |\boldsymbol{\xi}|^2. \tag{16.6}$$

Thus, ellipticity of the operator L is essentially equivalent to positive definiteness of the matrix $\mathbf{A}$.

The operator L is said to be *elliptic on the domain D* if and only if it is elliptic at every point of D. That is, L is elliptic on D if and only if a positive field $C(\mathbf{x})$ exists such that (16.6) is satisfied at every point of D for all real-valued N-vectors $\boldsymbol{\xi}$.

The operator L is said to be *uniformly elliptic* on the domain D if and only if it is elliptic at every point of D and a positive constant C_0 exists such that

$$C(\mathbf{x}) \geq C_0 \quad \forall\, \mathbf{x} \in D. \tag{16.7}$$

How does ellipticity differ from uniform ellipticity? While ellipticity (of L on D) requires the matrix $\mathbf{A}$ to be positive definite (i.e., its eigenvalues must be positive), its eigenvalues are allowed to be arbitrarily close to zero. In contrast, if L is uniformly elliptic on D, then the eigenvalues of $\mathbf{A}$ cannot be arbitrarily close to zero on D; they must have a positive lower limit C_0. Loosely, then, for L to be uniformly elliptic on D requires the eigenvalues of $\mathbf{A}$ to be "more positive" than mere ellipticity requires. The following exercises illustrate.

Exercise 16.2 Ellipticity of Operator *L*

Problem: Recall that by assumption $\mathbf{K}$ is symmetric and positive definite everywhere in D. Consider the operator L defined by (16.2) with $\mathbf{A}(\mathbf{x}) \equiv \mathbf{K}(\mathbf{x})$. Is L elliptic in D? Is L uniformly elliptic in D?

Solution: Because $\mathbf{K}$ is symmetric and positive definite everywhere in D, L is elliptic in D. However, L may or may not be uniformly elliptic in D; to make this determination we need information on the spatial variation of $\mathbf{K}(\mathbf{x})$. □

Exercise 16.3 Operator Properties

Problem: Consider the hydraulic conductivity field in one dimension:

$$K(x) = K_{\max} \frac{1 - e^{-x/x_{\max}}}{1 - e^{-1}} \quad \forall\, x \in D$$

where $K_{\max}$ and $x_{\max}$ are positive constants and $D \equiv \{x : 0 < x < x_{\max}\}$. Is $K(x)$ positive definite in D? Is the operator $L[v] \equiv d/dx\left[K(x)dv/dx\right]$ elliptic in D? Is $L[v]$ uniformly elliptic in D?

Solution: Although $K(x)$ is positive definite, (16.7) is not satisfied because

$$\lim_{x \to 0} K(x) = 0.$$

Therefore, although $L[v]$ is elliptic in D, it is not uniformly elliptic. □

Exercise 16.3 illustrates, by way of example, that conceptually even a medium whose hydraulic conductivity tensor is technically positive definite and spatially continuous can be effectively impermeable near a boundary. Requiring (16.7) to be satisfied for $\mathbf{A} = \mathbf{K}$ eliminates this possibility.

16.2. MAXIMUM PRINCIPLES

The development in this section is, in part, adapted from the discussion of maximum principles presented by *Protter and Weinberger* [1999].

In subsequent developments, we will encounter second-order differential inequalities involving uniformly elliptic operators, such as

$$\text{Strict}: L[f] > 0 \quad \forall\, \mathbf{x} \in G \tag{16.8a}$$

$$\text{Nonstrict}: L[f] \geq 0 \quad \forall\, \mathbf{x} \in G \tag{16.8b}$$

where (16.1) defines L, $\mathbf{A} \equiv \mathbf{K}(\mathbf{x})$, $f(\mathbf{x})$ is C^2 continuous, and G is a space domain (an open set). Sometimes when the function $f(\mathbf{x})$ satisfies such an inequality, one can apply a *maximum principle* to gain insight into the behavior of $f(\mathbf{x})$. We will use two such maximum principles, which we describe in the following subsections.

16.2.1. Strict Inequality

A maximum principle suitable for a strict inequality is *Hopf's maximum principle*, which can be stated as follows [*Protter and Weinberger*, 1999]:

Let $f(\mathbf{x})$ satisfy the differential inequality $L[f] > 0$ in a domain G where L is elliptic. Then f can't have a maximum in G.

That is, $f(\mathbf{x})$ attains its maximum value on the boundary of G rather than in G.

16.2.2. Nonstrict Inequality

A maximum principle suitable for a nonstrict inequality is the following [*Protter and Weinberger*, 1999]:

Let $f(\mathbf{x})$ satisfy the differential inequality $L[f] \geq 0$ in a domain G where L is uniformly elliptic. Suppose the coefficients a_{np} and b_n are uniformly bounded. If f attains a maximum $f_{\max}$ at a point of G, then $f = f_{\max}$ in G.

Under some circumstances one can relax the conditions of uniform ellipticity of the operator L and boundedness of the domain G; see *Protter and Weinberger* [1999] for details.

Exercise 16.4 Ellipticity of Laplacian Operator

Problem: Suppose the function $f(\mathbf{x})$ is C^2 continuous and satisfies Laplace's equation in a 3D space domain D:

$$\nabla^2 f(\mathbf{x}) = 0 \quad \forall\, \mathbf{x} \in D. \tag{16.9}$$

How is the operator $L[f]$ defined in this case? What is the domain G in this case? Is $L[f]$ elliptic, uniformly elliptic, or neither?

Solution: Define

$$\mathbf{A}(\mathbf{x}) \equiv \mathbf{I} \tag{16.10a}$$

$$\mathbf{b}(\mathbf{x}) \equiv \mathbf{0} \tag{16.10b}$$

where $\mathbf{I}$ is the 3×3 identity matrix. Substituting these results in (16.2) and then simplifying give

$$L[f] = \nabla^2 f(\mathbf{x}).$$

Also, the domain G is just the 3D space domain D. In this case $\mathbf{A}$ is symmetric and positive definite, so the operator L is elliptic. In fact, (16.7) is satisfied with $C_0 = 1$, so L is uniformly elliptic. □

Exercise 16.5 Extrema of Harmonic Functions

Problem: Suppose the function $f(\mathbf{x})$ is C^2 continuous and satisfies Laplace's equation (16.9) in a 3D space domain D. What, if anything, can you infer about the possible existence of extrema of the function $f(\mathbf{x})$ using maximum principles? Use the results of Exercise 16.4.

Solution: In Exercise 16.4 we found that L is uniformly elliptic. Therefore, the maximum principle for a nonstrict inequality applies. In this case it tells us that the function $f(\mathbf{x})$ has no maxima in D other than the trivial case $f(\mathbf{x}) = \text{constant}$. Repeating this analysis with the function $g(\mathbf{x}) \equiv -f(\mathbf{x})$ gives a similar result for $g(\mathbf{x})$; that is, the function $f(\mathbf{x})$ has no minima in D other than the trivial case $f(\mathbf{x}) = \text{constant}$. Therefore, the function $f(\mathbf{x})$ has neither maxima nor minima in D, other than the trivial case $f(\mathbf{x}) = \text{constant}$ in D. □

16.3. STATIONARY POINTS OF A_h AND B_h

Consider a particular harmonic constituent with frequency ω_m (m fixed). Suppose that with respect to this harmonic constituent the space domain is source free, i.e.,

$$A_u(\mathbf{x}; \omega_m) = 0 \tag{16.11a}$$

$$B_u(\mathbf{x}; \omega_m) = 0 \tag{16.11b}$$

$$\forall\, \mathbf{x} \in D.$$

Then the governing equations (5.10) for this constituent reduce to their homogenous forms, which we can express as

$$\nabla \cdot \left[\mathbf{K}(\mathbf{x})\nabla A_h(\mathbf{x}; \omega_m)\right] = -\omega_m S_s(\mathbf{x}) B_h(\mathbf{x}; \omega_m) \tag{16.12a}$$

$$\nabla \cdot \left[\mathbf{K}(\mathbf{x})\nabla B_h(\mathbf{x}; \omega_m)\right] = \omega_m S_s(\mathbf{x}) A_h(\mathbf{x}; \omega_m) \tag{16.12b}$$

$$\forall\, \mathbf{x} \in D.$$

16.3.1. Stationary Points of A_h

Consider (16.12a). We assume that both ω_m and S_s are positive, so the sign of the RHS is determined by the sign of $B_h(\mathbf{x}; \omega_m)$. We assume we can partition D into two disjoint subsets, one on which $B_h(\mathbf{x}; \omega_m)$ is nonpositive and the other on which $B_h(\mathbf{x}; \omega_m)$ is positive:

$$D_1 \cup D_2 = D \tag{16.13a}$$

$$D_1 \cap D_2 = \varnothing \tag{16.13b}$$

where

$$D_1 \equiv \{\mathbf{x} : \mathbf{x} \in D \text{ and } B_h(\mathbf{x}; \omega_m) \leq 0\} \tag{16.14a}$$

$$D_2 \equiv \{\mathbf{x} : \mathbf{x} \in D \text{ and } B_h(\mathbf{x}; \omega_m) > 0\}. \tag{16.14b}$$

In practice, making such a partition requires knowing the sign of $B_h(\mathbf{x}; \omega_m)$ throughout D.

Recall from Exercise 5.2 that if either $A_h(\mathbf{x}; \omega_m)$ or $B_h(\mathbf{x}; \omega_m)$ is uniform in D, then both $A_h(\mathbf{x}; \omega_m)$ and $B_h(\mathbf{x}; \omega_m)$ vanish everywhere in D. This case is trivial so we neglect it throughout the remainder of this chapter.

Combining the definitions (16.14) with (16.12a) then gives

$$\nabla\cdot\left[\mathbf{K}(\mathbf{x})\nabla A_\mathrm{h}(\mathbf{x};\omega_m)\right]\geq 0 \quad \forall\,\mathbf{x}\in D_1 \tag{16.15a}$$

$$\nabla\cdot\left[\mathbf{K}(\mathbf{x})\nabla A_\mathrm{h}(\mathbf{x};\omega_m)\right]< 0 \quad \forall\,\mathbf{x}\in D_2. \tag{16.15b}$$

The first of these corresponds to the nonstrict inequality (16.8b) with $f(\mathbf{x}) \equiv A_\mathrm{h}(\mathbf{x};\omega_m)$ and $G \equiv D_1$, while the second corresponds to the strict inequality (16.8a) with $f(\mathbf{x}) \equiv -A_\mathrm{h}(\mathbf{x};\omega_m)$ and $G \equiv D_2$. Applying the corresponding maximum principles then gives

$A_\mathrm{h}(\mathbf{x};\omega_m)$ has no (local or global) maximum on D_1. That is, if $A_\mathrm{h}(\mathbf{x};\omega_m)$ reaches a maximum, it does so on the boundary of D_1 rather than in D_1.

$A_\mathrm{h}(\mathbf{x};\omega_m)$ has no (local or global) minimum on D_2. That is, if $A_\mathrm{h}(\mathbf{x};\omega_m)$ attains a minimum, it does so on the boundary of D_2 rather than in D_2.

Thus, making inferences about the existence of local extrema of the coefficient function $A_\mathrm{h}(\mathbf{x};\omega_m)$ by applying maximum principles to the governing equation (16.12a) requires one to know the sign of the coefficient function $B_\mathrm{h}(\mathbf{x};\omega_m)$ throughout D.

Exercise 16.6 Extrema of Rectangular Form Coefficient Functions

Problem: Suppose the domain is 1D (i.e., $N = 1$) and $B_\mathrm{h}(x;\omega_m) > 0\,\forall\,x\in D$. Is it possible for $A_\mathrm{h}(x;\omega_m)$ to have more than one local maximum on D? Provide a simple geometric argument to support your answer.

Solution: No, it is not possible. In this case $D = D_2$. For $A_\mathrm{h}(x;\omega_m)$ to have more than one local maximum on D would require $A_\mathrm{h}(x;\omega_m)$ to have a local minimum at some intermediate location between each spatially adjacent pair of local maxima. The results obtained above prohibit this. □

Exercise 16.7 Generalization to Domains with Sources

Problem: Generalize the results obtained above to a domain with a nonzero source distribution.

Solution: Rearranging terms in the governing equation (5.10) gives

$$\nabla\cdot\left[\mathbf{K}(\mathbf{x})\nabla A_\mathrm{h}(\mathbf{x};\omega_m)\right] = -\omega_m S_\mathrm{s}(\mathbf{x})B_\mathrm{h}(\mathbf{x};\omega_m) - A_\mathrm{u}(\mathbf{x};\omega_m) \tag{16.16a}$$

$$\nabla\cdot\left[\mathbf{K}(\mathbf{x})\nabla B_\mathrm{h}(\mathbf{x};\omega_m)\right] = \omega_m S_\mathrm{s}(\mathbf{x})A_\mathrm{h}(\mathbf{x};\omega_m) - B_\mathrm{u}(\mathbf{x};\omega_m) \tag{16.16b}$$

$$\forall\,\mathbf{x}\in D.$$

These replace the homogeneous equations (16.12). Next, we generalize definitions (16.14) as

$$D_1 \equiv \left\{\mathbf{x} : \mathbf{x}\in D \text{ and } B_\mathrm{h}(\mathbf{x};\omega_m) \leq -\frac{A_\mathrm{u}(\mathbf{x};\omega_m)}{\omega_m S_\mathrm{s}(\mathbf{x})}\right\} \tag{16.17a}$$

$$D_2 \equiv \left\{\mathbf{x} : \mathbf{x}\in D \text{ and } B_\mathrm{h}(\mathbf{x};\omega_m) > -\frac{A_\mathrm{u}(\mathbf{x};\omega_m)}{\omega_m S_\mathrm{s}(\mathbf{x})}\right\}. \tag{16.17b}$$

Combining the definitions (16.17) with the governing equation (16.16a) gives, once again, inequalities (16.15). Applying the maximum principles then leads to the same basic conclusions as those obtained above for source-free domains. □

16.3.2. Stationary Points of B_h

The analysis of stationary points of $B_\mathrm{h}(\mathbf{x};\omega_m)$ is similar to that conducted in Section 16.3.1 for $A_\mathrm{h}(\mathbf{x};\omega_m)$, so rather than giving the details here we only summarize the applicable results.

We assume that we can partition the domain D as follows:

$$D_3 \cup D_4 = D \tag{16.18a}$$

$$D_3 \cap D_4 = \varnothing \tag{16.18b}$$

where

$$D_3 \equiv \{\mathbf{x} : \mathbf{x}\in D \text{ and } A_\mathrm{h}(\mathbf{x};\omega_m)\leq 0\} \tag{16.19a}$$

$$D_4 \equiv \{\mathbf{x} : \mathbf{x}\in D \text{ and } A_\mathrm{h}(\mathbf{x};\omega_m) > 0\}. \tag{16.19b}$$

Combining the definitions (16.19) with (16.12b) yields

$$\nabla\cdot\left[\mathbf{K}(\mathbf{x})\nabla B_\mathrm{h}(\mathbf{x};\omega_m)\right]\leq 0 \quad \forall\,\mathbf{x}\in D_3$$

$$\nabla\cdot\left[\mathbf{K}(\mathbf{x})\nabla B_\mathrm{h}(\mathbf{x};\omega_m)\right]> 0 \quad \forall\,\mathbf{x}\in D_4.$$

Applying the corresponding maximum principles to these inequalities yields the following:

$B_\mathrm{h}(\mathbf{x};\omega_m)$ has no (local or global) minimum on D_3. That is, if $B_\mathrm{h}(\mathbf{x};\omega_m)$ reaches a minimum, it does so on the boundary of D_3 rather than in D_3.

$B_\mathrm{h}(\mathbf{x};\omega_m)$ has no (local or global) maximum on D_4. That is, if $B_\mathrm{h}(\mathbf{x};\omega_m)$ attains a maximum, it does so on the boundary of D_4 rather than in D_4.

In practice, to make the partition (16.19), one must know the sign of $A_\mathrm{h}(\mathbf{x};\omega_m)$ throughout D. Thus, making inferences about the existence of local extrema of $B_\mathrm{h}(\mathbf{x};\omega_m)$ by applying maximum principles to the governing equation (16.12b) requires one to know the sign of $A_\mathrm{h}(\mathbf{x};\omega_m)$ throughout D.

16.4. CHAPTER SUMMARY

In this chapter we defined and described elliptic and uniformly elliptic operators and presented two maximum principles for differential inequalities involving elliptic operators. We showed how these principles could be applied to the analysis of stationary points of the rectangular form coefficient functions $A_{\mathrm{h}}(\mathbf{x};\omega_m)$ and $B_{\mathrm{h}}(\mathbf{x};\omega_m)$ of time-periodic flow. The results so obtained are not very useful because they generally require one to know the signs of $A_{\mathrm{h}}(\mathbf{x};\omega_m)$ and $B_{\mathrm{h}}(\mathbf{x};\omega_m)$ in D.

17

Stationary Points: Amplitude and Phase

In this chapter we conduct a theoretical analysis of stationary points of the hydraulic head amplitude and phase functions in source-free domains. The analysis effort illustrates application of the polar form of the space BVP (see Chapter 6) and its limitations. The analysis presented in this chapter relies heavily on the results derived in Chapter 6 and is based on the same assumptions (see Section 6.1).

17.1. PHYSICAL SIGNIFICANCE

Unlike stationary points of the rectangular form coefficients—i.e., $A_{\mathrm{h}}(\mathbf{x};\omega_m)$ and $B_{\mathrm{h}}(\mathbf{x};\omega_m)$—stationary points of the amplitude and phase functions have clear physical significance. Before we begin our analysis, we briefly examine some possible interpretations for such points.

Consider the amplitude function for a given hydraulic head harmonic constituent. Equation (25.2) implies that the stationary points of the amplitude function are zero points of the time-averaged, fluid mechanical energy flux density. Less generally, (25.1a) implies that in a medium that is homogeneous with respect to effective porosity, amplitude extrema correspond to extrema of the fluid mechanical energy density.

Additionally, based on the results of Exercises 14.6 and 14.7, we infer the following. If for a given harmonic constituent a stationary point of the amplitude function coincides with a stationary point of the phase function, then that point is a zero point of the hydraulic gradient harmonic constituent and therefore a stagnation point of the specific-discharge harmonic constituent. Chapter 18 discusses flow stagnation.

Finally, the rotation condition (15.72) indicates that stationary points of a harmonic constituent's amplitude and phase functions are points of zero rotation of the corresponding specific-discharge constituent. Exercise 17.1 addresses the converse proposition.

Exercise 17.1 Time Rotation and Stationary Points

Problem: Suppose that for some constituent (i.e., ω_m fixed) and some fixed location $\mathbf{x}$ we know that the specific-discharge harmonic constituent does not rotate with time. Is the location $\mathbf{x}$ a stationary point for the corresponding hydraulic head amplitude or phase functions?

Solution: The rotation condition (15.72) indicates two possibilities:

1. One or both of the corresponding amplitude and phase gradients vanishes at the point $\mathbf{x}$. In this case the point $\mathbf{x}$ is a stationary point for the amplitude function or the phase function or both.
2. Neither of the gradients vanishes, but they are collinear at the point $\mathbf{x}$. In this case the location $\mathbf{x}$ is not a stationary point for either the amplitude function or the phase function.

The information provided is insufficient to determine which of these applies, so the question posed is impossible to answer. □

17.2. GOVERNING AND BC EQUATIONS

If the space domain is free of sources and sinks, i.e.,

$$M_{\mathrm{u}}(\mathbf{x};\omega_m) = 0 \quad \forall\, \mathbf{x} \in D \quad (m = 1, 2, 3, \ldots)$$

then the governing equations (6.11) take the homogeneous forms

$$\begin{aligned}\nabla \cdot \left[\mathbf{K}(\mathbf{x})\nabla M_{\mathrm{h}}(\mathbf{x};\omega_m)\right] \\ - M_{\mathrm{h}}(\mathbf{x};\omega_m)\left|\mathbf{K}^{1/2}(\mathbf{x})\nabla\theta_{\mathrm{h}}(\mathbf{x};\omega_m)\right|^2 = 0\end{aligned} \tag{17.1a}$$

Hydrodynamics of Time-Periodic Groundwater Flow: Diffusion Waves in Porous Media, Geophysical Monograph 224,
First Edition. Joe S. Depner and Todd C. Rasmussen.

$$M_h(\mathbf{x};\omega_m)\nabla\cdot\left[\mathbf{K}(\mathbf{x})\nabla\theta_h(\mathbf{x};\omega_m)\right] + 2\left[\mathbf{K}(\mathbf{x})\nabla M_h(\mathbf{x};\omega_m)\right]\cdot\nabla\theta_h(\mathbf{x};\omega_m) - \omega_m S_s(\mathbf{x})M_h(\mathbf{x};\omega_m) = 0 \tag{17.1b}$$

$$\forall\, \mathbf{x} \in D_+(\omega_m); \quad m = 1, 2, 3, \ldots.$$

In addition, the boundary conditions (6.30) apply without modification.

Exercise 17.2 Spatially Uniform Amplitude Functions

Problem: Consider time-periodic flow in a source-free domain. Show that for a given harmonic constituent (i.e., ω_m fixed) space BVP solutions for which the hydraulic head amplitude function is both spatially uniform and nonzero are prohibited.

Solution: Suppose the amplitude function is both spatially uniform and positive; then the governing equations (17.1) are valid. Substituting zero for the amplitude gradient in (17.1a) and simplifying give

$$\nabla\theta_h(\mathbf{x};\omega_m) = \mathbf{0} \quad \forall\, \mathbf{x} \in D_+(\omega_m).$$

Substituting zero for the phase gradient in (17.1b) then yields

$$M_h(\mathbf{x};\omega_m) = 0 \quad \forall\, \mathbf{x} \in D_+(\omega_m).$$

This is a contradiction, because we defined the subregion $D_+(\omega_m)$ as that part of the space domain D for which the amplitude is positive. Consequently, our supposition must be false. Thus, solutions for which the amplitude function is both spatially uniform and nonzero are prohibited. □

Exercise 17.3 Spatially Uniform Phase Functions

Problem: Consider time-periodic flow in a source-free domain. Show that the polar form of the space BVP prohibits solutions for which the hydraulic head phase function is spatially uniform.

Solution: Suppose the phase function for a hydraulic head harmonic constituent is spatially uniform. Substituting zero for the phase gradient in the governing equations (17.1b) yields

$$M_h(\mathbf{x};\omega_m) = 0 \quad \forall\, \mathbf{x} \in D_+(\omega_m).$$

This is a contradiction, because we defined the subregion $D_+(\omega_m)$ as that part of the space domain D for which the amplitude is positive. Thus, our supposition must be false. Therefore, the governing equations prohibit solutions for which the phase function is spatially uniform. □

Neither of the results of Exercises 17.2 and 17.3 prohibits a priori the existence of continuous subregions within the domain D, throughout which $F_h(\mathbf{x};\omega_m) = 0$. We call such subregions, if any exist, *connected stagnation sets*. For a discussion of stagnation sets, see Chapter 18.

17.3. STATIONARY POINTS OF THE AMPLITUDE

Let $\mathbf{x} = \mathbf{x}_s$ be the coordinates of a stationary point of the amplitude function. Evaluating the governing equation (17.1b) at the stationary point leads to

$$\left\{\nabla\cdot\left[\mathbf{K}(\mathbf{x})\nabla\theta_h(\mathbf{x};\omega_m)\right]\right\}\Big|_{\mathbf{x}=\mathbf{x}_s} - \omega_m S_s(\mathbf{x}_s) = 0.$$

Neither the amplitude nor any of its spatial derivatives appears in this equation. Thus the equation provides no additional information about the amplitude function or its spatial variation at stationary points. Therefore, in this section we rely primarily on governing equation (17.1a) to make inferences about stationary points of the amplitude function.

17.3.1. Amplitude Maxima

In this section we use the governing equation (17.1a) and a maximum principle to investigate the possible existence of maxima of the amplitude function.

Solving the governing equation (17.1a) for the divergence term gives

$$\nabla\cdot\left[\mathbf{K}(\mathbf{x})\nabla M_h(\mathbf{x};\omega_m)\right] = M_h(\mathbf{x};\omega_m)\left|\mathbf{K}^{1/2}(\mathbf{x})\nabla\theta_h(\mathbf{x};\omega_m)\right|^2.$$

The RHS is nonnegative; we write this as

$$L_M\left[M_h(\mathbf{x};\omega_m)\right] \geq 0 \quad \forall\, \mathbf{x} \in D_+(\omega_m) \tag{17.2}$$

where L_M is the linear, second-order, differential operator defined as

$$L_M[f] \equiv \nabla\cdot\left[\mathbf{K}(\mathbf{x})\nabla f\right].$$

Here f, the operand, is a function of the space coordinates. The operator L_M can also be expressed as

$$L_M = \sum_{n=1}^{N}\sum_{p=1}^{N} a_{np}(\mathbf{x})\frac{\partial^2}{\partial x_n \partial x_p} + \sum_{n=1}^{N} b_n(\mathbf{x})\frac{\partial}{\partial x_n} \tag{17.3}$$

where

$$a_{np}(\mathbf{x}) = K_{np}(\mathbf{x}) \quad (n = 1, 2, \ldots, N;\; p = 1, 2, \ldots, N) \tag{17.4a}$$

$$b_n(\mathbf{x}) = \sum_{p=1}^{N} \frac{\partial K_{pn}(\mathbf{x})}{\partial x_p} \quad (n = 1, 2, \ldots, N). \tag{17.4b}$$

Because $\mathbf{K}$ is symmetric and positive definite in $D_+(\omega_m)$, the operator L_M is *elliptic*. From this point on, we will assume that the eigenvalues of $\mathbf{K}$ are bounded below by a positive constant in D, that is, that L_M is *uniformly elliptic* (see Section 16.1). Because the elements of $\mathbf{K}$ are C^1 continuous and uniformly bounded, the $a_{np}(\mathbf{x})$ and $b_n(\mathbf{x})$ coefficients are uniformly bounded.

Next we apply the maximum principle (see Section 16.2.2) to the nonstrict inequality (17.2). Let $G = D_+(\omega_m)$, $f = M_h(\mathbf{x}; \omega_m)$, and $L = L_M$; then all of the conditions of the maximum principle are satisfied. Therefore, the maximum principle applies to the amplitude function for each hydraulic head harmonic constituent; it tells us that the amplitude function attains a maximum value in $D_+(\omega_m)$ if and only if the amplitude is constant. However, the governing equations (17.1) imply that if the amplitude is constant it vanishes everywhere in $D_+(\omega_m)$, which is a contradiction. Therefore, the amplitude function cannot have a maximum in $D_+(\omega_m)$.

Exercise 17.4 Space-Periodic Amplitude Functions

Problem: Do the governing equations for time-periodic flow in a source-free domain admit solutions for which the amplitude function is both continuous and space periodic in $D_+(\omega_m)$?

Solution: No. A continuous, space-periodic solution would necessarily have multiple maxima of the amplitude function in $D_+(\omega_m)$. As shown above, amplitude maxima are prohibited. Therefore, space-periodic solutions are prohibited. □

In Section 25.2.2 we present a less abstract, more physical argument for the prohibition of amplitude maxima in source-free domains.

17.3.2. Amplitude Minima

In this section we examine particular examples that allow us to draw general conclusions about the existence of amplitude minima. Then we use the polar form of the governing equations to infer what we can about the nature of amplitude minima.

Numerous examples exist of flows in which no amplitude minimum appears. The following is a sampling of such flows:

- One dimension:

Finite domain: Cases a-2, a-3, m-2, m-3, n-2 (with $\nu = \ln 0.1$)

Infinite domain: Example 9.2

- Two dimensions: Example 12.6

Example 17.1 The Jacob-Ferris Solution on Finite Domain
Consider a finite-length, 1D, homogeneous, porous medium and a single harmonic constituent $h'(x, t; \omega_m)$ for which the boundary conditions are

$$h'(0, t; \omega_m) = M_h(0; \omega_m)\cos(\omega_m t)$$
$$h'(L, t; \omega_m) = M_h(L; \omega_m)\cos(\omega_m t - \Omega L)$$

where $M_h(0; \omega_m)$ is real valued and positive,

$$M_h(L; \omega_m) = M_h(0; \omega_m)e^{-\Omega L}$$

and

$$\Omega = \sqrt{\frac{\omega_m S_s}{2K}}.$$

The Jacob-Ferris solution (see Example 9.2) gives the steady periodic solution, valid for $0 < x < L$, as

$$h'(x, t; \omega_m) = M_h(x; \omega_m)\cos(\omega_m t - \Omega x).$$

The corresponding amplitude function is

$$M_h(x; \omega_m) = M_h(0; \omega_m)e^{-\Omega x}$$

which is a monotone decreasing function of x (see Figure 9.10). Therefore, in this example amplitude minima do not exist in the space domain. ◊

Example 17.2 One Dimension, Symmetric Dirichlet Boundary Conditions
Consider a finite-length, 1D, homogeneous, porous medium and a single harmonic constituent $h'(x, t; \omega_m)$ for which the boundary conditions are

$$h'(0, t; \omega_m) = h'(L, t; \omega_m) = M_0\cos(\omega_m t + \theta_0)$$

where M_0 is a real-valued, positive constant,

$$\Omega = \sqrt{\frac{\omega_m S_s}{2K}}$$

and

$$\tan\theta_0 = \frac{e^{-\Omega L}\sin(\Omega L)}{1 + e^{-\Omega L}\cos(\Omega L)}.$$

The steady periodic solution, valid for $0 < x < L$, is

$$h'(x, t; \omega_m) = C_0\Big\{e^{-\Omega x}\cos(\omega_m t - \Omega x) + e^{-\Omega(L-x)}\cos\big[\omega_m t - \Omega(L - x)\big]\Big\} \tag{17.5}$$

where

$$C_0 = \frac{M_h(0; \omega_m)}{\sqrt{1 + 2e^{-\Omega L}\cos(\Omega L) + e^{-2\Omega L}}}.$$

◊

Exercises 17.5– 17.8 pertain to Example 17.2.

Exercise 17.5 Symmetric Dirichlet BCs in One Dimension: Part 1 of 4

Problem: Review Example 17.2. Show that the corresponding amplitude function is

$$M_{\mathrm{h}}(x;\omega_m) = C_0\sqrt{e^{-2\Omega x} + e^{-2\Omega(L-x)} + 2e^{-\Omega L}\cos[\Omega(L-2x)]}. \tag{17.6}$$

Solution: Using trigonometric identity (B.2a), we can write the solution (17.5) as

$$h'(x,t;\omega_m) = C_0\Big[e^{-\Omega x}\Big\{\cos(\omega_m t)\cos(\Omega x) + \sin(\omega_m t)\sin(\Omega x)\Big\} + e^{-\Omega(L-x)}\Big\{\cos(\omega_m t)\cos\left[\Omega(L-x)\right] + \sin(\omega_m t)\sin\left[\Omega(L-x)\right]\Big\}\Big].$$

Grouping the coefficients for the $\cos(\omega_m t)$ and $\sin(\omega_m t)$ terms yields

$$h'(x,t;\omega_m) = C_0\Big[\Big\{e^{-\Omega x}\cos(\Omega x) + e^{-\Omega(L-x)}\times\cos\left[\Omega(L-x)\right]\Big\}\cos(\omega_m t) + \Big\{e^{-\Omega x}\sin(\Omega x) + e^{-\Omega(L-x)}\times\sin\left[\Omega(L-x)\right]\Big\}\sin(\omega_m t)\Big].$$

Therefore, the amplitude function satisfies

$$M_{\mathrm{h}}^2(x;\omega_m) = C_0^2\Big[\Big\{e^{-\Omega x}\cos(\Omega x) + e^{-\Omega(L-x)}\times\cos\left[\Omega(L-x)\right]\Big\}^2 + \Big\{e^{-\Omega x}\sin(\Omega x) + e^{-\Omega(L-x)}\times\sin\left[\Omega(L-x)\right]\Big\}^2\Big].$$

Squaring the terms on the RHS and using trigonometric identity (B.1) yield

$$M_{\mathrm{h}}^2(x;\omega_m) = C_0^2\Big[e^{-2\Omega x} + e^{-2\Omega(L-x)} + 2e^{-\Omega L}\Big\{\cos(\Omega x)\cos\left[\Omega(L-x)\right] + \sin(\Omega x)\sin\left[\Omega(L-x)\right]\Big\}\Big].$$

Using trigonometric identity (B.2b) to simplify the expression enclosed by the curved braces gives

$$M_{\mathrm{h}}^2(x;\omega_m) = C_0^2\Big\{e^{-2\Omega x} + e^{-2\Omega(L-x)} + 2e^{-\Omega L}\cos\left[\Omega(L-2x)\right]\Big\}.$$

Finally, taking the positive square root gives the result (17.6). □

Exercise 17.6 Symmetric Dirichlet BCs in One Dimension: Part 2 of 4

Problem: Derive an expression for the first derivative of the amplitude function with respect to x.

Solution

$$\frac{d}{dx}M_{\mathrm{h}}(x;\omega_m) = C_0\Omega\frac{e^{-2\Omega x} - e^{-2\Omega(L-x)} - e^{-\Omega L}\sin\left[\Omega(L-2x)\right]}{\sqrt{e^{-2\Omega x} + e^{-2\Omega(L-x)} + 2e^{-\Omega L}\cos\left[\Omega(L-2x)\right]}}.$$

□

Exercise 17.7 Symmetric Dirichlet BCs in One Dimension: Part 3 of 4

Problem: Show that the amplitude function has a stationary point at $x = L/2$.

Solution: By assumption the amplitude function is smooth, so its first derivative with respect to x vanishes at the stationary point(s). Let x_{s} denote the coordinate of the stationary point(s). Substituting $x = x_{\mathrm{s}}$ in the previous equation, substituting zero for the LHS, and simplifying give the transcendental equation in x_{s}:

$$0 = \frac{e^{-2\Omega x_{\mathrm{s}}} - e^{-2\Omega(L-x_{\mathrm{s}})} - e^{-\Omega L}\sin\left[\Omega(L-2x_{\mathrm{s}})\right]}{\sqrt{e^{-2\Omega x_{\mathrm{s}}} + e^{-2\Omega(L-x_{\mathrm{s}})} + 2e^{-\Omega L}\cos\left[\Omega(L-2x_{\mathrm{s}})\right]}}.$$

Substituting $x_{\mathrm{s}} = L/2$ in the previous equation gives the identity $0 = 0$, thus verifying that the amplitude has a stationary point at $x = L/2$. □

Exercise 17.8 Symmetric Dirichlet BCs in One Dimension: Part 4 of 4

Problem: Derive an expression for the value of the amplitude minimum at $x = L/2$.

Solution: Substituting $x = L/2$ in (17.6) and simplifying give

$$M_{\mathrm{h}}(L/2;\omega_m) = 2C_0 e^{-\Omega L/2}.$$

□

Examples 17.1 and 17.2 demonstrate that amplitude minima generally are neither required nor prohibited. Is it

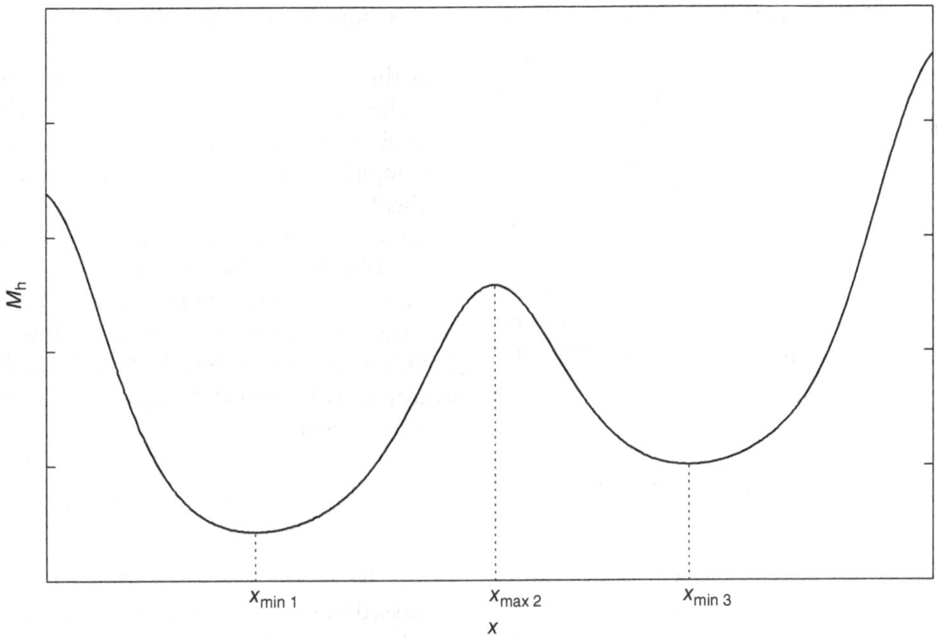

Figure 17.1 Prohibited 1D amplitude distribution.

possible for two or more isolated amplitude minima to exist at distinct locations in the space domain? We explore this question below.

Consider the 1D case. Suppose isolated amplitude minima exist at two distinct locations—the points with coordinates $x = x_{\mathrm{min}1}$ and $x = x_{\mathrm{min}3}$ where $x_{\mathrm{min}1} < x_{\mathrm{min}3}$. On a graph of amplitude (vertical axis) versus x (horizontal axis), these minima correspond to low points (see Figure 17.1). Recall that by assumption the amplitude function is smooth. Therefore, a high point (i.e., an amplitude maximum) necessarily appears at some point with coordinate $x = x_{\mathrm{max}2}$ lying between the two low points (i.e., $x_{\mathrm{min}1} < x_{\mathrm{max}2} < x_{\mathrm{min}3}$). This violates the general prohibition on amplitude maxima in source-free domains (see Section 17.3.1), so in the 1D case there can be at most one isolated amplitude minimum in the space domain.

Consider the 2D case. Suppose isolated amplitude minima exist at two distinct locations—the points with coordinates $\mathbf{x} = \mathbf{x}_{\mathrm{min}1}$ and $\mathbf{x} = \mathbf{x}_{\mathrm{min}3}$ where $|\mathbf{x}_{\mathrm{min}1} - \mathbf{x}_{\mathrm{min}3}| > 0$. Consider the variation of the amplitude along the line segment whose endpoints are the two stationary points. As in the 1D case, at some intermediate point ($\mathbf{x} = \mathbf{x}'$) on the segment the amplitude reaches a local maximum relative to the other points on the segment. Unlike the 1D case, in the 2D case that point need not correspond to an amplitude maximum on the space domain. For instance, it may correspond to a saddle point, which is not prohibited. Thus, the geometric argument used to rule out the existence of multiple minima in the 1D case does not carry over to the multidimensional case. In the multidimensional case we cannot rule out the possible existence of isolated amplitude minima at two or more distinct locations in the space domain.

Exercise 17.9 Use of Governing Equations: Amplitude Minima

Problem: Consider time-periodic flow in a bounded, source-free domain D. Suppose the hydraulic head amplitude is equal to its minimum value everywhere within a bounded N-dimensional subregion G, where $G \subset D$. Can you use the governing equations (17.1) to determine the value of the amplitude inside the subregion? Why or why not?

Solution: Yes, we can determine the value as follows. Inside G the amplitude is uniform. Therefore, both $\nabla M_{\mathrm{h}}(\mathbf{x};\omega_m) = \mathbf{0}$ and $\nabla \cdot [\mathbf{K}(\mathbf{x})\nabla M_{\mathrm{h}}(\mathbf{x};\omega_m)] = 0 \quad \forall \mathbf{x} \in G$. Suppose the minimum value is positive. Then the governing equations (17.1) are valid at every point in G. Substituting these in the governing equation (17.1a) and exploiting the positive definiteness of $\mathbf{K}$ lead to the following: Either $M_{\mathrm{h}}(\mathbf{x};\omega_m) = 0$ or $\nabla\theta_{\mathrm{h}}(\mathbf{x};\omega_m) = \mathbf{0}$ or both $\forall\ \mathbf{x} \in G$. Combining this restriction with the governing equation (17.1b) leads to the requirement that $M_{\mathrm{h}}(\mathbf{x};\omega_m) = 0 \quad \forall\ \mathbf{x} \in G$. This result contradicts our supposition, so the minimum value must be zero. □

Exercise 17.9 demonstrated that if the amplitude is uniform and equal to a minimum over a bounded N-dimensional subregion G in the N-dimensional space domain D, then that minimum is zero.

Exercise 17.10 Use of Governing Equations: Amplitude Minima

Problem: Consider time-periodic flow in a source-free, bounded, simply connected, multidimensional space domain (i.e., $N = 2$ or $N = 3$). Suppose the hydraulic head amplitude is equal to a minimum value everywhere within a bounded subregion G in the domain D. Further suppose that G is a lower-dimensional connected neighborhood of points (i.e., a continuous curve in two dimensions, a continuous surface in three dimensions). Can you use the governing equations (17.1) and the same argument that was used in Exercise 17.9 to determine the value of the amplitude inside G? Why or why not?

Solution: No, the value cannot be determined for the general case using only the governing equations. As in Exercise 17.9, $\nabla M_{\text{h}}(\mathbf{x};\omega_m) = \mathbf{0} \quad \forall\, \mathbf{x} \in G$. Unlike Exercise 17.9, however, in this exercise the second derivatives of the amplitude with respect to the space coordinates do not necessarily vanish in G, so the divergence term, $\nabla\cdot\left[\mathbf{K}(\mathbf{x})\nabla M_{\text{h}}(\mathbf{x};\omega_m)\right]$, of (17.1a) may or may not vanish. Therefore the governing equations, by themselves, do not require the amplitude to vanish in G. □

Exercise 17.10 demonstrated that the governing equations, by themselves, do not generally prohibit the existence of an amplitude minimum corresponding to a lower-dimensional subregion of a multidimensional space domain. Nor do they require that minimum amplitude to vanish.

Exercise 17.11 Amplitude Minima

Problem: Suppose the hydraulic head amplitude for a particular harmonic constituent (i.e., ω_m fixed) is a minimum at the isolated point with coordinates $\mathbf{x} = \mathbf{x}_{\text{min}}$, where $\mathbf{x}_{\text{min}} \in D$, and that minimum is positive. What, if anything, can one infer about the in-phase and quadrature components of the corresponding hydraulic gradient and specific-discharge vector harmonic constituents evaluated at the point with coordinates $\mathbf{x} = \mathbf{x}_{\text{min}}$?

Solution: The amplitude function is a minimum at the point with coordinates $\mathbf{x}_{\text{min}}$, so $\nabla M_{\text{h}}(\mathbf{x}_{\text{min}};\omega_m) = \mathbf{0}$. Substituting these results for the corresponding terms of (14.20a) and (14.22a), respectively, yields:

$$(\nabla h')_{\text{Ip}}(\mathbf{x}_{\text{min}}, t;\omega_m) = \mathbf{0}$$
$$\mathbf{q}'_{\text{Ip}}(\mathbf{x}_{\text{min}}, t;\omega_m) = \mathbf{0}.$$

That is, both of the in-phase component vectors vanish at $\mathbf{x} = \mathbf{x}_{\text{min}}$. □

17.3.3. Amplitude Saddle Points

In this section we use the governing equation (17.1a) and the second-derivative test of multivariate calculus to explore the following question: Are saddle points of the amplitude function generally required, prohibited, or neither?

Let $\mathbf{x} = \mathbf{x}_{\text{s}}$ be the coordinates of a stationary point of the amplitude function $M_{\text{h}}(\mathbf{x};\omega_m)$, where $\mathbf{x}_{\text{s}} \in D_{+}(\omega_m)$. Since $\mathbf{K}$ is symmetric and positive definite, it has positive eigenvalues. Therefore a linear coordinate transformation $\tilde{\mathbf{x}} = \mathbf{C}\mathbf{x}$ exists that renders the transformed medium locally isotropic with respect to hydraulic conductivity *at the stationary point*:

$$\tilde{\mathbf{K}}(\tilde{\mathbf{x}}_{\text{s}}) = \tilde{K}(\tilde{\mathbf{x}}_{\text{s}})\,\mathbf{I}$$

where the tilde symbol (i.e., "$\sim$") indicates a quantity expressed in terms of the transformed space coordinates. See Appendix C for details on linear transformations of the space coordinates.

Using the results presented in Appendix C, we can express the governing equation (17.1a) in terms of transformed quantities as

$$\tilde{\nabla}\cdot\left[\tilde{\mathbf{K}}(\tilde{\mathbf{x}})\tilde{\nabla}\tilde{M}_{\text{h}}(\tilde{\mathbf{x}};\omega_m)\right] = \tilde{M}_{\text{h}}(\tilde{\mathbf{x}};\omega_m)\left|\tilde{\mathbf{K}}^{1/2}(\tilde{\mathbf{x}})\tilde{\nabla}\tilde{\theta}_{\text{h}}(\tilde{\mathbf{x}};\omega_m)\right|^2. \tag{17.7}$$

Evaluating this equation at the stationary point ($\tilde{\mathbf{x}} = \tilde{\mathbf{x}}_{\text{s}}$) and simplifying give

$$\tilde{\nabla}^2\tilde{M}_{\text{h}}(\tilde{\mathbf{x}}_{\text{s}};\omega_m) = \tilde{M}_{\text{h}}(\tilde{\mathbf{x}}_{\text{s}};\omega_m)\left|\tilde{\nabla}\tilde{\theta}_{\text{h}}(\tilde{\mathbf{x}}_{\text{s}};\omega_m)\right|^2. \tag{17.8}$$

17.3.3.1. Multiple Dimensions

Next we show that in the multidimensional case (17.8) generally allows the existence of saddle points of the amplitude function. We do this by demonstrating their existence in the 2D case. We choose the 2D case for convenience; it is simpler and less notationally cumbersome than the 3D case.

The determinant of the transformed Hessian matrix (see Appendix C), evaluated at the stationary point, is

$$\det\tilde{\mathbf{H}}\left[\tilde{M}_{\text{h}}(\tilde{\mathbf{x}}_{\text{s}};\omega_m)\right] = \frac{\partial^2\tilde{M}_{\text{h}}}{\partial\tilde{x}_1^2}(\tilde{\mathbf{x}}_{\text{s}};\omega_m)\frac{\partial^2\tilde{M}_{\text{h}}}{\partial\tilde{x}_2^2}(\tilde{\mathbf{x}}_{\text{s}};\omega_m) - \left[\frac{\partial^2\tilde{M}_{\text{h}}}{\partial\tilde{x}_1\partial\tilde{x}_2}(\tilde{\mathbf{x}}_{\text{s}};\omega_m)\right]^2.$$

Solving (17.8) for the second derivative of the amplitude with respect to $\tilde{x}_2$ (evaluated at the stationary point) and substituting the result for the corresponding term in the previous equation lead to

$$\det \tilde{\mathbf{H}}\big[\tilde{M}_{\rm h}(\tilde{\mathbf{x}}_{\rm s};\omega_m)\big] = \frac{\partial^2 \tilde{M}_{\rm h}}{\partial \tilde{x}_1^2}(\tilde{\mathbf{x}}_{\rm s};\omega_m) \times \left\{ \tilde{M}_{\rm h}(\tilde{\mathbf{x}}_{\rm s};\omega_m)\left|\tilde{\nabla}\tilde{\theta}_{\rm h}(\tilde{\mathbf{x}}_{\rm s};\omega_m)\right|^2 - \frac{\partial^2 \tilde{M}_{\rm h}}{\partial \tilde{x}_1^2}(\tilde{\mathbf{x}}_{\rm s};\omega_m)\right\} - \left[\frac{\partial^2 \tilde{M}_{\rm h}}{\partial \tilde{x}_1 \partial \tilde{x}_2}(\tilde{\mathbf{x}}_{\rm s};\omega_m)\right]^2. \qquad (17.9)$$

Suppose the second derivative of the amplitude with respect to $\tilde{x}_1$ is negative at the stationary point (nothing prohibits this). Then the RHS of (17.9), and hence the determinant, is negative. Consequently the determinant of the corresponding *untransformed* Hessian matrix is negative (see Appendix C). That is, the Hessian matrix of the amplitude function, evaluated at the stationary point, is indefinite. By the second-derivative test of multivariate calculus [*Riddle*, 1974], the stationary point is thus a saddle point. Therefore, in the 2D case, (17.8) generally allows the existence of saddle points of the amplitude function.

For a specific instance of a time-periodic flow in a 2D, source-free domain for which no amplitude saddle points exist, refer to Section 13.8. That example demonstrates that multidimensional solutions exist for which the amplitude function has no saddle point. In contrast, Example 17.3 gives a multidimensional solution for which the amplitude function has at least one saddle point.

Example 17.3 Amplitude Saddle Point between Two Point Sources

Consider periodic flow in an infinite, 3D, homogeneous, porous medium with a single harmonic constituent of frequency ω_m (*m* fixed). If the medium contains two distinct, periodic point sources (call these point source 1 and point source 2) with identical frequencies ω_m and identical complex amplitudes $Q_0(\omega_m)$, then the midpoint between the two point sources is a saddle point for $M_{\rm h}(\mathbf{x};\omega_m)$ (i.e., an amplitude saddle point for the harmonic constituent with frequency ω_m). Exercises 17.12–17.16 explore some of the details. ◊

Exercise 17.12 Amplitude Function for Flow with a Two-Component FRF

Problem: Consider a periodic flow for which the hydraulic head FRF is given by the sum

$$F_{\rm h}(\mathbf{x};\omega_m) = \sum_{n=1}^{2} F^{(n)}(\mathbf{x};\omega_m) \qquad (17.10)$$

where the $F^{(n)}$ $(n = 1, 2)$ represent component FRFs (e.g., corresponding to two point sources). Writing each component FRF in terms of its corresponding amplitude and phase functions, we obtain

$$F^{(n)}(\mathbf{x};\omega_m) = M^{(n)}(\mathbf{x};\omega_m)\exp\left[i\theta^{(n)}(\mathbf{x};\omega_m)\right] \quad (n = 1, 2) \qquad (17.11)$$

where

$$M^{(n)}(\mathbf{x};\omega_m) = \left|F^{(n)}(\mathbf{x};\omega_m)\right| \qquad (17.12a)$$

$$\theta^{(n)}(\mathbf{x};\omega_m) = \arg F^{(n)}(\mathbf{x};\omega_m). \qquad (17.12b)$$

Show that the FRF amplitude can be written in terms of its component amplitude and phase functions as follows:

$$M_{\rm h}(\mathbf{x};\omega_m) = \sqrt{\left(M^{(1)}\right)^2 + \left(M^{(2)}\right)^2 + 2M^{(1)}M^{(2)}\cos\left(\theta^{(1)} - \theta^{(2)}\right)}. \qquad (17.13)$$

Solution: Substituting the RHS of (17.11) for $F^{(n)}(\mathbf{x};\omega_m)$, once with $n = 1$ and once with $n = 2$, in (17.10) and then grouping the real and imaginary parts give

$$F_{\rm h} = \left[M^{(1)}\cos\theta^{(1)} + M^{(2)}\cos\theta^{(2)}\right] + i\left[M^{(1)}\sin\theta^{(1)} + M^{(2)}\sin\theta^{(2)}\right].$$

Squaring the real and imaginary parts and then adding produce

$$\left|F_{\rm h}\right|^2 = M_{\rm h}^2 = \left[M^{(1)}\cos\theta^{(1)} + M^{(2)}\cos\theta^{(2)}\right]^2 + \left[M^{(1)}\sin\theta^{(1)} + M^{(2)}\sin\theta^{(2)}\right]^2.$$

Expanding the squared terms on the RHS leads to

$$M_{\rm h}^2 = \left(M^{(1)}\right)^2\cos^2\theta^{(1)} + \left(M^{(1)}\right)^2\sin^2\theta^{(1)} + \left(M^{(2)}\right)^2\cos^2\theta^{(2)} + \left(M^{(2)}\right)^2\sin^2\theta^{(2)} + 2M^{(1)}M^{(2)}\left(\cos\theta^{(1)}\cos\theta^{(2)} + \sin\theta^{(1)}\sin\theta^{(2)}\right).$$

Using the Pythagorean relation (B.1) and the angle-difference relation for cosine (B.2b) to simplify, we obtain

$$M_{\rm h}^2 = \left(M^{(1)}\right)^2 + \left(M^{(2)}\right)^2 + 2M^{(1)}M^{(2)}\cos\left(\theta^{(1)} - \theta^{(2)}\right). \qquad (17.14)$$

Taking the square root gives (17.13). □

Exercise 17.13 Amplitude Stationary Point between Two Point Sources

Problem: Consider the time-periodic flow described in Example 17.3. Show that the midpoint between the two sources is a stationary point of the amplitude function $M_{\rm h}(\mathbf{x};\omega_m)$.

Solution: Taking the gradient of (17.14) gives

$$\frac{1}{2}\nabla M_{\rm h}^2 = M^{(1)}\nabla M^{(1)} + \left[M^{(1)}\nabla M^{(2)} + M^{(2)}\nabla M^{(1)}\right]\cos\left(\theta^{(1)} - \theta^{(2)}\right) - M^{(1)}M^{(2)}\sin\left(\theta^{(1)} - \theta^{(2)}\right)\left(\nabla\theta^{(1)} - \nabla\theta^{(2)}\right) + M^{(2)}\nabla M^{(2)}. \qquad (17.15)$$

At the midpoint between the two sources we have

$$M^{(1)}(\mathbf{x}_{\text{midpt}}; \omega_m) = M^{(2)}(\mathbf{x}_{\text{midpt}}; \omega_m) \tag{17.16a}$$

$$\theta^{(1)}(\mathbf{x}_{\text{midpt}}; \omega_m) = \theta^{(2)}(\mathbf{x}_{\text{midpt}}; \omega_m) \tag{17.16b}$$

(where $\mathbf{x}_{\text{midpt}}$ denotes the space coordinates of the midpoint) because (1) the distance from the midpoint to source 1 is equal to the distance from the midpoint to source 2 and (2) as (12.14) and (12.15) show, the component amplitude and phase functions depend on the distance from the point to the source. Substituting (17.16) in (17.15) and grouping terms lead to

$$M_{\text{h}} \nabla M_{\text{h}} = \left(M^{(1)} + M^{(2)}\right) \times \left(\nabla M^{(1)} + \nabla M^{(2)}\right) \quad \text{at } \mathbf{x} = \mathbf{x}_{\text{midpt}}. \tag{17.17}$$

At the midpoint between the two sources we also have, due to symmetry,

$$\nabla M^{(1)}(\mathbf{x}_{\text{midpt}}; \omega_m) = -\nabla M^{(2)}(\mathbf{x}_{\text{midpt}}; \omega_m).$$

Substituting this result in (17.17) and simplifying produce

$$\nabla M_{\text{h}}(\mathbf{x}_{\text{midpt}}; \omega_m) = \mathbf{0}.$$

Therefore, the midpoint is a stationary point of the amplitude function $M_{\text{h}}(\mathbf{x}; \omega_m)$. □

Consider the time-periodic flow described in Example 17.3. Refer to the line that intersects the two point sources as the *line of symmetry*, because the distribution of sources, distribution of material properties, and boundary conditions are symmetric about this line. Consequently, the FRF $F_{\text{h}}(\mathbf{x}; \omega_m)$ is symmetric about this line. The midpoint between the two sources lies on the line of symmetry.

Similarly, refer to the plane that (1) is perpendicular to the line of symmetry and (2) contains the midpoint as the *plane of symmetry*, because the distribution of sources, distribution of material properties, and boundary conditions are symmetric about this plane. Consequently, the FRF $F_{\text{h}}(\mathbf{x}; \omega_m)$ is symmetric about this plane.

Exercise 17.14 Amplitude Saddle Point—Two Point Sources—Symmetry Plane

Problem: Consider the time-periodic flow described in Example 17.3. Consider the amplitude function $M_{\text{h}}(\mathbf{x}; \omega_m)$ on the plane of symmetry, that is, as a function whose domain is the 2D plane of symmetry. Is the amplitude function a local maximum, a local minimum, or neither at the midpoint?

Solution: On the plane of symmetry

$$M^{(1)}(\mathbf{x}; \omega_m) = M^{(2)}(\mathbf{x}; \omega_m) \tag{17.18a}$$

$$\theta^{(1)}(\mathbf{x}; \omega_m) = \theta^{(2)}(\mathbf{x}; \omega_m) \tag{17.18b}$$

because (1) at every point $\mathbf{x}$ on the plane of symmetry the distance from the point to source 1 is equal to the distance from the point to source 2 and (2) as (12.14) and (12.15) show, the component amplitude and phase functions depend on the distance from the point to the source. Substituting (17.18) into (17.13) and then simplifying yield

$$M_{\text{h}}(\mathbf{x}; \omega_m) = \sqrt{2} M^{(1)}(\mathbf{x}; \omega_m) \tag{17.19}$$

on the plane of symmetry. However, within the plane of symmetry, each component amplitude function $M^{(n)}$ ($n = 1, 2$) is a maximum at the midpoint, because this is the point of the plane that is nearest to the corresponding point source. Therefore, within the plane of symmetry the amplitude function M_{h} is a local *maximum* at the midpoint. □

Exercise 17.15 Amplitude Saddle Point—Two Point Sources—Symmetry Line

Problem: Consider the time-periodic flow described in Example 17.3. Consider the amplitude function $M_{\text{h}}(\mathbf{x}; \omega_m)$ on the line of symmetry, that is, as a function whose domain is the 1D line of symmetry. Is the amplitude function a local maximum, a local minimum, or neither at the midpoint?

Solution: The midpoint ($\mathbf{x} = \mathbf{x}_{\text{midpt}}$) lies on the plane of symmetry, where (17.18a), and (17.18b) are valid. Substituting (17.18a, b) into (17.14) with $\mathbf{x} = \mathbf{x}_{\text{midpt}}$ and then taking the square root yield

$$M_{\text{h}}(\mathbf{x}_{\text{midpt}}; \omega_m) = 2M^{(1)}(\mathbf{x}_{\text{midpt}}; \omega_m)$$

which is finite and positive. Consequently, on the line of symmetry, the amplitude function M_{h} increases from $2M^{(1)}(\mathbf{x}_{\text{midpt}}; \omega_m)$ at the midpoint to infinity as one approaches either point source (see 12.14). Therefore, within the line of symmetry the amplitude function M_{h} is a local *minimum* at the midpoint. □

Exercise 17.16 Amplitude Saddle Point between Two Point Sources

Problem: Consider the time-periodic flow described in Example 17.3. Use the results of Exercises 17.13, 17.14, and 17.15 to present a rational (not necessarily rigorous) argument for why the midpoint between the two sources is a saddle point of the amplitude function M_{h}.

Solution: (1) In Exercise 17.13 we showed that the midpoint is a stationary point of the amplitude function M_{h}. (2) In Exercise 17.14 we showed that within the plane of symmetry the amplitude function is a local *maximum* at the midpoint, so in three dimensions the amplitude function M_{h} cannot be a local minimum there. (3) In Exercise 17.15 we showed that within the line of symmetry the amplitude function is a local *minimum* at the midpoint, so in three dimensions the amplitude function M_h cannot be a local maximum there. Therefore, in three dimensions the amplitude function M_h is *neither* a local maximum nor a local minimum at the midpoint. This leaves only one other possibility—the midpoint is a saddle point of the amplitude function. □

17.3.3.2. One Dimension

Note 17.1 Saddle Points in One Dimension

In one dimension, a saddle point is both a stationary point and a point of inflection. The following conditions, when taken together, are both necessary and sufficient for saddle points:

- The derivative with respect to the space coordinate vanishes at the saddle point.
- The second derivative with respect to the space coordinate changes sign at the saddle point (i.e., is negative on one side of the saddle point and positive on the other).

That is, if the point with coordinate $x = x_s$ is a saddle point of the function $f(x)$, then both of the following must hold:

$$\left[\frac{d}{dx}f(x)\right]\Bigg|_{x=x_s} = 0 \tag{17.20a}$$

$$\left\{\left[\frac{d^2}{dx^2}f(x)\right]\Bigg|_{x=x_s-\delta}\right\}\left\{\left[\frac{d^2}{dx^2}f(x)\right]\Bigg|_{x=x_s+\delta}\right\} < 0 \tag{17.20b}$$

where δ is a positive but otherwise arbitrarily small increment. Inequality (17.20b) implies that at a saddle point the second derivative with respect to the space coordinate vanishes:

$$\left[\frac{d^2}{dx^2}f(x)\right]\Bigg|_{x=x_s} = 0. \tag{17.21}$$

Condition (17.21) is necessary but not sufficient to determine that a stationary point is a saddle point. △

Example 17.4 One Dimension: Point of Inflection versus Saddle Point

Consider the function $f(x) = x^4$. The point $x = 0$ is a stationary point of $f(x)$ because (17.20a), with $x_s = 0$, is satisfied. Equation (17.21), with $x_s = 0$, also is satisfied. However, (17.20b) is not satisfied, so the point with coordinate $x = 0$ is not a saddle point of $f(x)$. ◊

Exercise 17.17 Saddle Point of Amplitude Function in One Dimension

Problem: Show that if a saddle point of the amplitude function exists in a 1D, source-free domain, then it coincides with a stationary point of the phase function.

Solution: Let the point with coordinate $x = x_s$ be the saddle point of the amplitude function. Evaluating the governing equation (17.1a) at the saddle point gives

$$0 = M_h(x_s;\omega_m)\left\{\left[\frac{d}{dx}\theta_h(x;\omega_m)\right]\Bigg|_{x=x_s}\right\}^2.$$

By assumption $M_h(x_s;\omega_m) > 0$, so the derivative of the phase function is zero at the saddle point. Therefore, the saddle point of the amplitude function is a stationary point of the phase function. □

17.4. STATIONARY POINTS OF THE PHASE

Consider a stationary point of the phase function, with space coordinate $\mathbf{x} = \mathbf{x}_s$. Evaluating the governing equation (17.1a) at the stationary point leads to

$$\left\{\nabla\cdot\left[\mathbf{K}(\mathbf{x})\nabla M_h(\mathbf{x};\omega_m)\right]\right\}\Big|_{\mathbf{x}=\mathbf{x}_s} = 0.$$

Neither the phase nor any of its spatial derivatives appears in this equation. Thus the equation provides no additional information about the phase function or its spatial variation at stationary points. Therefore, in this section we rely primarily on governing equation (17.1b) to make inferences about stationary points of the phase function.

17.4.1. Phase Maxima

In this section we use the governing equations (17.1b) and a maximum principle to investigate the possible existence of maxima of the phase function.

Rearranging (17.1b) gives

$$\nabla\cdot\left[\mathbf{K}(\mathbf{x})\nabla\theta_h(\mathbf{x};\omega_m)\right] + 2\left[\mathbf{K}(\mathbf{x})\nabla Y_h(\mathbf{x};\omega_m)\right]\cdot\nabla\theta_h(\mathbf{x};\omega_m) = \omega_m S_s(\mathbf{x}) \tag{17.22}$$

where Y_h denotes the natural logarithm of the hydraulic head amplitude:

$$Y_h(\mathbf{x};\omega_m) \equiv \ln M_h(\mathbf{x};\omega_m).$$

The angular frequency and specific storage are positive by assumption, so the RHS of (17.22) is strictly positive; thus we can write it as

$$L_\theta\left[\theta_h(\mathbf{x};\omega_m)\right] > 0 \quad \forall\,\mathbf{x}\in D_+(\omega_m) \tag{17.23}$$

where L_θ is the linear, second-order, differential operator defined as

$$L_\theta[f] \equiv \nabla\cdot\left[\mathbf{K}(\mathbf{x})\nabla f\right] + 2\left[\mathbf{K}(\mathbf{x})\nabla Y_h(\mathbf{x};\omega_m)\right]\cdot\nabla f.$$

The operator L_θ can also be expressed as

$$L_\theta = \sum_{n=1}^{N}\sum_{p=1}^{N} a_{np}(\mathbf{x})\frac{\partial^2}{\partial x_n \partial x_p} + \sum_{n=1}^{N} b_n(\mathbf{x})\frac{\partial}{\partial x_n} \qquad (17.24)$$

where

$$a_{np}(\mathbf{x}) = K_{np}(\mathbf{x}) \quad (p = 1, 2, \ldots, N)$$

$$b_n(\mathbf{x}) = \sum_{p=1}^{N}\left[\frac{\partial K_{pn}(\mathbf{x})}{\partial x_p} + 2K_{np}(\mathbf{x})\frac{\partial Y_{\mathrm{h}}(\mathbf{x})}{\partial x_p}\right]$$

$$n = 1, 2, \ldots, N.$$

By assumption, the operator L_θ is *uniformly elliptic*, and the $a_{np}(\mathbf{x})$ and $b_n(\mathbf{x})$ coefficients are uniformly bounded, on $D_+(\omega_m)$ (see Section 17.3.1).

Next we apply Hopf's maximum principle (see Section 16.2.1) to the strict inequality (17.23). Let $G = D_+(\omega_m)$, $f = \theta_{\mathrm{h}}(\mathbf{x};\omega_m)$, and $L = L_\theta$; then all of the conditions of Hopf's maximum principle are satisfied. Therefore, Hopf's maximum principle applies to the phase function for each hydraulic head harmonic constituent; it tells us that the phase function cannot have a maximum value in $D_+(\omega_m)$.

Exercise 17.18 Application of Hopf's Maximum Principle

Problem: Consider porous media that generally may contain nonzero source distributions. Use Hopf's maximum principle and the governing equations (6.11b) to show the following: If for a given harmonic constituent (i.e., ω_m fixed)

$$M_{\mathrm{u}}(\mathbf{x};\omega_m) < \omega_m S_{\mathrm{s}}(\mathbf{x}) M_{\mathrm{h}}(\mathbf{x};\omega_m) \quad \forall\, \mathbf{x} \in D_+(\omega_m) \qquad (17.25)$$

then the corresponding hydraulic head phase function has no maximum in $D_+(\omega_m)$. What information is required to apply this rule? Is this rule of practical value?

Solution: Combining inequality (17.25) with the governing equation (6.11b) gives inequality (17.23). Following a line of reasoning similar to that used in the case of source-free domains (see above), Hopf's maximum principle tells us that the phase function has no maximum in $D_+(\omega_m)$. The use of this rule requires knowledge of the hydraulic head constituent amplitude function $M_{\mathrm{h}}(\mathbf{x};\omega_m)$ in $D_+(\omega_m)$; therefore, its practical value is limited. □

17.4.2. Phase Minima

In this section we examine particular examples that allow us to draw general conclusions about the existence of phase minima in source-free domains. Then we use the polar form of the governing equations to infer what we can about the nature of phase minima.

The examples presented in Chapter 9 for a 1D, finite-length, source-free domain filled with a homogeneous medium illustrate that phase minima generally are allowed but are not required. Can two or more isolated phase minima exist at distinct locations in the space domain? We explore this question below.

Exercise 17.19 Phase Minima in One Dimension

Problem: Consider 1D, source-free, porous media. Review the argument that was presented in Section 17.3.2 to prove that in that case there can be at most one isolated *amplitude* minimum in the space domain. Using a similar argument, show that if the amplitude is positive everywhere in the space domain, then there can be at most one *phase* minimum there.

Solution: Suppose isolated phase minima exist at two distinct locations — the points with coordinates $x = x_{\mathrm{min1}}$ and $x = x_{\mathrm{min3}}$ where $x_{\mathrm{min1}} < x_{\mathrm{min3}}$. On a graph of phase (vertical axis) versus x (horizontal axis), these minima correspond to low points (see Figure 17.2). Recall that by assumption the phase function is smooth. Therefore, a high point (i.e., a phase maximum) necessarily appears at some point with coordinate $x = x_{\mathrm{max2}}$ lying between the two low points (i.e., $x_{\mathrm{min1}} < x_{\mathrm{max2}} < x_{\mathrm{min3}}$). This violates the general prohibition on phase maxima in source-free domains (see Section 17.4.1), so in this case there can be at most one isolated phase minimum in the space domain. □

Consider the 2D case. Suppose isolated phase minima exist at two distinct locations—the points with coordinates $\mathbf{x} = \mathbf{x}_{\mathrm{min1}}$ and $\mathbf{x} = \mathbf{x}_{\mathrm{min3}}$ where $|\mathbf{x}_{\mathrm{min1}} - \mathbf{x}_{\mathrm{min3}}| > 0$. Consider the variation of the phase along the line segment whose endpoints are the two stationary points. As in the 1D case, at some intermediate point ($\mathbf{x} = \mathbf{x}'$) on the segment the phase reaches a local (possibly global) maximum *on the segment*. Unlike the 1D case, in the 2D case that point might not correspond to a phase maximum *on the space domain*. For instance, it may correspond to a saddle point, which is not prohibited. Therefore, we cannot rule out the possible existence of isolated phase minima at two or more distinct locations in a multidimensional space domain.

Exercise 17.20 Phase Minimum

Problem: Suppose the hydraulic head phase function for a particular harmonic constituent (i.e., ω_m fixed) is a minimum at the isolated point with coordinates $\mathbf{x} = \mathbf{x}_{\mathrm{min}}$. What, if anything, can one infer about the in-phase and quadrature components of the corresponding hydraulic gradient and specific-discharge vector harmonic constituents evaluated at the point with coordinates $\mathbf{x} = \mathbf{x}_{\mathrm{min}}$?

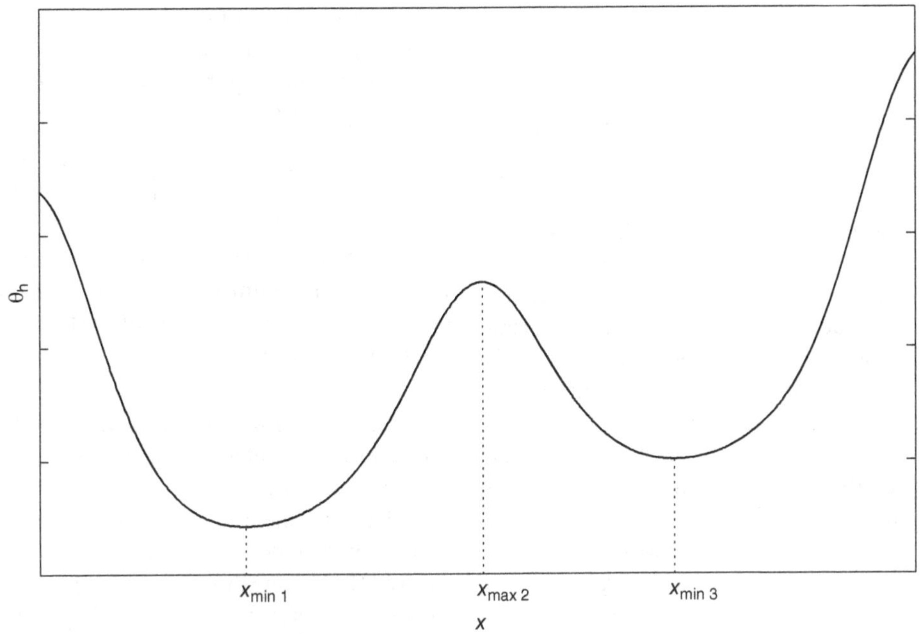

Figure 17.2 Prohibited 1D phase distribution.

Solution: The phase is a minimum at the point with coordinates $\mathbf{x}_{\min}$, so $\nabla\theta_{\mathrm{h}}(\mathbf{x}_{\min};\omega_m) = \mathbf{0}$. Substituting these results for the corresponding terms of (14.20b) and (14.22b), respectively, yields

$$(\nabla h')_{\mathrm{Qu}}(\mathbf{x}_{\min}, t; \omega_m) = \mathbf{0} \tag{17.26a}$$

$$\mathbf{q}'_{\mathrm{Qu}}(\mathbf{x}_{\min}, t; \omega_m) = \mathbf{0}. \tag{17.26b}$$

That is, both of the quadrature component vectors vanish at $\mathbf{x} = \mathbf{x}_{\min}$. □

17.4.3. Phase Saddle Points

In this section we use the governing equation (17.1b) and the second-derivative test of multivariate calculus to explore the following question: Are saddle points of the phase function required, prohibited, or neither? We follow a procedure similar to that of Section 17.3.3. The steps were described in some detail in that section so we provide only an abbreviated description here.

A linear coordinate transformation $\tilde{\mathbf{x}} = \mathbf{C}\mathbf{x}$ exists that renders the transformed medium locally isotropic with respect to hydraulic conductivity at the stationary point (see Section 17.3.3 for details). Using the results presented in Appendix C, we can express the governing equation (17.1b) in terms of transformed quantities as

$$\tilde{\nabla}\cdot\left[\tilde{\mathbf{K}}(\tilde{\mathbf{x}})\tilde{\nabla}\tilde{\theta}_{\mathrm{h}}(\tilde{\mathbf{x}};\omega_m)\right] + 2\left[\tilde{\mathbf{K}}(\tilde{\mathbf{x}})\tilde{\nabla}\tilde{Y}_{\mathrm{h}}(\tilde{\mathbf{x}};\omega_m)\right] \cdot\tilde{\nabla}\tilde{\theta}_{\mathrm{h}}(\tilde{\mathbf{x}};\omega_m) = \omega_m\tilde{S}_{\mathrm{s}}(\tilde{\mathbf{x}}) \tag{17.27}$$

where the tilde (i.e., "∼") denotes a quantity expressed relative to the transformed coordinates. Let $\mathbf{x} = \mathbf{x}_{\mathrm{s}}$ be the coordinates of a stationary point of the phase function $\theta_{\mathrm{h}}(\mathbf{x};\omega_m)$, where $\mathbf{x}_{\mathrm{s}} \in D_{+}(\omega_m)$. Evaluating the transformed governing equation (17.27) at the stationary point and simplifying give

$$\tilde{\nabla}^2\tilde{\theta}_{\mathrm{h}}(\tilde{\mathbf{x}}_{\mathrm{s}};\omega_m) = \frac{\omega_m\tilde{S}_{\mathrm{s}}(\tilde{\mathbf{x}}_{\mathrm{s}})}{\tilde{K}(\tilde{\mathbf{x}}_{\mathrm{s}})}. \tag{17.28}$$

17.4.3.1. Multiple Dimensions

Next we show that, in the multidimensional case, (17.28) generally allows the existence of saddle points of the phase function. We do this by demonstrating their existence in the 2D case. We choose the 2D case for convenience; it is simpler and less notationally cumbersome than the 3D case.

The determinant of the transformed Hessian matrix, evaluated at the stationary point, is

$$\det\tilde{\mathbf{H}}\left[\tilde{\theta}_{\mathrm{h}}(\tilde{\mathbf{x}}_{\mathrm{s}};\omega_m)\right] = \frac{\partial^2\tilde{\theta}_{\mathrm{h}}}{\partial\tilde{x}_1^2}(\tilde{\mathbf{x}}_{\mathrm{s}};\omega_m)\frac{\partial^2\tilde{\theta}_{\mathrm{h}}}{\partial\tilde{x}_2^2}(\tilde{\mathbf{x}}_{\mathrm{s}};\omega_m) - \left[\frac{\partial^2\tilde{\theta}_{\mathrm{h}}}{\partial\tilde{x}_1\partial\tilde{x}_2}(\tilde{\mathbf{x}}_{\mathrm{s}};\omega_m)\right]^2.$$

Solving (17.28) for the second derivative of the phase with respect to $\tilde{x}_2$ and substituting the result for the corresponding term in the previous equation lead to

$$\det \tilde{\mathbf{H}}\left[\tilde{\theta}_{\mathrm{h}}(\tilde{\mathbf{x}}_{\mathrm{s}};\omega_m)\right] = \frac{\partial^2\tilde{\theta}_{\mathrm{h}}}{\partial\tilde{x}_1^2}(\tilde{\mathbf{x}}_{\mathrm{s}};\omega_m)$$

$$\left[\frac{\omega_m\tilde{S}_{\mathrm{s}}(\tilde{\mathbf{x}}_{\mathrm{s}})}{\tilde{K}(\tilde{\mathbf{x}}_{\mathrm{s}})} - \frac{\partial^2\tilde{\theta}_{\mathrm{h}}}{\partial\tilde{x}_1^2}(\tilde{\mathbf{x}}_{\mathrm{s}};\omega_m)\right] - \left[\frac{\partial^2\tilde{\theta}_{\mathrm{h}}}{\partial\tilde{x}_1\partial\tilde{x}_2}(\tilde{\mathbf{x}}_{\mathrm{s}};\omega_m)\right]^2. \tag{17.29}$$

Suppose the second partial derivative of the phase with respect to $\tilde{x}_1$ is negative at the stationary point (nothing prohibits this). Then the RHS of (17.29), and hence the determinant, is negative. Consequently the determinant of the corresponding *untransformed* Hessian matrix is negative (see Appendix C). That is, the Hessian matrix of the phase function evaluated at the stationary point is indefinite. By the second-derivative test of multivariate calculus [*Riddle*, 1974], the stationary point is thus a saddle point. Therefore, saddle points of the phase function generally are allowed in the 2D case.

Exercise 17.21 describes another hypothetical case that allows for the existence of a saddle point of the phase function.

Exercise 17.21 Derivation

Problem: Suppose the second partial derivative of the phase with respect to $\tilde{x}_1$ is positive and the mixed second partial derivative is unknown at the stationary point. Derive a basic condition on the second partial derivative that guarantees that the determinant of the Hessian matrix is negative at the stationary point.

Solution: The condition must guarantee that the RHS of (17.29) is negative even when the mixed second partial derivative is zero:

$$\frac{\partial^2\tilde{\theta}_{\mathrm{h}}}{\partial\tilde{x}_1^2}(\tilde{\mathbf{x}}_{\mathrm{s}};\omega_m) > \frac{\omega_m\tilde{S}_{\mathrm{s}}(\tilde{\mathbf{x}}_{\mathrm{s}})}{\tilde{K}(\tilde{\mathbf{x}}_{\mathrm{s}})}.$$

□

For a specific instance of a time-periodic flow in a 2D, source-free domain for which no phase saddle points exist, refer to Section 13.8. That example demonstrates that in the multidimensional case solutions exist for which the phase function has no saddle point.

17.4.3.2. One Dimension

For basic information on saddle points in one dimension, see Note 17.1.

Exercise 17.22 Saddle Points of Phase Function in One Dimension

Problem: Show that saddle points of the phase function are prohibited in 1D, source-free domains.

Solution: Let the point with coordinate $x = x_{\mathrm{s}}$ be a stationary point of the phase function in a source-free, 1D medium. Evaluating the governing equation (17.1b) at the stationary point gives

$$\left[\frac{d^2}{dx^2}\theta_{\mathrm{h}}(x;\omega_m)\right]\bigg|_{x=x_{\mathrm{s}}} = \frac{\omega_m S_{\mathrm{s}}(x_{\mathrm{s}})}{K(x_{\mathrm{s}})}.$$

The RHS is positive, so by the second-derivative test the stationary point corresponds to a minimum. Consequently, saddle points of the phase function are prohibited in 1D, source-free domains. □

In Section 17.4.1 it was shown that maxima of the phase are prohibited, while Exercise 17.22 showed that in one dimension saddle points of the phase are prohibited. Thus, in one dimension every stationary point of the phase corresponds to a phase minimum. Exercise 17.19 showed that in one dimension there can be at most one distinct location at which the phase is a minimum. Therefore, in one dimension there can be at most one stationary point of the phase, and if such a point exists then it corresponds to a phase minimum.

Exercise 17.17 implies that in one dimension every saddle point of the amplitude corresponds to a stationary point of the phase (i.e., a phase minimum). Therefore, in 1D, source-free domains, at most one saddle point of the amplitude function exists, and if such a saddle point exists, then it corresponds to a phase minimum.

17.5. CHAPTER SUMMARY

In this chapter we presented the polar form of the space BVP for time-periodic groundwater flow in source-free domains. We presented exercises (see Section 17.2) to show that the following hold in three dimensions:

If the hydraulic head amplitude function is spatially uniform and equal to a minimum value in a subregion with positive volume, then the minimum value is zero.

The space BVP prohibits solutions for which the hydraulic head phase function is spatially uniform in a subregion with positive volume.

Similar results hold for a subregion with positive area in two dimensions and for a subregion with positive length in one dimension.

By manipulating the governing equations and applying maximum principles and the second-derivative test of multivariate calculus, we inferred basic rules about the existence and nature of stationary points of the hydraulic head amplitude and phase functions in source-free domains. We found that if the amplitude and phase functions vary smoothly over a subregion $D_+(\omega_m)$ in which the amplitude is positive, then the following conditions are satisfied in $D_+(\omega_m)$:

1. Both the amplitude and phase functions have the following characteristics:

(a) Maxima are prohibited.

(b) Minima are allowed but are not required.

(c) In the multidimensional case, multiple isolated minima at distinct locations are allowed but are not required.

(d) In the multidimensional case, saddle points are allowed but are not required.

2. In one dimension, the following hold:

(a) At most one stationary point of the phase exists. If such a point exists, then it corresponds to a phase minimum.

(b) At most two stationary points of the amplitude exist:

(i) At most one saddle point of the amplitude exists. If such a point exists, then it corresponds to a phase minimum.

(ii) At most one amplitude minimum exists. If such a point exists, then it may or may not correspond to a phase minimum. If an amplitude minimum exists and it corresponds to a phase minimum, then because of 2(a) and 2(b)(i), there is no amplitude saddle point.

18

Flow Stagnation

Flow stagnation is the absence of flow at one or more location(s) in the space domain (D). Ideally, complete characterization of a groundwater flow field requires the identification of any stagnant locations that may exist in that field. This chapter provides some theoretical background for understanding flow stagnation in the context of time-periodic flow.

Stagnation points are points in D where the specific-discharge vector vanishes at all times (i.e., locations where the flow is *stagnant*). A *stagnation set* is a collection of one or more stagnation point(s) in D. Thus, stagnation points are the fundamental units of stagnation sets.

18.1. STAGNATION POINTS

The point with coordinates $\mathbf{x} = \mathbf{x}_s$ is a stagnation point for the vector field $\mathbf{v}(\mathbf{x}, t)$ if and only if it satisfies

$$\mathbf{v}(\mathbf{x}_s, t) = \mathbf{0} \quad \forall\, t > 0, \text{ where } \mathbf{x}_s \in D. \tag{18.1}$$

One can define stagnation points with respect to various types of vector fields, including different types of specific-discharge vectors and the corresponding hydraulic gradient vectors:

- General: $\mathbf{q}(\mathbf{x}, t)$, $\nabla h(\mathbf{x}, t)$
- Periodic transient components: $\mathring{\mathbf{q}}(\mathbf{x}, t)$, $\nabla \mathring{h}(\mathbf{x}, t)$
- Harmonic constituents: $\mathbf{q}'(\mathbf{x}, t; \omega_m)$, $\nabla h'(\mathbf{x}, t; \omega_m)$
- In-phase components: $\mathbf{q}'_{\mathrm{Ip}}(\mathbf{x}, t; \omega_m)$, $\nabla (h')_{\mathrm{Ip}}(\mathbf{x}, t; \omega_m)$
- Quadrature components: $\mathbf{q}'_{\mathrm{Qu}}(\mathbf{x}, t; \omega_m)$, $\nabla (h')_{\mathrm{Qu}}(\mathbf{x}, t; \omega_m)$

Each of these specific-discharge vectors is related to the corresponding hydraulic gradient vector and to the hydraulic conductivity tensor $\mathbf{K}$ by Darcy's law. By assumption $\mathbf{K}$ is positive definite. Therefore, a point is a stagnation point for one of these specific-discharge vector fields if and only if it is a stagnation point for the corresponding hydraulic gradient vector field. Hence, the general results that we derive regarding stagnation points for any one of the specific-discharge vector fields listed above also apply to the corresponding hydraulic gradient vector field.

Exercise 18.1 Stagnation Points: Specific Discharge and Its Constituents

Problem: Show that if the point with coordinates $\mathbf{x} = \mathbf{x}_s$ is a stagnation point for every one of the specific-discharge vector harmonic constituents, i.e.,

$$\mathbf{q}'(\mathbf{x}_s, t; \omega_m) = \mathbf{0} \quad \forall\, t > 0, \text{ where } \mathbf{x}_s \in D;\ m = 1, 2, 3, \ldots \tag{18.2}$$

then the point is a stagnation point for the time-periodic component of the specific-discharge vector, $\mathring{\mathbf{q}}(\mathbf{x}, t)$.

Solution: Substituting $\mathbf{x} = \mathbf{x}_s$ and $\mathbf{q}'(\mathbf{x}_s, t; \omega_m) = \mathbf{0}$ (for $m = 1, 2, 3, \ldots$) in (4.21) gives

$$\mathring{\mathbf{q}}(\mathbf{x}_s, t) = \mathbf{0} \quad \forall\, t > 0, \text{ where } \mathbf{x}_s \in D$$

which is equivalent to (18.1) with the substitution $\mathbf{v}(\mathbf{x}, t) = \mathring{\mathbf{q}}(\mathbf{x}, t)$. Therefore, the point is a stagnation point for the time-periodic component of the specific-discharge vector. □

Exercise 18.2 Stagnation Points: Specific Discharge and Its Constituents

Problem: Show that if the point with coordinates $\mathbf{x} = \mathbf{x}_s$ is a stagnation point for the time-periodic component of the specific-discharge vector, then it is a stagnation point for every one of the corresponding vector harmonic constituents.

Hydrodynamics of Time-Periodic Groundwater Flow: Diffusion Waves in Porous Media, Geophysical Monograph 224,
First Edition. Joe S. Depner and Todd C. Rasmussen.

Solution: Substituting $\mathbf{x} = \mathbf{x}_s$ and $\mathring{\mathbf{q}}(\mathbf{x}_s, t) = \mathbf{0}$ in (3.10c) leads to

$$\mathbf{0} = \mathbf{K}(\mathbf{x}_s)\nabla\mathring{h}(\mathbf{x}, t) \quad \forall\, t > 0.$$

Substituting the RHS of (4.20) for the time-periodic hydraulic gradient vector yields

$$\mathbf{0} = \mathbf{K}(\mathbf{x}_s)\sum_{m=1}^{\infty}\nabla h'(\mathbf{x}_s, t; \omega_m) \quad \forall\, t > 0.$$

Substituting the RHS of (4.2a) for $h'(\mathbf{x}, t; \omega_m)$ and simplifying lead to

$$\mathbf{0} = \mathbf{K}(\mathbf{x}_s)\sum_{m=1}^{\infty}\Big[\nabla A_h(\mathbf{x}_s; \omega_m)\cos(\omega_m t) - \nabla B_h(\mathbf{x}_s; \omega_m)\sin(\omega_m t)\Big] \quad \forall\, t > 0.$$

Taking the inner products (as defined in Section B.2) of the previous equation with the functions $\cos(\omega_n t)$ and $\sin(\omega_n t)$, respectively, where n is a positive integer, gives

$$\begin{aligned}\mathbf{0} &= \mathbf{K}(\mathbf{x}_s)\nabla A_h(\mathbf{x}_s; \omega_n)\\ \mathbf{0} &= \mathbf{K}(\mathbf{x}_s)\nabla B_h(\mathbf{x}_s; \omega_n)\end{aligned} \qquad (18.3)$$

$$n = 1, 2, 3, \ldots.$$

Substituting these results for the corresponding terms of (4.24), with $\mathbf{x} = \mathbf{x}_s$, yields (18.2). □

We can summarize the results from Exercises 18.1 and 18.2 as follows: The point with coordinates $\mathbf{x} = \mathbf{x}_s$ is a stagnation point for the time-periodic component of the specific-discharge vector $\mathring{\mathbf{q}}(\mathbf{x}, t)$ if and only if it is a stagnation point for every one of the corresponding vector harmonic constituents $\mathbf{q}'(\mathbf{x}, t; \omega_m)$ $(m = 1, 2, 3, \ldots)$ [i.e., if and only if (18.2) is satisfied].

Exercise 18.3 Stagnation Points: Gradients of Rectangular Form Coefficients

Problem: Show that if the point with coordinates $\mathbf{x} = \mathbf{x}_s$ is a stagnation point for the vector harmonic constituent $\mathbf{q}'(\mathbf{x}, t; \omega_m)$, then the gradients of the amplitude coefficients, $\nabla A_h(\mathbf{x}; \omega_m)$ and $\nabla B_h(\mathbf{x}; \omega_m)$, vanish at that point.

Solution: Substituting $\mathbf{v}(\mathbf{x}, t) = \mathbf{q}'(\mathbf{x}, t; \omega_m)$ in (18.1) gives

$$\mathbf{q}'(\mathbf{x}_s, t; \omega_m) = \mathbf{0} \quad \forall\, t > 0, \text{ where } \mathbf{x}_s \in D.$$

Combining this equation with (4.24) gives

$$-\mathbf{K}(\mathbf{x}_s)\nabla A_h(\mathbf{x}_s; \omega_m)\cos(\omega_m t) + \mathbf{K}(\mathbf{x}_s)\nabla B_h(\mathbf{x}_s; \omega_m)\sin(\omega_m t) = \mathbf{0} \quad \forall\, t > 0. \qquad (18.4)$$

Multiplying by $\cos(\omega_m t)$ and $\sin(\omega_m t)$, respectively, and then integrating with respect to time, from $t = t_0$ to $t = t_0 + 2\pi$ rad$/\omega_m$, yield

$$\begin{aligned}\mathbf{K}(\mathbf{x}_s)\nabla A_h(\mathbf{x}_s; \omega_m) &= \mathbf{0}\\ \mathbf{K}(\mathbf{x}_s)\nabla B_h(\mathbf{x}_s; \omega_m) &= \mathbf{0}.\end{aligned}$$

Because $\mathbf{K}$ is positive definite, these imply

$$\nabla A_h(\mathbf{x}_s; \omega_m) = \mathbf{0} \qquad (18.5a)$$

$$\nabla B_h(\mathbf{x}_s; \omega_m) = \mathbf{0} \qquad (18.5b)$$

$$\mathbf{x}_s \in D.$$ □

Exercise 18.4 Stagnation Points: Gradients of Rectangular Form Coefficients

Problem: Show that if equations (18.5) are satisfied, then the point with coordinates $\mathbf{x} = \mathbf{x}_s$ is a stagnation point for the vector harmonic constituent $\mathbf{q}'(\mathbf{x}, t; \omega_m)$.

Solution: Substituting $\mathbf{x}_s$ for $\mathbf{x}$ and $\mathbf{0}$ for $\nabla A_h(\mathbf{x}_s; \omega_m)$ and for $\nabla B_h(\mathbf{x}_s; \omega_m)$ in (4.24) gives

$$\mathbf{q}'(\mathbf{x}_s, t; \omega_m) = \mathbf{0} \quad \forall\, t > 0, \text{ where } \mathbf{x}_s \in D \qquad (18.6)$$

which is equivalent to (18.1) with the substitution $\mathbf{v}(\mathbf{x}, t) = \mathbf{q}'(\mathbf{x}, t; \omega_m)$. Therefore, the point with coordinates $\mathbf{x} = \mathbf{x}_s$ is a stagnation point for $\mathbf{q}'(\mathbf{x}, t; \omega_m)$. □

We can summarize the results from Exercises 18.3 and 18.4 as follows: The point with coordinates $\mathbf{x} = \mathbf{x}_s$ is a stagnation point for the specific-discharge vector harmonic constituent $\mathbf{q}'(\mathbf{x}, t; \omega_m)$ if and only if the gradients of the amplitude coefficients, $\nabla A_h(\mathbf{x}; \omega_m)$ and $\nabla B_h(\mathbf{x}; \omega_m)$, vanish at that point [i.e., if and only if equations (18.5) are satisfied]. More abstractly, the point with coordinates $\mathbf{x} = \mathbf{x}_s$ is a stagnation point for the specific-discharge vector harmonic constituent $\mathbf{q}'(\mathbf{x}, t; \omega_m)$ if and only if it is a zero point for both of the gradients $\nabla A_h(\mathbf{x}; \omega_m)$ and $\nabla B_h(\mathbf{x}; \omega_m)$.

Exercise 18.5 Stagnation Points: Relationships between Gradients

Problem: The last sentence of the previous paragraph summarized the results from Exercises 18.3 and 18.4 in terms of the gradients $\nabla A_h(\mathbf{x}; \omega_m)$ and $\nabla B_h(\mathbf{x}; \omega_m)$. Express that result in terms of the gradient of the hydraulic head FRF, $\nabla F_h(\mathbf{x}; \omega_m)$.

Solution: Based on (7.6a), the equivalent statement is the following:

The point with coordinates $\mathbf{x} = \mathbf{x}_s$ is a stagnation point for the specific-discharge vector harmonic constituent $\mathbf{q}'(\mathbf{x}, t; \omega_m)$ if and only if it is a zero point for the gradient $\nabla F_h(\mathbf{x}; \omega_m)$. □

Exercise 18.6 Stagnation Points: Gradients of Rectangular Form Coefficients

Problem: Express the gradients $\nabla A_h(\mathbf{x};\omega_m)$ and $\nabla B_h(\mathbf{x};\omega_m)$ in terms of the amplitude and phase functions for the corresponding harmonic constituent. Are the resulting expressions valid at those points where the amplitude function is zero? Why or why not?

Solution: Taking the gradient of (4.4a) and (4.5a) gives

$$\nabla A_h(\mathbf{x};\omega_m) = \nabla[M_h(\mathbf{x};\omega_m)\cos\theta_h(\mathbf{x};\omega_m)]$$
$$\nabla B_h(\mathbf{x};\omega_m) = \nabla[M_h(\mathbf{x};\omega_m)\sin\theta_h(\mathbf{x};\omega_m)].$$

Expanding the terms on the RHSs leads to

$$\nabla A_h(\mathbf{x};\omega_m) = \nabla M_h\cos\theta_h - M_h\sin\theta_h\nabla\theta_h \quad (18.7a)$$
$$\nabla B_h(\mathbf{x};\omega_m) = \nabla M_h\sin\theta_h + M_h\cos\theta_h\nabla\theta_h \quad (18.7b)$$
$$\forall\,\mathbf{x}\in D_+(\omega_m).$$

Equations (18.7) are not valid at points where the amplitude function is zero because the phase function is not defined at such points. □

Exercise 18.7 Stagnation Points: Amplitude and Phase Gradients

Problem: Show that if the point with coordinates $\mathbf{x} = \mathbf{x}_s$, where $\mathbf{x}_s \in D_+(\omega_m)$, is a stagnation point for the vector harmonic constituent $\mathbf{q}'(\mathbf{x}, t;\omega_m)$, then

$$\nabla M_h(\mathbf{x}_s;\omega_m) = \mathbf{0} \quad (18.8a)$$
$$\nabla\theta_h(\mathbf{x}_s;\omega_m) = \mathbf{0} \quad (18.8b)$$

where $\mathbf{x}_s \in D_+(\omega_m)$.

Solution: Because $\mathbf{x} = \mathbf{x}_s$ is a stagnation point for the vector harmonic constituent $\mathbf{q}'(\mathbf{x}, t;\omega_m)$, equations (18.5) are satisfied. Substituting $\mathbf{x} = \mathbf{x}_s$, $\nabla A_h(\mathbf{x}_s;\omega_m) = \mathbf{0}$, and $\nabla B_h(\mathbf{x}_s;\omega_m) = \mathbf{0}$ in (18.7) leads to

$$\nabla M_h(\mathbf{x}_s;\omega_m)\cos\theta_h(\mathbf{x}_s;\omega_m) - M_h(\mathbf{x}_s;\omega_m)\sin\theta_h(\mathbf{x}_s;\omega_m)\nabla\theta_h(\mathbf{x}_s;\omega_m) = \mathbf{0} \quad (18.9a)$$
$$\nabla M_h(\mathbf{x}_s;\omega_m)\sin\theta_h(\mathbf{x}_s;\omega_m) + M_h(\mathbf{x}_s;\omega_m)\cos\theta_h(\mathbf{x}_s;\omega_m)\nabla\theta_h(\mathbf{x}_s;\omega_m) = \mathbf{0}. \quad (18.9b)$$

Eliminating $\theta_h(\mathbf{x}_s;\omega_m)$ and simplifying yield (18.8a) and (18.8b). □

Exercise 18.8 Stagnation Points: Amplitude and Phase Gradients

Problem: Show that if equations (18.8) are satisfied, then the point with coordinates $\mathbf{x} = \mathbf{x}_s$ is a stagnation point for the vector harmonic constituent $\mathbf{q}'(\mathbf{x}, t;\omega_m)$.

Solution: Substituting $\mathbf{x} = \mathbf{x}_s$ in (18.7) and implementing (18.8) lead to (18.5). Therefore, the point with coordinates $\mathbf{x} = \mathbf{x}_s$ is a stagnation point for the vector harmonic constituent $\mathbf{q}'(\mathbf{x}, t;\omega_m)$. □

We can summarize the results from Exercises 18.7 and 18.8 as follows: The point with coordinates $\mathbf{x} = \mathbf{x}_s$, where $\mathbf{x}_s \in D_+(\omega_m)$, is a stagnation point for the vector harmonic constituent $\mathbf{q}'(\mathbf{x}, t;\omega_m)$ if and only if the gradients of the corresponding amplitude and phase functions vanish at the point [i.e., if and only if equations (18.8) are satisfied]. More abstractly, the point with coordinates $\mathbf{x} = \mathbf{x}_s$, where $\mathbf{x}_s \in D_+(\omega_m)$, is a stagnation point for the specific-discharge vector harmonic constituent $\mathbf{q}'(\mathbf{x}, t;\omega_m)$ if and only if it is a zero point for both of the gradients $\nabla M_h(\mathbf{x};\omega_m)$ and $\nabla\theta_h(\mathbf{x};\omega_m)$.

Exercise 18.9 Stagnation Points on Domain Boundary

Problem: Is it possible to have a stagnation point on the boundary of a space domain D?

Solution: No. Although it is possible for the specific-discharge or hydraulic gradient vector harmonic constituents to vanish at a point on the boundary, such a point is not a stagnation point because it is not in the domain D [see equation (18.1)]. □

Exercise 18.10 Stagnation Points: In-Phase and Quadrature Components

Problem: Show that if the point with coordinates $\mathbf{x} = \mathbf{x}_s$, where $\mathbf{x}_s \in D_+(\omega_m)$, is a stagnation point for both the in-phase and quadrature components $\mathbf{q}'_{Ip}(\mathbf{x}, t;\omega_m)$ and $\mathbf{q}'_{Qu}(\mathbf{x}, t;\omega_m)$, then it is a stagnation point for the vector harmonic constituent $\mathbf{q}'(\mathbf{x}, t;\omega_m)$.

Solution: Substituting $\mathbf{x} = \mathbf{x}_s$ in (14.21) gives

$$\mathbf{q}'(\mathbf{x}_s, t;\omega_m) = \mathbf{q}'_{Ip}(\mathbf{x}_s, t;\omega_m) + \mathbf{q}'_{Qu}(\mathbf{x}_s, t;\omega_m).$$

Substituting $\mathbf{0}$ for $\mathbf{q}'_{Ip}(\mathbf{x}_s, t;\omega_m)$ and $\mathbf{q}'_{Qu}(\mathbf{x}_s, t;\omega_m)$ in the previous equation then yields

$$\mathbf{q}'(\mathbf{x}_s, t;\omega_m) = \mathbf{0} \quad \forall\, t, \text{ where } \mathbf{x}_s \in D_+(\omega_m).$$ □

Exercise 18.11 Stagnation Points: In-Phase and Quadrature Components

Problem: Show that if the point with coordinates $\mathbf{x} = \mathbf{x}_s$, where $\mathbf{x}_s \in D_+(\omega_m)$, is a stagnation point for the vector harmonic constituent $\mathbf{q}'(\mathbf{x}, t;\omega_m)$, then it is a stagnation point for both the in-phase and quadrature components $\mathbf{q}'_{Ip}(\mathbf{x}, t;\omega_m)$ and $\mathbf{q}'_{Qu}(\mathbf{x}, t;\omega_m)$.

Solution: Substituting the RHSs of (14.22) for the corresponding component vectors on the RHS of (14.21) gives

$$\mathbf{q}'(\mathbf{x}, t; \omega_m) = -\mathbf{K}(\mathbf{x})\nabla M_\mathrm{h}(\mathbf{x}; \omega_m)\cos\left[\omega_m t + \theta_\mathrm{h}(\mathbf{x}; \omega_m)\right] + M_\mathrm{h}(\mathbf{x}; \omega_m)\mathbf{K}(\mathbf{x})\nabla\theta_\mathrm{h}(\mathbf{x}; \omega_m) \times \sin\left[\omega_m t + \theta_\mathrm{h}(\mathbf{x}; \omega_m)\right].$$

Substituting $\mathbf{x} = \mathbf{x}_\mathrm{s}$ and $\mathbf{q}'(\mathbf{x}_\mathrm{s}, t; \omega_m) = \mathbf{0}$ in this equation leads to

$$\mathbf{0} = -\mathbf{K}(\mathbf{x}_\mathrm{s})\nabla M_\mathrm{h}(\mathbf{x}_\mathrm{s}; \omega_m)\cos\left[\omega_m t + \theta_\mathrm{h}(\mathbf{x}_\mathrm{s}; \omega_m)\right] + M_\mathrm{h}(\mathbf{x}_\mathrm{s}; \omega_m)\mathbf{K}(\mathbf{x}_\mathrm{s})\nabla\theta_\mathrm{h}(\mathbf{x}_\mathrm{s}; \omega_m)\sin\left[\omega_m t + \theta_\mathrm{h}(\mathbf{x}_\mathrm{s}; \omega_m)\right]. \tag{18.10}$$

Multiplying (18.10) by $\cos\left[\omega_m t + \theta_\mathrm{h}(\mathbf{x}; \omega_m)\right]$ and then integrating with respect to time, from $t = t_0$ to $t = t_0 + 2\pi\ \mathrm{rad}/\omega_m$, give

$$\mathbf{K}(\mathbf{x}_\mathrm{s})\nabla M_\mathrm{h}(\mathbf{x}_\mathrm{s}; \omega_m) = \mathbf{0}. \tag{18.11}$$

Similarly, multiplying (18.10) by $\sin[\omega_m t + \theta_\mathrm{h}(\mathbf{x}_\mathrm{s}; \omega_m)]$ and then integrating as above give

$$M_\mathrm{h}(\mathbf{x}_\mathrm{s}; \omega_m)\mathbf{K}(\mathbf{x}_\mathrm{s})\nabla\theta_\mathrm{h}(\mathbf{x}_\mathrm{s}; \omega_m) = \mathbf{0}. \tag{18.12}$$

Substituting the RHSs of (18.11) and (18.12) for the corresponding terms of (14.22) yields

$$\mathbf{q}'_\mathrm{Ip}(\mathbf{x}_\mathrm{s}, t; \omega_m) = \mathbf{0} \tag{18.13a}$$

$$\mathbf{q}'_\mathrm{Qu}(\mathbf{x}_\mathrm{s}, t; \omega_m) = \mathbf{0} \tag{18.13b}$$

$$\forall\ t, \text{ where } \mathbf{x}_\mathrm{s} \in D_+(\omega_m).$$

These are equivalent to (18.1) with the substitutions $\mathbf{v}(\mathbf{x}, t) = \mathbf{q}'_\mathrm{Ip}(\mathbf{x}, t; \omega_m)$ and $\mathbf{v}(\mathbf{x}, t) = \mathbf{q}'_\mathrm{Qu}(\mathbf{x}, t; \omega_m)$, respectively. □

We can summarize the results from Exercises 18.10 and 18.11 as follows: The point with coordinates $\mathbf{x} = \mathbf{x}_\mathrm{s}$, where $\mathbf{x}_\mathrm{s} \in D_+(\omega_m)$, is a stagnation point for the vector harmonic constituent $\mathbf{q}'(\mathbf{x}, t; \omega_m)$ if and only if it is a stagnation point for both the in-phase and quadrature components $\mathbf{q}'_\mathrm{Ip}(\mathbf{x}, t; \omega_m)$ and $\mathbf{q}'_\mathrm{Qu}(\mathbf{x}, t; \omega_m)$ [i.e., if and only if equations (18.13) are satisfied].

Section Summary

The most important points of this section are as follows:

1. The following statements are logically equivalent:
   - The point with coordinates $\mathbf{x} = \mathbf{x}_\mathrm{s}$, where $\mathbf{x}_\mathrm{s} \in D$, is a stagnation point for $\mathbf{q}'(\mathbf{x}, t; \omega_m)$.
   - $\mathbf{q}'(\mathbf{x}_\mathrm{s}, t; \omega_m) = \mathbf{0} \quad \forall\ t$, where $\mathbf{x}_\mathrm{s} \in D$.
   - $\nabla A_\mathrm{h}(\mathbf{x}_\mathrm{s}; \omega_m) = \mathbf{0} = \nabla B_\mathrm{h}(\mathbf{x}_\mathrm{s}; \omega_m)$, where $\mathbf{x}_\mathrm{s} \in D$.
   - $\nabla F_\mathrm{h}(\mathbf{x}_\mathrm{s}; \omega_m) = \mathbf{0}$, where $\mathbf{x}_\mathrm{s} \in D$.
   - $\nabla M_\mathrm{h}(\mathbf{x}_\mathrm{s}; \omega_m) = \mathbf{0} = \nabla\theta_\mathrm{h}(\mathbf{x}_\mathrm{s}; \omega_m)$, where $\mathbf{x}_\mathrm{s} \in D$.
   - $\mathbf{q}'_\mathrm{Ip}(\mathbf{x}_\mathrm{s}, t; \omega_m) = \mathbf{0} = \mathbf{q}'_\mathrm{Qu}(\mathbf{x}_\mathrm{s}, t; \omega_m) \quad \forall\ t$, where $\mathbf{x}_\mathrm{s} \in D$.

2. There are two types of stagnation points for the vector harmonic constituent $\mathbf{q}'(\mathbf{x}, t; \omega_m)$:

Type 1 $M_\mathrm{h}(\mathbf{x}_\mathrm{s}; \omega_m) > 0$. Then the point with coordinates $\mathbf{x} = \mathbf{x}_\mathrm{s}$ is a stagnation point if and only if $\nabla M_\mathrm{h}(\mathbf{x}_\mathrm{s}; \omega_m) = \nabla\theta_\mathrm{h}(\mathbf{x}_\mathrm{s}; \omega_m) = \mathbf{0}$.

Type 2 $M_\mathrm{h}(\mathbf{x}_\mathrm{s}; \omega_m) = 0$, so the phase function is undefined at $\mathbf{x} = \mathbf{x}_\mathrm{s}$. Then the point is a stagnation point if and only if $\nabla A_\mathrm{h}(\mathbf{x}_\mathrm{s}; \omega_m) = \nabla B_\mathrm{h}(\mathbf{x}_\mathrm{s}; \omega_m) = \mathbf{0}$.

18.1.1. One Dimension

This section presents exercises dealing with stagnation in 1D flows through source-free domains filled with homogeneous media. The exercises frequently reference the material presented in Chapter 9.

Exercise 18.12 Stagnation Point in Finite, Source-Free, Ideal Medium

Problem: Consider time-periodic, 1D flow in a source-free, finite-length domain filled with a homogeneous medium. Suppose we have a Dirichlet condition at the left boundary and a Neumann condition at the right boundary. Express the requirements for existence of a stagnation point in terms of the constants a_m and b_m and the eigenvalue λ_m.

Solution: Suppose a stagnation point exists and its coordinate is $x = x_\mathrm{s}$. Then from Exercise 18.5 we have

$$\left[\frac{d}{dx}F_\mathrm{h}(x; \omega_\mathrm{m})\right]\Bigg|_{x=x_\mathrm{s}} = 0.$$

Substituting the RHS of (9.7) for F_h gives

$$a_m\lambda_m e^{\lambda_m x_\mathrm{s}} - b_m\lambda_m e^{-\lambda_m x_\mathrm{s}} = 0.$$

Solving for x_s yields

$$x_\mathrm{s} = \frac{1}{2\lambda_m}\ln\left(\frac{b_m}{a_m}\right). \tag{18.14}$$

The conditions for existence of a stagnation point are that the RHS of (18.14) must (a) be defined and real valued and (b) satisfy $0 < x_\mathrm{s} < L$. For instance, condition (a) prohibits the existence of a stagnation point in those cases where either $a_m = 0$ or $b_m = 0$ and in those cases where the ratio b_m/a_m is real valued. □

Exercise 18.13 Type 1 and Type 2 Stagnation Points

Problem: Consider the results obtained in Exercise 18.12. If such a stagnation point exists, then is it of type 1 or type 2?

Solution: Substituting $x = x_\mathrm{s}$ in (9.7) gives

$$F_\mathrm{h}(x_\mathrm{s}; \omega_\mathrm{m}) = a_m e^{\lambda_m x_\mathrm{s}} + b_m e^{-\lambda_m x_\mathrm{s}}.$$

Substituting the RHS of (18.14) for x_s and simplifying yield

$$F_h(x_s;\omega_m) = 2(a_m b_m)^{1/2}. \qquad (18.15)$$

Both a_m and b_m must be nonzero for a stagnation point to exist, so $M_h(x_s;\omega_m) = 2|a_m b_m|^{1/2} > 0$. Therefore, in 1D flows through source-free domains filled with homogeneous media, if a stagnation point exists, then it is of type 1. Also see Exercise 18.15. □

Exercise 18.14 Type 1 and Type 2 Stagnation Points

Problem: Consider time-periodic flow in a 1D homogeneous medium with the Dirichlet conditions

$$F_h(L;\omega_m) = F_h(0;\omega_m).$$

Find the x coordinate of the stagnation point. Also, express the value of the hydraulic head FRF at the stagnation point in terms of $F_h(0;\omega_m)$, λ_m, and L. What type of stagnation point is it?

Solution: Substituting $F_h(L;\omega_m) = F_h(0;\omega_m)$ in (9.11) gives

$$a_m = \frac{F_h(0;\omega)[1 - e^{-\lambda_m L}]}{2\sinh(\lambda_m L)} \qquad (18.16a)$$

$$b_m = \frac{F_h(0;\omega)[e^{\lambda_m L} - 1]}{2\sinh(\lambda_m L)}. \qquad (18.16b)$$

Substituting the RHSs for a_m and b_m in (18.14) and then simplifying yield $x_s = L/2$. Substituting this result for x and the RHSs of (18.16) for a_m and b_m in (9.7) and then simplifying lead to

$$\frac{F_h(x_s;\omega_m)}{F_h(0;\omega_m)} = \frac{2\sinh(\lambda_m L/2)}{\sinh(\lambda_m L)}.$$

where $F_h(x_s;\omega_m)$ is nonzero, so the amplitude is positive at the stagnation point. Therefore, it is a type 1 stagnation point. □

Exercise 18.15 Type 2 Stagnation Points

Problem: Consider time-periodic, 1D flow in a source-free domain filled with a homogeneous medium. Prove that type 2 stagnation points are prohibited in such flows.

Solution: Suppose a zero point of the hydraulic head FRF exists and that it has the space coordinate $x = x_0$:

$$F_h(x_0;\omega_m) = 0.$$

Substituting $x = x_0$ in (9.7) and equating the result to zero give

$$a_m e^{\lambda_m x_0} + b_m e^{-\lambda_m x_0} = 0.$$

Solving for x_0 yields

$$x_0 = \frac{1}{2\lambda_m} \ln\left(\frac{-b_m}{a_m}\right). \qquad (18.17)$$

Comparing the RHS to that of (18.14) reveals that $x_0 \neq x_s$. Thus, if a stagnation point exists, then it cannot be a zero point of the hydraulic head FRF. Therefore, type 2 stagnation points are prohibited in 1D flows through source-free domains filled with homogeneous media. Also see Exercise 18.13. □

Exercise 18.16 FRF Zero Points and Stagnation Points

Problem: Consider time-periodic, 1D flow in a source-free domain filled with a homogeneous medium. Show that in such flows there can be a zero point of the hydraulic head FRF or a stagnation point but not both.

Solution: Write (18.17) as

$$\begin{aligned} x_0 &= \frac{1}{2\lambda_m} \ln\left(e^{i\pi}\frac{b_m}{a_m}\right) \\ &= \frac{i\pi}{2\lambda_m} + \frac{1}{2\lambda_m} \ln\left(\frac{b_m}{a_m}\right). \end{aligned}$$

Substituting the RHS of (9.5) for λ_m in the first term on the RHS and using (18.14) to rewrite the second term give

$$x_0 = \frac{\pi\sqrt{2}}{4\sqrt{\alpha}}(1+i) + x_s.$$

This equation cannot be satisfied if both x_0 and x_s are real valued. Therefore, in 1D flows through source-free domains filled with homogeneous media there can be a zero point of the hydraulic head FRF or a stagnation point but not both. □

18.2. STAGNATION SETS

A *stagnation set* for a vector field $\mathbf{v}(\mathbf{x}, t)$ is a collection of one or more stagnation point(s) for $\mathbf{v}(\mathbf{x}, t)$.

We define the *global stagnation set* for a vector field $\mathbf{v}(\mathbf{x}, t)$ as

$$G_{\text{global}}[\mathbf{v}] \equiv \{\mathbf{x} : \mathbf{x} \in D \text{ and } \mathbf{v}(\mathbf{x}, t) = \mathbf{0} \quad \forall\, t > 0\}.$$

That is, the global stagnation set for the vector field $\mathbf{v}(\mathbf{x}, t)$ is the set containing all of the stagnation points for $\mathbf{v}(\mathbf{x}, t)$ and no other points. Just as we can define stagnation points with respect to various vector fields, we can define global stagnation sets with respect to those same vector fields.

Exercise 18.17 Global Stagnation Set

Problem: Write a simple equation that concisely relates the global stagnation set for $\mathring{\mathbf{q}}(\mathbf{x}, t)$ to the global stagnation sets of the corresponding harmonic constituents $\mathbf{q}'(\mathbf{x}, t; \omega_m)$ $(m = 1, 2, 3, \ldots)$.

Solution: The definition of the global stagnation set given earlier in this section and the results of Exercises 18.1 and 18.2 imply

$$G_{\text{global}}[\mathring{\mathbf{q}}] = \bigcap_m G_{\text{global}}[\mathbf{q}'(\mathbf{x}, t; \omega_m)]. \tag{18.18}$$

□

Equation (18.18) implies that the global stagnation set $G_{\text{global}}[\mathring{\mathbf{q}}]$ is empty if and only if the global stagnation sets of the corresponding harmonic constituents $G_{\text{global}}[\mathbf{q}'(\mathbf{x}, t; \omega_m)]$ $(m = 1, 2, 3, \ldots)$ are disjoint.

We can partition every nonempty global stagnation set into one or more mutually disjoint subsets of two basic types—unconnected stagnation sets and connected stagnation sets. The following sections discuss unconnected and connected stagnation sets.

18.2.1. Unconnected Stagnation Sets

An *unconnected stagnation set* is a set that contains only isolated stagnation points. An *isolated stagnation point* is a stagnation point that is completely surrounded by points that are not stagnation points. A more formal definition is as follows:

Suppose the point with coordinates $\mathbf{x} = \mathbf{x}_s$ is an isolated stagnation point for the vector field $\mathbf{v}(\mathbf{x}, t)$. Then there exists a ball of positive radius centered at $\mathbf{x} = \mathbf{x}_s$ for which $|\mathbf{v}(\mathbf{x}, t)| > 0$ at every interior point of the ball except the center.

In three dimensions, ball means sphere. In two dimensions, ball means disk. In one dimension *ball of positive radius* means segment of positive length.

Example 18.1 Isolated, Type 1 Stagnation Point

A homogeneous, source-free, 1D porous medium of finite length $(0 < x < L)$ is subjected to time-periodic forcing consisting of a single harmonic constituent applied at both endpoints. The amplitude of the forcing is identical at the two endpoints, as is the phase, as per case b-0 (see Table 9.1 and Figure 9.2). The midpoint $(x = L/2)$ is a stagnation point for the vector harmonic constituent $\mathbf{q}'(\mathbf{x}, t; \omega_m)$ because

$$\left[\frac{d}{dx} M_h(x; \omega_m)\right]\Bigg|_{x=L/2} = 0$$

$$\left[\frac{d}{dx} \theta_h(x; \omega_m)\right]\Bigg|_{x=L/2} = 0.$$

The stagnation point at $x = L/2$ is type 1 because

$$M_h(L/2; \omega_m) > 0.$$

◊

18.2.2. Connected Stagnation Sets

Informally, a *connected stagnation set* is a stagnation set the points of which are spatially contiguous. For example, in one dimension a connected stagnation set is a line segment, a ray, or a line. In two dimensions a connected stagnation set includes a continuous trajectory with a positive arc length, a region of positive area, a spatially connected combination of these geometric elements.

Example 18.2 Disjoint, Connected Stagnation Sets in 1D Medium

Consider time-periodic flow through a hypothetical, 1D, porous medium $D \equiv \{x | 0 < x < 10\}$ due to a single harmonic constituent ω_m. Suppose the global stagnation set for the specific-discharge vector harmonic constituent consists of two unconnected, finite, line segments:

$$A \equiv \{x | 1 < x < 2\}$$
$$B \equiv \{x | 7 < x < 9\}.$$

Each line segment is a connected stagnation set because it is spatially continuous. However, the global stagnation set, which is $A \cup B$, is not a connected stagnation set because A and B are not spatially connected to one another. ◊

Exercise 18.18 Connected Stagnation Sets

Problem: Consider time-periodic flow through a hypothetical, 2D, porous medium in which the space domain is the right half-plane:

$$D \equiv \{(x_1, x_2) | x_1 > 0\}.$$

Suppose the global stagnation set for the specific-discharge vector harmonic constituent is

$$G \equiv A \cup B$$

where the subsets are

$$A \equiv \{(x_1, x_2) | (x_1 - 5)^2 + x_2^2 < 9\}$$
$$B \equiv \{(x_1, x_2) | 8 < x_1 < 9 \text{ and } x_2 = 4\}.$$

Is subset A a connected stagnation set? Is subset B a connected stagnation set? Is the global stagnation set a connected stagnation set? Why or why not?

Solution: Each of the subsets A and B is a connected set because it is spatially continuous. However, the global stagnation set G is not a connected stagnation set because A and B are not spatially connected to one another. □

We can partition every connected stagnation set into one or more mutually disjoint subsets of two basic types, which we call *stagnation zones* and *stagnation divides*. To understand the distinction between stagnation zones and stagnation divides, it is useful to define the *N-volume* of a connected set: In three dimensions, the N-volume of a connected set is simply its volume. In two dimensions the N-volume of a connected set is its area, and in one dimension the N-volume of a connected set is its length. We use the notation $|G|$ to denote the N-volume of the connected set G. It follows that $|G| \geq 0$.

Informally, the difference between a stagnation zone and a stagnation divide is as follows. Suppose G is a connected stagnation set. If $|G| > 0$, then G contains at least one stagnation zone. Alternatively, if $|G| = 0$, then G contains no stagnation zones and G is a stagnation divide. The following sections describe stagnation zones and stagnation divides in more detail.

18.2.2.1. Stagnation Zones

Suppose the set G is a connected stagnation set for the vector field $\mathbf{v}(\mathbf{x}, t)$. Then G is a *stagnation zone* for the vector field $\mathbf{v}(\mathbf{x}, t)$ if and only if it satisfies the following:

For every point $\mathbf{x} \in G$ there exists a ball of nonzero radius centered at the point $\mathbf{x}$ for which every interior point is a stagnation point of $\mathbf{v}(\mathbf{x}, t)$.

In three dimensions, ball means sphere. In two dimensions, ball means disk. In one dimension *ball of nonzero radius* means segment of nonzero length. It follows that if a subregion G is a stagnation zone for the vector field $\mathbf{v}(\mathbf{x}, t)$, then G is an open set and $|G| > 0$. The set G need not be simply connected.

Exercise 18.19 Stagnation Zones

Problem: Consider a 3D domain that contains a stagnation set defined as follows:

$$G \equiv \{\mathbf{x} : |\mathbf{x} - \mathbf{a}|^2 < r^2\} \tag{18.19}$$

where $\mathbf{x}$ is the Cartesian space coordinate vector, $\mathbf{a}$ is the space coordinate vector of a fixed center point, and r is a positive, real number. Provide a concise (one-sentence) verbal description of the basic geometry of the set G. Determine whether G is (1) an open or closed set and (2) a stagnation zone, a stagnation divide, or neither. No proof is necessary.

Solution: The stagnation set G is the interior of a sphere, it is an open set, and it is a stagnation zone. □

Exercise 18.20 Stagnation Zones

Problem: Suppose the time-periodic component of the specific-discharge vector has a single stagnation zone G in D. Suppose each of the vector harmonic constituents $\mathbf{q}'(\mathbf{x}, t; \omega_m)$ $(m = 1, 2, 3, \ldots)$ also has a single stagnation zone $G(\omega_m)$ in D. Write the basic relation that exists between the set G and the sets $G(\omega_m)$ $(m = 1, 2, 3, \ldots)$ in this case.

Solution: In this case G is the global stagnation set for $\mathring{\mathbf{q}}(\mathbf{x}, t)$ and the $G(\omega_m)$ $(m = 1, 2, 3, \ldots)$ are the global stagnation sets for the corresponding harmonic constituents $\mathbf{q}'(\mathbf{x}, t; \omega_m)$ $(m = 1, 2, 3, \ldots)$. Therefore, the relation is (18.18), with $G_{\text{global}}[\mathring{\mathbf{q}}] = G$ and $G_{\text{global}}[\mathbf{q}'(\mathbf{x}, t; \omega_m)] = G(\omega_m)$:

$$G = \bigcap_m G(\omega_m).$$

□

Combining the results (18.5) with the definition of stagnation zone given above yields the following, equivalent statement: A connected subregion $G(\omega_m)$, with $G(\omega_m) \subset D$ and $|G(\omega_m)| > 0$, is a stagnation zone for the vector harmonic constituent $\mathbf{q}'(\mathbf{x}, t; \omega_m)$ if and only if the following are satisfied:

$$\nabla A_{\text{h}}(\mathbf{x}; \omega_m) = \mathbf{0} \tag{18.20a}$$

$$\nabla B_{\text{h}}(\mathbf{x}; \omega_m) = \mathbf{0} \tag{18.20b}$$

$$\forall\, \mathbf{x} \in G(\omega_m).$$

Exercise 18.21 Stagnation Zones

Problem: Suppose $G(\omega_m)$ is a connected subregion with $G(\omega_m) \subset D$ and $|G(\omega_m)| > 0$. Show that if $M_{\text{h}}(\mathbf{x}; \omega_m) = 0 \quad \forall\, \mathbf{x} \in G(\omega_m)$, then $G(\omega_m)$ is a stagnation zone for $\mathbf{q}'(\mathbf{x}, t; \omega_m)$.

Solution: By assumption the amplitude function vanishes in $G(\omega_m)$:

$$M_{\text{h}}(\mathbf{x}; \omega_m) = 0 \quad \forall\, \mathbf{x} \in G(\omega_m).$$

The previous equation and (4.6a) imply

$$A_{\text{h}}(\mathbf{x}; \omega_m) = 0$$

$$B_{\text{h}}(\mathbf{x}; \omega_m) = 0$$

$$\forall\, \mathbf{x} \in G(\omega_m).$$

From these it follows that equations (18.20) are satisfed. Hence, $G(\omega_m)$ is a stagnation zone for the vector harmonic constituent $\mathbf{q}'(\mathbf{x}, t; \omega_m)$. □

In Section 17.2 it was shown that space BVP solutions for which the amplitude function is both constant and positive in a subregion with positive N-volume are prohibited in source-free domains. This result, together with the result of Exercise 18.21, implies the following:

In source-free domains, if the connected stagnation set $G(\omega_m)$ is a stagnation zone for the vector harmonic constituent $\mathbf{q}'(\mathbf{x}, t; \omega_m)$, then $M_{\text{h}}(\mathbf{x}; \omega_m) = 0 \quad \forall\, \mathbf{x} \in G(\omega_m)$.

That is, if a stagnation zone exists, then its points consist of type 2 stagnation points.

Exercise 18.22 Stagnation Zones

Problem: Suppose the subregion $G(\omega_m)$ is a stagnation zone for the vector harmonic constituent $\mathbf{q}'(\mathbf{x}, t; \omega_m)$. Summarize the basic relations that exist between the set $G(\omega_m)$ and each of the sets $D_o(\omega_m)$ and $D_+(\omega_m)$.

Solution: The basic relations are

$$G(\omega_m) \cap D_o(\omega_m) = G(\omega_m) \quad [\text{alternatively, } G(\omega_m) \subset D_o(\omega_m)]$$

$$G(\omega_m) \cap D_+(\omega_m) = \emptyset.$$

□

18.2.2.2. Stagnation Divides

A subregion G is a *stagnation divide* for a vector field $\mathbf{v}(\mathbf{x}, t)$ if and only if G is a connected stagnation set for $\mathbf{v}$ and no point $\mathbf{x}$ in G is an element of any stagnation zone for $\mathbf{v}$.

Technically, in one dimension stagnation divides do not exist. The 1D equivalent of a stagnation zone is an isolated stagnation point, but an isolated stagnation point is not a connected set so it cannot be a stagnation divide. If G is a stagnation divide, then one of the following must be true:

- In two dimensions, G would be a continuous trajectory, either straight or curved, with positive arc length, i.e., a 1D geometric object. Examples include a line or line segment and a circular arc.
- In three dimensions, G would be one of the following:
  - A trajectory with positive arc length (i.e., a 1D object)
  - A surface, either flat or curved, of positive area (i.e., a 2D object)
  - A spatially connected combination of trajectories and surfaces

It follows that $|G| = 0$. Note that G need not be geometrically regular or simply connected.

Exercise 18.23 Stagnation Divides

Problem: Consider a 3D domain that contains a stagnation set defined as follows:

$$G \equiv \{\mathbf{x} : |\mathbf{x} - \mathbf{a}|^2 = r^2\}$$

where $\mathbf{x}$ is the Cartesian space coordinate vector, $\mathbf{a}$ is the space coordinate vector of a fixed center point, and r is a positive, real number. Provide a concise (one-sentence) verbal description of the basic geometry of the set G. Determine whether G is (1) an open or closed set and (2) a stagnation zone, a stagnation divide, or neither. No proof is necessary.

Solution: The stagnation set G is a sphere (spherical shell), it is a closed set, and it is a stagnation divide. □

Exercise 18.24 Stagnation Divides versus Stagnation Zones

Problem: Consider a 3D domain that contains a stagnation set defined as follows:

$$G \equiv \{\mathbf{x} : |\mathbf{x} - \mathbf{a}|^2 \leq r^2\}$$

where $\mathbf{x}$ is the Cartesian space coordinate vector, $\mathbf{a}$ is the space coordinate vector of a fixed center point, and r is a positive, real number. Provide a concise (one-sentence) verbal description of the basic geometry of the set G. Determine whether G is (1) an open or closed set and (2) a stagnation zone, a stagnation divide, or neither. No proof is necessary.

Solution: The stagnation set G consists of a sphere (spherical shell) and the enclosed interior, it is a closed set, and it is neither a stagnation zone nor a stagnation divide but contains both a stagnation zone (the sphere interior) and a stagnation divide (the spherical shell). □

No flow occurs across a stagnation divide. Hence the feature locally *divides* the flow field much as an impermeable barrier would if present at the same location. However, in multidimensional, time-periodic flows there is an important difference between stagnation divides and impermeable surfaces. Although the component of flux perpendicular to an impermeable surface necessarily vanishes (i.e., $\mathbf{q} \cdot \hat{\mathbf{n}} = 0$), the component(s) parallel to the surface may or may not vanish. Thus, points adjacent to an impermeable surface may or may not be stagnation points.

18.2.2.3. Closed-Curve Stagnation Divides and Stagnation Zones

Consider time-periodic flow in a source-free, 2D medium that contains a stagnation divide consisting of a closed curve (e.g., an ellipse or rectangle). Let G_1 denote the stagnation divide and let G_2 denote the 2D subdomain it encloses. Then G_1 and G_2 are closed and open sets, respectively. The stagnation divide G_1 defines the boundary of the subdomain G_2, so the boundary of G_2 is subject to the no-flow condition:

$$\mathbf{q}' \cdot \hat{\mathbf{n}} = 0 \quad \forall\, \mathbf{x} \in G_1$$

where $\hat{\mathbf{n}}$ denotes the outward unit-normal vector for the subdomain G_2. Therefore, the FRF solution on the enclosed subdomain is

$$F_h(\mathbf{x}; \omega_m) = C(\omega_m) \quad \forall\, \mathbf{x} \in G_2$$

where $\nabla C(\omega_m) = \mathbf{0}$. This and the result from Exercise 18.5 then imply that the enclosed subdomain is a stagnation zone.

Exercise 18.25 Type 1 and Type 2 Stagnation Zones

Problem: Referring to the case of time-periodic flow in a 2D medium described at the beginning of this section, we saw that the enclosed subdomain G_2 is stagnation zone. Is G_2 a type 1 stagnation zone, a type 2 stagnation zone, or neither?

Solution: The result of Exercise 17.2 indicates that

$$F_{\mathrm{h}}(\mathbf{x};\omega_m) = 0 \quad \forall\, \mathbf{x} \in G_2.$$

Therefore, all of the points in the enclosed subdomain are type 2 stagnation points. The enclosed subdomain is a type 2 stagnation zone. □

Similarly, in a source-free, 3D medium, if a stagnation divide consists of a closed surface (e.g., an ellipsoid or rectangular box), then the enclosed volume is a stagnation zone.

18.2.3. Partially Connected Stagnation Sets

A *partially connected* stagnation set is a stagnation set that contains two or more subsets, at least one of which is a connected set, and at least one of the subsets is unconnected to the other subsets. It follows that every stagnation set is either connected, unconnected, or partially connected.

Example 18.3 Partially Connected Stagnation Set in 2D Medium
Consider time-periodic flow through a hypothetical, 2D, infinite porous medium,

$$D \equiv \{(x_1, x_2) \;:\; x_1 \in \mathbb{R}, x_2 \in \mathbb{R}\}$$

in which the hydraulic head field consists of a single harmonic constituent (i.e., ω_m fixed). Suppose we defined the global stagnation set for the specific-discharge vector harmonic constituent as

$$G \equiv G_1 \cup G_2$$

where we define the subsets G_1 and G_2, respectively, as

$$G_1 \equiv \{(x_1, x_2) \;:\; x_1 = 0, x_2 \in \mathbb{R}\}$$
$$G_2 \equiv \{(x_1, x_2) \;:\; x_1 = 6, x_2 = 0\}.$$

The set G_1, which consists of the x_2 axis, is a stagnation divide and therefore a connected stagnation set. The set G_2, which consists of the single point whose coordinates are $(x_1, x_2) = (6, 0)$, is an isolated stagnation point and therefore an unconnected stagnation set. The subsets G_1 (connected) and G_2 (unconnected) are spatially unconnected to one another. Therefore, the global stagnation set G is a partially connected stagnation set. ◊

Exercise 18.26 Connectivity of Stagnation Sets

Problem: Consider time-periodic flow through a hypothetical, 2D, infinite porous medium

$$D \equiv \{(x_1, x_2) \;:\; x_1 \in \mathbb{R}, x_2 \in \mathbb{R}\}$$

in which the hydraulic head field consists of a single harmonic constituent (i.e., ω_m fixed). Suppose we defined the global stagnation set for the specific-discharge vector harmonic constituent as

$$G \equiv G_1 \cup G_2$$

where we define the subsets G_1 and G_2, respectively, as

$$G_1 \equiv \{(x_1, x_2) \;:\; x_1 = 0, x_2 \in \mathbb{R}\}$$
$$G_2 \equiv \{(x_1, x_2) \;:\; x_1 = 11, x_2 \in \mathbb{R}\}.$$

For each of the stagnation sets G_1, G_2, and G, answer the following question: Is the set connected, unconnected, or partially connected?

Solution: The set G_1, which consists of the x_2 axis, is a stagnation divide and therefore a connected stagnation set. The set G_2, which consists of the line given by $x_1 = 11$, also is a stagnation divide and therefore a connected stagnation set. The subsets G_1 (connected) and G_2 (connected) are spatially unconnected to one another. Therefore, the global stagnation set G is a partially connected stagnation set. □

18.3. CHAPTER SUMMARY

In this chapter, we described the concept of flow stagnation in the context of time-periodic groundwater flow. We formally defined stagnation sets and introduced a system for classifying them based on spatial connectedness—unconnected, connected, and partially connected. We further classified connected stagnation sets into two categories—stagnation zones and stagnation divides. Our classification system is intuitive and helps relate the basic geometric and topologic properties of various types of stagnation sets to mathematical properties of the flow field, such as the stationary points of the amplitude and phase functions for hydraulic head harmonic constituents.

Part VI
Wave Propagation

In this part we explore the interpretation of time-periodic groundwater flow as steady-state propagation of spatially modulated (damped) diffusion waves, i.e., as traveling, harmonic, hydraulic head waves.

Hydrodynamics of Time-Periodic Groundwater Flow: Diffusion Waves in Porous Media, Geophysical Monograph 224,
First Edition. Joe S. Depner and Todd C. Rasmussen.

19

Harmonic, Hydraulic Head Waves

In this chapter we explain basic concepts of wave propagation (e.g., wave propagation velocity) in continuous media and then present two similar but different interpretations of time-periodic groundwater flow as steady-state propagation of spatially modulated (damped), diffusion waves, i.e, as traveling, harmonic, hydraulic head waves.

19.1. INTRODUCTION

Perhaps surprisingly, it is not particularly easy to clearly define a *wave*. *Scales and Snieder* [1999] define a wave as "an organized propagating imbalance." This definition sounds vague because of its generality, but it will suffice unless and until we need a more specific definition.

We expect that most readers are at least somewhat familiar with *ordinary waves*, which appear in physical systems governed by wave-type (hyperbolic) PDEs. Ordinary waves transmit energy but do not transport mass [*Salazar*, 2006]. Ordinary waves also exhibit reflection and refraction phenomena at material interfaces [*Mandelis et al.*, 2001].

In the literature, waves that appear in physical systems governed by a diffusion-type (parabolic) PDE rather than a wave-type PDE are commonly referred to as *diffusion waves*. For a general discussion of diffusion waves, see, the work of *Mandelis* [2000] and *Davis et al.* [2001]. The use of the term diffusion waves implies a broadening of, and therefore a departure from, the conventional meaning of the term *waves* (i.e., ordinary waves) in physics; consequently its use is somewhat controversial.

Unlike ordinary waves, the harmonic, hydraulic head waves described in this chapter propagate in systems governed by a diffusion-type PDE—the groundwater flow equation (2.5). Later, in Chapter 22, we show that harmonic, hydraulic head waves satisfy wave-type PDEs as well.

19.2. WAVES IN CONTINUOUS MEDIA

To describe the wave interpretation of time-periodic groundwater flow, this section establishes consistent terminology and explains the concept of wave propagation velocity in continuous media.

19.2.1. Terminology

The wave-related terminology used in this book (e.g., wave phase, wavefront, phase velocity, propagation velocity, wave vector, wave number) generally is consistent with established conventions in other fields of physical wave propagation.

Consider a generic *wave field*:

$$f(\mathbf{x}, t; \omega_m) = M(\mathbf{x}; \omega_m) \cos \xi(\mathbf{x}, t; \omega_m) \tag{19.1}$$

where we define the *wave phase* (not to be confused with the phase function, θ) as

$$\xi(\mathbf{x}, t; \omega_m) \equiv \omega_m t + \theta(\mathbf{x}; \omega_m). \tag{19.2}$$

The RHS of (19.1) is the product of two terms—a unit-amplitude sinusoid, $\cos \xi$, which we call the *unit-amplitude wave field*, and a *wave amplitude function* (also, *wave amplitude*), M. The unit-amplitude sinusoid has a constant amplitude (i.e., unity) and a frequency-dependent phase, θ, that varies continuously in space. The cosine function is a continuous function of its argument, so the unit-amplitude wave field varies continuously in both time and space. At any fixed location the unit-amplitude

Hydrodynamics of Time-Periodic Groundwater Flow: Diffusion Waves in Porous Media, Geophysical Monograph 224,
First Edition. Joe S. Depner and Todd C. Rasmussen.

sinusoid varies sinusoidally with time. The wave amplitude, which is frequency dependent, is time invariant and spatially modulates the unit-amplitude sinusoid. We assume that the wave amplitude also varies continuously in space.

We say that (19.1) represents a *harmonic, traveling wave* (also *wave train*) because it describes a field for which both of the following hold:

1. At any fixed location the field varies sinusoidally in time.
2. The field represents or can be interpreted as a disturbance that travels through the porous medium.

The meaning of condition 1 is clear enough, and Section 19.2.2 will clarify the meaning of condition 2.

19.2.2. Wave Propagation Velocity

In this section we define a wave propagation velocity vector field corresponding to a single harmonic traveling wave, and derive a general expression for it.

Define a *wavefront* as a surface of constant wave phase (ξ = constant). Then the *phase velocity* (also *propagation velocity*) is the velocity at which such a wavefront moves (propagates) through the medium. Let $\mathbf{v}(\mathbf{x};\omega_m)$ denote the phase velocity of the wave.

Consider an imaginary, massless tracer particle that travels passively through the fictitious continuum, always at the location of a moving wavefront. We define the phase velocity as the velocity of such a particle subject to the following two conditions:

1. The wave phase, as observed continuously with time at the (moving) location of the tracer particle, is time invariant:

$$\left.\frac{D\xi}{Dt}\right|_{\mathbf{v}} = 0 \tag{19.3}$$

where we define the substantial derivative for the velocity field $\mathbf{v}$ as

$$\left.\frac{D}{Dt}\right|_{\mathbf{v}} \equiv \frac{\partial}{\partial t} + \mathbf{v}\cdot\nabla. \tag{19.4}$$

Although this condition restricts the location of the tracer particle to the traveling wavefront, it does not prohibit the particle from moving arbitrarily within that moving surface.

2. The tracer particle always moves parallel to the gradient of the wave phase:

$$\mathbf{v}(\mathbf{x};\omega_m) = \pm c(\mathbf{x};\omega_m)\frac{\nabla\xi}{|\nabla\xi|}, \quad |\nabla\xi(\mathbf{x},t;\omega_m)| > 0 \tag{19.5}$$

where c is the *propagation speed* (i.e., $c \equiv |\mathbf{v}|$). This condition prohibits the tracer particle from moving in any direction parallel to the wavefront. [The phase velocity is single valued; the plus-or-minus sign ($\pm$) in (19.5) indicates a sign ambiguity that we will resolve later in our development (see Exercise 19.2).]

In summary, the phase velocity is the velocity of an imaginary, massless particle that moves passively through the medium while constantly at the location of a moving wavefront and moving perpendicular to it.

Exercise 19.1 Wave Propagation Speed

Problem: Express the propagation speed, $c(\mathbf{x};\omega_m)$, compactly in terms of the constituent frequency (ω_m) and the phase gradient ($\nabla\theta$).

Solution: Substituting the RHS of (19.5) for the phase velocity in (19.3) and simplifying yield

$$\frac{\partial}{\partial t}\xi(\mathbf{x},t;\omega_m) \pm c(\mathbf{x};\omega_m)\left|\nabla\xi(\mathbf{x},t;\omega_m)\right| = 0. \tag{19.6}$$

Differentiating (19.2) with respect to time and the space coordinates, respectively, gives

$$\tfrac{\partial}{\partial t}\xi(\mathbf{x},t;\omega_m) = \omega_m \tag{19.7a}$$

$$\nabla\xi(\mathbf{x},t;\omega_m) = \nabla\theta(\mathbf{x};\omega_m). \tag{19.7b}$$

Substituting the RHSs for the derivative terms of the wave phase in (19.6), noting that the propagation speed is nonnegative, and then solving for the propagation speed give

$$c(\mathbf{x};\omega_m) = \frac{\omega_m}{\left|\nabla\theta(\mathbf{x};\omega_m)\right|}, \quad \left|\nabla\theta(\mathbf{x};\omega_m)\right| > 0. \tag{19.8}$$

□

Equation (19.8) implies that the propagation speed is nominally proportional to the constituent frequency and inversely proportional to the magnitude of the phase gradient. (We say *nominally* here because the phase function and hence its gradient generally depend in part on the constituent frequency.) Consequently, the phase velocity is undefined at the zero points of the phase gradient, if any exist; at all other points the phase velocity is finite.

The constituent frequency is independent of the material hydrologic properties $\mathbf{K}$ and S_s. Also, the phase function θ depends partly on the boundary conditions and the distribution of sources and not solely on the material hydrologic properties. Therefore, (19.8) implies that the propagation speed is not an intrinsic physical property of the porous medium.

Equation (19.8) expresses the requirement that for a given frequency ω_m the wave propagation speed c must decrease wherever the magnitude of the phase gradient increases, and vice versa, to maintain the constant frequency of the harmonic variation. Stated another way, in those areas where the phase gradient increases or decreases with distance, respectively, the wave propagation speed decreases or increases proportionally to compensate for the spatial variation of the phase gradient.

Exercise 19.2 Phase Velocity

Problem: Derive an expression for the phase velocity, $\mathbf{v}(\mathbf{x};\omega_m)$, in terms of the constituent frequency and the phase gradient. Is your result consistent with the definition of propagation speed given in (19.8)?

Solution: Substituting the RHSs of (19.7b) and (19.8) for the gradient of the wave phase and the propagation speed, respectively, in (19.5) gives

$$\mathbf{v}(\mathbf{x};\omega_m) = \frac{\pm\omega_m \nabla\theta(\mathbf{x};\omega_m)}{\left|\nabla\theta(\mathbf{x};\omega_m)\right|^2}.$$

Taking the magnitude gives (19.8), so this result is consistent with that equation. Imposing condition (19.3) resolves the sign ambiguity, giving

$$\mathbf{v}(\mathbf{x};\omega_m) = \frac{-\omega_m \nabla\theta(\mathbf{x};\omega_m)}{\left|\nabla\theta(\mathbf{x};\omega_m)\right|^2}, \quad \left|\nabla\theta(\mathbf{x};\omega_m)\right| > 0. \quad (19.9)$$

□

Equation (19.9) indicates that the phase velocity is directed opposite the phase gradient. That is, each wave travels in the direction in which its corresponding phase *decreases*.

19.3. TWO WAVE INTERPRETATIONS

Consider a single hydraulic head harmonic constituent (4.3a):

$$h'(\mathbf{x},t;\omega_m) = M_h(\mathbf{x};\omega_m)\cos\xi_h(\mathbf{x},t;\omega_m) \quad (19.10)$$

where

$$\xi_h(\mathbf{x},t;\omega_m) \equiv \omega_m t + \theta_h(\mathbf{x};\omega_m). \quad (19.11)$$

Compare (19.10) and (19.11) to (19.1) and (19.2), respectively. The RHS of (19.10) resembles the expression for a spatially attenuated, harmonic, traveling wave, which suggests that one might interpret time-periodic groundwater flow as the propagation of harmonic, hydraulic head waves. We present two such interpretations in Sections 19.3.1 and 19.3.2.

19.3.1. Constituent Wave Interpretation

In the *constituent wave interpretation*, at any given point in the space domain each hydraulic head harmonic constituent $h'(\mathbf{x};\omega_m)$ (m fixed) represents one traveling harmonic wave, which we call a *constituent wave*. Sometimes we call the harmonic constituent, $h'(\mathbf{x};\omega_m)$, the *constituent wave field*.

We call ξ_h the *constituent wave phase* (not to be confused with the phase function, θ_h). As per the definition given in Section 19.2.1, the corresponding wave propagation velocity (19.9) is

$$\mathbf{v}_h(\mathbf{x};\omega_m) = \frac{-\omega_m \nabla\theta_h(\mathbf{x};\omega_m)}{\left|\nabla\theta_h(\mathbf{x};\omega_m)\right|^2} \forall\, \mathbf{x} \in D_+(\omega_m), \quad \left|\nabla\theta_h(\mathbf{x};\omega_m)\right| > 0 \quad (19.12)$$

and the corresponding propagation speed is

$$c_h(\mathbf{x};\omega_m) = \frac{\omega_m}{\left|\nabla\theta_h(\mathbf{x};\omega_m)\right|} \forall\, \mathbf{x} \in D_+(\omega_m), \quad \left|\nabla\theta_h(\mathbf{x};\omega_m)\right| > 0. \quad (19.13)$$

We assume that the phase function θ_h and its first and second derivatives with respect to the space coordinates are continuous and finite throughout the space domain $D_+(\omega_m)$. Then the phase velocity is finite and nonzero everywhere in $D_+(\omega_m)$ except possibly at stationary points (if any) of the phase function.

The constituent wave interpretation exhibits the following conceptual indiosyncrasies regarding stationary points of the phase θ_h:

- At stationary points the corresponding constituent wave propagation velocity is undefined.
- Stationary points that correspond to local minima of the phase function are terminal points for some or all of the corresponding constituent wave ray paths, depending on the problem geometry and boundary conditions.
- In some cases where stationary points exist, the corresponding constituent wave undergoes *negative attenuation*. See Section 20.5.1 for a definition of negative attenuation.

These characteristics are perhaps most easily visualized in flows through 1D media. In Section 17.4 we found that in 1D media at most one stationary point of the phase exists and that if such a point exists then it corresponds to a phase minimum. For example, in each of the following 1D cases (see Table 10.1) a phase minimum exists inside the space domain:

b-2, b-3, c-2, c-3, d-2, d-3, n-2, n-3, p-2, p-3, q-2, q-3

Each of these cases displays the following characteristics:

- The space domain, which is finite, is a continuous interval because the amplitude function has no zero point, i.e., $D_+(\omega_m) = D$.
- There are two oppositely directed constituent wave ray paths, one on either side of the stationary point, and both terminate at the stationary point.
- Negative attenuation occurs throughout a finite subinterval. This is clearly recognizable from the graphs of the corresponding amplitude and phase functions.

Some multidimensional flows, such as the 3D, radial flow in a spherical shell in Example 12.5 (see Figure 12.3) and the 2D, radial flow in an annulus in Example 12.12 (see Figure 12.6), exhibit similar behavior.

19.3.2. Component Wave Interpretation

In Part III we saw numerous examples in which we can represent the hydraulic head FRF as the weighted sum of two linearly independent FREs, each of which satisfies the governing equation of the space BVP. Generally we can represent the FRF as the weighted sum of one or more FREs:

$$F_{\mathrm{h}}(\mathbf{x};\omega_m) = \sum_{n=1}^{N} a_m^{(n)} F^{(n)}(\mathbf{x},t;\omega_m)$$

$$\forall\, \mathbf{x} \in D \quad (m = 1, 2, \ldots)$$

where $N = N(\omega_m)$ is a positive integer, the $a_m^{(n)}$ $(n = 1, 2, \ldots, N)$ are complex constants (dimensions: L), the $F^{(n)}(\mathbf{x};\omega_m)$ $(n = 1, 2, \ldots, N)$ represent the linearly independent FREs (dimensionless) for the mth harmonic constituent, and

$$\nabla \cdot \left[\mathbf{K}(\mathbf{x})\nabla F^{(n)}(\mathbf{x};\omega_m)\right] - iS_{\mathrm{s}}(\mathbf{x})F^{(n)}(\mathbf{x};\omega_m) = 0$$

$$\forall\, \mathbf{x} \in D; \quad n = 1, 2, \ldots, N.$$

We can represent the hydraulic head field for each constituent as the superposition of one or more linearly independent, *component hydraulic head fields*,

$$h'(\mathbf{x},t;\omega_m) = \sum_{n=1}^{N} h^{(n)}(\mathbf{x},t;\omega_m) \quad \forall\, \mathbf{x} \in D \quad (m = 1, 2, \ldots) \tag{19.14}$$

where

$$h^{(n)}(\mathbf{x},t;\omega_m) = \mathrm{Re}\left[a_m^{(n)} F^{(n)}(\mathbf{x};\omega_m) e^{\mathrm{i}\omega_m t}\right] \quad (n = 1, 2, \ldots, N). \tag{19.15}$$

Each component field is purely periodic in time with frequency ω_m,

$$h^{(n)}(\mathbf{x},t;\omega_m) = M^{(n)}(\mathbf{x};\omega_m)\cos\left[\omega_m t + \theta^{(n)}(\mathbf{x};\omega_m)\right] \quad (n = 1, 2, \ldots, N) \tag{19.16}$$

where

$$M^{(n)}(\mathbf{x};\omega_m) \equiv \left|a_m^{(n)}\right| \left|F^{(n)}(\mathbf{x};\omega_m)\right| \tag{19.17a}$$

$$\theta^{(n)}(\mathbf{x};\omega_m) \equiv \arg F^{(n)}(\mathbf{x};\omega_m) + \arg a_m^{(n)}. \tag{19.17b}$$

Thus, each component field represents a traveling harmonic wave, which we call a *component wave*. Importantly, we can represent every hydraulic head harmonic constituent field $h'(\mathbf{x},t;\omega_m)$ (ω_m fixed) as the superposition of one or more traveling, harmonic component waves of frequency ω_m.

Each component field satisfies the homogeneous, space-time governing equation for its corresponding harmonic constituent [i.e., equation (5.25a) with $u' \equiv 0$]:

$$\nabla \cdot \left[\mathbf{K}(\mathbf{x})\nabla h^{(n)}(\mathbf{x},t;\omega_m)\right] = S_{\mathrm{s}}(\mathbf{x})\frac{\partial h^{(n)}(\mathbf{x},t;\omega_m)}{\partial t} \tag{19.18}$$

$$\forall\, \mathbf{x} \in D; \quad n = 1, 2, \ldots, N.$$

If for some particular harmonic constituent (i.e., ω_m fixed) we have $N(\omega_m) = 1$, then there is only one component wave. In that case the component wave field is identical to the constituent wave field, i.e., $F^{(1)}(\mathbf{x};\omega_m) = F(\mathbf{x};\omega_m)$ and $h^{(1)}(\mathbf{x},t;\omega_m) = h'(\mathbf{x},t;\omega_m)$. Accordingly, the constituent wave and the component wave are one and the same.

Exercise 19.3 Governing Equations for Component FRFs

Problem: Express (19.14) and (19.18) in terms of the corresponding hydraulic head FRFs.

Solution: Substituting the RHS of (7.2a) for $h'(\mathbf{x},t;\omega_m)$ and the RHS of (19.15) for $h^{(n)}(\mathbf{x},t;\omega_m)$ and then dividing by $e^{i\omega_m t}$ yield

$$F_{\mathrm{h}}(\mathbf{x};\omega_m) = \sum_{n=1}^{N} a_m^{(n)} F^{(n)}(\mathbf{x},t;\omega_m) \quad \forall\, \mathbf{x} \in D \quad (m = 1, 2, \ldots) \tag{19.19}$$

and

$$\nabla \cdot \left[\mathbf{K}(\mathbf{x})\nabla F^{(n)}(\mathbf{x};\omega_m)\right] - iS_{\mathrm{s}}(\mathbf{x})F^{(n)}(\mathbf{x};\omega_m) = 0 \tag{19.20}$$

$$\forall\, \mathbf{x} \in D; \quad n = 1, 2, \ldots, N.$$

□

The examples presented in Section 19.4 support and illustrate the component wave interpretation as it applies to time-periodic flows in some particular hypothetical porous media.

19.4. EXAMPLES

19.4.1. Examples: One Dimension

This section presents examples and exercises to illustrate the component wave interpretation of time-periodic, 1D flows.

Recall that in time-periodic, 1D flows the hydraulic head FRF and its FREs are functions of the single space coordinate x. We represent them symbolically as

$$F_{\mathrm{h}}(\mathbf{x};\omega_m) = G(x;\omega_m)$$

$$F^{(n)}(\mathbf{x};\omega_m) = G^{(n)}(x;\omega_m) \quad (n = 1, 2).$$

Combining the second of these with (19.15) allows us to write the component hydraulic head fields as

$$h^{(n)}(x,t;\omega_m) = \mathrm{Re}\left[a_m^{(n)} G^{(n)}(x;\omega_m)\exp\left(i\omega_m t\right)\right] \quad (n=1,2) \tag{19.21}$$

where the $a_m^{(n)}$ are complex constants.

19.4.1.1. Plane Waves in Exponential Media

In Chapter 10 we saw that for time-periodic, 1D flow in exponential media (which includes the special case of ideal media), we can write the FREs as

$$G^{(n)}(x;\omega_m) = \exp\left(\lambda_m^{(n)} x\right) \quad (n=1,2)$$

where the eigenvalues $\lambda_m^{(n)}$ are given by (10.11). Substituting the RHS for $G^{(n)}$ in (19.21) leads to

$$\begin{aligned} h^{(n)}(x,t;\omega_m) = {} & \left|a_m^{(n)}\right| \exp\left(\mathrm{Re}\,\lambda_m^{(n)} x\right) \\ & \cos\left[\omega_m t + \left(\mathrm{Im}\,\lambda_m^{(n)}\right)x + \arg a_m^{(n)}\right] \end{aligned} \tag{19.22}$$

$$n = 1,2.$$

which is the equation of a harmonic plane wave traveling parallel to the x axis.

Exercise 19.4 Component Waves for 1D Flow in Exponential Medium

Problem: Consider time-periodic, 1D flow through a source-free domain filled with an exponential medium. For each of the component waves, express the wave amplitude and phase function in terms of x, the eigenvalues $\lambda_m^{(n)}$, and the constants $a_m^{(n)}$.

Solution: Comparing (19.22) to the definitions (19.1) and (19.2), evidently the nth component wave has the following wave amplitude and phase function, respectively:

$$M^{(n)}(x;\omega_m) = \left|a_m^{(n)}\right| \exp\left(\mathrm{Re}\,\lambda_m^{(n)} x\right) \tag{19.23a}$$

$$\theta^{(n)}(x;\omega_m) = \left(\mathrm{Im}\,\lambda_m^{(n)}\right)x + \arg a_m^{(n)} \tag{19.23b}$$

$$n = 1,2.$$

□

Example 19.1 Plane Component Waves in Ideal Medium

Consider time-periodic, 1D flow in a source-free, finite-length (ℓ) domain filled with an ideal medium. For plotting, it is convenient to scale the FREs as

$$G_\mathrm{D}^{(1)}(X;\eta_m) \equiv \frac{G^{(1)}(x;\omega_m)}{G^{(1)}(0;\omega_m)}$$

$$G_\mathrm{D}^{(2)}(X;\eta_m) \equiv \frac{G^{(2)}(x;\omega_m)}{G^{(2)}(l;\omega_m)}$$

where

$$X \equiv \frac{x}{l}$$

$$\eta_m \equiv \sqrt{\alpha_m} l.$$

and $G^{(1)}(0;\omega_m) = G^{(2)}(l;\omega_m)$, so in this case the scaling is the same for the two FRFs. Then define the scaled amplitude and phase functions as

$$M_\mathrm{D}^{(n)}(X;\eta_m) \equiv \left|G_\mathrm{D}^{(n)}(X;\eta_m)\right|$$

$$\theta_\mathrm{D}^{(n)}(X;\eta_m) \equiv \arg G_\mathrm{D}^{(n)}(X;\eta_m).$$

Figure 19.1 shows the FRFs for case b-0 (see Table 9.1) with $\eta_m = 5$. Notice that the amplitude and phase functions have no stationary points, so the component waves travel completely through the medium in opposite directions without undergoing negative attenuation. Compare the corresponding constituent wave amplitude and phase functions (Figure 9.2). ◊

Exercise 19.5 Component Waves: Propagation Velocity and Speed

Problem: Consider time-periodic, 1D flow through a source-free domain filled with an exponential medium. For each of the component waves, express the propagation velocity and propagation speed in terms of $\hat{\mathbf{x}}$, ω_m, and the eigenvalues $\lambda_m^{(n)}$. In words, concisely describe the relationship between each component wave's eigenvalue and its direction of travel.

Solution: Taking the gradient of (19.23b) gives

$$\nabla\theta^{(n)}(x;\omega_m) = \left(\mathrm{Im}\,\lambda_m^{(n)}\right)\hat{\mathbf{x}} \quad (n=1,2). \tag{19.24}$$

Substituting the RHS for the phase gradient in (19.9) and using inequalities (10.12b) and (10.12d) yield the propagation velocity

$$\mathbf{v}^{(n)}(x;\omega_m) = \frac{-\omega_m}{\mathrm{Im}\,\lambda_m^{(n)}}\hat{\mathbf{x}} \quad (n=1,2) \tag{19.25}$$

and the propagation speed

$$c^{(n)}(x;\omega_m) = \frac{\omega_m}{\left|\mathrm{Im}\,\lambda_m^{(n)}\right|} \quad (n=1,2). \tag{19.26}$$

For $\mathrm{Im}\,\lambda_m^{(n)} > 0$ the component wave travels in the direction of decreasing x, and for $\mathrm{Im}\,\lambda_m^{(n)} < 0$ the wave travels in the direction of increasing x. □

Inequalities (10.12) imply

$$\left(\mathrm{Re}\,\lambda_m^{(n)}\right)\left(\mathrm{Im}\,\lambda_m^{(n)}\right) > 0 \quad (n=1,2). \tag{19.27}$$

Thus, for fixed n the real and imaginary parts of the eigenvalue are nonzero and have the same sign. Consequently, the waves described by (19.22) are attenuated exponentially with travel distance.

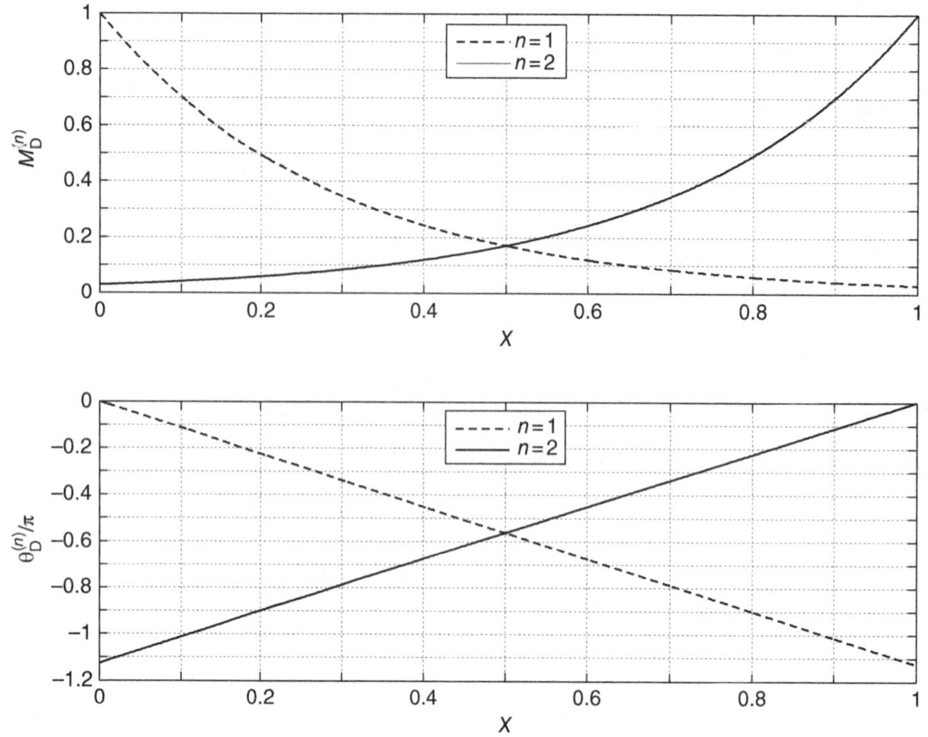

Figure 19.1 FRFs for plane component waves in ideal medium (context: see Example 19.1).

19.4.1.2. Plane Waves in Power Law Media

In Chapter 11 we saw that for time-periodic, 1D flow in power law media we can write the FREs as

$$G^{(n)}(x;\omega_m) = \left|\frac{x}{x_c}\right|^{p_m^{(n)}} \quad (n = 1, 2)$$

where the eigenvalues $p_m^{(n)}$ are given by (11.12) and $|x| > 0$. Substituting the RHS for $G^{(n)}$ in (19.21) leads to

$$h^{(n)}(x, t; \omega_m) = \left|a_m^{(n)}\right|\left|\frac{x}{x_c}\right|^{\mathrm{Re}\,p_m^{(n)}} \cos\left[\omega_m t + \left(\mathrm{Im}\,p_m^{(n)}\right)\ln\left|\frac{x}{x_c}\right| + \arg a_m^{(n)}\right] \quad (19.28)$$

$$n = 1, 2$$

which is the equation of a harmonic plane wave traveling parallel to the x axis.

Exercise 19.6 Component Waves for 1D Flow in Power Law Medium

Problem: Consider time-periodic, 1D flow through a source-free domain filled with a power law medium. For each of the component waves, express the wave amplitude and phase function in terms of x, x_c, the eigenvalues $p_m^{(n)}$, and the constants $a_m^{(n)}$.

Solution: Comparing (19.28) to the definitions (19.1) and (19.2), evidently the nth component wave has the following wave amplitude and phase function, respectively:

$$M^{(n)}(x;\omega_m) = \left|a_m^{(n)}\right|\left|\frac{x}{x_c}\right|^{\mathrm{Re}\,p_m^{(n)}} \quad (19.29a)$$

$$\theta^{(n)}(x;\omega_m) = \left(\mathrm{Im}\,p_m^{(n)}\right)\ln\left|\frac{x}{x_c}\right| + \arg a_m^{(n)} \quad (19.29b)$$

$$n = 1, 2.$$

□

Example 19.2 Plane Component Waves in Power Law Medium

Consider time-periodic, 1D flow in a source-free domain filled with a power law medium. Define the scaled amplitude and phase functions as

$$M_D^{(n)}(X;f_m) \equiv \left|G^{(n)}(X;f_m)\right| \quad (19.30a)$$

$$\theta_D^{(n)}(X;f_m) \equiv \arg G^{(n)}(X;f_m) \quad (19.30b)$$

where (11.4) defines the dimensionless space coordinate X and (11.5) defines the dimensionless frequency f_m. Figure 19.2 shows the component wave amplitude and phase functions for the particular case where $b = 2$ and $f_m = 1$. Notice that the amplitude and phase functions do

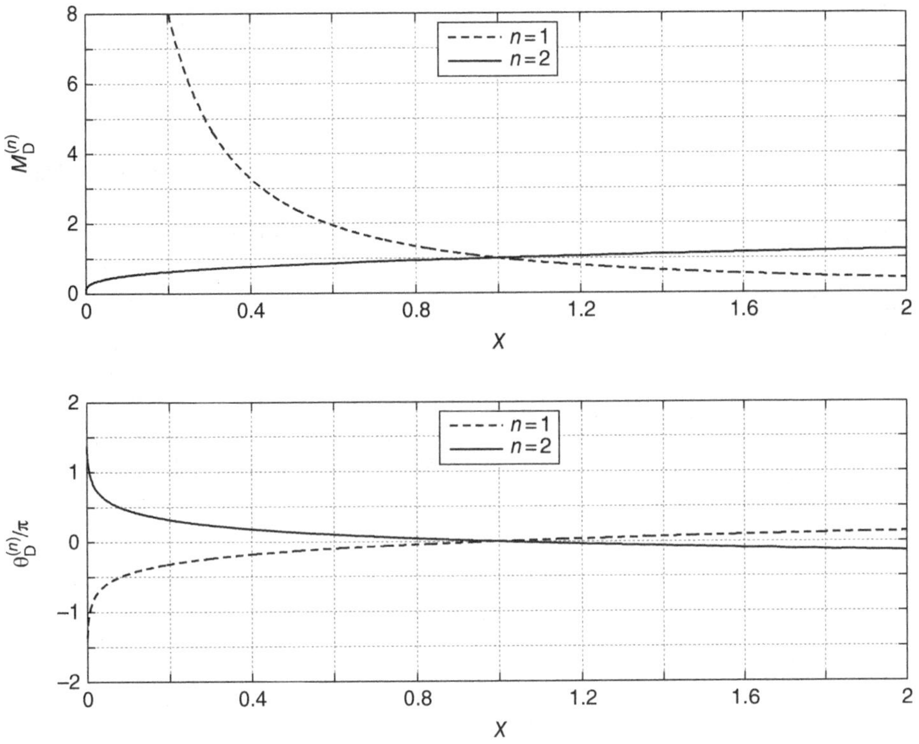

Figure 19.2 FRFs for plane component waves in power law medium (b = 2, f_m = 1) (context: see Example 19.2).

not have stationary points, so the component waves travel completely through the medium in opposite directions without undergoing negative attenuation. ◊

Exercise 19.7 Component Waves: Propagation Velocity and Speed

Problem: Consider time-periodic, 1D flow through a source-free domain filled with a power law medium. For each of the component waves, express the propagation velocity and propagation speed in terms of x, $\hat{\mathbf{x}}$, ω_m, and the eigenvalues $p_m^{(n)}$. In words, concisely describe the relationship between each component wave's eigenvalue and its direction of travel.

Solution: Taking the gradient of (19.29b) gives

$$\nabla\theta^{(n)}(x;\omega_m) = \left(\operatorname{Im} p_m^{(n)}\right)x^{-1}\hat{\mathbf{x}} \quad (n = 1, 2). \tag{19.31}$$

Substituting the RHS for the phase gradient in (19.9) and using inequalities (11.13b) and (11.13d) yield the propagation velocity

$$\mathbf{v}^{(n)}(x;\omega_m) = \frac{-\omega_m}{\operatorname{Im} p_m^{(n)}}x\hat{\mathbf{x}} \quad (n = 1, 2)$$

and the propagation speed

$$c^{(n)}(x;\omega_m) = \frac{\omega_m}{\left|\operatorname{Im} p_m^{(n)}\right|}|x| \quad (n = 1, 2). \tag{19.32}$$

For $\operatorname{Im} p_m^{(n)} > 0$ the component wave travels in the direction of decreasing $|x|$, and for $\operatorname{Im} p_m^{(n)} < 0$ the wave travels in the direction of increasing $|x|$. □

By combining (11.14b) and (19.32) we obtain

$$c^{(1)}(x;\omega_m) = c^{(2)}(x;\omega_m). \tag{19.33}$$

That is, at any given point x the two component waves travel with the same propagation speed.

Inequalities (11.13) imply

$$\left(\operatorname{Re} p_m^{(n)}\right)\left(\operatorname{Im} p_m^{(n)}\right) > 0 \quad (n = 1, 2). \tag{19.34}$$

Thus, for fixed n the real and imaginary parts of the eigenvalue are nonzero and have the same sign. Consequently, the waves described by (19.28) are attenuated with travel distance.

Exercise 19.8 Component Waves for 1D Flow in Power Law Medium

Problem: Consider 1D flow in a finite, source-free domain filled with a power law medium. Suppose that for a particular harmonic constituent (i.e., ω_m fixed) the n = 2 component wave amplitude function takes the form

$$M^{(2)}(x;\omega_m) = \left|a_m^{(2)}\right|\left|\frac{x}{x_c}\right|. \tag{19.35}$$

Derive an equation that compactly expresses the parameter f_m in terms of the parameter b in this case. What form does the $n = 1$ component wave amplitude function take in this case?

Solution: Comparing the RHSs of (19.35) and (19.29b), we infer that

$$\mathrm{Re}\, p_m^{(2)} = 1. \tag{19.36}$$

Substituting the RHS of (11.12), with $n = 2$, for $p_m^{(2)}$ leads to

$$\sqrt{\frac{\sqrt{(1-b)^4 + 16f_m^2} + (1-b)^2}{2}} = 1 + b.$$

Squaring, multiplying by 2, and rearranging produce

$$\sqrt{(1-b)^4 + 16f_m^2} = 1 + 6b + b^2.$$

Squaring and then solving for f_m yield

$$f_m = (b+1)\sqrt{b}, \quad b > 0. \tag{19.37}$$

This completes the first part of the exercise. Combining (11.14a) with (19.36) yields

$$\mathrm{Re}\, p_m^{(1)} = -b.$$

Substituting the RHS for $\mathrm{Re}\, p_m^{(1)}$ in (19.29a), with $n = 1$, then gives for this case

$$M^{(1)}(x;\omega_m) = |a_m^{(1)}| \left| \frac{x}{x_c} \right|^{-b}.$$

□

Example 19.3 Plane Component Waves in Power Law Medium
Consider time-periodic, 1D flow in a source-free domain filled with a power law medium. Figure 19.3 shows the component wave FRFs for the case where $b = 1$ and $f_m = 2$, where we used the dimensionless variables defined in Example 19.2. In this case $M^{(2)}(x;\omega_m)$ increases linearly with the space coordinate, so b and f_m jointly satisfy (19.37). ◊

19.4.2. Examples: Multiple Dimensions

This section presents examples and exercises to illustrate the component wave interpretation of time-periodic, multidimensional flows.

Recall that in time-periodic, radial spherical flows and axisymmetric flows the hydraulic head FRF and its FREs are functions of the single space coordinate r. We represent them symbolically as

$$F_h(\mathbf{x};\omega_m) = G(r;\omega_m)$$

$$F^{(n)}(\mathbf{x};\omega_m) = G^{(n)}(r;\omega_m) \quad (n = 1, 2).$$

Combining the second of these with (19.15) allows us to write the component hydraulic head fields as

$$h^{(n)}(r,t;\omega_m) = \mathrm{Re}\left[a_m^{(n)} G^{(n)}(r;\omega_m) \exp(i\omega_m t)\right] \quad (n = 1, 2) \tag{19.38}$$

where the $a_m^{(n)}$ are complex constants.

19.4.2.1. Spherical Waves in Ideal Media

In Chapter 12 we saw that for time-periodic, radial spherical flow in ideal media we can write the FREs as

$$G^{(1)}(r;\omega_m) = \frac{e^{-\lambda_m r}}{\lambda_m r} \tag{19.39a}$$

$$G^{(2)}(r;\omega_m) = \frac{\sinh(\lambda_m r)}{\lambda_m r}. \tag{19.39b}$$

Substituting the RHSs for the $G^{(n)}$ $(n = 1, 2)$ and the RHS of (12.4) for λ_m in (19.38) leads to

$$h^{(1)}(r,t;\omega_m) = |a_m^{(1)}| \frac{e^{-\sqrt{\alpha_m/2}\,r}}{\sqrt{\alpha_m}\,r} \times \cos\left[\omega_m t - \sqrt{\frac{\alpha_m}{2}}\,r - \frac{\pi}{4} + \arg a_m^{(1)}\right] \tag{19.40a}$$

$$h^{(2)}(r,t;\omega_m) = |a_m^{(2)}| \frac{|\sinh(e^{i\pi/4}\sqrt{\alpha_m}\,r)|}{\sqrt{\alpha_m}\,r} \times \cos\left[\omega_m t + \arg\sinh(e^{i\pi/4}\sqrt{\alpha_m}\,r) - \frac{\pi}{4} + \arg a_m^{(2)}\right]. \tag{19.40b}$$

These are the equations of harmonic, spherical waves that respectively diverge from and converge on the origin.

Exercise 19.9 Component Waves for Radial Spherical Flow in Ideal Medium

Problem: Consider time-periodic, radial spherical flow through a source-free domain filled with an ideal medium. For each of the component waves, express the wave amplitude and phase function in terms of r, α_m, and the constants $a_m^{(n)}$.

Solution: Comparing (19.40a) to the definitions (19.1) and (19.2) for the $n = 1$ component wave we have

$$M^{(1)}(r;\omega_m) = |a_m^{(1)}| \frac{e^{-\sqrt{\alpha_m/2}\,r}}{\sqrt{\alpha_m}\,r} \tag{19.41a}$$

$$\theta^{(1)}(r;\omega_m) = -\sqrt{\frac{\alpha_m}{2}}\,r - \frac{\pi}{4}\,\mathrm{rad} + \arg a_m^{(1)}. \tag{19.41b}$$

Comparing (19.40b) to the definitions (19.1) and (19.2), evidently the $n = 2$ component wave has the following wave amplitude and phase function, respectively:

$$M^{(2)}(r;\omega_m) = |a_m^{(2)}| \frac{|\sinh(e^{i\pi/4}\sqrt{\alpha_m}\,r)|}{\sqrt{\alpha_m}\,r} \tag{19.42a}$$

$$\theta^{(2)}(r;\omega_m) = \arg\sinh(e^{i\pi/4}\sqrt{\alpha_m}\,r) - \frac{\pi}{4}\,\mathrm{rad} + \arg a_m^{(2)}. \tag{19.42b}$$

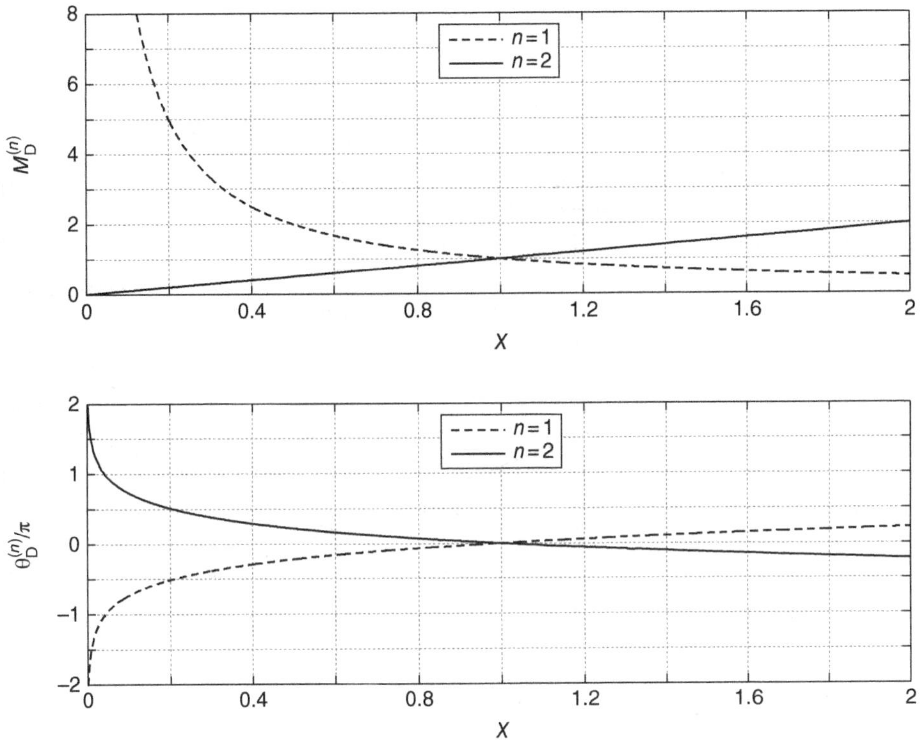

Figure 19.3 FRFs for plane component waves in power law medium (b = 1, f_m = 2) (context: see Example 19.3).

Using identities (12.21) to express the modulus and argument of the hyperbolic sine in terms of real-valued quantities leads to

$$M^{(2)}(r;\omega_m) = |a_m^{(2)}|\frac{\sqrt{\cosh(\sqrt{2\alpha_m}r) - \cos(\sqrt{2\alpha_m}r)}}{\sqrt{2\alpha_m}r} \tag{19.43a}$$

$$\theta^{(2)}(r;\omega_m) = \arctan\left[\coth\left(\sqrt{\frac{\alpha_m}{2}}r\right)\tan\left(\sqrt{\frac{\alpha_m}{2}}r\right)\right] - \frac{\pi}{4}\text{ rad} + \arg a_m^{(2)}. \tag{19.43b}$$

□

Example 19.4 Spherical Component Waves in Ideal Medium
Consider time-periodic, radial spherical flow in a spherical shell filled with an ideal medium due to time-periodic forcing at the inner and outer boundaries. For plotting, it is convenient to scale the FREs as

$$G_D^{(1)}(R;\eta_m) \equiv \frac{G^{(1)}(r;\omega_m)}{G^{(1)}(r_{\min};\omega_m)}$$

$$G_D^{(2)}(R;\eta_m) \equiv \frac{G^{(2)}(r;\omega_m)}{G^{(2)}(r_{\max};\omega_m)}$$

where

$$R \equiv \frac{r}{r_{\max}}$$

$$\eta_m \equiv \sqrt{\alpha_m} r_{\max}.$$

Define the corresponding amplitude and phase functions as

$$M_D^{(n)}(X;\eta_m) \equiv \left|G_D^{(n)}(X;\eta_m)\right| \tag{19.44a}$$

$$\theta_D^{(n)}(X;\eta_m) \equiv \arg G_D^{(n)}(X;\eta_m). \tag{19.44b}$$

Figure 19.4 shows the FREs for the case where $r_{\min}/r_{\max}$ = 0.2 and η_m = 5. The component amplitude and phase functions have no stationary points, so the component waves travel completely through the medium in opposite directions while undergoing normal (positive) attenuation. Compare the corresponding constituent wave amplitude and phase functions (Figure 12.3). ◊

For the n = 1 component wave, we can easily express the propagation velocity in terms of $\hat{\mathbf{r}}$, ω_m, and α_m. Taking the gradient of (19.41b) gives

$$\nabla\theta^{(1)}(r;\omega_m) = -\sqrt{\frac{\alpha_m}{2}}\hat{\mathbf{r}}. \tag{19.45}$$

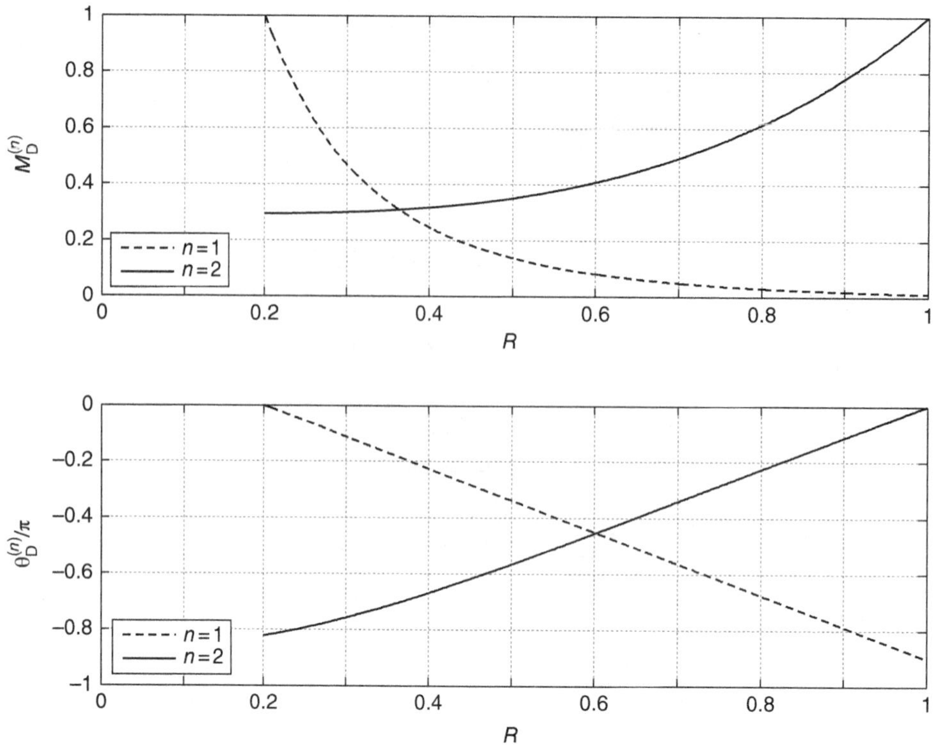

Figure 19.4 FRFs for spherical component waves in ideal medium (context: see Example 19.4).

Substituting the RHS for the phase gradient in (19.9) produces

$$\mathbf{v}^{(1)}(r;\omega_m) = \omega_m\sqrt{\frac{2}{\alpha_m}}\hat{\mathbf{r}}. \tag{19.46}$$

Thus, the $n = 1$ component wave travels outward, away from the origin, at constant speed.

For the $n = 2$ component wave we can express the propagation velocity in terms of r, $\hat{\mathbf{r}}$, ω_m, and α_m (see Exercise 19.11). However, based on (19.9) and Figure 19.4 alone, we know that the propagation velocity $\mathbf{v}^{(2)}(r;\omega_m)$ is a vector that points in the direction of decreasing r, and its magnitude increases as r decreases. That is, the $n = 2$ component wave travels inward, toward the origin, and its speed increases as the wavefront approaches the origin.

Exercise 19.10 Component Waves for Radial Spherical Flow in Ideal Medium

Problem: Consider time-periodic, radial spherical flow through a source-free domain filled with an ideal medium. For the $n = 2$ component wave, express the phase gradient, $\nabla\theta^{(2)}(r;\omega_m)$, in terms of r, $\hat{\mathbf{r}}$, ω_m, and α_m.

Solution: Define the auxiliary variable

$$u \equiv \sqrt{\frac{\alpha_m}{2}}r. \tag{19.47}$$

Then we can rewrite (19.43b) as

$$\theta^{(2)}(r;\omega_m) = \arctan\left[\coth(u)\tan(u)\right] - \pi/4 \text{ rad} + \arg a_m^{(2)}.$$

Taking the gradient yields

$$\nabla\theta^{(2)}(r;\omega_m) = \frac{d}{du}\left\{\arctan\left[\coth(u)\tan(u)\right]\right\}\sqrt{\frac{\alpha_m}{2}}\hat{\mathbf{r}}. \tag{19.48}$$

To evaluate the derivative on the RHS, use the chain rule and the following identities from *Roy and Olver* [2016]:

$$\frac{d}{du}\arctan(u) = \frac{1}{1+u^2}$$
$$\frac{d}{du}\tan(u) = \sec^2(u)$$
$$\frac{d}{du}\coth(u) = -\text{csch}^2(u).$$

The result is

$$\frac{d}{du}\left\{\arctan\left[\coth(u)\tan(u)\right]\right\} = \frac{\coth(u)\sec^2(u) - \text{csch}^2(u)\tan(u)}{1 + \left\{\arctan\left[\coth(u)\tan(u)\right]\right\}^2}.$$

Substituting the RHS for the derivative in (19.48) and substituting the RHS of (19.47) for u lead to

$$\nabla\theta^{(2)}(r;\omega_m) = \frac{\left[\begin{array}{l}\coth\left(\sqrt{\frac{\alpha_m}{2}}r\right)\sec^2\left(\sqrt{\frac{\alpha_m}{2}}r\right) \\ -\operatorname{csch}^2\left(\sqrt{\frac{\alpha_m}{2}}r\right)\tan\left(\sqrt{\frac{\alpha_m}{2}}r\right)\end{array}\right]}{1+\left\{\arctan\left[\coth\left(\sqrt{\frac{\alpha_m}{2}}r\right)\tan\left(\sqrt{\frac{\alpha_m}{2}}r\right)\right]\right\}^2}\sqrt{\frac{\alpha_m}{2}}\hat{\mathbf{r}}. \tag{19.49}$$

□

Exercise 19.11 Component Waves for Radial Spherical Flow in Ideal Medium

Problem: Consider time-periodic, radial spherical flow through a source-free domain filled with an ideal medium. For the $n = 2$ component wave, express the propagation velocity and propagation speed in terms of r, $\hat{\mathbf{r}}$, ω_m, and α_m.

Solution: Substituting the RHS of (19.49) for the phase gradient in (19.9) produces

$$\mathbf{v}^{(2)}(r;\omega_m) = -\omega_m\sqrt{\frac{2}{\alpha_m}} \times \frac{1+\left\{\arctan\left[\coth\left(\sqrt{\frac{\alpha_m}{2}}r\right)\tan\left(\sqrt{\frac{\alpha_m}{2}}r\right)\right]\right\}^2}{\begin{array}{l}\coth\left(\sqrt{\frac{\alpha_m}{2}}r\right)\sec^2\left(\sqrt{\frac{\alpha_m}{2}}r\right) \\ -\operatorname{csch}^2\left(\sqrt{\frac{\alpha_m}{2}}r\right)\tan\left(\sqrt{\frac{\alpha_m}{2}}r\right)\end{array}}\hat{\mathbf{r}}.$$

The denominator of the second (long) term on the RHS is positive, so the phase speed is

$$c^{(2)}(r;\omega_m) = \omega_m\sqrt{\frac{2}{\alpha_m}} \times \frac{1+\left\{\arctan\left[\coth\left(\sqrt{\frac{\alpha_m}{2}}r\right)\tan\left(\sqrt{\frac{\alpha_m}{2}}r\right)\right]\right\}^2}{\begin{array}{l}\coth\left(\sqrt{\frac{\alpha_m}{2}}r\right)\sec^2\left(\sqrt{\frac{\alpha_m}{2}}r\right) \\ -\operatorname{csch}^2\left(\sqrt{\frac{\alpha_m}{2}}r\right)\tan\left(\sqrt{\frac{\alpha_m}{2}}r\right)\end{array}}.$$

□

19.4.2.2. Cylindrical Waves in Ideal Media

In Chapter 12 we saw that for time-periodic, axisymmetric flow in ideal media we can write the FREs as

$$G^{(1)}(r;\omega_m) = N_0(\sqrt{\alpha_m}r)e^{i\phi_0(\sqrt{\alpha_m}r)} \tag{19.50a}$$

$$G^{(2)}(r;\omega_m) = M_0(\sqrt{\alpha_m}r)e^{i\theta_0(\sqrt{\alpha_m}r)} \tag{19.50b}$$

where M_0 and N_0 are the order-zero Kelvin modulus functions and θ_0 and ϕ_0 are the order-zero Kelvin phase functions. Readers who have not yet done so should read Appendix E. Substituting the RHSs of (19.50) for $G^{(n)}$ in (19.38) leads to

$$h^{(1)}(r,t;\omega_m) = \left|a_m^{(1)}\right|N_0(\sqrt{\alpha_m}r)\cos\left[\omega_m t+\phi_0(\sqrt{\alpha_m}r)+\arg a_m^{(1)}\right] \tag{19.51a}$$

$$h^{(2)}(r,t;\omega_m) = \left|a_m^{(2)}\right|M_0(\sqrt{\alpha_m}r)\cos\left[\omega_m t+\theta_0(\sqrt{\alpha_m}r)+\arg a_m^{(2)}\right]. \tag{19.51b}$$

These are the equations of harmonic, cylindrical waves that respectively diverge from and converge on the axial line $r = 0$.

Exercise 19.12 Component Waves for Axisymmetric Flow in Ideal Medium

Problem: Consider time-periodic, axisymmetric flow through a source-free, annular domain filled with an ideal medium. For each of the component waves, express the wave amplitude and phase function in terms of r, α_m, and the constants $a_m^{(n)}$. Use the Kelvin modulus and phase functions.

Solution: Comparing (19.51) to the definitions (19.1) and (19.2), evidently the $n = 1$ component wave has the following amplitude and phase function:

$$M^{(1)}(r;\omega_m) = \left|a_m^{(1)}\right|N_0(\sqrt{\alpha_m}r) \tag{19.52a}$$

$$\theta^{(1)}(r;\omega_m) = \phi_0(\sqrt{\alpha_m}r)+\arg a_m^{(1)}. \tag{19.52b}$$

Similarly, for the $n = 2$ component wave we have

$$M^{(2)}(r;\omega_m) = \left|a_m^{(2)}\right|M_0(\sqrt{\alpha_m}r) \tag{19.53a}$$

$$\theta^{(2)}(r;\omega_m) = \theta_0(\sqrt{\alpha_m}r)+\arg a_m^{(2)}. \tag{19.53b}$$

□

Example 19.5 Cylindrical Component Waves in Ideal Medium

Consider time-periodic, axisymmetric flow in an annular domain filled with an ideal medium due to time-periodic forcing at the inner and outer boundaries. For plotting, it is convenient to scale the FREs as

$$G_D^{(1)}(R;\eta_m) \equiv \frac{G^{(1)}(r;\omega_m)}{G^{(1)}(r_{min};\omega_m)}$$

$$G_D^{(2)}(R;\eta_m) \equiv \frac{G^{(2)}(r;\omega_m)}{G^{(2)}(r_{max};\omega_m)}$$

where

$$R \equiv \frac{r}{r_{max}}$$

$$\eta_m \equiv \sqrt{\alpha_m}r_{max}.$$

Define the corresponding amplitude and phase functions as

$$M_{\rm D}^{(n)}(X;\eta_m) \equiv \left|G_{\rm D}^{(n)}(X;\eta_m)\right|$$
$$\theta_{\rm D}^{(n)}(X;\eta_m) \equiv \arg G_{\rm D}^{(n)}(X;\eta_m).$$

Figure 19.5 shows the FRFs for the case $\eta_m = 5$. Notice that the amplitude and phase functions do not have stationary points, so the component waves travel completely through the medium in opposite directions without undergoing negative attenuation. Compare the corresponding constituent wave amplitude and phase functions (Figure 12.6). ◊

For each of the cylindrical component waves, we can express the propagation velocity in terms of r, the constituent frequency, and α_m. Consider the case $n = 1$. Taking the gradient of (19.52b) gives

$$\nabla\theta^{(1)}(r;\omega_m) = \sqrt{\alpha_m}\left[\frac{d}{du}\phi_0(u)\right]\Bigg|_{u=\sqrt{\alpha_m}r}\hat{\mathbf{r}}.$$

Using the second of equations 10.68.13 of *Olver and Maximon* [2016], we can write this as

$$\nabla\theta^{(1)}(r;\omega_m)$$
$$= \sqrt{\alpha_m}\frac{N_1(\sqrt{\alpha_m}r)}{N_0(\sqrt{\alpha_m}r)}\sin\left[\phi_1(\sqrt{\alpha_m}r) - \phi_0(\sqrt{\alpha_m}r) - \pi/4\right]\hat{\mathbf{r}}$$

Substituting the RHS for the phase gradient in (19.9) and using the monotonicity characterstics of the Kelvin phase functions (see Appendix E.5) lead to

$$\mathbf{v}^{(1)}(r;\omega_m) = \frac{-\omega_m}{\sqrt{\alpha_m}}\frac{N_0(\sqrt{\alpha_m}r)}{N_1(\sqrt{\alpha_m}r)}\frac{1}{\left|\sin\left[\phi_1(\sqrt{\alpha_m}r) - \phi_0(\sqrt{\alpha_m}r) - \pi/4\right]\right|}\hat{\mathbf{r}}. \quad (19.54)$$

Thus, the $n = 1$ component wave travels outward, away from the axis of symmetry ($r = 0$), and its speed varies with r.

Exercise 19.13 Component Waves for Axisymmetric Flow in Ideal Medium

Problem: Consider time-periodic, axisymmetric flow through a source-free domain filled with an ideal medium. Express the propagation velocity of the $n = 2$ component wave in terms of r, $\hat{\mathbf{r}}$, ω_m, and α_m. Use the Kelvin modulus and phase functions. In what direction does the $n = 2$ component wave travel relative to the axis of symmetry ($r = 0$)? Does its speed vary with location?

Solution: Taking the gradient of (19.53b) gives

$$\nabla\theta^{(2)}(r;\omega_m) = \sqrt{\alpha_m}\left[\frac{d}{du}\theta_0(u)\right]\Bigg|_{u=\sqrt{\alpha_m}r}\hat{\mathbf{r}}.$$

Using the second of equations 10.68.13 of *Olver and Maximon* [2016], we can write this as

$$\nabla\theta^{(2)}(r;\omega_m) = \sqrt{\alpha_m}\frac{M_1(\sqrt{\alpha_m}r)}{M_0(\sqrt{\alpha_m}r)}\sin\left[\theta_1(\sqrt{\alpha_m}r) - \theta_0(\sqrt{\alpha_m}r) - \pi/4\right]\hat{\mathbf{r}}. \quad (19.55)$$

Substituting the RHS for the phase gradient in (19.9) and using the monotonicity characterstics of the Kelvin phase functions (see Section E.5) lead to

$$\mathbf{v}^{(2)}(r;\omega_m) = \frac{\omega_m}{\sqrt{\alpha_m}}\frac{M_0(\sqrt{\alpha_m}r)}{M_1(\sqrt{\alpha_m}r)}\frac{1}{\left|\sin\left[\theta_1(\sqrt{\alpha_m}r) - \theta_0(\sqrt{\alpha_m}r) - \pi/4\right]\right|}\hat{\mathbf{r}}. \quad (19.56)$$

Thus, the $n = 2$ component wave travels inward, toward the axis of symmetry, and its speed varies with r. □

19.4.2.3. Oblique Plane Waves in Exponential Media

In Chapter 13 we saw that for time-periodic, uniform-gradient flow in exponential media we can write the FREs as

$$F^{(n)}(\mathbf{x};\omega_m) = \exp\left[\mathbf{w}^{(n)}(\omega_m)\cdot\mathbf{x}\right] \quad (n = 1,2) \quad (19.57)$$

where the $\mathbf{w}^{(n)}(\omega_m)$ are complex-valued eigenvectors that satisfy the characteristic equation (13.11). Substituting the RHS for the $F^{(n)}$ in (19.15) leads to

$$h^{(n)}(\mathbf{x},t;\omega_m) = \left|a_m^{(n)}\right|\exp\left[\mathrm{Re}\,\mathbf{w}^{(n)}(\omega_m)\cdot\mathbf{x}\right] \times\cos\left[\omega_m t + \mathrm{Im}\,\mathbf{w}^{(n)}(\omega_m)\cdot\mathbf{x} + \arg a_m^{(n)}\right] \quad (n = 1,2) \quad (19.58)$$

which is the equation of a particular type of traveling, harmonic plane wave.

Exercise 19.14 Component Waves for Uniform-Gradient Flow

Problem: Consider time-periodic, uniform-gradient flow through a source-free domain filled with an exponential medium. For each of the component waves, express the wave amplitude and phase function in terms of x, the eigenvectors $\mathbf{w}^{(n)}(\omega_m)$, and the constants $a_m^{(n)}$.

Solution: Comparing (19.58) to the definitions (19.1) and (19.2), evidently the nth component wave has the following wave amplitude and phase function, respectively:

$$M^{(n)}(\mathbf{x};\omega_m) = \left|a_m^{(n)}\right|\exp\left[\mathrm{Re}\,\mathbf{w}^{(n)}(\omega_m)\cdot\mathbf{x}\right] \quad (19.59a)$$
$$\theta^{(n)}(\mathbf{x};\omega_m) = \mathrm{Im}\,\mathbf{w}^{(n)}(\omega_m)\cdot\mathbf{x} + \arg a_m^{(n)} \quad (19.59b)$$
$$n = 1,2.$$

□

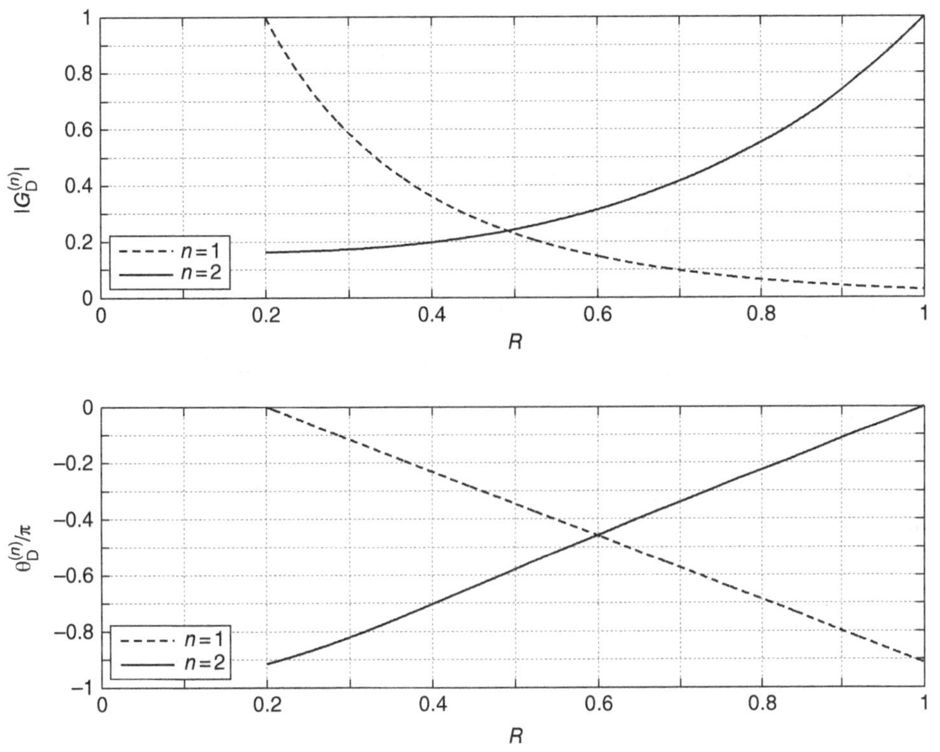

Figure 19.5 FRFs for cylindrical component waves in ideal medium (context: see Example 19.5).

Exercise 19.15 Component Waves for Uniform-Gradient Flow

Problem: Consider time-periodic, uniform-gradient flow through a source-free domain filled with an exponential medium. For each of the component waves, express the propagation velocity and propagation speed in terms of $\hat{\mathbf{x}}$, ω_m, and the eigenvectors $\mathbf{w}^{(n)}(\omega_m)$. In words, concisely describe the relationship between each component wave's eigenvector and its direction of travel.

Solution: Taking the gradient of (19.59b) gives

$$\nabla\theta^{(n)}(x;\omega_m) = \operatorname{Im}\mathbf{w}^{(n)}(\omega_m) \quad (n = 1, 2).$$

Substituting the RHS for the phase gradient in (19.9) yields the propagation velocity

$$\mathbf{v}^{(n)}(x;\omega_m) = \frac{-\omega_m}{\left|\operatorname{Im}\mathbf{w}^{(n)}(\omega_m)\right|^2}\operatorname{Im}\mathbf{w}^{(n)}(\omega_m) \quad (n = 1, 2)$$

and the propagation speed

$$c^{(n)}(x;\omega_m) = \frac{\omega_m}{\left|\operatorname{Im}\mathbf{w}^{(n)}(\omega_m)\right|} \quad (n = 1, 2).$$

The nth component wave travels in the direction opposite the vector $\operatorname{Im}\mathbf{w}^{(n)}(\omega_m)$. □

The eigenvector relation (13.17) implies that $\operatorname{Im}\mathbf{w}^{(2)}(\omega_m) = -\operatorname{Im}\mathbf{w}^{(1)}(\omega_m)$. Consequently, the two component waves travel in opposite directions. The phase gradient $\nabla\theta^{(n)}$ and the log conductivity gradient $\boldsymbol{\mu}$ need not be collinear; this fact motivates our use of the term *oblique plane waves*.

19.5. DISCUSSION: CONSTITUENT WAVES VERSUS COMPONENT WAVES

In the constituent wave interpretation, we view each hydraulic head harmonic constituent field $h'(\mathbf{x}, t; \omega_m)$ (m fixed) at each point in the porous medium as a single traveling wave. In the component wave interpretation, we view each harmonic constituent field $h'(\mathbf{x}, t; \omega_m)$ (m fixed) as an interference pattern that results from the superposition of the corresponding component waves $h^{(n)}(\mathbf{x}, t; \omega_m)$ $(n = 1, 2, \ldots, N)$.

Both wave interpretations present conceptual challenges in those cases where stationary points of the corresponding phase functions (θ_h or $\theta^{(n)}$) appear, including the following:

- Waves terminate in the space domain at stationary point(s).
- Wave propagation velocities grow unbounded at stationary point(s).

For example, consider 3D (or 2D) radial flow in a spherical (or cylindrical) domain filled with an ideal medium due to time-periodic forcing of the outer boundary, as in Example 12.4 (Example 12.11). In these cases the center of the domain ($r = 0$) is a stationary point, and each harmonic constituent corresponds to a single component wave that propagates from the outer boundary to the center, where it terminates with unbounded velocity.

With constituent waves additional problems can arise. For instance, in many 1D cases (see Section 19.3.1 for a list) and in some multidimensional cases (e.g., see Examples 12.5 and 12.12) the constituent wave phase function exhibits one or more stationary points in the space domain, while the corresponding component waves exhibit no stationary points. That is, while the constituent wave enters the medium and then terminates at the stationary point(s), the component waves travel completely through the medium, entering at one boundary and exiting at another. Perhaps more importantly, in these cases the constituent wave interpretation implies that the constituent wave undergoes negative attenuation in the medium adjacent to the stationary point(s). We are unable to conceive of a credible physical rationale for this negative attenuation, so we consider the constituent wave interpretation deficient in this regard.

The component wave interpretation seems more physically intuitive than the constituent wave interpretation and therefore more useful. For this reason we favor the component wave interpretation.

19.6. MORE WAVE KINEMATICS

In Section 19.2.2 we defined the phase velocity of a harmonic wave traveling in a continuous medium. The phase velocity is a concept of *wave kinematics*, the subject that deals with the motion (e.g., speed, direction, travel path) of waves without regard to the forces that cause the motion. In this section we develop a few additional concepts of wave kinematics in the context of component waves.

19.6.1. Wave Vector, Wave Number, and Wavelength

Define the *wave vector* as

$$\begin{aligned} \mathbf{k}^{(n)}(\mathbf{x};\omega_m) &\equiv -\nabla\xi^{(n)}(\mathbf{x},t;\omega_m) \\ &= -\nabla\theta^{(n)}(\mathbf{x};\omega_m). \end{aligned} \tag{19.60}$$

and the *wave number* as the magnitude of the wave vector, i.e.,

$$\begin{aligned} k^{(n)}(\mathbf{x};\omega_m) &\equiv \left|\mathbf{k}^{(n)}(\mathbf{x};\omega_m)\right| \\ &= \left|\nabla\theta^{(n)}(\mathbf{x};\omega_m)\right|. \end{aligned} \tag{19.61}$$

Both the wave vector and the wave number have the dimensions of reciprocal length and the units of radians per unit length. These allow us to write (19.9) and (19.8), respectively, as

$$\mathbf{v}^{(n)}(\mathbf{x};\omega_m) = \frac{\omega_m \mathbf{k}^{(n)}(\mathbf{x};\omega_m)}{(k^{(n)})^2(\mathbf{x};\omega_m)} \tag{19.62a}$$

$$c^{(n)}(\mathbf{x};\omega_m) = \frac{\omega_m}{k^{(n)}(\mathbf{x};\omega_m)} \tag{19.62b}$$

$$k^{(n)}(\mathbf{x};\omega_m) > 0.$$

Equations (19.62) show that the phase velocity and propagation speed are steady-state, spatially variable, frequency-dependent fields.

Exercise 19.16 Plane Waves in Ideal Media: Propagation Velocity

Problem: Consider plane waves in a source-free domain filled with an ideal medium. Assume the phase function is given by

$$\theta^{(n)}(\mathbf{x};\omega_m) = -\mathbf{k}^{(n)}(\omega_m)\cdot\mathbf{x}$$

where $\mathbf{k}^{(n)}(\omega_m)$ denotes a space-invariant wave vector. Evaluate the phase velocity and propagation speed. What can you say about their spatial variation?

Solution: In this case the phase gradient is spatially uniform, i.e.,

$$\nabla\theta^{(n)}(\mathbf{x};\omega_m) = -\mathbf{k}^{(n)}(\omega_m).$$

Substituting the RHS for the wave vector in (19.62b) gives the following for the phase velocity and propagation speed, respectively:

$$\mathbf{v}^{(n)}(\mathbf{x};\omega_m) = \frac{\omega_m \mathbf{k}^{(n)}(\omega_m)}{(k^{(n)})^2(\omega_m)} \tag{19.63a}$$

$$c^{(n)}(\mathbf{x};\omega_m) = \frac{\omega_m}{k^{(n)}(\omega_m)} \tag{19.63b}$$

where $k^{(n)}(\omega_m) \equiv |\mathbf{k}^{(n)}(\omega_m)|$. The phase velocity and propagation speed are spatially uniform, depending only on the frequency. □

Define the *local wavelength* of the nth component wave, $\lambda^{(n)}$, in terms of the corresponding wave number, $k^{(n)}$, as

$$\lambda^{(n)}(\mathbf{x};\omega_m) \equiv \frac{2\pi \text{ rad}}{k^{(n)}(\mathbf{x};\omega_m)}. \tag{19.64}$$

The local wavelength has the dimensions of length.

Exercise 19.17 Local Wavelength, Frequency, and Propagation Speed

Problem: Derive a simple, general equation relating the local wavelength, $\lambda^{(n)}(\mathbf{x};\omega_m)$, to the angular frequency, ω_m, and the wave propagation speed, $c^{(n)}(\mathbf{x};\omega_m)$.

Solution: Solving (19.62b) for the wave number gives

$$\frac{1}{k^{(n)}(\mathbf{x};\omega_m)} = \frac{c^{(n)}(\mathbf{x};\omega_m)}{\omega_m}.$$

Substituting the RHS for $1/k^{(n)}$ in (19.64) and then multiplying the resulting equation by ω_m yield

$$\omega_m \lambda^{(n)}(\mathbf{x};\omega_m) = 2\pi c^{(n)}(\mathbf{x};\omega_m). \qquad (19.65)$$

□

Equation (19.65) states that for a given constituent frequency ω_m the wave propagation speed $c^{(n)}$ must be proportional to the wavelength $\lambda^{(n)}$ to maintain the constant frequency of the harmonic variation.

19.6.2. Ray Path, Travel Distance, and Travel Time

We define a *ray path* as the continuous trajectory followed by an imaginary, massless tracer particle that travels passively through the fictitious continuum, with velocity equal to the phase velocity. Ray paths may be straight lines or curved trajectories.

We define the *travel distance* of the nth component wave, $s^{(n)}$, as the curvilinear coordinate of the tracer particle measured along the ray path, starting from the particle's initial position ($s^{(n)} = 0$ at $t = 0$) and increasing in the direction of the phase velocity. Because the tracer particle is always at the location of the traveling wavefront, $s^{(n)}$ is the travel distance of the wavefront along the ray path. Geometrically, $s^{(n)}$ is equivalent to an arc length parameter of the particle trajectory.

Let $\hat{s}^{(n)}$ denote the unit vector parallel to the ray path of the nth component wave. We will call $\hat{s}^{(n)}$ the *ray path vector*. The ray path vector is parallel to the phase velocity vector, so by (19.9)

$$\hat{s}^{(n)}(\mathbf{x};\omega_m) \equiv \frac{\mathbf{v}^{(n)}}{\left|\mathbf{v}^{(n)}\right|} = \frac{-\nabla\theta^{(n)}}{\left|\nabla\theta^{(n)}\right|}. \qquad (19.66)$$

Consequently we have the following:

$$\mathbf{k}^{(n)} = k^{(n)}\hat{s}^{(n)} \qquad (19.67a)$$

$$\mathbf{v}^{(n)} = c^{(n)}\hat{s}^{(n)}. \qquad (19.67b)$$

Exercise 19.18 Phase Velocity at No-Flow Boundary

Problem: Consider a time-periodic flow for which a particular harmonic constituent (i.e., ω_m fixed) consists of a single component wave [i.e., $N(\omega_m) = 1$] and the point $\mathbf{x} = \mathbf{x}_0$ lies on a no-flow boundary. Then

$$\mathbf{q}'(\mathbf{x}_0, t;\omega_m) \cdot \hat{n}(\mathbf{x}_0) = 0, \quad \forall\, t$$

where $\mathbf{x}_0 \in \Gamma$ and $\hat{n}(\mathbf{x})$ denotes the unit vector that is outward normal to the boundary at the point $\mathbf{x}$. Further suppose that the medium is locally isotropic there:

$$\mathbf{K}(\mathbf{x}_0) = K(\mathbf{x}_0)\,\mathbf{I} \qquad (19.68)$$

where $K > 0$. What can you infer about the phase velocity of the constituent wave at the point $\mathbf{x} = \mathbf{x}_0$?

Solution: Substituting the RHS of (19.68) for $\mathbf{K}$ in (14.24b) yields

$$\nabla\theta^{(n)}(\mathbf{x}_0;\omega_m) \cdot \hat{n}(\mathbf{x}_0) = 0.$$

Combining this with (19.66) and (19.67b) gives

$$\mathbf{v}^{(n)}(\mathbf{x}_0;\omega_m) \cdot \hat{n}(\mathbf{x}_0) = 0.$$

Therefore, at the point $\mathbf{x} = \mathbf{x}_0$ the phase velocity vector either vanishes or is parallel to the boundary. □

Next, we express the relationship between the variation of the phase along a ray path to the travel distance of the wavefront. The variation of the phase with travel distance (i.e., in the direction of the ray path) is given by the directional derivative:

$$\frac{\partial\theta^{(n)}}{\partial s^{(n)}} = \nabla\theta^{(n)} \cdot \hat{s}^{(n)}.$$

At any fixed time the wavefronts coincide with the level surfaces of the phase function, so the phase varies only in the direction perpendicular to the wavefront. In other words, the phase varies only in the direction of the ray path, so we can write the LHS as an ordinary derivative:

$$\frac{d\theta^{(n)}}{ds^{(n)}} = -\mathbf{k}^{(n)} \cdot \hat{s}^{(n)}.$$

Thus, along a fixed ray path the differentials are related by

$$d\theta^{(n)} = -k^{(n)}\, \mathrm{d}s^{(n)}. \qquad (19.69)$$

Exercise 19.19 Wave Travel Time

Problem: Let $\Delta T^{(n)} \equiv T^{(n)}(s^{(n)};\omega_m) - T^{(n)}(s_0^{(n)};\omega_m)$ denote the time for the nth component wavefront to travel along a fixed ray path between the points whose travel distance coordinates are $s_0^{(n)}$ (initial) and $s^{(n)}$. We call $\Delta T^{(n)}$ the *wave travel time* for the nth component wave. Derive a simple expression relating $d\theta^{(n)}$ to $dT^{(n)}$, $k^{(n)}$, and ω_m for a wavefront traveling along the ray path.

Solution: The travel distance and travel time differentials are related by

$$ds^{(n)} = c^{(n)}\; dT^{(n)}.$$

Substituting the RHS for $ds^{(n)}$ in (19.69) and using (19.62b) to express the propagation speed in terms of the wave number and the frequency give

$$d\theta^{(n)} = -\omega_m\; dT^{(n)}.$$

□

19.7. CHAPTER SUMMARY

In this chapter we interpreted time-periodic groundwater flow as the steady-state propagation of spatially modulated (damped), traveling diffusion waves, i.e., harmonic, hydraulic head waves. We presented and discussed two distinct wave interpretations, which we call the *constituent wave interpretation* and the *component wave interpretation*. We concluded that the component wave interpretation is preferable to the constituent wave interpretation because it is more intuitive. The component wave interpretation views each hydraulic head harmonic constituent field as an interference pattern that results from the superposition of the component waves corresponding to that frequency.

We presented hypothetical examples with analytical solutions of time-periodic flows in particular media, both one dimensional and multidimensional, to demonstrate the wave interpretation. We also presented some basic concepts of wave kinematics, including wave propagation velocity, wave vector, wave number, wavelength, ray path, travel distance, and travel time.

20
Wave Distortion

In this chapter we define and describe various types of wave distortion. We do this in the context of harmonic, hydraulic head waves.

20.1. INTRODUCTION

A *wave packet* is a complex wave formed by the superposition of two or more harmonic waves (either constituent waves or component waves) with distinct frequencies generally traveling in the same or nearly the same direction. Consequently a wave packet is an *anharmonic wave*. *Wave distortion* is the change in shape of a wave or wave packet as it propagates through a medium.

20.2. DISPERSIVE AND NONDISPERSIVE DISTORTION

We distinguish between two basic types of wave distortion here—dispersive and nondispersive—and describe both in the following sections.

20.2.1. Dispersive Distortion

In Section 19.2.2 we showed that the phase velocity of a harmonic, hydraulic head wave depends on the constituent frequency and position. At any given location within the medium, then, waves with different frequencies travel with different velocities. This causes the relative phase relationships between the various waves to change with travel distance, thus changing the shape of the wave packet as it travels, even in a homogeneous medium for which spatial attenuation is negligible. We call the resulting wave distortion *dispersive distortion* (i.e., *wave dispersion*). The concept of wave dispersion is meaningful only in the context of a traveling wave *packet*.

Conceptually, the phase velocity can exhibit two types of frequency dependence, thus giving rise to two types of dispersion—frequency dispersion and geometric dispersion; we briefly describe both below.

Frequency dispersion (also *physical dispersion*) is the distortion that arises because the propagation speeds of constituent waves depend intrinsically on their respective frequencies. In Section 19.2.2 we showed that the propagation speed of a harmonic, hydraulic head wave depends on the constituent frequency. Therefore, hydraulic head wave propagation exhibits frequency dispersion. Because frequency dispersion arises from the frequency dependence of the phase *speed*, rather than the frequency dependence of the phase velocity *direction*, it occurs even in 1D flow (e.g., see Section 19.4.1). We discuss frequency dispersion further in Section 20.3.

Geometric dispersion is an apparent dispersion that can appear in 2D and 3D flows when waves of different frequency travel along different ray paths (e.g., due to frequency dependence of the boundary conditions and source distributions). In 1D flows geometric dispersion does not occur because the ray paths of the various constituent waves coincide. We will not discuss geometric dispersion further in this book.

20.2.2. Nondispersive Distortion

Nondispersive distortion is the change in shape of a single harmonic wave as it propagates due to mechanisms other than dispersion. Nondispersive distortion includes *wave attenuation* and *wave dilation/contraction*. Wave attenuation is the decrease of wave amplitude with change in position. Wave dilation/contraction is the increase or decrease, respectively, of the wavelength with change in position. We discuss wave attenuation

Hydrodynamics of Time-Periodic Groundwater Flow: Diffusion Waves in Porous Media, Geophysical Monograph 224, First Edition. Joe S. Depner and Todd C. Rasmussen.

and wave dilation/contraction in Sections 20.5 and 20.6, respectively.

20.3. FREQUENCY DISPERSION AND GROUP VELOCITY

Consider the hydraulic head field corresponding to the superposition of two component wave fields, the frequencies of which are distinct but very nearly equal:

$$h'(\mathbf{x}, t) = h^{(1)}(\mathbf{x}, t; \omega) + h^{(1)}(\mathbf{x}, t; \omega + \Delta\omega) \tag{20.1}$$

where

$$h^{(1)}(\mathbf{x}, t; \omega) = M^{(1)}(\mathbf{x}; \omega) \cos\left[\omega t + \theta^{(1)}(\mathbf{x}; \omega)\right] \tag{20.2a}$$

$$h^{(1)}(\mathbf{x}, t; \omega + \Delta\omega) = M^{(1)}(\mathbf{x}; \omega + \Delta\omega) \cos\left[(\omega + \Delta\omega)t + \theta^{(1)}(\mathbf{x}; \omega + \Delta\omega)\right] \tag{20.2b}$$

and

$$|\Delta\omega| \ll \omega. \tag{20.3}$$

Define the amplitude and phase increments associated with $\Delta\omega$ as

$$\Delta M \equiv M^{(1)}(\mathbf{x}; \omega + \Delta\omega) - M^{(1)}(\mathbf{x}; \omega) \tag{20.4a}$$

$$\Delta\theta \equiv \theta^{(1)}(\mathbf{x}; \omega + \Delta\omega) - \theta^{(1)}(\mathbf{x}; \omega). \tag{20.4b}$$

Then we can rewrite (20.1) as

$$\begin{aligned} h'(\mathbf{x}, t) = M^{(1)}(\mathbf{x}; \omega)\Big\{ &\cos\left[\omega t + \theta^{(1)}\right] \\ &+ \cos\left[(\omega + \Delta\omega)t + (\theta^{(1)} + \Delta\theta)\right]\Big\} \\ &+ \Delta M \cos\left[(\omega + \Delta\omega)t + (\theta^{(1)} + \Delta\theta)\right] \end{aligned} \tag{20.5}$$

We assume that the frequency increment $\Delta\omega$ is sufficiently small that the corresponding amplitude increment is relatively small, i.e., $|\Delta M| \ll M^{(1)}$, so that we can approximate (20.5) as

$$\begin{aligned} h'(\mathbf{x}, t) \approx M^{(1)}(\mathbf{x}; \omega)\Big\{ &\cos\left[\omega t + \theta^{(1)}\right] \\ &+ \cos\left[(\omega + \Delta\omega)t + (\theta^{(1)} + \Delta\theta)\right]\Big\}. \end{aligned} \tag{20.6}$$

Using trigonometric identity (B.3) to express the RHS of (20.6) as a product of cosines gives

$$h'(\mathbf{x}, t) \approx 2M^{(1)} \cos\left(\frac{\Delta\omega t + \Delta\theta}{2}\right) \cos\left(\frac{\sigma_\omega t + \sigma_\theta}{2}\right) \tag{20.7}$$

where

$$\sigma_\omega \equiv 2\omega + \Delta\omega \tag{20.8a}$$

$$\sigma_\theta \equiv \theta^{(1)}(\mathbf{x}; \omega + \Delta\omega) + \theta^{(1)}(\mathbf{x}; \omega). \tag{20.8b}$$

We can also write (20.7) as

$$h'(\mathbf{x}, t) \approx 2M^{(1)} \cos\varphi \cos\left(\frac{\sigma_\omega t + \sigma_\theta}{2}\right) \tag{20.9}$$

where

$$\varphi \equiv \frac{\Delta\omega t + \Delta\theta}{2}.$$

The $\cos\varphi$ term in (20.9) represents a relatively low frequency traveling envelope; the other cosine term represents a relatively high frequency carrier wave that is modulated by that envelope.

The *group velocity* is the velocity of the traveling envelope; it satisfies the following:

$$\left.\frac{D\varphi}{Dt}\right|_{\mathbf{v}=\mathbf{v}_{\text{group}}} = 0 \tag{20.10a}$$

$$\mathbf{v}_{\text{group}}(\mathbf{x}; \omega_m) = \pm|\mathbf{v}_{\text{group}}(\mathbf{x}; \omega_m)|\frac{\nabla\varphi}{|\nabla\varphi|}, \quad |\nabla\varphi(\mathbf{x}, t; \omega_m)| > 0. \tag{20.10b}$$

Substituting the RHS of (20.10b) for $\mathbf{v}_{\text{group}}$ in (20.10a) and using the definition of the substantial derivative (19.4) lead to

$$\frac{\partial\varphi}{\partial t} \pm |\mathbf{v}_{\text{group}}(\mathbf{x}; \omega_m)||\nabla\varphi(\mathbf{x}, t; \omega_m)| = 0.$$

Carrying out the differentiation of φ and then solving for the group speed give

$$|\mathbf{v}_{\text{group}}(\mathbf{x}; \omega)| = \left|\frac{\Delta\omega}{\nabla\Delta\theta(\mathbf{x}; \omega)}\right|.$$

Substituting the RHS for the group speed in (20.10b), expressing the gradient $\nabla\varphi$ in terms of $\nabla\theta_{\text{h}}$, and simplifying give the group velocity as

$$\mathbf{v}_{\text{group}}\mathbf{x}; \omega = \frac{\pm\nabla(\Delta\theta/\Delta\omega)}{\left|\nabla(\Delta\theta/\Delta\omega)\right|^2}.$$

Imposing (20.10a) resolves the sign ambiguity, giving

$$\mathbf{v}_{\text{group}}(\mathbf{x}; \omega) = \frac{-\nabla(\Delta\theta/\Delta\omega)}{\left|\nabla(\Delta\theta/\Delta\omega)\right|^2}.$$

In the limit as $\Delta\omega \to 0$, this becomes

$$\mathbf{v}_{\text{group}}(\mathbf{x}; \omega) = \frac{-\nabla(\partial\theta/\partial\omega)}{\left|\nabla(\partial\theta/\partial\omega)\right|^2}.$$

Reversing the order of differentiation with respect to the frequency and the space coordinates gives the alternative form

$$\mathbf{v}_{\text{group}}(\mathbf{x}; \omega) = \frac{-\partial/\partial\omega(\nabla\theta)}{\left|\partial/\partial\omega(\nabla\theta)\right|^2}.$$

Using the definition of wave vector (19.60), we can express this compactly as

$$\mathbf{v}_{\text{group}}(\mathbf{x};\omega) = \frac{\partial \mathbf{k}/\partial\omega}{\left|\partial \mathbf{k}/\partial\omega\right|^2}.$$

The *group speed* is the magnitude of the group velocity vector:

$$c_{\text{group}}(\mathbf{x};\omega) = \left|\frac{\partial \mathbf{k}}{\partial\omega}\right|^{-1}.$$

The derivations of expressions for the group velocity and group speed presented above assume that the frequencies of the harmonic waves in the traveling wave packet differ relatively little from one another. That is, they assume that (20.3) is satisfied. If the frequencies differ substantially, then (20.3) is violated and the concept of group velocity is meaningless. We can calculate the relative difference between two frequencies $\omega_{\min}$ and $\omega_{\max}$ as

$$\frac{\Delta\omega}{\bar{\omega}} \equiv \frac{\omega_{\max} - \omega_{\min}}{\left(\omega_{\max} + \omega_{\min}\right)/2} = 2\left(\frac{\omega_{\max} - \omega_{\min}}{\omega_{\max} + \omega_{\min}}\right). \quad (20.11)$$

Exercise 20.1 Group Velocity of Particular Diurnal Wave Packet

Problem: Analysis of groundwater-level fluctuations in a coastal aquifer indicates that the following three tidal constituents contribute approximately 99% of the spectral energy in the diurnal frequency band:
Lunar diurnal (O_1): $\omega_1 \approx 6.76 \times 10^{-5}$rad s^{-1}
Solar diurnal (P_1): $\omega_2 \approx 7.25 \times 10^{-5}$rad s^{-1}
Lunar diurnal (K_1): $\omega_3 \approx 7.29 \times 10^{-5}$rad s^{-1}
Consequently, these three constituents alone might be used to approximate the diurnal band of the hydraulic head spectrum. Define a diurnal wave packet as the superposition of the corresponding three hydraulic head harmonic constituents. Is the concept of group velocity meaningful for such a wave packet?

Solution: Using (20.11) to calculate the relative difference between the highest and lowest constituent frequencies in the packet gives

$$\frac{\Delta\omega}{\bar{\omega}} = 2\left(\frac{\omega_3 - \omega_1}{\omega_3 + \omega_1}\right) \approx 0.075 \ll 1. \quad (20.12)$$

In this case (20.3) is satisfied, so the concept of group velocity is meaningful. □

Exercise 20.2 Group Velocity of a Diurnal/Semidiurnal Wave Packet

Problem: Suppose a wave packet results from the superposition of multiple diurnal and semidiurnal hydraulic head harmonic constituents in the frequency range $\omega_{\min} \leq \omega_m \leq \omega_{\max}$, where the minimum and maximum frequencies correspond to the following tidal constituents:
Larger elliptic diurnal ($2Q_1$): $\omega_{\min} \approx 6.2 \times 10^{-5}$rad s^{-1}
Shallow-water semidiurnal ($2SM_2$): $\omega_{\max} \approx 1.5 \times 10^{-4}$rad s^{-1}
Is the concept of group velocity meaningful for such a wave packet?

Solution: Using (20.11) to calculate the relative difference between the highest and lowest constituent frequencies in the packet yields

$$\frac{\Delta\omega}{\bar{\omega}} = 2\left(\frac{\omega_{\max} - \omega_{\min}}{\omega_{\max} + \omega_{\min}}\right) \approx 0.83 \approx 1. \quad (20.13)$$

In this case (20.3) is violated, so the concept of group velocity is *not* meaningful. □

20.4. LONGITUDINAL AND TRANSVERSE GRADIENTS

Consider a space field $f(\mathbf{x};\omega_m)$ associated with a particular component wave (i.e., n fixed) and the corresponding ray path. Examples of such a field include the wave amplitude ($f = M^{(n)}$), the wave phase ($f = \theta^{(n)}$), and the local wavelength ($f = \lambda^{(n)}$). Let $s^{(n)}$ denote the ray path travel distance (arc length) and let $\hat{\mathbf{s}}^{(n)}$ denote the ray path vector [see equation (19.66)]. We can resolve the gradient of f into two component vectors that are parallel and perpendicular to the corresponding ray path as

$$\nabla f = (\nabla f)_{||s} + (\nabla f)_{\perp s} \quad (20.14)$$

where we define the longitudinal and transverse component vectors, respectively, as

$$(\nabla f)_{||s} \equiv \left(\nabla f \cdot \hat{\mathbf{s}}^{(n)}\right)\hat{\mathbf{s}}^{(n)} \quad (20.15a)$$

$$(\nabla f)_{\perp s} \equiv \nabla f - \left(\nabla f \cdot \hat{\mathbf{s}}^{(n)}\right)\hat{\mathbf{s}}^{(n)}. \quad (20.15b)$$

Then we define the longitudinal rate of change of the field f as

$$\frac{\partial f}{\partial s^{(n)}} = \nabla f \cdot \hat{\mathbf{s}}^{(n)}. \quad (20.16)$$

We use a partial derivative here because generally f also may vary in the direction(s) perpendicular to the ray path (i.e., transversely).

Exercise 20.3 Longitudinal and Transverse Gradients

Problem: Using the definitions (20.15), verify that $(\nabla f)_{||s}$ and $(\nabla f)_{\perp s}$ satisfy (20.14). Also, verify that $(\nabla f)_{||s} \cdot (\nabla f)_{\perp s} = 0$. □

Exercise 20.4 Longitudinal Rate of Change

Problem: Use (20.16) to express $\partial\theta^{(n)}/\partial s^{(n)}$ in terms of $\nabla\theta^{(n)}$.

Solution: Substituting $f = \theta^{(n)}$ and the RHS of (19.66) for $\hat{\mathbf{s}}^{(n)}$ in (20.16) yields

$$\frac{\partial \theta^{(n)}}{\partial s^{(n)}} = -|\nabla \theta^{(n)}|. \tag{20.17}$$

□

20.5. WAVE ATTENUATION

20.5.1. Positive and Negative Attenuation

At a point $\mathbf{x}$ we define the local wave attenuation to be positive or negative according to the criterion

$$\frac{\partial}{\partial s^{(n)}} M^{(n)}(\mathbf{x}; \omega_m) \begin{cases} < 0, & \text{normal (positive) attenuation} \\ > 0, & \text{negative (reverse) attenuation} \end{cases} \tag{20.18}$$

where (20.16) defines the longitudinal rate of change, $\partial/\partial s^{(n)}$. That is, at those points where the wave amplitude function decreases in the direction of wave travel, the wave is said to be *normally (positively) attenuated.* Conversely, at those points where the wave amplitude function increases in the direction of wave travel, the wave is said to be *negatively (reversely) attenuated.*

Exercise 20.5 Normal/Reverse Attenuation in Terms of Gradients

Problem: Define normal and reverse attenuation in terms of the wave amplitude and phase gradients by rewriting (20.18).

Solution: Equation (20.16) gives

$$\frac{\partial M^{(n)}}{\partial s^{(n)}} = \nabla M^{(n)} \cdot \hat{\mathbf{s}}^{(n)}.$$

Substituting the RHS of (19.66) for the ray path vector leads to

$$\frac{\partial M^{(n)}}{\partial s^{(n)}} = \frac{-\nabla M^{(n)} \cdot \nabla \theta^{(n)}}{|\nabla \theta^{(n)}|}.$$

Substituting the RHS for $\partial M^{(n)}/\partial s^{(n)}$ in (20.18) and then simplifying the result give

$$\nabla M^{(n)}(\mathbf{x}; \omega_m) \cdot \nabla \theta^{(n)}(\mathbf{x}; \omega_m) \begin{cases} > 0, & \text{normal (positive) attenuation} \\ < 0, & \text{negative (reverse) attenuation.} \end{cases} \tag{20.19}$$

□

In each of the following 1D cases (see Chapter 10), constituent waves undergo negative attenuation over a finite subinterval:

b-2, b-3, c-2, c-3, d-2, d-3, n-2, n-3, p-2, p-3, q-2, q-3.

This is clearly recognizable from the graphs of the corresponding amplitude and phase functions. Additionally, in each of the following multidimensional flows constituent waves undergo negative attenuation over a finite subregion:

- 3D, radial flow in a spherical shell, Example 12.5 (see Figure 12.3)
- 2D, radial flow in an annulus, Example 12.12 (see Figure 12.6)

Exercise 20.6 Flow in Spherical Shell: Negative Attenuation

Problem: Consider the time-periodic flow in a spherical shell described in Example 12.5. For the case $\eta_m = 5$ describe the subregion in which the constituent wave undergoes negative attenuation in terms of the dimensionless radius R. Use Figure 12.3 to visually estimate the relevant numerical values of R.

Solution: From Figure 12.3, the constituent wave's amplitude and phase gradients vanish at $R \approx 0.36$ and $R \approx 0.50$, respectively. For R values between these two, the amplitude and phase gradients are of opposite sign. Therefore, the region in which the constituent wave undergoes negative attenuation is given by (approximately)

$$0.36 < R < 0.50$$

which is a spherical shell contained wholly within the space domain. □

Exercise 20.7 Flow in Annulus: Negative Attenuation

Problem: Consider the time-periodic flow in an annulus described in Example 12.12. For the case $\eta_m = 5$ describe the subregion in which the constituent wave undergoes negative attenuation in terms of the dimensionless radius R. Use Figure 12.6 to visually estimate the relevant numerical values of R.

Solution: From Figure 12.6, $R \approx 0.44$ and $R \approx 0.55$, respectively, are the values of the dimensionless radius where the constituent wave amplitude and phase gradients vanish. For R values between these two, the amplitude and phase gradients are of opposite sign. Therefore, the region in which the constituent wave undergoes negative attenuation is given by (approximately)

$$0.44 < R < 0.55$$

which is an annulus contained wholly within the space domain. □

20.5.2. Geometric and Nongeometric Attenuation

Conceptually, we can think of wave attenuation as the result of two more or less independent contributions—*nongeometric* (or *intrinsic*) attenuation and *geometric* attenuation. These contributions combine to produce the total (net) attenuation.

Nongeometric attenuation results from mechanisms that are independent of ray path geometry, so it occurs even when the ray paths are straight lines with no geometric spreading. For instance, wave attenuation that occurs in 1D flows is necessarily nongeometric. We expect harmonic, hydraulic head waves propagating in real media to exhibit some degree of nongeometric attenuation.

Geometric attenuation arises purely as a result of ray path geometry, independent of the mechanisms that normally cause wave attenuation in nondiverging ray geometries. At locations where wave ray paths diverge, geometric attenuation is positive so total or net attenuation is greater than nongeometric attenuation alone. Similarly, at locations where ray paths converge, geometric attenuation is negative so total attenuation is less than nongeometric attenuation alone. Equation (20.61) (see Exercise 20.34) provides some support for this view.

One dimensionless measure of the geometric spreading of the ray paths is the *ray path geometric divergence*:

$$\Gamma^{(n)}(\mathbf{x}; \omega_m) \equiv \frac{1}{k^{(n)}} \nabla \cdot \hat{\mathbf{s}}^{(n)}. \tag{20.20}$$

The geometric divergence represents the degree of local *angular* divergence of the ray paths. Positive values indicate ray path spreading; negative values indicate ray path convergence; and a value of zero indicates straight, parallel ray paths, i.e., neither convergence nor divergence. If at any location $\mathbf{x}$ the value of $\Gamma^{(n)}$ is negative, positive, or zero, then we expect the contribution of geometric attenuation to the net attenuation to be negative, positive, or zero, respectively.

Exercise 20.8 Plane Waves: Geometric Divergence and Attenuation

Problem: Consider plane waves in a 3D, semi-infinite domain filled with an ideal medium due to a vertically oriented, single-constituent plane source of infinite extent located at $x_1 = 0$ (see Example 9.3). Suppose the constituent frequency is ω_m (m fixed). What is the geometric divergence of such waves? What can you infer about the contribution of geometric attenuation to the total or net attenuation in this case?

Solution: Combining the expression for the phase gradient obtained in Exercise 20.19 with (19.66) gives $\hat{\mathbf{s}}^{(n)} = \hat{\mathbf{e}}_1$. Therefore $\nabla \cdot \hat{\mathbf{s}}^{(n)} = 0$. Substituting this result for the divergence on the RHS of (20.20) gives $\Gamma^{(n)} = 0$. That is, for plane waves the unit ray path vector is spatially uniform, so the geometric divergence vanishes. Therefore, there is no geometric attenuation in this case. □

Example 20.1 Forced Outer Cylindrical Boundary—Geometric Divergence

Consider a small oceanic island that is circular in outline and underlain by a homogeneous and isotropic (i.e., ideal), confined aquifer of uniform thickness. Due to ocean tides, cylindrical waves are propagated from the aquifer's seaward boundary inward toward the center of the island. Let r denote the radial coordinate, as measured in the horizontal plane with the origin ($r = 0$) at the center of the island and $r = r_{\max}$ at the seaward boundary. Assume the wave number $k^{(n)}(r)$ is finite and positive for $0 < r < r_{\max}$ and varies continuously with r. What can we infer about the geometric divergence of such waves? What does this imply about the contribution of geometric attenuation to the total (net) attenuation?

This problem was discussed in Section 12.4.3. In this case the constituent wave ray path vector is given by $\hat{\mathbf{s}}^{(n)} = -\hat{\mathbf{r}}$, where $\hat{\mathbf{r}}$ is the unit vector pointing radially outward from the origin (see Figure 12.5). Therefore,

$$\nabla \cdot \hat{\mathbf{s}}^{(n)} = \nabla \cdot (-\hat{\mathbf{r}}) = \frac{1}{r}\frac{\partial}{\partial r}(-r) = \frac{-1}{r}, \qquad r > 0.$$

Substituting this result for the divergence on the RHS of (20.20) gives

$$\Gamma^{(n)} = \frac{-1}{k^{(n)} r}, \qquad r > 0.$$

Thus, the geometric divergence is negative in this case. From Figure 12.5 we also infer that $|\nabla\theta^{(n)}|$, and therefore $k^{(n)}(r)$, is an increasing function of r, so $\Gamma^{(n)}$ decreases in absolute value as r increases.

Since $\Gamma^{(n)} < 0$ and decreases in absolute value with increasing r, the contribution of geometric attenuation to the total attenuation is negative and diminishes in magnitude with distance from the center of the island. ◊

Exercise 20.9 Cylinder Waves: Geometric Divergence and Attenuation

Problem: Consider cylindrical waves in a 3D, infinite domain filled with an ideal medium due to a vertically oriented, infinitely long, uniform line source located at $(x_1, x_2, x_3) = (0, 0, x_3)$. Let r denote the radial coordinate, as measured in the horizontal (x_1x_2) plane, with the origin ($r = 0$) located at the line source. What can you infer about the geometric divergence of such waves? What can you infer about the contribution of geometric attenuation to the total or net attenuation in this case?

Solution: This problem was discussed in Section 12.4.2. In this case the constituent wave ray path vector is given by $\hat{\mathbf{s}}^{(n)} = \hat{\mathbf{r}}$, where $\hat{\mathbf{r}}$ is the unit vector pointing radially outward from the origin (see Figure 12.4). Therefore,

$$\nabla \cdot \hat{\mathbf{s}}^{(n)} = \nabla \cdot \hat{\mathbf{r}} = \frac{1}{r}\frac{\partial}{\partial r} r = \frac{1}{r}, \quad r > 0.$$

Substituting this result for the divergence on the RHS of (20.20) gives

$$\Gamma^{(n)} = \frac{1}{k^{(n)} r}, \quad r > 0.$$

Thus, the geometric divergence is positive in this case. From Figure 12.4 we also infer that $|\nabla\theta^{(n)}|$, and therefore $k^{(n)}(r)$, is a decreasing function of r but does not decrease faster than $1/r$. Therefore, $\Gamma^{(n)}$ decreases in absolute value as r increases.

Since $\Gamma^{(n)} > 0$ and decreases with increasing r, the contribution of geometric attenuation to the total attenuation is positive and diminishes with distance from the line source. □

Exercise 20.10 Spherical Waves: Geometric Divergence and Attenuation

Problem: Consider spherical waves in a 3D, infinite domain filled with an ideal medium due to a single-constituent point source at the origin (see Example 12.1). Suppose the constituent frequency is ω_m (m fixed). Express the geometric divergence in terms of α_m and the radial space coordinate r. What can you infer about the contribution of geometric attenuation to the total or net attenuation in this case?

Solution: Combining the expression obtained for the phase gradient in Exercise 20.15 with (19.66) gives $\hat{\mathbf{s}}^{(n)} = \hat{\mathbf{r}}$. Therefore,

$$\nabla \cdot \hat{\mathbf{s}}^{(n)} = \nabla \cdot \hat{\mathbf{r}} = \frac{1}{r^2}\frac{\partial}{\partial r} r^2 = \frac{2}{r}, \quad r > 0.$$

Substituting this result for the divergence on the RHS of (20.20) gives

$$\Gamma^{(n)} = \frac{2}{k^{(n)} r}, \quad r > 0. \tag{20.21}$$

Combining the expression obtained for the phase gradient in Exercise 20.15 with the definition of the wave number (19.61) yields $k^{(n)} = \sqrt{\alpha_m/2}$. Substituting the RHS for the wave number in (20.21) gives

$$\Gamma^{(n)} = \frac{2\sqrt{2}}{\sqrt{\alpha_m} r}, \quad r > 0.$$

Since $\Gamma^{(n)} > 0$ and decreases with increasing r, the contribution of geometric attenuation to the total attenuation is positive and diminishes with distance from the point source. □

20.5.3. Wave Attenuation Measures

In this section we present several local measures of wave attenuation.

20.5.3.1. Dimensionless Wave Attenuation

We define the *longitudinal* and *transverse, dimensionless attenuation rate vectors* as $-(\nabla M^{(n)})_{\parallel s}$ and $-(\nabla M^{(n)})_{\perp s}$, respectively. We also define the *longitudinal, dimensionless attenuation rate* (scalar) as $-\partial M^{(n)}/\partial s^{(n)}$. A positive attenuation rate corresponds to a decrease in the wave amplitude as the wave progresses along its ray path.

20.5.3.2. Relative Wave Attenuation

Let $Y^{(n)}$ denote the logarithm of the wave amplitude:

$$Y^{(n)}(\mathbf{x}; \omega_m) \equiv \ln M^{(n)}(\mathbf{x}; \omega_m).$$

Then we define the *longitudinal* and *transverse, relative attenuation rate vectors* as $-(\nabla Y^{(n)})_{\parallel s}$ and $-(\nabla Y^{(n)})_{\perp s}$, respectively. We also define the *longitudinal, relative attenuation rate* (scalar) as $-\partial Y^{(n)}/\partial s^{(n)}$. It has the dimensions of reciprocal length.

Exercise 20.11 Longitudinal, Relative Attenuation Rate

Problem: Express the longitudinal, relative attenuation rate in terms of the log amplitude and the phase.

Solution: The longitudinal, relative attenuation is given by substituting $f = Y^{(n)}$ in (20.16) and then multiplying by -1:

$$-\frac{\partial Y^{(n)}}{\partial s^{(n)}} = -\nabla Y^{(n)} \cdot \hat{\mathbf{s}}^{(n)}.$$

Using (19.66) to express the unit vector in terms of the phase gradient then gives

$$-\frac{\partial Y^{(n)}}{\partial s^{(n)}} = \frac{\nabla Y^{(n)} \cdot \nabla\theta^{(n)}}{|\nabla\theta^{(n)}|}. \tag{20.22}$$

□

20.5.3.3. Spatial Attenuation Scale

We define the *spatial attenuation scale* (also *attenuation scale*) as

$$\delta^{(n)}(\mathbf{x}; \omega_m) \equiv \frac{1}{|\nabla Y^{(n)}(\mathbf{x}; \omega_m)|}, \quad |\nabla Y^{(n)}(\mathbf{x}; \omega_m)| > 0. \tag{20.23}$$

The local attenuation scale has the dimensions of length. It gives the distance over which the wave amplitude would decrease by a factor of $1/e$ in the direction opposite the amplitude gradient *if the log amplitude gradient was spatially uniform and equal in magnitude to its value at that point* ($\mathbf{x}$). Because of the reciprocal relationship (20.23), greater (lesser) values of the scale indicate lower (higher) rates of relative wave attenuation.

Exercise 20.12 Radial Spherical Flow: Amplitude and Phase Functions

Problem: Consider the solution for radial spherical flow due to a point source in an infinite, 3D domain filled with an ideal medium (see Example 12.1). Give expressions for the corresponding amplitude and phase functions.

Solution: In this case there is only one component wave, so it is identical to the constituent wave, i.e., $h' = h^{(1)}$. The amplitude and phase, respectively, are given by the modulus and argument of the RHS of (12.10):

$$M^{(1)}(\mathbf{x};\omega_m) = \frac{|Q_0(\omega_m)|}{4\pi K r}\exp\left(-\sqrt{\frac{\alpha_m}{2}}r\right) \quad (20.24a)$$

$$\theta^{(1)}(\mathbf{x};\omega_m) = -\sqrt{\frac{\alpha_m}{2}}r \quad (20.24b)$$

$$r > 0.$$

The amplitude and phase functions are undefined at $r = 0$ because the FRF is undefined there. □

Exercise 20.13 Axisymmetric Flow: Amplitude and Phase Functions

Problem: Consider the solution for 2D, axisymmetric flow due to an infinitely long, uniform line source in an infinite, 3D domain filled with an ideal medium (see Section 12.4.2). Express the corresponding amplitude and phase functions, in dimensional form, in terms of the Kelvin modulus and phase functions.

Solution: In this case there is only one component wave, so it is identical to the constituent wave, i.e., $h' = h^{(1)}$. Equations (12.40) give the dimensionless forms of the FRF amplitude and phase functions. Using the definitions (12.38) to rewrite these in dimensional form yields

$$M^{(1)}(\mathbf{x};\omega_m) = \frac{|\Delta Q_0(\omega_m)|}{2\pi K}N_0(\sqrt{\alpha_m}r) \quad (20.25a)$$

$$\theta^{(1)}(\mathbf{x};\omega_m) = \phi_0(\sqrt{\alpha_m}r) + \arg\Delta Q_0(\omega_m) \quad (20.25b)$$

$$r > 0$$

where N_0 and ϕ_0 are the Kelvin modulus and phase functions defined by (E.7). The FRF amplitude and phase functions are undefined at $r = 0$ because the FRF is undefined there. □

Exercise 20.14 Rectangular Flow: Amplitude and Phase Functions

Problem: Consider the solution for rectangular flow due to a uniform plane source of infinite extent in an infinite, 3D domain filled with an ideal medium (see Example 9.3). Give expressions for the corresponding amplitude and phase functions.

Solution: In this case there is only one component wave, so it is identical to the constituent wave, i.e., $h' = h^{(1)}$. The amplitude and phase, respectively, are given by the modulus and argument of the RHS of (9.53):

$$M^{(1)}(\mathbf{x};\omega_m) = \frac{|\Delta Q_0(\omega_m)|}{2K\sqrt{\alpha_m}}\exp\left(-\sqrt{\frac{\alpha_m}{2}}|x|\right) \quad (20.26a)$$

$$\theta^{(1)}(\mathbf{x};\omega_m) = -\frac{\pi}{4}\text{ rad} - \sqrt{\frac{\alpha_m}{2}}|x| \quad (20.26b)$$

$$|x| > 0.$$

The amplitude and phase functions are undefined at $|x| = 0$ because the FRF is undefined there. □

Exercise 20.15 Local Attenuation Scale for Radial Spherical Flow

Problem: Consider the solution for radial spherical flow due to a point source in an infinite, 3D domain filled with an ideal medium (see Example 12.1). Find expressions for the gradients of the amplitude, phase, and log amplitude functions. Also, find an expression for the local attenuation scale and use it to evaluate its near-source ($r \to 0$) and far-field ($r \to \infty$) limiting values.

Solution: In this case there is only one component wave, so it is identical to the constituent wave, i.e., $h' = h^{(1)}$. In spherical coordinates, $\nabla f(r) = (\partial f/\partial r)\hat{\mathbf{r}}$, where $\hat{\mathbf{r}}$ denotes the unit radius vector that points outward from the origin. Differentiating (20.24) then gives

$$\nabla M^{(1)}(\mathbf{x};\omega_m) = -\frac{|Q_0(\omega_m)|}{4\pi K r}\exp\left(-\sqrt{\frac{\alpha_m}{2}}r\right)\left(\frac{1}{r}+\sqrt{\frac{\alpha_m}{2}}\right)\hat{\mathbf{r}} \quad (20.27a)$$

$$\nabla\theta^{(1)}(\mathbf{x};\omega_m) = -\sqrt{\frac{\alpha_m}{2}}\,\hat{\mathbf{r}} \quad (20.27b)$$

$$r > 0.$$

Combining the RHSs of (20.24a) and (20.27a) yields

$$\nabla Y^{(1)}(\mathbf{x};\omega_m) = -\left(\frac{1}{r}+\sqrt{\frac{\alpha_m}{2}}\right)\hat{\mathbf{r}}, \quad r > 0. \quad (20.28)$$

Notice that both $\nabla\theta^{(1)}$ and $\nabla Y^{(1)}$ are independent of the point source amplitude Q_0. Substituting the RHS of (20.28) for $\nabla Y^{(1)}$ in (20.23) leads to

$$\delta^{(1)}(\mathbf{x};\omega_m) = \frac{1}{1/r + \sqrt{\alpha_m/2}}, \quad r > 0.$$

Taking the appropriate limits of the RHS gives the near-source and far-field limits:

$$\lim_{r\to 0}\delta^{(1)}(\mathbf{x};\omega_m) = 0 \quad (20.29a)$$

$$\lim_{r\to\infty}\delta^{(1)}(\mathbf{x};\omega_m) = \sqrt{\frac{2}{\alpha_m}}. \quad (20.29b)$$

□

Exercise 20.16 Axisymmetric Flow: Amplitude and Phase Gradients

Problem: Consider the solution for 2D, axisymmetric flow due to an infinitely long, uniform line source in an infinite, 3D domain filled with an ideal medium (see Section 12.4.2). Express the gradients of the amplitude, phase, and log amplitude functions in terms of the Kelvin modulus and phase functions.

Solution: In this case there is only one component wave, so it is identical to the constituent wave, i.e., $h' = h^{(1)}$. In cylindrical coordinates, $\nabla f(r) = (\partial f/\partial r)\hat{\mathbf{r}}$, where $\hat{\mathbf{r}}$ denotes the unit radius vector that points outward from the origin. Differentiating (20.25) then gives

$$\nabla M^{(1)}(\mathbf{x};\omega_m) = \frac{|\Delta Q_0(\omega_m)|\sqrt{\alpha_m}}{2\pi K}\left[\frac{d}{dR}N_0(R)\right]\bigg|_{R=\sqrt{\alpha_m}r}\hat{\mathbf{r}} \tag{20.30a}$$

$$\nabla\theta^{(1)}(\mathbf{x};\omega_m) = \sqrt{\alpha_m}\left[\frac{d}{dR}\phi_0(R)\right]\bigg|_{R=\sqrt{\alpha_m}r}\hat{\mathbf{r}} \tag{20.30b}$$

$$r > 0.$$

Using the equations from Section 10.68 of *Olver and Maximon* [2016] to differentiate the Kelvin functions on the RHSs leads to

$$\nabla M^{(1)}(\mathbf{x};\omega_m) = \frac{|\Delta Q_0(\omega_m)|\sqrt{\alpha_m}}{2\pi K}N_1(\sqrt{\alpha_m}r) \times \cos\left[\phi_1(\sqrt{\alpha_m}r) - \phi_0(\sqrt{\alpha_m}r) - \frac{\pi}{4}\right]\hat{\mathbf{r}} \tag{20.31a}$$

$$\nabla\theta^{(1)}(\mathbf{x};\omega_m) = \sqrt{\alpha_m}\frac{N_1(\sqrt{\alpha_m}r)}{N_0(\sqrt{\alpha_m}r)} \times \sin\left[\phi_1(\sqrt{\alpha_m}r) - \phi_0(\sqrt{\alpha_m}r) - \frac{\pi}{4}\right]\hat{\mathbf{r}} \tag{20.31b}$$

$$r > 0.$$

□

Exercise 20.17 Local Attenuation Scale for Axisymmetric Flow

Problem: Consider the solution for 2D, axisymmetric flow due to an infinitely long, uniform line source in an infinite, 3D domain filled with an ideal medium (see Section 12.4.2). Find an expression for the local attenuation scale and use it to evaluate its far-field ($r \to \infty$) limiting value.

Solution: Combining the RHSs of (20.25a) and (20.31a) yields

$$\nabla Y^{(1)}(\mathbf{x};\omega_m) = \sqrt{\alpha_m}\frac{N_1(\sqrt{\alpha_m}r)}{N_0(\sqrt{\alpha_m}r)}\cos\left[\phi_1(\sqrt{\alpha_m}r) - \phi_0(\sqrt{\alpha_m}r) - \frac{\pi}{4}\right]\hat{\mathbf{r}}, \quad r > 0. \tag{20.32}$$

Notice that both gradients, $\nabla Y^{(1)}$ and $\nabla\theta^{(1)}$, are independent of the plane source amplitude ΔQ_0. Substituting the RHS of (20.32) for $\nabla Y^{(1)}$ in (20.23) leads to

$$\delta^{(1)}(\mathbf{x};\omega_m) = \frac{N_0(\sqrt{\alpha_m}r)}{\sqrt{\alpha_m}N_1(\sqrt{\alpha_m}r)\left|\cos\left[\phi_1(\sqrt{\alpha_m}r) - \phi_0(\sqrt{\alpha_m}r) - \frac{\pi}{4}\right]\right|}, \quad r > 0. \tag{20.33}$$

To evaluate the far-field limit, use the asymptotic expansions for the large argument presented in Section 10.68(iii) of *Olver and Maximon* [2016] to obtain

$$\lim_{r\to\infty}\delta^{(1)}(\mathbf{x};\omega_m) = \sqrt{\frac{2}{\alpha_m}}.$$

□

Exercise 20.18 Local Attenuation Scale for Axisymmetric Flow

Problem: Consider the solution for 2D, axisymmetric flow due to an infinitely long, uniform line source in an infinite, 3D domain filled with an ideal medium (see Section 12.4.2). Equation (20.33) expresses the local attenuation scale in terms of Kelvin modulus and phase functions. Find the near-source ($r \to 0$) limiting value of the local attenuation scale.

Solution: We can write (20.33) in abbreviated form as

$$\delta^{(1)}(\mathbf{x};\omega_m) = \frac{N_0(R)}{\sqrt{\alpha_m}N_1(R)\left|\cos\left[\phi_1(R) - \phi_0(R) - \pi/4\right]\right|}, \quad R > 0 \tag{20.34}$$

where $R \equiv \sqrt{\alpha_m}r$. Equation (E.14) leads to

$$\cos\left[\phi_1(0) - \phi_0(0) - \pi/4\right] = -1. \tag{20.35}$$

From the definitions of the k-type Kelvin modulus functions (e.g., see Appendix E) we have

$$\frac{N_0(R)}{N_1(R)} = \left|\frac{K_0(z)}{K_1(z)}\right|$$

where $z \equiv e^{i\pi/4}R$. Consequently,

$$\lim_{R\to 0}\frac{N_0(R)}{N_1(R)} = \lim_{z\to 0}\left|\frac{K_0(z)}{K_1(z)}\right|.$$

Using Equations 10.30.2 and 10.30.3 of *Olver and Maximon* [2016] to evaluate the RHS gives

$$\lim_{R\to 0}\frac{N_0(R)}{N_1(R)} = 0. \tag{20.36}$$

Using the results in (20.35) and (20.36) to evaluate the limit of the RHS of (20.34) yields

$$\lim_{r \to 0} \delta^{(1)}(\mathbf{x}; \omega_m) = 0, \quad r > 0.$$

□

Exercise 20.19 Local Attenuation Scale for Rectangular Flow

Problem: Consider the solution for rectangular flow due to a uniform plane source of infinite extent in an infinite, 3D domain filled with an ideal medium (see Example 9.3). Find expressions for the gradients of the amplitude, phase, and log amplitude functions and the local attenuation scale.

Solution: In this case there is only one component wave, so it is identical to the constituent wave, i.e., $h' = h^{(1)}$. Differentiating (20.26) and noting that $\nabla|x| = \mathrm{sign}(x)\hat{\mathbf{x}}$ (valid for $|x| > 0$) yield

$$\nabla M^{(1)}(\mathbf{x}; \omega_m) = -\mathrm{sign}(x)\frac{|\Delta Q_0(\omega_m)|}{2\sqrt{2}K}\exp\left(-\sqrt{\frac{\alpha_m}{2}}|x|\right)\hat{\mathbf{x}} \tag{20.37a}$$

$$\nabla\theta^{(1)}(\mathbf{x}; \omega_m) = -\mathrm{sign}(x)\sqrt{\frac{\alpha_m}{2}}\hat{\mathbf{x}} \tag{20.37b}$$

$$|x| > 0.$$

Combining the RHSs of (20.26a) and (20.37a) yields

$$\nabla Y^{(1)}(\mathbf{x}; \omega_m) = -\mathrm{sign}(x)\sqrt{\frac{\alpha_m}{2}}\hat{\mathbf{x}}, \quad |x| > 0. \tag{20.38}$$

Notice that $\nabla Y^{(1)} = \nabla\theta^{(1)}$, and both gradients are independent of the plane source amplitude ΔQ_0. Substituting the RHS of (20.38) for $\nabla Y^{(1)}$ in (20.23) leads to

$$\delta^{(1)}(\mathbf{x}; \omega_m) = \sqrt{\frac{2}{\alpha_m}}, \quad |x| > 0.$$

Notice that the local attenuation scale is space invariant in this case. □

20.5.3.4. Penetration Depth

Consider time-periodic flow due to boundary forcing with a single component wave in a source-free porous medium. Suppose that for some point on the boundary the component wave's amplitude and phase gradients are perpendicular to the boundary and point outward, i.e.,

$$\frac{\nabla M^{(n)}}{|\nabla M^{(n)}|} = \frac{\nabla\theta^{(n)}}{|\nabla\theta^{(n)}|} = \hat{\mathbf{n}}(\mathbf{x}) \quad \text{for some point } \mathbf{x} \in \Gamma$$

where $\hat{\mathbf{n}}$ is the unit, outward normal vector to the boundary (see Section 2.1). That is, the component wave penetrates the medium in the direction perpendicular to the boundary, the wave is attenuated as it penetrates the medium, and the direction of maximum wave attenuation is perpendicular to the boundary. Then the *penetration depth*, $L^{(n)}$, if it exists, is the distance over which the component wave amplitude decreases to $1/e$ of its value at the boundary. It satisfies

$$\frac{M^{(n)}(\mathbf{x} - L^{(n)}\hat{\mathbf{n}}; \omega_m)}{M^{(n)}(\mathbf{x}; \omega_m)} = \frac{1}{e} \quad \left(\mathbf{x} \in \Gamma,\ \mathbf{x} - L^{(n)}\hat{\mathbf{n}} \in D\right). \tag{20.39}$$

The penetration depth has the dimensions of length and depends on position (on the boundary) and constituent frequency, i.e., $L^{(n)} = L^{(n)}(\mathbf{x}; \omega_m)$.

Example 20.2 Penetration Depths for 1D Flow in Finite Medium

Consider time-periodic flow in a 1D, finite medium. Let $n = 1$ and $n = 2$ denote component waves that travel in the directions of decreasing and increasing x, respectively (e.g., as in Example 19.1). Rewrite the penetration depth condition (20.39) specifically for this problem geometry, once for each of the two component waves.

The space domain and its boundary, respectively, are given by

$$D \equiv \{x : 0 < x < l\}$$

$$\Gamma \equiv \{x : x = 0 \text{ or } x = l\}.$$

The component waves with $n = 1$ and $n = 2$ enter the medium at the boundary points with coordinates $x = l$ and $x = 0$, respectively. The outward unit normal vectors at the two boundaries are

$$\hat{\mathbf{n}}(0) = -\hat{\mathbf{x}} \tag{20.40a}$$

$$\hat{\mathbf{n}}(l) = +\hat{\mathbf{x}} \tag{20.40b}$$

where $\hat{\mathbf{x}}$ is the unit vector pointing in the direction of increasing x. Combining each of these with (20.39) yields

$$\frac{M^{(1)}(l - L^{(1)}; \omega_m)}{M^{(1)}(l; \omega_m)} = \frac{1}{e} \tag{20.41a}$$

$$\frac{M^{(2)}(L^{(2)}; \omega_m)}{M^{(2)}(0; \omega_m)} = \frac{1}{e} \tag{20.41b}$$

where the penetration depths, if they exist, satisfy

$$0 < L^{(1)}(0; \omega_m) < l \tag{20.42a}$$

$$0 < L^{(2)}(l; \omega_m) < l. \tag{20.42b}$$

◊

Exercise 20.20 Penetration Depth for 1D Flow in Ideal Medium

Problem: Consider time-periodic, 1D flow through a semi-infinite, source-free domain filled with an ideal medium as described by the Jacob-Ferris solution (see Section 9.4.1). In this case there is only one component wave, and it travels in the direction of increasing x. Let $L^{(1)}$ denote its penetration depth. Express $L^{(1)}$ compactly in terms of α_m and then in terms of K, S_s, and ω_m. How does $L^{(1)}$ depend on the amplitude at the boundary, i.e., on $M^{(1)}(0; \omega_m)$?

Solution: In this case we define the space domain and its boundary by

$$D \equiv \{x : x > 0\} \tag{20.43a}$$

$$\Gamma \equiv \{x : x = 0\} \tag{20.43b}$$

and the outward unit normal vector is $\hat{\mathbf{n}}(0) = -\hat{\mathbf{x}}$. Using these in (20.39) gives

$$\frac{M^{(1)}(L^{(1)};\omega_m)}{M^{(1)}(0;\omega_m)} = \frac{1}{e} \tag{20.44}$$

provided $L^{(1)}$ exists. Combining (9.48a) and (9.47b) gives, for the amplitude function,

$$M^{(1)}(x;\omega_m) = M^{(1)}(0;\omega_m)\exp\left(-\sqrt{\frac{\alpha_m}{2}}x\right).$$

Substituting $x = L^{(1)}$ and then substituting the RHS for $M^{(1)}(L^{(1)};\omega_m)$ in (20.44) and finally solving for $L^{(1)}$ yield

$$L^{(1)} = \sqrt{\frac{2}{\alpha_m}} = \sqrt{\frac{2K}{\omega_m S_s}} \tag{20.45}$$

where we have used (5.31). Therefore, the penetration depth is independent of $M^{(1)}(0;\omega_m)$. □

Exercise 20.21 Penetration Depth for 1D Flow in Exponential Medium

Problem: Consider time-periodic, 1D flow through a semi-infinite, source-free domain filled with an exponential medium. In this case there is only one component wave, and it travels in the direction of increasing x. Let $L^{(1)}$ denote its penetration depth. Express $L^{(1)}$ compactly in terms of α_m and μ. How does $L^{(1)}$ depend on the amplitude at the boundary, i.e., on $M^{(1)}(0;\omega_m)$? Are your results consistent with those obtained in Exercise 20.20?

Solution: In this case we define the space domain and its boundary by

$$D \equiv \{x : x > 0\} \tag{20.46a}$$

$$\Gamma \equiv \{x : x = 0\} \tag{20.46b}$$

and the outward unit normal vector is $\hat{\mathbf{n}}(0) = -\hat{\mathbf{x}}$. Using these in (20.39) gives

$$\frac{M^{(1)}(L^{(1)};\omega_m)}{M^{(1)}(0;\omega_m)} = \frac{1}{e} \tag{20.47}$$

provided $L^{(1)}$ exists. Combining (10.80a), (10.76b), and (10.76c) gives, for the amplitude function,

$$M^{(1)}(x;\omega_m) = M^{(1)}(0;\omega_m) \times \exp\left\{-\left[\frac{\mu}{2} + \sqrt{\frac{\sqrt{\mu^4 + 16\alpha_m^2} + \mu^2}{8}}\right]x\right\}.$$

Substituting $x = L^{(1)}$ and then substituting the RHS for $M^{(1)}(L^{(1)};\omega_m)$ in (20.47) and finally solving for $L^{(1)}$ yield

$$L^{(1)} = \left[\frac{\mu}{2} + \sqrt{\frac{\sqrt{\mu^4 + 16\alpha_m^2} + \mu^2}{8}}\right]^{-1}. \tag{20.48}$$

The penetration depth is independent of $M^{(1)}(0;\omega_m)$.

Consider the special case where the medium is homogeneous. Substituting $\mu = 0$ on the RHS of (20.48) yields (20.45), so this result is consistent with that obtained in Exercise 20.20. □

Exercise 20.22 Penetration Depth Approximation

Problem: Consider time-periodic, 1D flow through a semi-infinite, source-free domain filled with an exponential medium. Using (20.48), derive an approximate expression for $L^{(1)}$ valid for the special case where

$$\mu \gg 2\sqrt{\alpha_m}. \tag{20.49}$$

How does $L^{(1)}$ depend on ω_m in this case?

Solution: Raising both sides of inequality (20.49) to the fourth power yields

$$\mu^4 \gg 16\alpha_m^2$$

which suggests the approximation

$$\sqrt{\mu^4 + 16\alpha_m^2} \approx \mu^2.$$

Substituting the RHS for $\sqrt{\mu^4 + 16\alpha_m^2}$ in (20.48) and then simplifying lead to

$$L^{(1)} \approx \mu^{-1} \quad \text{for } \mu \gg 2\sqrt{\alpha_m} > 0.$$

Under these conditions $L^{(1)}$ is approximately independent of ω_m. □

Exercise 20.23 Comparison: Penetration Depths for 1D Component Waves

Problem: Consider time-periodic, 1D flow through a finite, source-free domain filled with an ideal medium. In this case there are two component waves, one traveling in the direction of decreasing x and the other traveling in the direction of increasing x. Denote these with $n = 1$ and $n = 2$, respectively, as in Example 19.1. Express the penetration depths $L^{(1)}$ and $L^{(2)}$ compactly in terms of α_m and l. How does $L^{(1)}$ depend on the amplitude of the component wave at the boundary where it enters the medium, i.e., on $M^{(1)}(l;\omega_m)$? How does $L^{(2)}$ depend on $M^{(2)}(0;\omega_m)$? How do your results compare with the penetration depth of a component wave in a 1D, homogeneous, semi-infinite medium (see Exercise 20.20)?

Solution: The results of Example 20.2 apply directly here, so the penetration depths satisfy (20.41). The component wave amplitude functions are given by (19.23a), where the eigenvalues ($\lambda_m^{(n)}$) are given by (10.11) with $\mu = 0$. Using these to solve (20.41) for $L^{(1)}$ and $L^{(2)}$ yields

$$L^{(1)}(l;\omega_m) = L^{(2)}(0;\omega_m) = \sqrt{\frac{2}{\alpha_m}}.$$

The penetration depth is independent of the component wave amplitudes at the boundaries. Also, each component wave's penetration depth is the same as that for a semi-infinite medium [see equation (20.45)]. □

As Exercises 20.19–20.23 demonstrate, basic theory implies that the local attenuation scale and the penetration depth are decreasing functions of the constituent frequency. This has important consequences for time-periodic hydraulic testing of nonhomogeneous materials, both in the field and in the laboratory. The basic ideas is this: If the relationship between local attenuation scale and constituent frequency is known, then by controlling the frequency one might exert some control over the volume of material substantially affected by a hydraulic test [*Rigord et al.*, 1993; *Song and Renner*, 2006, 2007].

Example 20.3 Penetration Depth for Radial Spherical Flow
Consider 3D, radial flow through a source-free, spherical domain filled with an ideal medium due to time-periodic forcing at the outer boundary (see Exercise 12.6). In this case there is only one component wave, which we denote with $n = 1$, and it travels into the medium from the outer boundary toward the center. Is it possible to express the penetration depth $L^{(1)}$ compactly in terms of α_m and $r_{\max}$ in this case?

From (12.24a) we deduce that the amplitude function is

$$M^{(1)}(r;\omega_m) = M^{(1)}(r_{\max};\omega_m)\frac{\sqrt{\cosh(\sqrt{2}\eta_m R) - \cos(\sqrt{2}\eta_m R)}}{R\sqrt{\cosh(\sqrt{2}\eta_m) - \cos(\sqrt{2}\eta_m)}}$$

where (12.20b) and (12.20c), respectively, define the dimensionless radius (R) and dimensionless frequency (η_m). Using the expression on the RHS to evaluate (20.39) leads to the transcendental equation

$$\cosh y - \cos y - cy^2 = 0 \qquad (20.50)$$

where we define the dimensionless variable y and the dimensionless constant c, respectively, as

$$y \equiv \sqrt{2}\eta_m\left(1 - \frac{L^{(1)}}{r_{\max}}\right) \qquad (20.51a)$$

$$c \equiv \frac{\cosh(\sqrt{2}\eta_m) - \cos(\sqrt{2}\eta_m)}{2e^2\eta_m^2}. \qquad (20.51b)$$

Apparently then, it is impossible to obtain a closed-form solution for the penetration depth in this problem. ◊

Exercise 20.24 Spherical Radial Flow: Estimation of Penetration Depth

Problem: Consider 3D, radial flow through a source-free, spherical domain filled with an ideal medium due to time-periodic forcing at the outer boundary. In Example 20.3 we found that it is impossible to obtain a closed-form solution for the penetration depth in this problem. Use the graphs in Figure 12.2 to visually estimate the dimensionless penetration depth, $L^{(1)}/r_{\max}$, for the cases (a) $\eta_m = 5$ and (b) $\eta_m = 10$. Comparing your results for the two cases, how does the dimensionless penetration depth vary with dimensionless frequency?

Solution: Use the dimensionless amplitude graphs in Figure 12.2 and note that $e^{-1} \approx 0.368$.
Case (a) $\eta_m = 5$ The value $|G_D| = 0.368$ gives $R \approx 0.53$, so $L^{(1)}/r_{\max} \approx 1 - 0.53 = 0.47$.
Case (b) $\eta_m = 10$ The value $|G_D| = 0.368$ gives $R \approx 0.83$, so $L^{(1)}/r_{\max} \approx 1 - 0.83 = 0.17$.
The dimensionless penetration depth is a decreasing function of the dimensionless frequency. □

Exercise 20.25 Axisymmetric Flow: Estimation of Penetration Depth

Problem: Consider 2D, axisymmetric flow in a source-free, cylindrical domain filled with an ideal medium due to time-periodic forcing at the outer boundary. Use the graphs in Figure 12.5 to visually estimate the dimensionless penetration depth, $L^{(1)}/r_{\max}$, for the cases (a) $\eta_m = 5$ and (b) $\eta_m = 10$. Comparing your results for the two cases, how does the dimensionless penetration depth vary with dimensionless frequency?

Solution: Use the dimensionless amplitude graphs in Figure 12.5 and note that $e^{-1} \approx 0.368$.
Case (a) $\eta_m = 5$ The value $|G_D| = 0.368$ gives $R \approx 0.66$, so $L^{(1)}/r_{\max} \approx 1 - 0.66 = 0.34$.
Case (b) $\eta_m = 10$ The value $|G_D| = 0.368$ gives $R \approx 0.85$, so $L^{(1)}/r_{\max} \approx 1 - 0.85 = 0.15$.

The dimensionless penetration depth is a decreasing function of the dimensionless frequency. □

20.5.3.5. Phase-Relative Wave Attenuation

We can formulate a dimensionless, local measure of relative, longitudinal attenuation that compensates for a spatially variable phase gradient. We call it the *phase-relative wave attenuation*:

$$\frac{\partial Y^{(n)}}{\partial \theta^{(n)}} = \left(\frac{\partial Y^{(n)}}{\partial s^{(n)}}\right)\frac{ds^{(n)}}{d\theta^{(n)}} = \frac{\left(\partial Y^{(n)}/\partial s^{(n)}\right)}{\left(d\theta^{(n)}/ds^{(n)}\right)}. \tag{20.52}$$

Here, $\partial Y^{(n)}/\partial \theta^{(n)}$ measures the relative rate of change of the amplitude per change of phase along the ray path. The ray path is opposite the phase gradient [see equation (19.69)], so

$$\frac{d\theta^{(n)}}{ds^{(n)}} = -\left|\nabla \theta^{(n)}\right|.$$

Substituting the RHS for the denominator in (20.52) gives

$$\frac{\partial Y^{(n)}}{\partial \theta^{(n)}} = \frac{-1}{\left|\nabla \theta^{(n)}\right|}\frac{\partial Y^{(n)}}{\partial s^{(n)}} \tag{20.53}$$

which shows that the phase-relative spatial attenuation and the longitudinal, relative attenuation have the same sign.

To express $\partial Y^{(n)}/\partial \theta^{(n)}$ solely in terms of the log amplitude and phase gradients, substitute the RHS of (20.22) for $-\partial Y^{(n)}/\partial s^{(n)}$ to obtain

$$\frac{\partial Y^{(n)}}{\partial \theta^{(n)}} = \frac{\nabla Y^{(n)} \cdot \nabla \theta^{(n)}}{\left|\nabla \theta^{(n)}\right|^2}. \tag{20.54}$$

Exercise 20.26 Plane Waves: Longitudinal and Phase-Relative Attenuation

Problem: Consider plane waves in a 3D, infinite domain filled with an ideal medium due to a vertically oriented, single-constituent plane source of infinite extent located at $x = 0$ (see Example 9.3). Consider a single component wave, i.e., $h' \equiv h^{(n)}$ (n fixed). Find compact expressions for the longitudinal, relative attenuation and the phase-relative spatial attenuation.

Solution: Combining the phase gradient expression from Exercise 20.19 with (19.66) gives $\hat{\mathbf{s}}^{(n)} = \hat{\mathbf{x}}$, so $s^{(n)} = x$. Combining this result with those from Exercise 20.11 yields

$$-\frac{\partial Y^{(n)}}{\partial s^{(n)}}(\mathbf{x};\omega_m) = \sqrt{\frac{\alpha_m}{2}}, \quad x > 0.$$

Combining (20.37b) and (20.38) gives $\nabla Y^{(n)} = \nabla \theta^{(n)}$. Substituting this result in (20.54) leads to

$$\frac{\partial Y^{(n)}}{\partial \theta^{(n)}} = 1, \quad x > 0.$$

Notice that the RHS is independent of the space coordinates, the frequency, and the material hydrologic properties. □

Exercise 20.27 Spherical Waves: Longitudinal and Phase-Relative Attenuation

Problem: Consider spherical waves in a 3D, infinite domain filled with an ideal medium due to a single-constituent point source located at the origin (see Example 12.1). Consider a single component wave, i.e., $h' = h^{(n)}$ (n fixed). Find compact expressions for the longitudinal, relative attenuation and the phase-relative spatial attenuation. Give low-order approximations for both, valid in the far field (i.e., distant from the point source, where $r \gg 1/\sqrt{\alpha_m}$).

Solution: Combining the phase gradient expression from Exercise 20.15 with (19.66) gives $\hat{\mathbf{s}}^{(n)} = \hat{\mathbf{r}}$. Therefore $s^{(n)} = r$. Combining this result with the results from Exercise 20.11 yields

$$-\frac{\partial Y^{(n)}}{\partial s^{(n)}}(\mathbf{x};\omega_m) = \sqrt{\frac{\alpha_m}{2}} + \frac{1}{r}, \quad r > 0.$$

Substituting the expressions for the gradients of the log amplitude and phase (see Exercise 20.15) for the corresponding terms of (20.54) gives

$$\frac{\partial Y^{(n)}}{\partial \theta^{(n)}}(\mathbf{x};\omega_m) = 1 + \sqrt{\frac{2}{\alpha_m}}\frac{1}{r}, \quad r > 0.$$

In the far-field limit we have, respectively, the approximations:

$$-\left.\frac{\partial Y^{(n)}}{\partial s^{(n)}}\right|_{r \gg 1/\sqrt{\alpha_m}} \approx \lim_{r\to\infty}\left(-\frac{\partial Y^{(n)}}{\partial s^{(n)}}\right) = \sqrt{\frac{\alpha_m}{2}}$$

$$\left.\frac{\partial Y^{(n)}}{\partial \theta^{(n)}}\right|_{r \gg 1/\sqrt{\alpha_m}} \approx \lim_{r\to\infty}\frac{\partial Y^{(n)}}{\partial \theta^{(n)}} = 1.$$

□

Exercise 20.27 shows that for spherical waves due to a single-constituent point source in a domain filled with an ideal medium the longitudinal, relative attenuation is bounded below but not bounded above. The lower bound, which corresponds to the far-field limit, equals the value for uniform plane waves (see Exercise 20.26). For these spherical waves the phase-relative spatial attenuation exceeds that of uniform plane waves by the factor $\sqrt{2}/(\sqrt{\alpha_m}r)$.

In Exercises 20.26 and 20.27, the spatial attenuation vector and the ray path vector were parallel throughout the space domain. More generally, the spatial attenuation vector and the ray path vector may be nonparallel locally (e.g., see Example 20.4); this suggests that under some circumstances it may be possible to have $\partial Y^{(n)}/\partial \theta^{(n)} < 1$ locally.

Example 20.4 Wave Propagation Not Parallel to Attenuation Direction
Consider an infinite, source-free domain filled with an ideal medium in which plane waves propagate in a direction other than the direction of maximum spatial attenuation. Suppose the hydraulic head FRF is a uniform-gradient function, as described in Chapter 13 (also see Section 13.8). In this case we can derive rough bounds on $\partial Y^{(n)}/\partial\theta^{(n)}$.

Substituting the RHS of (13.71b) for the numerator on the RHS of (20.54) leads to

$$\frac{\partial Y^{(n)}}{\partial\theta^{(n)}} = \frac{\alpha_m}{2\left|\nabla\theta^{(n)}\right|^2}.$$

Comparing the RHS to (13.72), we obtain

$$\frac{\partial Y^{(n)}}{\partial\theta^{(n)}} = \cos\epsilon_m$$

where ϵ_m is the positive angle between $\nabla Y^{(n)}$ and $\nabla\theta^{(n)}$. By assumption the gradients are nonparallel, so the RHS is less than unity. Also, (13.71b) indicates that $\cos\epsilon_m > 0$. Therefore,

$$0 < \frac{\partial Y^{(n)}}{\partial\theta^{(n)}} < 1.$$

Note that $\partial Y^{(n)}/\partial\theta^{(n)}$ is space invariant in this case. ◊

20.5.4. Attenuation Distortion

Earlier in this section we discussed how attenuation changes the shapes of harmonic traveling waves, both constituent waves and component waves. The spatial attenuation rate generally increases with frequency, so as a wave packet travels through a porous medium its higher-frequency constituents are more highly attenuated than its lower-frequency constituents. Consequently, after some travel distance the relative proportions of the various constituent amplitudes change, thus changing the shape of the wave packet. We call this type of distortion, which is conceptually distinct from frequency dispersion, *attenuation distortion*. Like the concept of wave dispersion, the concept of attenuation distortion is meaningful only in the context of a traveling wave packet.

20.6. WAVE DILATION/CONTRACTION

Generally harmonic, hydraulic head waves undergo dilation (lengthening) and/or contraction (shortening) as they propagate through a porous medium. By this we mean that the wavelength increases or decreases as the wave progresses along its path. An important exception is the special case where the phase gradient is spatially uniform; then no dilation or contraction occurs.

Consider the unit-amplitude component wave field:

$$V^{(n)}(\mathbf{x}, t; \omega_m) = \cos\left[\omega_m t + \theta^{(n)}(\mathbf{x}; \omega_m)\right]. \tag{20.55}$$

Suppose we fix the constituent frequency and the time and then plot the variation of $V^{(n)}$ with travel distance $s^{(n)}$ along one of the corresponding ray paths. Generally we would get a distorted sinusoidal curve with unit amplitude. The curve would be distorted in the sense that at some locations it would be stretched or shrunken relative to others.

A simple, dimensionless measure of the local stretching of the wave with respect to its progress along the ray path is the *longitudinal wave dilation*:

$$\Lambda^{(n)}(\mathbf{x}; \omega_m) \equiv \frac{1}{2\pi}\frac{\partial\lambda^{(n)}}{\partial s^{(n)}} \tag{20.56}$$

where $\lambda^{(n)}$ is the local wavelength (19.64) and $s^{(n)}$ is the wave's travel distance. Positive and negative values of $\Lambda^{(n)}$ correspond to dilation and contraction, respectively. Substituting the RHS of (19.64) for the wavelength in (20.56) expresses the wave dilation in terms of wave number rather than local wavelength:

$$\Lambda^{(n)}(\mathbf{x}; \omega_m) = \frac{-1}{(k^{(n)})^2}\frac{\partial k^{(n)}}{\partial s^{(n)}}. \tag{20.57}$$

Exercise 20.28 Plane Waves in Ideal Media: Wave Dilation

Problem: Consider plane waves in a 3D, semi-infinite domain filled with an ideal medium due to a vertically oriented, single-constituent plane source of infinite extent located at $x = 0$ (see Example 9.3). Evaluate the wave dilation.

Solution: Combining the expression obtained for the phase gradient in Exercise 20.19 with the definition of the wave number (19.61) yields

$$k^{(n)} = \sqrt{\frac{\alpha_m}{2}}. \tag{20.58}$$

Substituting the RHS for $k^{(n)}$ in (20.57) leads to $\Lambda^{(n)} = 0$. Thus, in this particular case the wave dilation is zero. □

Exercise 20.29 Plane Waves in Exponential Media: Wave Dilation

Problem: Consider time-periodic, 1D flow through a semi-infinite domain filled with an exponential medium, so that the FRF solution described in Section 10.4 is valid. Consider a single constituent wave whose frequency is ω_m (m fixed). Evaluate the wave dilation.

Solution: Equation (10.80b) indicates that the phase varies linearly with the space coordinate, so the phase gradient is uniform. It follows that the wave number is spatially uniform. Therefore, the wave dilation is zero. □

Exercise 20.30 Spherical Waves in Ideal Media: Wave Dilation

Problem: Consider spherical waves in a 3D, infinite domain filled with an ideal medium due to a single-constituent point source located at $\mathbf{x} = \mathbf{0}$ (see Example 12.1). Evaluate the wave dilation.

Solution: Combining the expression obtained for the phase gradient in Exercise 20.15 with the definition of the wave number (19.61) yields $k^{(n)} = \sqrt{\alpha_m}$. Substituting the RHS for $k^{(n)}$ in (20.57) leads to $\Lambda^{(n)} = 0$. That is, in the case of spherical waves propagating from a point source in a ideal medium, the wave dilation is zero. □

Exercise 20.31 Wave Dilation in Terms of Propagation Speed

Problem: Consider time-periodic flow in a porous medium with a single harmonic constituent ω_m (m fixed). Express the wave dilation in terms of the variation of propagation speed along the ray path. For a given constituent frequency, does the propagation speed increase or decrease in regions where the wave undergoes dilation?

Solution: Solving (19.62b) for the wave number gives

$$k^{(n)}(\mathbf{x};\omega_m) = \frac{\omega_m}{c^{(n)}(\mathbf{x};\omega_m)}.$$

Substituting the RHS for $k^{(n)}$ in (20.57) and then simplifying the result yield

$$\Lambda^{(n)}(\mathbf{x};\omega_m) = \frac{1}{\omega_m}\frac{\partial c^{(n)}}{\partial s^{(n)}}. \tag{20.59}$$

In regions where the wave undergoes dilation the propagation speed increases, and where the wave contracts the propagation speed decreases. That way the frequency of the hydraulic head variation remains constant. □

Exercise 20.32 Wave Dilation and Relative Convective Dilation

Problem: Express the wave dilation compactly in terms of ω_m, $\lambda^{(n)}$ [see equation (19.64)], and $D\lambda^{(n)}/Dt$ (with $\mathbf{v}^{(n)}$ equal to the phase velocity). Interpret your result.

Solution: Substituting $\hat{\mathbf{s}}^{(n)} \cdot \nabla\lambda^{(n)}$ for $\partial\lambda^{(n)}/\partial s^{(n)}$ on the RHS of (20.56) gives

$$\begin{aligned}\Lambda^{(n)}(\mathbf{x};\omega_m) &= \frac{1}{2\pi}\hat{\mathbf{s}}^{(n)} \cdot \nabla\lambda^{(n)} \\ &= \frac{1}{2\pi}\left(c^{(n)}\right)^{-1}\frac{D\lambda^{(n)}}{Dt} \\ &= \frac{1}{2\pi}\left(\frac{2\pi}{\omega_m\lambda^{(n)}}\right)\frac{D\lambda^{(n)}}{Dt} \\ &= \frac{1}{\omega_m\lambda^{(n)}}\frac{D\lambda^{(n)}}{Dt}.\end{aligned}$$

Thus, the wave dilation is directly proportional (proportionality constant ω_m^{-1}) to the relative convective dilation, which is the relative rate of dilation of a traveling wave as observed in a frame that moves with the wavefront. □

Exercise 20.33 Wave Distortion: An Identity

Problem: Consider a wave-based interpretation of the Laplacian of the component wave phase function ($\theta^{(n)}$). Show that the Laplacian of the phase function is proportional to the difference between the wave dilation and the ray path geometric divergence:

$$\nabla^2\theta^{(n)} = \left(k^{(n)}\right)^2\left[\Lambda^{(n)}(\mathbf{x};\omega_m) - \Gamma^{(n)}(\mathbf{x};\omega_m)\right]. \tag{20.60}$$

Solution: Begin by writing the Laplacian of the phase function in terms of the wave vector:

$$\begin{aligned}\nabla^2\theta^{(n)} &= -\nabla \cdot (-\nabla\theta^{(n)}) \\ &= -\nabla \cdot \mathbf{k}^{(n)} \\ &= -\nabla \cdot \left(k^{(n)}\hat{\mathbf{s}}^{(n)}\right) \\ &= -\nabla k^{(n)} \cdot \hat{\mathbf{s}}^{(n)} - k^{(n)}\nabla \cdot \hat{\mathbf{s}}^{(n)} \\ &= -\frac{\partial k^{(n)}}{\partial s^{(n)}} - k^{(n)}\nabla \cdot \hat{\mathbf{s}}^{(n)}.\end{aligned}$$

Factoring $(k^{(n)})^2$ from the RHS yields

$$\nabla^2\theta^{(n)} = (k^{(n)})^2\left[-\frac{1}{(k^{(n)})^2}\frac{\partial k^{(n)}}{\partial s^{(n)}} - \frac{1}{k^{(n)}}\nabla \cdot \hat{\mathbf{s}}^{(n)}\right].$$

Using (20.57) and (20.20), respectively, to rewrite the first and second terms within the brackets on the RHS gives (20.60). □

Exercise 20.34 Wave Distortion in Ideal Media: An Identity

Problem: Consider the propagation of harmonic, hydraulic head waves due to time-periodic flow through an ideal medium. Use (20.60) and the polar form governing equation (6.11b) to show that in this case the phase-relative attenuation is related to the ray path geometric divergence, the wave dilation, the wave number, and α_m as

$$\frac{\partial Y^{(n)}}{\partial\theta^{(n)}} = \frac{1}{2}\left(\Gamma^{(n)} - \Lambda^{(n)}\right) + \frac{\alpha_m}{2(k^{(n)})^2}. \tag{20.61}$$

Solution: Substituting $M_h = M^{(n)}$, $\theta_h = \theta^{(n)}$, $\mathbf{K} = K\mathbf{I}$ with $\nabla K = \mathbf{0}$, and $S_s(\mathbf{x}) = S_s$ with $\nabla S_s = \mathbf{0}$ in the polar form governing equation (6.11b) and then simplifying the result yield

$$2M^{(n)}\nabla M^{(n)} \cdot \nabla\theta + (M^{(n)})^2\nabla^2\theta^{(n)} - \alpha_m(M^{(n)})^2 = 0$$

where we have used the definition (5.31). Dividing by $(M^{(n)})^2$ and writing the amplitude in terms of the log amplitude, and substituting the RHS of (20.60) for $\nabla^2\theta^{(n)}$ give

$$2\nabla Y^{(n)} \cdot \nabla\theta^{(n)} + (k^{(n)})^2\left(\Lambda^{(n)} - \Gamma^{(n)}\right) - \alpha_m = 0.$$

Dividing by $-2(k^{(n)})^2$ and using (20.54) to express the scalar product of the gradients in terms of the phase-relative attenuation produce (20.61). □

20.7. CHAPTER SUMMARY

In this chapter we defined wave distortion as the change in shape of a harmonic, hydraulic head wave or wave packet as it travels through a porous medium. The definition applies to both constituent waves and component waves.

We defined two types of distortion—dispersive and nondispersive. Dispersive distortion is the distortion of a wave packet that results from the frequency dependence of the propagation velocity. Dispersive distortion includes conceptually distinct sub categories—physical dispersion and geometric dispersion. The group velocity is the velocity at which a wave packet travels; the concept of group velocity is meaningful only if the wave packet's constituent frequencies differ relatively little from one another. We showed that under some conditions the concept of group velocity is valid for traveling wave packets.

Nondispersive distortion includes wave attenuation and wave dilation. Wave attenuation has two distinct components—intrinsic (nongeometric) and geometric. The intrinsic component is that which arises purely from dissipation, independent of ray path geometry, and the geometric component is that which arises purely as a result of ray path geometry, independent of dissipation. The geometric divergence of a ray path measures its degree of local angular divergence. We listed examples from earlier chapters showing that constituent waves sometimes exhibit negative attenuation and we presented several measures of wave attenuation.

The along-ray wave dilation is the local lengthening (stretching) of a constituent wave with distance traveled along the ray path. We also explained that harmonic, hydraulic head waves traveling through nonhomogeneous media generally undergo wave dilation and/or contraction.

21

Waves in One Dimension

In Section 19.4.1 we presented examples of component wave propagation in 1D flows. In this chapter we further explore the component wave interpretation of time-periodic groundwater flow (see Section 19.3.2) as it applies to 1D flows.

21.1. LINEARLY INDEPENDENT FRFS

Exercise 21.1 Complex-Form Governing Equation for 1D Flow

Problem: Consider time-periodic flow in a source-free, 1D, generally nonhomogeneous, porous medium. Adapt the governing equation (7.17a) for such flows.

Solution: Substituting $\mathbf{x} = x$, $\mathbf{K} = K\mathbf{I}$, and $F_{\mathrm{u}} = 0$ in (7.17a) yields

$$\frac{d}{dx}\left[K(x)\frac{dF_{\mathrm{h}}}{dx}\right] - i\omega_m S_{\mathrm{s}}(x) F_{\mathrm{h}}(x;\omega_m) = 0.$$

Expanding the derivative term and dividing by $K(x)$ give

$$\frac{d^2F_{\mathrm{h}}}{dx^2} + \left(\frac{d\ln K}{dx}\right)\frac{dF_{\mathrm{h}}}{dx} - i\left(\frac{\omega_m S_{\mathrm{s}}(x)}{K(x)}\right)F_{\mathrm{h}}(x;\omega_m) = 0. \tag{21.1}$$

□

Equation (21.1) is a homogeneous, second-order, linear ODE with (assumed) continuously variable coefficients. Such an equation has at most two linearly independent solutions [*Korn and Korn*, 2000], so we write the general solution as

$$F_{\mathrm{h}}(x;\omega_m) = \sum_{n=1}^{2} a_m^{(n)} F^{(n)}(x;\omega_m) \tag{21.2}$$

where the $a_m^{(n)}$ are complex constants and the two FREs $F^{(n)}(x;\omega_m)$ are linearly independent solutions of the governing equation (21.1):

$$\frac{d^2F^{(n)}}{dx^2} + \left(\frac{d\ln K}{dx}\right)\frac{dF^{(n)}}{dx} - i\left(\frac{\omega_m S_{\mathrm{s}}(x)}{K(x)}\right)F^{(n)}(x;\omega_m) = 0 \quad (n = 1,2). \tag{21.3}$$

Note 21.1 Linear Independence of Functions

The following informal definitions of linearly dependent and linearly independent functions are based loosely on *Weisstein* [2016g].

Consider a set of N functions $F^{(n)}(\mathbf{x};\omega_m)$ $(n = 1,2,\ldots,N)$, where N is a finite integer and $N \geq 2$, defined on a finite, continuous domain D. The set is either *linearly independent* or *linearly dependent* on D. The set is linearly independent if and only if no single one of the functions can be expressed as a linear combination of the remaining $N-1$ functions. Alternatively, if we can express any one of the functions as a linear combination of the remaining $N-1$ functions everywhere on D, e.g.,

$$F^{(p)}(\mathbf{x};\omega_m) = \sum_{n\neq p} b^{(n)} F^{(n)}(\mathbf{x};\omega_m) \quad \forall\, \mathbf{x} \in D$$

where $1 \leq p \leq N$ and $\nabla b^{(n)} = \mathbf{0}$, then the set of N functions is *linearly dependent.*

In the case where $N = 2$, the two functions $F^{(1)}(x;\omega_m)$ and $F^{(2)}(x;\omega_m)$ defined on a finite interval D are linearly dependent if and only if we can express one of them as a constant multiple of the other on D:

$$F^{(1)}(x;\omega_m) = b^{(2)} F^{(2)}(x;\omega_m) \quad \forall\, x \in D$$

where $b^{(2)}$ is space invariant. △

Recall that the relevant polar form governing equations for time-periodic flow in source-free, 1D media are (6.21). Define

Hydrodynamics of Time-Periodic Groundwater Flow: Diffusion Waves in Porous Media, Geophysical Monograph 224, First Edition. Joe S. Depner and Todd C. Rasmussen.

$$\zeta^{(n)}(x;\omega_m) \equiv \frac{d}{dx}\theta^{(n)}(x;\omega_m). \tag{21.4}$$

Exercise 21.2 Linearly Independent, Complex-Valued Functions

Problem: Consider two complex-valued functions $F^{(n)}(x;\omega_m)$ $(n = 1,2)$ defined on the continuous space domain D with

$$|F^{(n)}(x;\omega_m)| > 0 \quad \forall\, x \in D \quad (n = 1,2)$$

and for which the phase derivatives $\zeta^{(n)}(x;\omega_m)$ $(n = 1,2)$ are distinct (nonidentical) functions on D. Show that the functions $F^{(n)}(x;\omega_m)$ $(n = 1,2)$ are linearly independent.

Solution: Because the $F^{(n)}$ are complex valued we can write

$$\frac{F^{(1)}(x;\omega_m)}{F^{(2)}(x;\omega_m)} = M(x;\omega_m)\exp\left[i\theta(x;\omega_m)\right]$$

where

$$M(x;\omega_m) \equiv \frac{|F^{(1)}(x;\omega_m)|}{|F^{(2)}(x;\omega_m)|} \tag{21.5a}$$

$$\theta(x;\omega_m) \equiv \arg F^{(1)}(x;\omega_m) - \arg F^{(2)}(x;\omega_m). \tag{21.5b}$$

Differentiating the RHS of (21.5b) with respect to x yields

$$\frac{d}{dx}\theta(x;\omega_m) = \zeta^{(1)}(x;\omega_m) - \zeta^{(2)}(x;\omega_m).$$

The space derivatives $\zeta^{(n)}(x;\omega_m)$ $(n = 1,2)$ are distinct functions, so the LHS does not vanish everywhere on D, and thus $\theta(x;\omega_m)$ is not constant on D. Consequently, regardless of whether or not M is space invariant, the ratio $F^{(1)}/F^{(2)}$ is not space invariant on D. Therefore, the functions $F^{(n)}(x;\omega_m)$ $(n = 1,2)$ are linearly independent. □

Exercise 21.3 Linearly Independent Amplitude Functions

Problem: Consider two complex-valued functions $F^{(n)}(x;\omega_m)$ $(n = 1,2)$ for which the corresponding amplitude functions $|F^{(n)}(x;\omega_m)|$ $(n = 1,2)$ are linearly independent. Show that the functions $F^{(n)}(x;\omega_m)$ $(n = 1,2)$ are linearly independent.

Solution: Because the $F^{(n)}$ are complex valued we can write

$$\frac{F^{(1)}(x;\omega_m)}{F^{(2)}(x;\omega_m)} = M(x;\omega_m)\exp\left[i\theta(x;\omega_m)\right] \tag{21.6}$$

where

$$M(x;\omega_m) \equiv \frac{|F^{(1)}(x;\omega_m)|}{|F^{(2)}(x;\omega_m)|}$$

$$\theta(x;\omega_m) \equiv \arg F^{(1)}(x;\omega_m) - \arg F^{(2)}(x;\omega_m).$$

Regardless of whether $\theta(x;\omega_m)$ is constant or variable on D, $M(x;\omega_m)$ is not constant on D because $|F^{(1)}(x;\omega_m)|$ and $|F^{(2)}(x;\omega_m)|$ are linearly independent. Consequently, the ratio $F^{(1)}/F^{(2)}$ is not space invariant on D. Therefore, the functions $F^{(n)}(x;\omega_m)$ $(n = 1,2)$ are linearly independent. □

We can summarize the results of Exercises 21.2 and 21.3 as follows. Consider two complex-valued functions $F^{(n)}(x;\omega_m)$ $(n = 1,2)$ for which the corresponding amplitude functions $|F^{(n)}(x;\omega_m)|$ are nonzero on D. If either (a) the amplitude functions are linearly independent or (b) the phase derivatives $\zeta^{(n)}(x;\omega_m)$ are distinct functions, then the functions $F^{(n)}(x;\omega_m)$ $(n = 1,2)$ are linearly independent.

Exercise 21.4 Linear Independence: Amplitude and Phase Functions

Problem: Show that if two complex-valued functions $F^{(n)}(x;\omega_m)$ $(n = 1,2)$ are linearly independent on D, then at least one of the following must be true:

(a) The corresponding amplitude functions $M^{(n)}(x;\omega_m)$ are linearly independent on D.

(b) The corresponding phase derivatives $\zeta^{(n)}(x;\omega_m)$ are distinct functions on D.

Solution: Because the $F^{(n)}$ are complex valued we can write

$$\frac{F^{(1)}(x;\omega_m)}{F^{(2)}(x;\omega_m)} = M(x;\omega_m)\exp\left[i\theta(x;\omega_m)\right] \tag{21.7}$$

where

$$M(x;\omega_m) \equiv \frac{|F^{(1)}(x;\omega_m)|}{|F^{(2)}(x;\omega_m)|} \tag{21.8a}$$

$$\theta(x;\omega_m) \equiv \arg F^{(1)}(x;\omega_m) - \arg F^{(2)}(x;\omega_m). \tag{21.8b}$$

The LHS of (21.7) is not space invariant because the $F^{(n)}(x;\omega_m)$ are linearly independent. Therefore, either $M(x;\omega_m)$ or $\theta(x;\omega_m)$ is space variable or both are space variable. The first of these requires the amplitude functions $|F^{(n)}(x;\omega_m)|$ to be linearly independent on D, the second requires the phase derivatives $\zeta^{(n)}(x;\omega_m)$ to be distinct functions on D, and the third requires both. □

Exercise 21.5 Linearly Independent FRF Solutions

Problem: Consider time-periodic flow in a source-free, 1D, generally nonhomogeneous, porous medium. Suppose the $F^{(n)}$ $(x;\omega_m)$ $(n = 1,2)$ are linearly independent solutions of the governing equation (21.1) and the component phase derivatives are nonzero throughout the space domain:

$$|\zeta^{(n)}(x;\omega_m)| > 0 \quad \forall\, x \in D \quad (n = 1,2). \tag{21.9}$$

Show that it must *not* be true that

$$\zeta^{(1)}(x;\omega_m) = \zeta^{(2)}(x;\omega_m) \quad \forall\, x \in D. \tag{21.10}$$

That is, show that the phase derivatives $\zeta^{(n)}(x;\omega_m)$ $(n = 1,2)$ are distinct (nonidentical) functions on the space domain D. Use (6.23b).

Solution: Because of (21.3) and (21.9), (6.23b) is valid when we replace M_h and ζ_h with $M^{(n)}$ and $\zeta^{(n)}$, respectively:

$$\frac{M^{(n)}(x;\omega_m)}{M^{(n)}(x_0;\omega_m)} = \sqrt{\frac{K(x_0)\zeta^{(n)}(x_0;\omega_m)}{K(x)\zeta^{(n)}(x;\omega_m)}} \times \exp\left(\frac{\omega_m}{2}\int_{x_0}^{x}\frac{S_s(v)}{K(v)\zeta^{(n)}(v;\omega_m)}\,dv\right). \tag{21.11}$$

We will use the method of proof by contradiction. Suppose (21.10) is valid. Using it twice in (21.11), once with $n = 1$ and once with $n = 2$, leads to

$$M^{(1)}(x;\omega_m) = MM^{(2)}(x;\omega_m) \quad \forall\, x \in D \tag{21.12}$$

where M is real valued and space invariant:

$$M \equiv \frac{M^{(1)}(x_0;\omega_m)}{M^{(2)}(x_0;\omega_m)}.$$

Substituting $x = v$ in (21.10) and then integrating with respect to v, from $v = x_0$ to $v = x$, produce

$$\theta^{(1)}(x;\omega_m) = \theta^{(2)}(x;\omega_m) + \theta \quad \forall\, x \in D \tag{21.13}$$

where θ is real valued and space invariant:

$$\theta \equiv \theta^{(1)}(x_0;\omega_m) - \theta^{(2)}(x_0;\omega_m).$$

Using (21.12) and (21.13) leads to

$$\frac{F^{(1)}(x;\omega_m)}{F^{(2)}(x;\omega_m)} = Me^{i\theta}.$$

The RHS is complex valued and space invariant, so the $F^{(n)}$ $(n = 1,2)$ are linearly dependent. This contradicts our assumption that they are linearly independent, so our supposition that (21.10) is valid must be false. That is, (21.10) must be false. The component phase derivatives are nonidentical. □

Exercise 21.5 and (19.9), taken together, imply that if we can represent a 1D flow as two component waves, then the two waves have distinct phase velocity profiles. However, the two corresponding phase *speed* profiles may or may not be distinct.

Exercise 21.6 Sign of Phase Derivative and Direction of Wave Propagation

Problem: Consider time-periodic flow in a source-free, 1D, generally nonhomogeneous, porous medium. Suppose a particular hydraulic head harmonic constituent (i.e., ω_m fixed) consists of a single, harmonic traveling wave, with

$$|\zeta_h(x;\omega_m)| > 0 \quad \forall\, x \in D$$

where (6.22) defines $\zeta_h(x;\omega_m)$ and equations (6.23) are valid. In (6.23b), how is the sign of the exponent related to the sign of ζ_h? Describe how the sign of the exponent is related to the direction of wave propagation.

Solution: In (6.23b), ω_m, $S_s(x)$, and $K(x)$ are positive. Consequently, the exponent and $\zeta_h(x;\omega_m)$ have the same sign.

For waves traveling in the direction of increasing x, $\zeta_h(x;\omega_m) < 0$, while for waves traveling in the direction of decreasing x, $\zeta_h(x;\omega_m) > 0$. Therefore, the exponent is negative or positive, respectively, for waves traveling in the direction of decreasing or increasing x. □

21.2. COMPONENT WAVE HYPOTHESES

This section presents basic hypotheses about 1D flows in which we can represent the hydraulic head harmonic constituent as the superposition of two component waves. These hypotheses are based on the examples of wave propagation in 1D flows discussed in Section 19.4.1. Although the particular examples presented in this book support these hypotheses, we have not proven the hypotheses generally.

21.2.1. Through-Travel Hypothesis

In Chapter 17 we saw that for 1D flows the hydraulic head phase function $\theta_h(x;\omega_m)$ has at most one stationary point, and if such a stationary point exists, then it is a minimum (see Section 17.5). This is true for each component phase function $\theta^{(n)}(x;\omega_m)$ $(n = 1,2)$ as well, because the FREs also solve the governing equation of the space BVP for the FRF. Consequently, for each FRE one and only one of the following conditions is satisfied:

1. If the corresponding phase function, $\theta^{(n)}(x;\omega_m)$, has no stationary point in D, then

$$\left|\zeta^{(n)}(x;\omega_m)\right| > 0 \quad \forall\, x \in D. \tag{21.14}$$

2. If the corresponding phase function, $\theta^{(n)}(x;\omega_m)$, has a minimum at the point $x = x_s$, where $x_s \in D$, then

$$\zeta^{(n)}(x;\omega_m) \begin{cases} < 0, & x < x_s \text{ and } x \in D \\ = 0, & x = x_s \\ > 0, & x > x_s \text{ and } x \in D. \end{cases} \tag{21.15}$$

The *through-travel hypothesis* for component waves in 1D flows is represented mathematically by inequality (21.14). If the through-travel hypothesis is valid, then neither of the two component phase functions has a stationary point in D. Assuming $\zeta^{(n)}(x;\omega_m)$ is continuous in x, this hypothesis implies that either

$$\zeta^{(n)}(x;\omega_m) > 0 \quad \forall\, x \in D$$

or

$$\zeta^{(n)}(x;\omega_m) < 0 \quad \forall\, x \in D.$$

The definition of wave propagation speed (19.8) implies that if the through-travel hypothesis is valid, then each of the two component waves travels through the entire space domain from one boundary to the other. The ray path of each component wave has one starting point and one terminal point, and these lie on the boundary at opposite ends of the domain. The hypothesis does not specify whether the two component waves travel in the same or opposite directions.

Exercise 21.7 Through-Travel Hypothesis for Constituent Waves

Problem: Consider the through-travel hypothesis for constituent waves, rather than for component waves, in one dimension:

$$0 < \left|\zeta_{\text{h}}(x;\omega_m)\right| < \infty \quad \forall\, x \in D \quad (m = 1, 2, 3, \dots).$$

Compare this to inequality (21.14). Is such a hypothesis generally valid? Why or why not?

Solution: No, the through-travel hypothesis for constituent waves is not generally valid. The numerous 1D cases listed in Section 19.3.1 (i.e., b-2, b-3, c-2, c-3, d-2, d-3, n-2, n-3, p-2, p-3, q-2, q-3) serve as counterexamples. □

21.2.2. Opposing-Velocity Hypothesis

The *opposing-velocity hypothesis* for component waves in 1D flows is as follows:

$$\zeta^{(1)}(x;\omega_m)\zeta^{(2)}(x;\omega_m) < 0 \quad \forall\, x \in D. \qquad (21.16)$$

If the opposing-velocity hypothesis is valid, then neither of the two component phase functions has a stationary point in D, and the two component waves travel in opposite directions. Notice that the opposing-velocity hypothesis implies the through-travel hypothesis.

21.2.3. Positive-Attenuation Hypothesis

The *positive-attenuation hypothesis* for component waves in 1D flows is as follows:

$$\zeta^{(n)}(x;\omega_m)\frac{d}{dx}M^{(n)}(x;\omega_m) > 0 \quad (\forall\, x \in D;\ n = 1, 2). \qquad (21.17)$$

If the positive-attenuation hypothesis is valid, then the following are satisfied in D:

- Each of the two component waves undergoes positive attenuation as it travels.
- Neither of the two component amplitude functions has a stationary point.
- Neither of the two component phase functions has a stationary point.

Notice that the positive-attenuation hypothesis implies the through-travel hypothesis. Like the through-travel hypothesis, the positive-attenuation hypothesis does not specify whether the two component waves travel in the same or opposite directions.

Exercise 21.8 Component Waves and Their Amplitude Functions

Problem: Consider 1D flows in which we can represent the harmonic constituent as the superposition of two component waves and suppose the positive-attenuation and opposing-velocity hypotheses are valid. Are the derivatives of the component wave amplitude functions $dM^{(1)}/dx$ and $dM^{(2)}/dx$ of the same sign or of opposite sign? Is it possible for either of the component amplitude functions to have a stationary point in D?

Solution: Inequality (21.16) implies that either (a) $\zeta^{(1)}(x;\omega_m) < 0$ and $\zeta^{(2)}(x;\omega_m) > 0$ or (b) $\zeta^{(1)}(x;\omega_m) > 0$ and $\zeta^{(2)}(x;\omega_m) < 0$. Consider case (a) first. Inequality (21.17) gives

$$\frac{d}{dx}M^{(1)}(x;\omega_m) < 0 \qquad (21.18a)$$

$$\frac{d}{dx}M^{(2)}(x;\omega_m) > 0 \qquad (21.18b)$$

$$\forall\, x \in D.$$

Together these give

$$\frac{d}{dx}M^{(1)}(x;\omega_m)\frac{d}{dx}M^{(2)}(x;\omega_m) < 0 \quad \forall\, x \in D. \qquad (21.19)$$

A similar analysis for case (b) gives

$$\frac{d}{dx}M^{(1)}(x;\omega_m) > 0$$

$$\frac{d}{dx}M^{(2)}(x;\omega_m) < 0$$

$$\forall\, x \in D.$$

Together, these also yield inequality (21.19). Therefore, if the positive-attenuation and opposing-velocity hypotheses are valid, then the derivatives of the component wave amplitude functions are of opposite sign. Inequality (21.19) implies that neither amplitude function has a stationary point in D. □

21.3. UNIFORM-VELOCITY WAVES

21.3.1. General Analysis

In Section 19.4.1 we showed that we can represent some time-periodic flows in source-free, 1D media as the superposition of uniform-velocity waves that are attenuated as they travel. These component waves are of the form (19.16)

$$h^{(n)}(\mathbf{x}, t; \omega_m) = M^{(n)}(\mathbf{x}; \omega_m) \cos\left[\omega_m t + \theta^{(n)}(\mathbf{x}; \omega_m)\right] \tag{21.20}$$

where

$$\theta^{(n)}(x; \omega_m) = \theta^{(n)}(x_0; \omega_m) + (-1)^{n+1} k^{(n)}(x - x_0) \tag{21.21a}$$

$$\zeta^{(n)}(x; \omega_m) = (-1)^{n+1} k^{(n)} \tag{21.21b}$$

$$n = 1, 2$$

for fixed ω_m, $x_0(\omega_m)$ and the $\theta^{(n)}(x_0; \omega_m)$ are real-valued constants, and

$$k^{(n)}(\omega_m) > 0 \quad (n = 1, 2). \tag{21.22}$$

Can such waves propagate in flows through other 1D media? In this section we explore the general conditions under which such waves propagate in source-free, 1D media.

Substituting the RHS of (21.21b) for ζ_h in (6.24a) and then multiplying by -1 and simplifying lead to

$$4\left(k^{(n)}\right)^4 + C\left(k^{(n)}\right)^2 + D^{(n)}k^{(n)} + E = 0 \qquad \forall\, x \in D \quad (n = 1, 2) \tag{21.23}$$

where we define the coefficients as

$$C(x) \equiv \left(\frac{d \ln K}{dx}\right)^2 + 2\frac{d}{dx}\left(\frac{d \ln K}{dx}\right) \tag{21.24a}$$

$$D^{(n)}(x; \omega_m) \equiv (-1)^n\, 2\frac{d}{dx}\left(\frac{\omega_m S_s}{K}\right) \tag{21.24b}$$

$$E(x; \omega_m) \equiv -\left(\frac{\omega_m S_s}{K}\right)^2. \tag{21.24c}$$

Equation (21.23) is quartic in $k^{(n)}$, which is space invariant. Only those porous media and harmonic constituents for which (21.23) is satisfied will admit uniform-velocity waves.

Substituting $M^{(n)}$ for M_h and the RHS of (21.21b) for ζ_h in (6.23b) gives

$$\frac{M^{(n)}(x; \omega_m)}{M^{(n)}(x_0; \omega_m)} = \sqrt{\frac{K(x_0)}{K(x)}} \exp\left(\frac{(-1)^{n+1}\omega_m}{2k^{(n)}(\omega_m)} \int_{x_0}^{x} \frac{S_s(v)}{K(v)}\, dv\right) \quad (n = 1, 2). \tag{21.25}$$

If (21.23) is satisfied and $k^{(n)}$ is known, then we can use (21.25) to infer the component wave's amplitude function. In (21.25), when $n = 1$ the exponent is positive so the exponential term is an increasing function of x. Conversely, when $n = 2$ the exponent is negative so the exponential term is a decreasing function of x.

Exercise 21.9 1D Flow in Exponential Medium

Problem: Consider time-periodic, 1D flow through a domain filled with an exponential medium. Derive simplified expressions for the coefficients C, $D^{(n)}$, and E and solve (21.23) for $k^{(n)}$. What does the expression for $k^{(n)}$ reduce to in the special case where the medium is homogeneous?

Solution: Substituting the following in (21.24)

$$K(x) = K(0)e^{\mu x}$$
$$S_s(x) = S_s(0)e^{\mu x}$$

leads to

$$C(x) \equiv \mu^2$$
$$D^{(n)}(x; \omega_m) \equiv 0$$
$$E(x; \omega_m) \equiv -\alpha_m^2.$$

Substituting the RHSs for the corresponding coefficients in (21.23) yields

$$4\left(k^{(n)}\right)^4 + \mu^2\left(k^{(n)}\right)^2 - \alpha_m^2 = 0 \quad \forall\, x \in D \quad (n = 1, 2)$$

which is quadratic in $\left(k^{(n)}\right)^2$. Applying the quadratic formula gives

$$\left(k^{(n)}\right)^2 = \frac{-\mu^2 \pm \sqrt{(\mu^2)^2 - 4(4)(-\alpha_m^2)}}{2(4)}. \tag{21.26}$$

The square of the wave number cannot be negative, so we replace the plus-or-minus sign with a positive sign. Then, taking the positive square root gives

$$k^{(n)} = \frac{1}{2}\sqrt{\frac{\sqrt{\mu^4 + 16\alpha_m^2} - \mu^2}{2}} \quad (n = 1, 2). \tag{21.27}$$

In the special case where the medium is homogeneous, $\mu = 0$ so this reduces to

$$k^{(n)} = \sqrt{\frac{\alpha_m^{(1)}}{2}} \quad (n = 1, 2).$$

□

Exercise 21.10 1D Flow in Exponential Medium

Problem: Consider time-periodic flow through a source-free, 1D domain filled with an exponential medium. Suppose we can represent the flow as the superposition of two component waves described by (19.16). Using (21.25), express the wave amplitude, $M^{(n)}(x; \omega_m)$, explicitly in terms of $\alpha_m^{(1)}$, μ, $k(\omega_m)$, $M^{(n)}(x_0; \omega_m)$, x, and x_0.

Solution: Equations (10.1) and (10.2) give $\sqrt{K(x_0)/K(x)} = e^{-\mu x/2}$ and $\omega_m S_s/K = \alpha_m^{(1)}$. Substituting these and $x_0 = 0$ in (21.25) gives

$$M^{(n)}(x; \omega_m) = M^{(n)}(x_0; \omega_m) \exp\left[-\frac{1}{2}\left(\mu + \frac{(-1)^n \alpha_m^{(1)}}{k^{(n)}(\omega_m)}\right)(x - x_0)\right] \quad (n = 1, 2). \tag{21.28}$$

□

21.3.2. Space-Invariant Coefficients

One way (perhaps the only way) that (21.23) could be satisfied is if all three of the coefficients C, $D^{(n)}$, and E defined by (21.24) are space invariant. In this section we explore that scenario.

Exercise 21.11 Space-Invariant Wave number and Hydraulic Diffusivity

Problem: Prove that if both $k^{(1)}$ and the hydraulic diffusivity are space invariant, then the coefficient C is space invariant.

Solution: Equation (21.23) is satisfied because $k^{(1)}$ is space invariant. Equations (21.24b) and (21.24c) indicate, respectively, that $D^{(n)} = 0$ $(n = 1, 2)$ and $E = -\alpha_m^2$ [where we have used the definition of $\alpha_m^{(1)}$ in equation (5.31)], because the hydraulic diffusivity is space invariant. Substituting these for $D^{(n)}$ and E in (21.23) then gives

$$4\left(k^{(1)}\right)^4 + C\left(k^{(1)}\right)^2 - \alpha_m^2 = 0 \quad \forall\, x \in D.$$

Solving for C yields

$$C = \frac{\alpha_m^2 - 4\left(k^{(1)}\right)^4}{\left(k^{(1)}\right)^2}.$$

The RHS is space invariant, so C is space invariant. □

Equation (21.24c) indicates that if the coefficent E is to be space invariant, then the hydraulic diffusivity must be space invariant [i.e., $E = -\alpha_m^2$, where we have used the definition of $\alpha_m^{(1)}$ in equation (5.31)]. Then (21.24b) indicates that $D^{(n)} = 0$ $(n = 1, 2)$. Substituting these values for the coefficients in the polynomial (21.23) gives

$$4\left(k^{(n)}\right)^4 + C\left(k^{(n)}\right)^2 - \alpha_m^2 = 0 \quad (n = 1, 2). \tag{21.29}$$

This equation is quadratic in $\left(k^{(n)}\right)^2$. Using the quadratic formula to solve it for $\left(k^{(n)}\right)^2$ and then taking the positive square root (because k is nonnegative by definition) lead to

$$k^{(n)} = \frac{\sqrt{\sqrt{C^2 + 16\alpha_m^2} - C}}{2\sqrt{2}}. \tag{21.30}$$

For any real value C, this gives $k^{(n)} > 0$. Therefore, if the coefficients C, $D^{(n)}$, and E are space invariant, then there is one and only one possible value for the wave numbers $k^{(n)}$. Consequently, the wave numbers of the component waves in (19.16) are identical:

$$k^{(1)}(\omega_m) = k^{(2)}(\omega_m) = k(\omega_m).$$

What forms can the hydraulic conductivity distribution take in this special case where the coefficients C, $D^{(n)}$, and E are space invariant? We can write (21.24a) as

$$2\frac{dg}{dx} + g^2 = C \tag{21.31}$$

where C is understood to be a real constant and we define $g(x)$ as

$$g(x) \equiv \frac{d \ln K}{dx}.$$

Rearranging (21.31) leads to

$$\frac{dg}{C - g^2} = \frac{dx}{2}.$$

Substituting $x = v$ and $g(x) = u(v)$ and then integrating from $v = 0$ to $v = x$ give

$$\int_{g(0)}^{g(x)} \frac{du}{C - u^2} = \frac{1}{2}x. \tag{21.32}$$

The result obtained by evaluating the integral on the LHS depends on whether the constant C is positive, negative, or zero. We consider each of these three cases separately in the following exercises.

Exercise 21.12 Space-Invariant Coefficients

Problem: Suppose all three of the coefficients C, $D^{(n)}$, and E defined by (21.24) are space invariant and $C > 0$. Show that in this case the log conductivity gradient is given by

$$g(x) \equiv \frac{d \ln K}{dx} = \sqrt{C}\frac{Pe^{\sqrt{C}x} - 1}{Pe^{\sqrt{C}x} + 1} \tag{21.33}$$

where

$$P \equiv \frac{\sqrt{C} + g(0)}{\sqrt{C} - g(0)} \tag{21.34}$$

and $g(0) \neq \sqrt{C}$.

Solution: Using equation 66 of *Beyer* [1984] to evaluate the integral on the LHS of (21.32) produces

$$\frac{1}{2\sqrt{C}}\left[\ln \frac{\sqrt{C} + u}{\sqrt{C} - u}\right]\Bigg|_{u=g(0)}^{u=g(x)} = \frac{1}{2}x.$$

Evaluating the term on each side at the indicated limits and then multiplying by $2\sqrt{C}$ give

$$\ln \frac{\sqrt{C} + g(x)}{\sqrt{C} - g(x)} - \ln P = \sqrt{C}x.$$

Exponentiating and then rearranging give

$$\frac{\sqrt{C} + g(x)}{\sqrt{C} - g(x)} = Pe^{\sqrt{C}x}.$$

Solving for $g(x)$ then yields (21.33). □

Exercise 21.13 Space-Invariant Coefficients

Problem: Suppose all three of the coefficients C, $D^{(n)}$, and E defined by (21.24) are space invariant and $C > 0$. Derive a closed-form expression for $K(x)$ by integrating (21.33).

Solution: We can write (21.33) as

$$\frac{d \ln K}{dx} = \sqrt{C}\left(1 - \frac{2}{Pe^{\sqrt{C}x} + 1}\right).$$

Substituting $x = v$ and then integrating with respect to v from $v = 0$ to $v = x$ give

$$\int_0^x \left[\frac{d \ln K(v)}{dv}\right] dv = \sqrt{C}\int_0^x \left(1 - \frac{2}{1 + Pe^{\sqrt{C}v}}\right) dv.$$

Using equation 4.10.9 of *Roy and Olver* [2016] to evaluate the second term of the integral on the RHS leads to

$$[\ln K(v)]\big|_0^x = \sqrt{C}\left[-v + \frac{2}{\sqrt{C}}\ln\left(1 + Pe^{\sqrt{C}v}\right)\right]\Bigg|_0^x.$$

Evaluating each term at the indicated limits produces

$$\ln\left[\frac{K(x)}{K(0)}\right] = 2\ln\left[\frac{1 + Pe^{\sqrt{C}x}}{1 + P}\right] - \sqrt{C}x.$$

Exponentiating yields

$$\frac{K(x)}{K(0)} = \left[\frac{1 + Pe^{\sqrt{C}x}}{1 + P}\right]^2 e^{-\sqrt{C}x} \tag{21.35}$$

where (21.34) defines P. □

Exercise 21.14 Restrictions on Space Domain

Problem: Suppose $K(x)$ is given by (21.35). Is it necessary to restrict the range of x so that $K(x)$ is defined everywhere in the space domain? If so, specify such restrictions.

Solution: Equation (21.35) implies that $K(x)$ is defined for all x if $P \neq -1$. Equation (21.34) implies that this requires $|C| > 0$, which is true because $C > 0$ in this case. Therefore, $K(x)$ is defined everywhere in the space domain; no restriction on x is necessary. □

Exercise 21.15 Restrictions on Space Domain

Problem: Suppose $K(x)$ is given by (21.35). Is it necessary to restrict the values of x so that $K(x) > 0$ everywhere in the space domain? If so, specify such restrictions. Assume $K(0) > 0$.

Solution: The RHS of (21.35) is never negative, so we need only consider possible zero points. Let x_0 denote the zero point of $K(x)$. Substituting $x = x_0$ and $K(x_0) = 0$ in (21.35) gives

$$1 + Pe^{\sqrt{C}x_0} = 0.$$

Solving for x_0 gives

$$x_0 = \frac{1}{\sqrt{C}}\ln\left(\frac{-1}{P}\right). \tag{21.36}$$

The coordinate x_0 must be real valued, so such a zero point exists only if $P < 0$.

Substituting the RHS of (21.34) for P gives

$$\frac{\sqrt{C} + g(0)}{\sqrt{C} - g(0)} < 0. \tag{21.37}$$

Here we consider two cases, depending on whether the denominator is positive or negative:

Case 1: The denominator is positive [i.e., $g(0) < \sqrt{C}$]. Multiplying inequality (21.37) by the denominator then gives $g(0) < -\sqrt{C}$. Therefore, $g(0) < -\sqrt{C}$.

Case 2: The denominator is negative [i.e., $g(0) > \sqrt{C}$]. Multiplying inequality (21.37) by the denominator then gives $g(0) > -\sqrt{C}$. Therefore, $g(0) > \sqrt{C}$.

Conclusion: If $|g(0)| > \sqrt{C}$, then a zero point, defined by (21.36), potentially exists. The space domain must not include this point. On the other hand, if $|g(0)| < \sqrt{C}$, then no zero point exists. □

Exercise 21.16 Space-Invariant Coefficients

Problem: Suppose all three of the coefficients C, $D^{(n)}$, and E defined by (21.24) are space invariant and $C = 0$. Show that in this case

$$K(x) \equiv K(0)\left[1 + \frac{g(0)}{2}x\right]^2 \tag{21.38a}$$

$$S_s(x) = S_s(0)\left[1 + \frac{g(0)}{2}x\right]^2. \tag{21.38b}$$

Solution: Substituting $C = 0$ in (21.32) gives

$$\int_{g(0)}^{g(x)} \frac{du}{u^2} = \frac{-1}{2}x. \tag{21.39}$$

Evaluating the integral leads to

$$\left.\frac{1}{u}\right|_{g(0)}^{g(x)} = \frac{1}{2}x. \tag{21.40}$$

Evaluating the term on the LHS at the indicated limits and then solving for $g(x)$ yields

$$g(x) = \frac{g(0)}{1 + [g(0)/2]x}. \tag{21.41}$$

Substituting $x = v$ and $g(x) = d \ln K(v)/dv$ and then integrating with respect to v from $v = 0$ to $v = x$ give

$$\ln \frac{K(x)}{K(0)} = 2 \ln\left[1 + \frac{g(0)}{2}x\right]. \tag{21.42}$$

Exponentiating produces (21.38a). Then, because E is space invariant, the specific storage takes the form in (21.38b). □

Exercise 21.17 Restrictions on Space Domain

Problem: Suppose $K(x)$ is given by (21.38a). Is it necessary to restrict the values of x so that $K(x) > 0$ everywhere in the space domain? If so, specify such restrictions. Assume $K(0) > 0$.

Solution: The RHS of (21.38a) is never negative, so we need only consider possible zero points. Let x_0 denote the zero point of $K(x)$. Substituting $x = x_0$ and $K(x_0) = 0$ in (21.38a) and then solving for x_0 give

$$x_0 = \frac{-2}{g(0)}.$$

The space domain must not include this point. □

Exercise 21.18 Space-Invariant Coefficients

Problem: Suppose all three of the coefficients C, $D^{(n)}$, and E defined by (21.24) are space invariant and $C < 0$. Show that in this case the log conductivity gradient is given by

$$g(x) \equiv \sqrt{|C|}\tan\left(r_0 - \frac{\sqrt{|C|}}{2}x\right) \tag{21.43}$$

where

$$r_0 \equiv \tan^{-1}\frac{g(0)}{\sqrt{|C|}}. \tag{21.44}$$

Solution: In this case we can write $C = -\left(\sqrt{|C|}\right)^2$. Substituting this in (21.32) gives

$$\int_{g(0)}^{g(x)} \frac{du}{\left(\sqrt{|C|}\right)^2 + u^2} = \frac{-1}{2}x.$$

Evaluating the integral yields

$$\left.\left[\tan^{-1}\frac{u}{\sqrt{|C|}}\right]\right|_{g(0)}^{g(x)} = \frac{-\sqrt{|C|}}{2}x.$$

Evaluating the term on the LHS at the indicated limits and rearranging lead to

$$\tan^{-1}\frac{g(x)}{\sqrt{|C|}} = \tan^{-1}\frac{g(0)}{\sqrt{|C|}} - \frac{\sqrt{|C|}}{2}x.$$

Taking the tangent then yields (21.43). □

Exercise 21.19 Space-Invariant Coefficients

Problem: Suppose all three of the coefficients C, $D^{(n)}$, and E defined by (21.24) are space invariant and $C < 0$. Derive a closed-form expression for $K(x)$ by integrating (21.43).

Solution: Substituting $x = v$ in (21.43) gives

$$\frac{d \ln K(v)}{dv} = \sqrt{|C|}\tan\left(r_0 - \frac{\sqrt{|C|}}{2}v\right).$$

Integrating with respect to v from $v = 0$ to $v = x$ leads to

$$\int_0^x \left[\frac{d \ln K(v)}{dv}\right] dv = \sqrt{C}\int_0^x \tan\left(r_0 - \frac{\sqrt{|C|}}{2}v\right) dv.$$

Letting $w = r_0 - \sqrt{|C|}v/2$ in the integral on the RHS gives $dv = -2dw/\sqrt{|C|}$. Substituting these in the RHS produces

$$\left.\left[\ln K(v)\right]\right|_0^x = -2\int_{r_0}^{r_0-\sqrt{|C|}x/2} \tan w \, dw.$$

Using equation 4.26.3 of *Roy and Olver* [2016] to evaluate the integral on the RHS, we obtain

$$\left.\left[\ln K(v)\right]\right|_0^x = 2\left.\left[\ln \cos w\right]\right|_{r_0}^{r_0-\sqrt{|C|}x/2}.$$

Evaluating both sides at the indicated limits gives

$$\ln \frac{K(x)}{K(0)} = 2\left[\ln \cos\left(r_0 - \frac{\sqrt{|C|}}{2}x\right) - \ln \cos r_0\right].$$

Exponentiating yields

$$\frac{K(x)}{K(0)} = \left[\frac{\cos\left(r_0 - \sqrt{|C|}x/2\right)}{\cos r_0}\right]^2.$$

Using the angle-difference relation for the cosine (B.2b) to expand the numerator and then (21.44) to eliminate r_0 produces

$$\frac{K(x)}{K(0)} = \left[\cos\left(\frac{\sqrt{|C|}}{2}x\right) + \frac{g(0)}{\sqrt{|C|}}\sin\left(\frac{\sqrt{|C|}}{2}x\right)\right]^2. \tag{21.45}$$

□

Exercise 21.20 Restrictions on Space Domain

Problem: Suppose $K(x)$ is given by (21.45). Is it necessary to restrict the range of x so that $K(x)$ is defined everywhere in the space domain? If so, specify such restrictions.

Solution: Equation (21.45) implies that $K(x)$ is undefined only if $C = 0$. By assumption $C < 0$ in this case, so no restriction on x is necessary to ensure that $K(x)$ is defined everywhere in the space domain. □

Exercise 21.21 Restrictions on Space Domain

Problem: Suppose $K(x)$ is given by (21.45). Is it necessary to restrict the range of x so that $K(x) > 0$ everywhere in the space domain? If so, specify such restrictions. Assume $K(0) > 0$.

Solution: The RHS of (21.45) is never negative, so we need only consider its possible zero points. Let x_n ($n \in \mathbb{Z}$) denote the zero point(s) of $K(x)$. Substituting $x = x_n$ and $K(x_n) = 0$ in (21.45) and simplifying give

$$\tan\left(\frac{\sqrt{|C|}}{2}x_n\right) = \frac{-\sqrt{|C|}}{g(0)}.$$

Solving for x_n yields

$$x_n = \frac{-2}{\sqrt{|C|}}\left[\text{Arctan}\frac{\sqrt{|C|}}{g(0)} + n\pi \text{ rad}\right] \quad \forall n \in \mathbb{Z} \tag{21.46}$$

where Arctan denotes the principal value of the inverse-tangent function. We consider the cases where $g(0) > 0$ and $g(0) < 0$ separately below:

Case 1: $g(0) > 0$ Then the argument of the arctangent function in (21.46) is positive, so [*Eves*, 1984]

$$x_n = \frac{2}{\sqrt{|C|}}\left[\text{Arctan}\frac{g(0)}{\sqrt{|C|}} - \left(n + \frac{1}{2}\right)\pi \text{ rad}\right] \quad \forall n \in \mathbb{Z}. \tag{21.47}$$

Case 2: $g(0) < 0$ Then the argument of the arctangent function in (21.46) is negative, so [*Eves* 1984]

$$x_n = \frac{-2}{\sqrt{|C|}}\left[\text{Arctan}\frac{|g(0)|}{\sqrt{|C|}} + \left(n - \frac{1}{2}\right)\pi \text{ rad}\right] \quad \forall n \in \mathbb{Z}. \tag{21.48}$$

In either case the space domain must not include any of the corresponding zero points. Additionally, the space domain must exclude some finite interval on either side of each zero point. Otherwise, if a boundary coincides with a zero point, then the medium is effectively impermeable in the space domain adjacent to the boundary. □

Exercise 21.22 Restrictions on Space Domain

Problem: In Exercise 21.21 we found that the function defined by the RHS of (21.45) has infinitely many zero points on the real line. If the hydraulic conductivity is to be nonzero throughout the space domain, then the domain must exclude such points. Do the results obtained in Exercise 21.21 effectively place an upper limit on the length of the space domain? If so, what is that limit?

Solution: The zero points are spaced at even intervals along the x axis. The distance between adjacent zero points is $|\Delta x| \equiv |x_{n+1} - x_n|$, where $n \in \mathbb{Z}$. Therefore, if the space domain is to be continuous, then its length (l) must satisfy $l < |\Delta x|$. Using (21.46) to express $|\Delta x|$ in terms of $\sqrt{|C|}$ then yields

$$l < \frac{2\pi}{\sqrt{|C|}}.$$

□

21.4. NONUNIFORM-VELOCITY WAVES

In Section 21.3 we interpreted time-periodic, 1D flows in some particular media as the superposition of two harmonic waves traveling in opposite directions with space-invariant speeds. In this section we consider the more general case of time-periodic, 1D flows through media that are continuously, but otherwise more or less arbitrarily, nonhomogeneous. We interpret the hydraulic head field of such a flow as the superposition of harmonic waves traveling with generally *nonuniform* (i.e., space-varying) speeds. The example and exercises presented below illustrate the concept.

21.4.1. Waves in 1D Flow through Power Law Media

We defined power law media in Section 11.1. In this case, $|d\zeta^{(n)}/dx| > 0$ for both $n = 1$ and $n = 2$, so the component waves propagate at spatially nonuniform speeds.

Exercise 21.23 Component wave Propagation Speeds in Power Law Media

Problem: Consider time-periodic, 1D flow through a source-free domain filled with a power law medium. Show that the following equation is satisfied:

$$2\frac{d}{dx}\left(\frac{\omega_m S_s}{K}\right)\zeta^{(1)} - 4\left(\frac{\omega_m S_s}{K}\right)\frac{d\zeta^{(1)}}{dx} = 0. \qquad (21.49)$$

Using the uncoupled governing equation (6.24a), relate this result to the observation [see (19.33)] that the two component waves have identical propagation speeds at any fixed location.

Solution: Substituting $n = 1$ in (19.29b) and then differentiating with respect to x give

$$\zeta^{(1)}(x;\omega_m) \equiv \frac{d\theta^{(1)}}{dx} = \left(\operatorname{Im} p_m^{(1)}\right)x^{-1}.$$

Substituting the RHS for $\zeta^{(1)}$ and the RHSs of (11.1) and (11.2) for K and S_s, respectively, in the LHS of (21.49) lead to

$$2\frac{d}{dx}\left(\frac{\omega_m S_s}{K}\right)\zeta^{(1)} - 4\left(\frac{\omega_m S_s}{K}\right)\frac{d\zeta^{(1)}}{dx}$$
$$= \left(\operatorname{Im} p_m^{(n)}\right)(-4x^{-4} + 4x^{-4}) = 0.$$

Thus, (21.49) is satisfied. Using this result, the governing equation (6.24a) reduces to

$$\left(\frac{\omega_m S_s}{K}\right)^2 - \left[\left(\frac{d\ln K}{dx}\right)^2 + 2\frac{d}{dx}\left(\frac{d\ln K}{dx}\right)\right]\zeta_h^2$$
$$-4\zeta_h^4 + 3\left(\frac{d\zeta_h}{dx}\right)^2 - 2\zeta_h\frac{d^2\zeta_h}{dx^2} = 0.$$

The LHS is even in ζ_h, so if a phase derivative $\zeta^{(1)}$ satisfies this governing equation, then its opposite, $\zeta^{(2)} = -\zeta^{(1)}$, also satisfies the equation. Consequently, at any fixed location the propagation speeds of the two component waves are identical. □

Exercise 21.24 Component Waves in Power Law Media

Problem: Consider the component wave fields $G^{(n)}(x;\omega_m)$ $(n = 1, 2)$ for the flow whose FRF is given in (11.11). Express the local wavelength $\lambda(x;\omega_m)$ and the local wave dilation $\Lambda(x;\omega_m)$, for both component wave fields in terms of b, f_m, and x.

Solution: Review (19.31). $\operatorname{Im}\lambda^{(1)} > 0$, so the wave numbers, ray path vectors, and ray parameters for the two component waves are, respectively,

$$k^{(n)} = (-1)^n\left(\operatorname{Im} p_m^{(n)}\right)x^{-1} \qquad (21.50a)$$
$$\hat{\mathbf{s}}^{(n)} = (-1)^{n+1}\hat{\mathbf{x}} \qquad (21.50b)$$
$$s^{(n)} = (-1)^{n+1}x \qquad (21.50c)$$
$$n = 1, 2.$$

Equation (21.50c) gives $ds^{(n)} = (-1)^{n+1}dx$. Using these results to evaluate the RHSs of (19.64) and (20.56) produces, respectively,

$$\lambda^{(n)}(x;\omega_m) = \frac{(-1)^n 2\pi}{\operatorname{Im} p_m^{(n)}}x$$
$$\Lambda^{(n)}(x;\omega_m) = \frac{-1}{\operatorname{Im} p_m^{(n)}}.$$

Using the RHS of (11.12) to express $\operatorname{Im} p_m^{(1)}$ in terms of b and f_m then yields

$$\lambda^{(n)}(x;\omega_m) = \frac{4\pi\sqrt{2}}{\sqrt{\sqrt{(1-b)^4 + 16f_m^2} - (1-b)^2}}x$$
$$\Lambda^{(n)}(x;\omega_m) = \frac{(-1)^{n+1}2\sqrt{2}}{\sqrt{\sqrt{(1-b)^4 + 16f_m^2} - (1-b)^2}}$$
$$n = 1, 2.$$

□

21.5. CHAPTER SUMMARY

In this chapter we interpreted time-periodic flow in source-free, 1D media as the steady-state propagation of spatially modulated (damped), traveling diffusion waves, i.e., harmonic, hydraulic head waves. We developed conditions on the space distributions of material hydrologic properties for waves that travel at spatially uniform (i.e., space-invariant) speeds and presented examples with analytical solutions. We also presented a hypothetical example, with an analytical solution, of a 1D flow that we can interpret as the superposition of two component waves traveling in opposite directions and at nonuniform speeds.

22
Wave Equation

In this chapter we show that each component wave field satisfies a generalized wave equation with unique coefficients.

22.1. GENERALIZED WAVE EQUATION

Recall that the governing equation, i.e., the groundwater flow equation (2.5), is of the parabolic (diffusion) type rather than the hyperbolic (wave) type (e.g., see Exercise 7.3). However, it is easy to show (see Exercise 22.1) that each hydraulic head component wave field $h^{(n)}(\mathbf{x};\omega_m)$ $(n = 1, 2, \ldots, N)$ satisfies the generalized wave equation:

$$\nabla^2 h^{(n)} = \frac{1}{(c^{(n)})^2(\mathbf{x};\omega_m)} \frac{\partial^2 h^{(n)}}{\partial t^2} + \gamma^{(n)}(\mathbf{x};\omega_m)\frac{\partial h^{(n)}}{\partial t} + \beta^{(n)}(\mathbf{x};\omega_m)h^{(n)} \qquad (22.1)$$

where (19.62b) gives the propagation speed, $c^{(n)}$, and we define the coefficients $\gamma^{(n)}$ and $\beta^{(n)}$ as

$$\gamma^{(n)}(\mathbf{x};\omega_m) \equiv \frac{\nabla \cdot [(M^{(n)})^2 \nabla\theta^{(n)}]}{\omega_m (M^{(n)})^2} \quad (\text{dimensions: } L^{-2}t) \qquad (22.2a)$$

$$\beta^{(n)}(\mathbf{x};\omega_m) \equiv \frac{\nabla^2 M^{(n)}}{M^{(n)}} \quad (\text{dimensions: } L^{-2}). \qquad (22.2b)$$

For fixed n, equation (22.1) is a second-order, homogeneous PDE with real-valued coefficients that depend on the space coordinates and the constituent frequency.

Although neither the source term nor the hydraulic conductivity tensor **K** appears explicitly in (22.1), the equation is valid in the general case where the source distribution is nonzero and the medium is nonhomogeneous and anisotropic.

Exercise 22.1 Generalized Wave Equation: Component Waves

Problem: Using (19.16), express each of $\partial h^{(n)}/\partial t$, $\partial^2 h^{(n)}/\partial t^2$, $\nabla h^{(n)}$, and $\nabla^2 h^{(n)}$ in terms of ω_m, t, $M^{(n)}$, and $\theta^{(n)}$. Use these results to show that the component wave field $h^{(n)}(\mathbf{x};\omega_m)$ satisfies the generalized wave equation (22.1) where (19.62b) gives the propagation speed, $c^{(n)}$, and (22.2) defines the coefficients $\gamma^{(n)}$ and $\beta^{(n)}$.

Solution: Differentiating the RHS of (19.16) with respect to time gives

$$\frac{\partial h^{(n)}}{\partial t} = -\omega_m M^{(n)} \sin[\omega_m t + \theta^{(n)}] \qquad (22.3a)$$

$$\frac{\partial^2 h^{(n)}}{\partial t^2} = -\omega_m^2 M^{(n)} \cos[\omega_m t + \theta^{(n)}]. \qquad (22.3b)$$

Taking the gradient of the RHS of (19.16) leads to

$$\nabla h^{(n)} = \cos[\omega_m t + \theta^{(n)}]\nabla M^{(n)} - M^{(n)} \sin[\omega_m t + \theta^{(n)}]\nabla\theta^{(n)}.$$

Taking the divergence of this equation gives the Laplacian of the RHS of (19.16):

$$\nabla^2 h^{(n)} = -\cos[\omega_m t + \theta^{(n)}]\left[M^{(n)}|\nabla\theta^{(n)}|^2 - \nabla^2 M^{(n)}\right] - \sin[\omega_m t + \theta^{(n)}]\left[2\nabla M^{(n)} \cdot \nabla\theta^{(n)} + M^{(n)}\nabla^2\theta^{(n)}\right].$$

Rearranging the RHS yields

$$\nabla^2 h^{(n)} = \nabla^2 M^{(n)} \cos[\omega_m t + \theta^{(n)}] - |\nabla\theta^{(n)}|^2 M^{(n)} \cos[\omega_m t + \theta^{(n)}] - \frac{\nabla \cdot [(M^{(n)})^2 \nabla\theta^{(n)}]}{M^{(n)}} \sin[\omega_m t + \theta^{(n)}].$$

Hydrodynamics of Time-Periodic Groundwater Flow: Diffusion Waves in Porous Media, Geophysical Monograph 224, First Edition. Joe S. Depner and Todd C. Rasmussen.

Using (19.16), (22.3b), and (22.3a), respectively, to rewrite the first, second, and third terms on the RHS and then rearranging terms give

$$\nabla^2 h^{(n)} = \frac{|\nabla\theta^{(n)}|^2}{\omega_m^2}\frac{\partial^2 h^{(n)}}{\partial t^2} + \frac{\nabla\cdot[(M^{(n)})^2\nabla\theta^{(n)}]}{\omega_m (M^{(n)})^2}\frac{\partial h^{(n)}}{\partial t} + \frac{\nabla^2 M^{(n)}}{M^{(n)}} h^{(n)}.$$

Combining this equation with the phase speed definition (19.62b) and the coefficient definitions (22.2) gives (22.1). □

Exercise 22.2 Generalized Wave Equation: Constituent Waves

Problem: Using (19.10) and (19.11), express each of $\partial h'/\partial t$, $\partial^2 h'/\partial t^2$, $\nabla h'$, and $\nabla^2 h'$ in terms of ω_m, t, M_h, and θ_h. Use these results to show that the constituent wave field $h'(\mathbf{x};\omega_m)$ satisfies the generalized wave equation:

$$\nabla^2 h' = \frac{1}{c_h^2(\mathbf{x};\omega_m)}\frac{\partial^2 h'}{\partial t^2} + \gamma_h(\mathbf{x};\omega_m)\frac{\partial h'}{\partial t} + \beta_h(\mathbf{x};\omega_m)h' \quad \forall\, \mathbf{x} \in D_+(\omega_m) \tag{22.4}$$

where (19.8) gives the propagation speed, c_h, and we define the coefficients γ_h and β_h as

$$\gamma_h(\mathbf{x};\omega_m) \equiv \frac{\nabla\cdot[M_h^2\nabla\theta_h]}{\omega_m M_h^2} \quad (\text{dimensions: } L^{-2}t) \tag{22.5a}$$

$$\beta_h(\mathbf{x};\omega_m) \equiv \frac{\nabla^2 M_h}{M_h} \quad (\text{dimensions: } L^{-2}). \tag{22.5b}$$

Solution: Differentiating the RHS of (19.10) with respect to time and using (19.11) give

$$\frac{\partial h'}{\partial t} = -\omega_m M_h \sin[\omega_m t + \theta_h] \tag{22.6a}$$

$$\frac{\partial^2 h'}{\partial t^2} = -\omega_m^2 M_h \cos[\omega_m t + \theta_h]. \tag{22.6b}$$

Taking the gradient of the RHS of (19.10) and using (19.11) lead to

$$\nabla h' = \cos[\omega_m t + \theta_h]\nabla M_h - M_h \sin[\omega_m t + \theta_h]\nabla\theta_h.$$

Taking the divergence of this equation gives the Laplacian of the RHS of (19.10):

$$\nabla^2 h' = -\cos[\omega_m t + \theta_h]\Big[M_h|\nabla\theta_h|^2 - \nabla^2 M_h\Big] - \sin[\omega_m t + \theta_h]\Big[2\nabla M_h\cdot\nabla\theta_h + M_h\nabla^2\theta_h\Big].$$

Rearranging yields

$$\nabla^2 h' = \nabla^2 M_h \cos[\omega_m t + \theta_h] - |\nabla\theta_h|^2 M_h \cos[\omega_m t + \theta_h] - \frac{\nabla\cdot[M_h^2\nabla\theta_h]}{M_h}\sin[\omega_m t + \theta_h].$$

Using (19.10), (22.6b), and (22.6a), respectively, to rewrite the first, second, and third terms on the RHS and then rearranging terms give

$$\nabla^2 h' = \frac{|\nabla\theta_h|^2}{\omega_m^2}\frac{\partial^2 h'}{\partial t^2} + \frac{\nabla\cdot[M_h^2\nabla\theta_h]}{\omega_m M_h^2}\frac{\partial h'}{\partial t} + \frac{\nabla^2 M_h}{M_h}h'.$$

Combining this equation with the phase speed (19.13) and the coefficient definitions (22.5) gives (22.4). □

We can summarize the results obtained in this section as follows. If each generic wave field, f, can represent either a constituent wave field ($f = h'$) or a component wave field ($f = h^{(n)}$), then f satisfies the generalized wave equation:

$$\nabla^2 f = \frac{1}{c^2(\mathbf{x};\omega_m)}\frac{\partial^2 f}{\partial t^2} + \gamma(\mathbf{x};\omega_m)\frac{\partial f}{\partial t} + \beta(\mathbf{x};\omega_m)f \quad \forall\, \mathbf{x} \in D_+(\omega_m) \tag{22.7}$$

where (19.8) gives the propagation speed, c, and we define the coefficients γ and β as

$$\gamma(\mathbf{x};\omega_m) \equiv \frac{\nabla\cdot[M^2\nabla\theta]}{\omega_m M^2} \quad (\text{dimensions: } L^{-2}t) \tag{22.8a}$$

$$\beta(\mathbf{x};\omega_m) \equiv \frac{\nabla^2 M}{M} \quad (\text{dimensions: } L^{-2}). \tag{22.8b}$$

22.2. INTERPRETATION

For a given harmonic constituent (i.e., ω_m fixed), (22.7) is a homogeneous, second-order PDE with space-variable coefficients. Equation (22.7) superficially resembles the damped wave equations of mathematical physics, such as those encountered in elasticity and electrodynamics. It is also a form of the *telegraph* (or *telegraphers'*) *equation* with space-variable coefficients.

For the purpose of discussion we will refer to the terms of (22.7) as follows:

$$\nabla^2 f \equiv \text{Laplacian term}$$

$$\frac{1}{c^2}\frac{\partial^2 f}{\partial t^2} \equiv \text{order-2 term}$$

$$\gamma\frac{\partial f}{\partial t} \equiv \text{order-1 term}$$

$$\beta f \equiv \text{order-0 term}.$$

One way to understand the roles of the various terms in the wave equation (22.7) is by analogy with other physical systems that are governed by similar equations. Example 22.1 illustrates such an analogy.

Example 22.1 Wave Equation: Vibrating-String Analogy
Consider the 1D transverse deflection of an elastic string undergoing small-amplitude vibrations and subject to a viscous damping force and a Hookean restoring force:

$$\frac{\partial^2 u}{\partial t^2} = c^2 \frac{\partial^2 u}{\partial x^2} - c^2 G \frac{\partial u}{\partial t} - c^2 B u \tag{22.9}$$

where $u = u(x, t)$ represents the vertical deflection (distance) of the string from its horizontal, equilibrium position ($u = 0$) and G and B are spatially variable coefficient functions. This equation is a statement of vertical momentum balance. The LHS represents the vertical acceleration of the string at the point whose horizontal space coordinate is x. The equation states that the acceleration is equal to the sum of three accelerations, represented by the three terms on the RHS:

- The term $c^2(\partial^2 u/\partial x^2)$ represents the acceleration due to the elastic restoring force of the deflected string.
- At those points where $G > 0$ the term $-c^2 G(\partial u/\partial t)$ gives the acceleration due to a viscous damping force that acts opposite to the string velocity $\partial u/\partial t$. Conversely, at those points, if any, where $G < 0$ this term represents the effect of a viscous force that tends to amplify the deflection rather than to dampen it.
- At those points where $B > 0$ the term $-c^2 Bu$ represents the acceleration due to a restoring force that acts proportional and opposite to the string deflection u. Conversely, at those points, if any, where $B < 0$ this term represents the effect of a destabilizing rather than restoring force.

Compare the wave equation (22.7) for time-periodic, 1D flow:

$$\frac{\partial^2 f}{\partial t^2} = c^2 \frac{\partial^2 f}{\partial x^2} - c^2 \gamma \frac{\partial f}{\partial t} - c^2 \beta f. \tag{22.10}$$

Equations (22.10) and (22.9) are identical in form, with f replacing u and γ and β replacing G and B, respectively. Thus, if γ and β are positive, then the order-1 and order-0 terms of the wave equation (22.7) represent damping "forces" that act to reduce the amplitude of the hydraulic head fluctuation. ◊

22.3. COEFFICIENT FUNCTIONS γ AND β

In Example 22.1 we used the analogy of a vibrating elastic string to show that the order-1 term of the wave equation (22.7) represents damping of the hydraulic head fluctuations at points where γ is positive and amplification at points where γ is negative. We also showed that similar results apply for the order-0 term and the coefficient function β. Intuitively, it would seem to follow that in a source free domain the coefficient functions γ and β cannot both be negative everywhere, because that would appear to require a net input of energy.

Exercise 22.3 Wave Equation Coefficient Functions

Problem: Consider a source-free domain filled with a medium that is homogeneous and isotropic with respect to hydraulic conductivity. What can you infer about the signs of the coefficient functions γ and β in this case? Based on your findings, what is the significance of the order-1 and order-0 terms of the wave equation (22.7) in this case?

Solution: Substituting M for M_{h}, θ for θ_{h}, $M_{\mathrm{u}} = 0$, and K (with $\nabla K = \mathbf{0}$) for $\mathbf{K}$ in (6.11) leads to

$$\begin{aligned} \beta &= |\nabla\theta|^2 \\ \gamma &= \frac{S_{\mathrm{s}}(\mathbf{x})}{K} \end{aligned} \tag{22.11}$$

from which we conclude

$$\begin{aligned} \beta &\geq 0 \\ \gamma &> 0. \end{aligned} \tag{22.12}$$

Thus, in the special case where the domain is source free and the porous medium is homogeneous and isotropic with respect to hydraulic conductivity, the order-1 and order-0 terms of the wave equation (22.7) generally represent damping, rather than amplification, of the hydraulic head fluctuations. Possible exceptions are stationary points of the phase function; at such points (if any) $\beta = 0$ and the order-0 term of the wave equation vanishes. □

Exercise 22.4 Wave Equation Coefficient Function

Problem: Consider time-periodic, 1-D flow through a semi-infinite, source-free domain filled with an exponential medium. Suppose the FRF solution described in Section 10.4 is valid. Evaluate the coefficient function $\gamma(\mathbf{x}; \omega_m)$. What can you say about the spatial variation and sign of $\gamma(\mathbf{x}; \omega_m)$ in this case?

Solution: Section 10.4 indicates that in this case there is only one component wave, so $f = h' = h^{(1)}$. Expressing the RHS of (22.8a) in terms of the log amplitude gives

$$\gamma(\mathbf{x}; \omega_m) = \frac{1}{\omega_m}\left(2\nabla Y \cdot \nabla\theta + \nabla^2\theta\right)$$

where $Y \equiv \ln M$. Equation (10.80b) indicates that the phase varies linearly with the space coordinate, so $\nabla^2\theta = 0$. Substituting this in the previous equation yields

$$\gamma(\mathbf{x};\omega_m) = \frac{2}{\omega_m}\nabla Y \cdot \nabla\theta.$$

Using the results of Exercise 10.22 we can express this as

$$\gamma(\mathbf{x};\omega_m) = \frac{2}{\omega_m}\left[\frac{\mu}{2} + \sqrt{\frac{\sqrt{\mu^4+16}+\mu^2}{8}}\right] \times \sqrt{\frac{\sqrt{\mu^4+16}-\mu^2}{8}} \quad (22.13)$$

which implies that $\gamma(\mathbf{x};\omega_m)$ is spatially uniform in this case.

Also, because $\sqrt{\mu^4+16} > \mu^2$, we have

$$\frac{\mu}{2} + \sqrt{\frac{\sqrt{\mu^4+16}+\mu^2}{8}} > 0$$

$$\sqrt{\frac{\sqrt{\mu^4+16}-\mu^2}{8}} > 0.$$

Using these to evaluate the sign of the RHS of (22.13) gives $\gamma(\mathbf{x};\omega_m) > 0$. □

Exercise 22.5 Wave Equation Coefficient Function

Problem: Consider time-periodic, 1D flow through a semi-infinite, source-free domain filled with an exponential medium. Suppose the FRF solution described in Section 10.4 is valid. Evaluate the coefficient function $\beta(\mathbf{x};\omega_m)$. What can you say about the spatial variation and sign of $\beta(\mathbf{x};\omega_m)$ in this case?

Solution: Section 10.4 indicates that in this case there is only one component wave, so $f = h' = h^{(1)}$. Expressing the RHS of (22.8b) in terms of the log amplitude gives

$$\beta(\mathbf{x};\omega_m) = |\nabla Y|^2 + \nabla^2 Y$$

where $Y \equiv \ln M$. Equation (10.80a) indicates that the log amplitude varies linearly with the space coordinate, so $\nabla^2 Y = 0$. Substituting this in the previous equation yields

$$\beta(\mathbf{x};\omega_m) = |\nabla Y|^2 \quad (22.14)$$

which implies that $\beta(\mathbf{x};\omega_m)$ is nonnegative. Using the results of Exercise 10.22 we can express this as

$$\beta(\mathbf{x};\omega_m) = \left[\frac{\mu}{2} + \sqrt{\frac{\sqrt{\mu^4+16}+\mu^2}{8}}\right]^2 \quad (22.15)$$

which implies that $\beta(\mathbf{x};\omega_m)$ is spatially uniform (and nonnegative) in this case.

Also, because $\sqrt{\mu^4+16} > \mu^2$, we have

$$\frac{\mu}{2} + \sqrt{\frac{\sqrt{\mu^4+16}+\mu^2}{8}} > 0.$$

Using this to evaluate the sign of the RHS of (22.15) gives $\beta(\mathbf{x};\omega_m) > 0$. □

Exercise 22.6 Wave Equation Coefficient Functions

Problem: Consider the general case of time-periodic groundwater flow in a nonhomogeneous, anisotropic medium. Do the coefficients of the generalized wave equation (22.7) (i.e., c^{-2}, γ, β) depend either explicitly or implicitly on the boundary conditions? Do they depend on the source distribution? Do they depend on the material hydrologic properties $\mathbf{K}$ and S_s? Explain.

Solution: The hydraulic head constituent amplitude and phase functions depend directly, albeit implicitly, on the boundary conditions, the source distribution, and the hydraulic properties via the polar form of the space BVP [see (6.11) and (6.30)]. Therefore, by (22.8), the coefficients depend directly, but implicitly, on the boundary conditions, the source distribution, and the hydraulic properties. □

22.4. DISCUSSION

That the coefficients of the generalized wave equation (22.7) are known functions of the amplitude and phase has the following important consequences:

1. The coefficients generally are unknown unless the amplitude and phase functions are known, that is, unless the solution of the space BVP [e.g., the system of equations (7.17)] is known.
2. The coefficients depend implicitly not only on the material hydrologic properties $\mathbf{K}$ and S_s but also on the boundary conditions and the source distribution (see Exercise 22.6). Therefore, the coefficients do not directly represent intrinsic physical properties of the medium.
3. Superposition is not valid for the generalized wave equation (22.7).

These consequences limit the usefulness of the generalized wave equation.

Although we have shown that the wave field f satisfies the generalized wave equation (22.7), we do not claim that the equation has any particular physical significance in this context. In fact, the wave field f also may satisfy numerous other wave equations that differ from one another in nontrivial ways. Perhaps the most important point of this chapter is simply that the wave field f satisfies a wave equation and it is not difficult to write such an equation.

22.5. CHAPTER SUMMARY

In this chapter we showed that each generic hydraulic head wave field satisfies a generalized wave equation (22.7), which is a homogenous, second-order PDE with real-valued coefficients that depend on the space coordinates and the constituent frequency. We expressed the coefficients explicitly in terms of the amplitude and phase functions and the frequency of the generic wave.

Part VII
Energy Transport

In this part we explore the transport of fluid mechanical energy by time-periodic groundwater flow in elastically deformable, saturated porous media under isothermal conditions. Our examination is based on assumptions that are consistent with those of the groundwater flow equation (2.5).

Throughout our discussion of energy, we assume that the fluid velocity is sufficiently low that the fluid's kinetic energy is negligible relative to its nonkinetic mechanical energy.

Hydrodynamics of Time-Periodic Groundwater Flow: Diffusion Waves in Porous Media, Geophysical Monograph 224,
First Edition. Joe S. Depner and Todd C. Rasmussen.

23
Mechanical Energy of Groundwater

In this chapter we begin our examination of the transport of fluid mechanical energy by groundwater flow under isothermal conditions. We develop differential and integral equations describing the instantaneous conservation of fluid mechanical energy, and we derive expressions for fluid mechanical energy density and its rate of change, flux density, and dissipation rate.

23.1. BASIC DEFINITIONS

Define the effective fluid *mechanical energy density* as

$$\epsilon(\mathbf{x}, t) \equiv \phi_e \rho_w g h \tag{23.1}$$

where ρ_w is the mass density of resident groundwater and ϕ_e is the effective (interconnected) porosity of the porous medium. The effective fluid mechanical energy density is the amount of nonkinetic mechanical energy stored in the resident groundwater per unit volume of porous medium. It has the dimensions of energy per unit volume, or $ML^{-1}t^{-2}$.

Our use of the qualifier *effective* in the term *effective fluid mechanical energy* reflects our assumption that the fluid kinetic energy is negligible relative to the fluid nonkinetic mechanical energy. Henceforth, for brevity we will drop the qualifier and instead use the term *fluid mechanical energy*.

Define the fluid mechanical energy *flux density* as

$$\mathbf{J}(\mathbf{x}, t) \equiv \epsilon \mathbf{q}. \tag{23.2}$$

The flux density gives the instantaneous rate of flow of fluid mechanical energy per unit cross-sectional area due to fluid flow. It has the dimensions of Mt^{-3}.

Define the *nominal seepage velocity* as

$$\mathbf{v}(\mathbf{x}, t) \equiv \frac{\mathbf{q}}{\phi_e}. \tag{23.3}$$

The nominal seepage velocity is the average particle velocity of groundwater flowing through a fictitious porous medium whose specific discharge is $\mathbf{q}$ and whose effective porosity is ϕ_e. We can also write it as

$$\mathbf{v} = \mathbf{v}^{(s)} + \mathbf{v}^{(r)} \tag{23.4}$$

where $\mathbf{v}^{(s)}$ is the average velocity of the solid matrix and $\mathbf{v}^{(r)}$ represents the average velocity of the groundwater *relative to* the solid matrix, i.e.,

$$\mathbf{v}^{(r)} \equiv \mathbf{v} - \mathbf{v}^{(s)}.$$

We can write the flux density in terms of the velocities $\mathbf{v}^{(s)}$ and $\mathbf{v}^{(r)}$ by combining (23.2), (23.3), and (23.4) as follows:

$$\mathbf{J} = \epsilon \phi_e \left(\mathbf{v}^{(s)} + \mathbf{v}^{(r)}\right). \tag{23.5}$$

23.2. CONSERVATION EQUATIONS

In this section we develop equations describing the conservation of fluid mechanical energy for a deforming porous medium in three dimensions. *Oosterbaan et al.* [1996] provide a less formal development, applicable to groundwater flow in an unconfined aquifer.

Let $V(t)$ denote a closed, N-dimensional control volume within the space domain D. The control volume moves (deforms) with the solid matrix. We will assume that the deformation is such that the shape and total volume of $V(t)$ change continuously with time. Let $S(t)$ denote the boundary of $V(t)$. Then $S(t)$ is a control surface whose position and shape change continuously with time.

The total amount of fluid mechanical energy within $V(t)$ at time t is

$$E[t; V(t)] \equiv \int_{V(t)} \epsilon \, dV \tag{23.6}$$

Hydrodynamics of Time-Periodic Groundwater Flow: Diffusion Waves in Porous Media, Geophysical Monograph 224, First Edition. Joe S. Depner and Todd C. Rasmussen.

where dV is a differential element of volume (dimensions: L^3). We assume that the fluid mechanical energy satisfies the following general balance law:

$$\frac{dE}{dt} = -\int_{S(t)} \epsilon\phi_e \mathbf{v}^{(r)} \cdot \hat{\mathbf{n}}\, dS - \int_{V(t)} \eta\, dV + \int_{V(t)} \sigma\, dV \quad (23.7)$$

where

$\hat{\mathbf{n}}$ is the outward unit-normal vector to the bounding surface S (dimensionless);

dS is a differential element of surface area (dimensions: L^2);

$\eta(\mathbf{x}, t)$ is the local per-volume rate of mechanical energy dissipation (e.g., viscous dissipation) (dimensions: $ML^{-1}t^{-3}$); and

$\sigma(\mathbf{x}, t)$ is the local per-volume rate of mechanical energy delivery by internal sources (dimensions: $ML^{-1}t^{-3}$).

Equation (23.7) is an integral equation for the conservation of fluid mechanical energy. It states that the instantaneous time rate of change of the total amount of fluid mechanical energy within the moving control volume $V(t)$ is the sum of three contributions, which we describe below.

The first term on the RHS of (23.7) represents the energy flux imbalance. This surface integral term, which we also refer to as the *flux term*, corresponds to the total (net) rate of flow of fluid mechanical energy into the moving control volume $V(t)$ across its boundary.

The second term on the RHS of (23.7) represents energy conversion. This volume integral term, which we refer to as the *dissipation term*, corresponds to the net rate of dissipation ($\eta > 0$) and/or accumulation ($\eta < 0$) of fluid mechanical energy by means other than energy flux imbalance and other than internal source generation. For more on dissipation, see Section 23.5.

The third term on the RHS of (23.7) represents the net rate of energy delivery by sources. This volume integral term, which we refer to as the energy *source term*, corresponds to the rate of delivery ($\sigma > 0$) or removal ($\sigma < 0$) of mechanical energy by internal sources. For more on energy sources, see Section 25.4.2. In source-free domains $\sigma = 0$.

We can write the RHS of (23.7) as a single volume integral by applying the divergence theorem to the surface integral:

$$\frac{dE}{dt} = \int_{V(t)} \left[-\nabla \cdot (\epsilon\phi_e \mathbf{v}^{(r)}) - \eta + \sigma \right] dV. \quad (23.8)$$

Applying the Reynolds transport theorem [*Aris*, 1962] to (23.6) gives

$$\frac{dE}{dt} = \int_{V(t)} \left[\frac{\partial \epsilon}{\partial t} + \nabla \cdot (\epsilon\phi_e \mathbf{v}^{(s)}) \right] dV. \quad (23.9)$$

Equating the RHSs of (23.8) and (23.9), rearranging, and using (23.5) yield

$$\int_{V(t)} \left[\frac{\partial \epsilon}{\partial t} + \nabla \cdot \mathbf{J} + \eta - \sigma \right] dV = 0.$$

This equation is valid for every control volume within the space domain, so it is valid in the limit as the size of the control volume decreases to zero:

$$\frac{\partial \epsilon}{\partial t} = -\nabla \cdot \mathbf{J} - \eta + \sigma. \quad (23.10)$$

Equation (23.10) is a differential equation for the conservation of fluid mechanical energy. It states that the local time rate of change of the fluid mechanical energy density is the sum of three contributions, which we describe below:

The first contribution, corresponding to the term $-\nabla \cdot \mathbf{J}$, is the local per-volume rate of mechanical energy accumulation due to flux density convergence.

The second contribution, corresponding to the term $-\eta$, is the local per-volume rate of mechanical energy accumulation/dissipation by means other than flux convergence and other than internal source generation. The conditions $\eta > 0$ and $\eta < 0$, respectively, represent dissipation and accumulation.

The third contribution, corresponding to the term σ, is the local per-volume rate of mechanical energy delivery/removal by internal sources. The conditions $\sigma > 0$ and $\sigma < 0$, respectively, represent mechanical energy delivery and removal. In source-free domains $\sigma = 0$.

23.3. RATE OF CHANGE OF ENERGY DENSITY

In this section we express the local rate of change of the fluid mechanical energy density in terms of the hydraulic head and its time derivative.

Using the definition (23.1) for the energy density ϵ and assuming the acceleration of gravity is time invariant, we can write

$$\frac{\partial \epsilon}{\partial t} = \rho_w g h \frac{\partial \phi_e}{\partial t} + \phi_e g h \frac{\partial \rho_w}{\partial t} + \phi_e \rho_w g \frac{\partial h}{\partial t}. \quad (23.11)$$

Notice that the derivatives with respect to time in the first and second terms on the RHS are not yet expressed in terms of the hydraulic head.

Note 23.1 Spatiotemporal Variations of Effective Porosity and Void Ratio

We can write the effective porosity as

$$\phi_e = \frac{V_v}{V_s + V_v}$$

where V_v denotes the volume of interconnected voids and V_s denotes the combined volume of solids and isolated voids in a finite-volume sample of porous medium. The effective porosity satisfies $0 \leq \phi_e \leq 1$. Dividing both the numerator and the denominator on the RHS by V_s gives

$$\phi_e = \frac{e}{1+e} \tag{23.12}$$

where e denotes the *void ratio*, which we define as

$$e \equiv \frac{V_v}{V_s}, \quad V_s > 0.$$

The void ratio satisfies $e \geq 0$. Thus, the effective porosity is a univariate function of the void ratio. Taking the total differential of (23.12) yields

$$d\phi_e = \frac{de}{(1+e)^2}. \tag{23.13}$$

△

Note 23.2 Void Ratio, Effective Stress, and Water Pressure
Assume the void ratio depends on effective stress, σ_e, and on position:

$$e = f[\sigma_e(\mathbf{x}, t), \mathbf{x}].$$

If the grains of the porous matrix are incompressible, so that all of the bulk dilation/contraction is accommodated directly by commensurate changes in effective porosity, then the medium's *bulk compressibility* (α) is given by

$$\alpha \equiv \frac{-1}{1+e}\frac{\partial e}{\partial \sigma_e}.$$

Solving for the derivative gives

$$\frac{\partial e}{\partial \sigma_e} = -(1+e)\alpha.$$

Assuming that incremental increases in the effective stress are offset by corresponding decreases in the water pressure and vice versa, i.e.,

$$d\sigma_e = -dp$$

yields

$$\frac{\partial e}{\partial p} = (1+e)\alpha. \tag{23.14}$$

△

Exercise 23.1 Derivative of Effective Porosity with Respect to Pressure

Problem: Derive a compact expression for $\partial\phi_e/\partial p$ in terms of the effective porosity and the bulk compressibility.

Solution: Assume the effective porosity depends on the void ratio and the position, as

$$\phi_e(\mathbf{x}, t) = f[e(\mathbf{x}, t), \mathbf{x}]$$

and that the void ratio is a function of the pressure and the position, as

$$e(\mathbf{x}, t) = F[p(\mathbf{x}, t), \mathbf{x}].$$

Differentiating the effective porosity with respect to the pressure and using the chain rule give

$$\frac{\partial \phi_e}{\partial p} = \left(\frac{\partial \phi_e}{\partial e}\right)\left(\frac{\partial e}{\partial p}\right).$$

Using (23.13) to rewrite the first term on the RHS, substituting the RHS of (23.14) for the second term on the RHS, and then using (23.12) to express the void ratio in terms of the effective porosity yield

$$\frac{\partial \phi_e}{\partial p} = (1-\phi_e)\alpha. \tag{23.15}$$

□

Exercise 23.2 Derivative of Effective Porosity with Respect to Time

Problem: Express $\partial\phi_e/\partial t$ in terms of $\partial p/\partial t$ and then in terms of $\partial h/\partial t$.

Solution: Assume the effective porosity depends on both the water pressure and the position, as

$$\phi_e(\mathbf{x}, t) = f[p(\mathbf{x}, t), \mathbf{x}].$$

Differentiating with respect to time and using the chain rule give

$$\frac{\partial \phi_e}{\partial t} = \left(\frac{\partial \phi_e}{\partial p}\right)\left(\frac{\partial p}{\partial t}\right).$$

Substituting the RHS of (23.15) for $\partial\phi_e/\partial p$ yields

$$\frac{\partial \phi_e}{\partial t} = (1-\phi_e)\alpha\frac{\partial p}{\partial t}.$$

Using (2.3) with the substitution $d\mathbf{x} = \mathbf{0}$ leads to

$$\frac{\partial \phi_e}{\partial t} = (1-\phi_e)\alpha\rho_w g\frac{\partial h}{\partial t}. \tag{23.16}$$

□

Using (23.16), we can write the first term on the RHS of (23.11) as

$$\rho_w g h\frac{\partial \phi_e}{\partial t} = (1-\phi_e)\alpha\rho_w^2 g^2 h\frac{\partial h}{\partial t}. \tag{23.17}$$

Note 23.3 Spatiotemporal Variations of Groundwater Density and Pressure
Assuming that fresh groundwater is a chemically pure or nearly pure fluid, its density is a function of its pressure and its temperature, so that its equation of state is of the form

$$\rho_w = f(p, T)$$

where p and T denote water pressure and water temperature, respectively. The total differential of the water density is then

$$d\rho_{\mathrm{w}} = \left(\frac{\partial \rho_{\mathrm{w}}}{\partial p}\right) \mathrm{d}p + \left(\frac{\partial \rho_{\mathrm{w}}}{\partial T}\right) dT$$

where p and T denote pressure and temperature, respectively. If the groundwater system is isothermal or approximately isothermal, then $dT = 0$, giving

$$d\rho_{\mathrm{w}} = \beta_{\mathrm{w}} \rho_{\mathrm{w}}\, dp \tag{23.18}$$

where β_{w} is the isothermal compressibility of water, i.e.,

$$\beta_{\mathrm{w}} \equiv \frac{1}{\rho_{\mathrm{w}}} \left(\frac{\partial \rho_{\mathrm{w}}}{\partial p}\right)_{\mathrm{T}}.$$

△

Using (2.3), we can rewrite (23.18) as

$$d\rho_{\mathrm{w}} = \beta_{\mathrm{w}} \rho_{\mathrm{w}}^2 g (dh - dx_3).$$

Using this result, we can write the second term on the RHS of (23.11) as

$$\phi_{\mathrm{e}} g h \frac{\partial \rho_{\mathrm{w}}}{\partial t} = \phi_{\mathrm{e}} \beta_{\mathrm{w}} \rho_{\mathrm{w}}^2 g^2 h \frac{\partial h}{\partial t}. \tag{23.19}$$

Substituting the RHSs of (23.17) and (23.19) for the first and second terms, respectively, on the RHS of (23.11) and grouping terms give

$$\frac{\partial \epsilon}{\partial t} = \left(\alpha - \phi_{\mathrm{e}} \alpha + \phi_{\mathrm{e}} \beta_{\mathrm{w}}\right) \rho_{\mathrm{w}}^2 g^2 h \frac{\partial h}{\partial t} + \phi_{\mathrm{e}} \rho_{\mathrm{w}} g \frac{\partial h}{\partial t}. \tag{23.20}$$

23.4. MECHANICAL ENERGY FLUX DENSITY

In this section we express the fluid mechanical energy flux density and its divergence in terms of the hydraulic head and its derivatives.

23.4.1. General Results

Using (23.5), we can write the flux density as

$$\mathbf{J} = \epsilon \mathbf{q}^{(r)} + \phi_{\mathrm{e}} \epsilon \mathbf{v}^{(\mathrm{s})} \tag{23.21}$$

where $\mathbf{q}^{(r)}$ represents the volumetric specific discharge of groundwater as observed in the moving frame of the deforming solid matrix, i.e.,

$$\mathbf{q}^{(r)} = \phi_{\mathrm{e}} \mathbf{v}^{(r)}. \tag{23.22}$$

Combining Darcy's law, (2.4), with (23.21) gives

$$\mathbf{J} = -\epsilon \mathbf{K} \nabla h + \phi_{\mathrm{e}} \epsilon \mathbf{v}^{(\mathrm{s})}. \tag{23.23}$$

Note 23.4 Mass Conservation for Solid Matrix
Mass conservation for the solid matrix requires [*Freeze and Cherry*, 1979]

$$-\nabla \cdot \left[(1 - \phi_{\mathrm{e}}) \rho_{\mathrm{s}} \mathbf{v}^{(\mathrm{s})}\right] = \frac{\partial}{\partial t}\left[(1 - \phi_{\mathrm{e}}) \rho_{\mathrm{s}}\right]$$

where ρ_{s} represents the density of the solid matrix (including isolated voids) of the porous medium. Assuming the solid matrix is spatially homogeneous with respect to mass density and incompressible (i.e., $|\nabla \rho_{\mathrm{s}}| = \partial \rho_{\mathrm{s}}/\partial t = 0$), the continuity equation yields

$$\nabla \cdot (\phi_{\mathrm{e}} \mathbf{v}^{(\mathrm{s})}) = \nabla \cdot \mathbf{v}^{(\mathrm{s})} - \frac{\partial \phi_{\mathrm{e}}}{\partial t}. \tag{23.24}$$

△

23.4.2. Divergence of Energy Flux Density

Taking the divergence of (23.23) gives

$$\nabla \cdot \mathbf{J} = -\nabla \cdot (\epsilon \mathbf{K} \nabla h) + \nabla \cdot (\phi_{\mathrm{e}} \epsilon \mathbf{v}^{(\mathrm{s})}). \tag{23.25}$$

Expanding the first term on the RHS leads to

$$-\nabla \cdot (\epsilon \mathbf{K} \nabla h) = -(\mathbf{K} \nabla h) \cdot \nabla \epsilon - \epsilon \nabla \cdot (\mathbf{K} \nabla h).$$

Using the groundwater flow equation (2.5) to rewrite the divergence on the RHS gives

$$-\nabla \cdot (\epsilon \mathbf{K} \nabla h) = -(\mathbf{K} \nabla h) \cdot \nabla \epsilon - \epsilon S_{\mathrm{s}} \frac{\partial h}{\partial t}. \tag{23.26}$$

Expanding the second term on the RHS of (23.25) gives

$$\nabla \cdot (\phi_{\mathrm{e}} \epsilon \mathbf{v}^{(\mathrm{s})}) = \phi_{\mathrm{e}} \mathbf{v}^{(\mathrm{s})} \cdot \nabla \epsilon + \epsilon \nabla \cdot (\phi_{\mathrm{e}} \mathbf{v}^{(\mathrm{s})}).$$

Substituting the RHS of (23.24) for the divergence on the RHS yields

$$\nabla \cdot (\phi_{\mathrm{e}} \epsilon \mathbf{v}^{(\mathrm{s})}) = \phi_{\mathrm{e}} \mathbf{v}^{(\mathrm{s})} \cdot \nabla \epsilon + \epsilon \nabla \cdot \mathbf{v}^{(\mathrm{s})} - \epsilon \frac{\partial \phi_{\mathrm{e}}}{\partial t}.$$

Using the approximation $\nabla \cdot \mathbf{v}^{(\mathrm{s})} = \alpha\, \partial p/\partial t$ (for details see *Freeze and Cherry* [1979]), this becomes

$$\nabla \cdot (\phi_{\mathrm{e}} \epsilon \mathbf{v}^{(\mathrm{s})}) = \phi_{\mathrm{e}} \mathbf{v}^{(\mathrm{s})} \cdot \nabla \epsilon + \alpha \epsilon \frac{\partial p}{\partial t} - \epsilon \frac{\partial \phi_{\mathrm{e}}}{\partial t}.$$

Combining this result with (23.25) and (23.26) leads to

$$\nabla \cdot \mathbf{J} = -(\mathbf{K} \nabla h) \cdot \nabla \epsilon + \phi_{\mathrm{e}} \mathbf{v}^{(\mathrm{s})} \cdot \nabla \epsilon + \epsilon \left(- S_{\mathrm{s}} \frac{\partial h}{\partial t} + \alpha \frac{\partial p}{\partial t} - \frac{\partial \phi_{\mathrm{e}}}{\partial t} \right). \tag{23.27}$$

We will assume that

$$|\phi_{\mathrm{e}} \mathbf{v}^{(\mathrm{s})} \cdot \nabla \epsilon| \ll \left| \frac{\partial \epsilon}{\partial t} \right|$$

so that the second term on the RHS of (23.27) makes a negligible contribution to the RHS of (23.10). Therefore, (23.27) is approximated as

$$\nabla \cdot \mathbf{J} = -(\mathbf{K}\nabla h) \cdot \nabla \epsilon + \epsilon\left(- S_s \frac{\partial h}{\partial t} + \alpha \frac{\partial p}{\partial t} - \frac{\partial \phi_e}{\partial t} \right).$$

Using (23.1) to express the energy density in terms of the hydraulic head and using (2.3) and (23.16) to express the derivatives with respect to time of the pressure and the effective porosity in terms of the hydraulic head and simplifying yield

$$\nabla \cdot \mathbf{J} = -(\mathbf{K}\nabla h) \cdot \nabla(\phi_e \rho_w g h) + \phi_e \rho_w g(\rho_w g \phi_e \alpha - S_s) h \frac{\partial h}{\partial t}. \tag{23.28}$$

23.5. DISSIPATION

Possible mechanisms for the dissipation of fluid mechanical energy include the following:

- conversion to heat by viscous dissipation,
- conversion to heat in the solid matrix by anelastic deformation, and
- conversion to elastic potential energy stored in the solid matrix.

We will assume that the work done by the fluid to deform the solid matrix is negligible in comparison to the amount of mechanical energy lost by viscous dissipation, so that the dissipation term effectively represents viscous dissipation. Consequently, the dissipation will be positive at those points and times for which the specific discharge is nonzero:

$$\operatorname{sign}\eta = \operatorname{sign}|\mathbf{q}|$$

so that

$$\eta \geq 0 \quad \forall \mathbf{x} \in D, \forall t. \tag{23.29}$$

Solving (23.10) for the per-volume, mechanical energy dissipation rate gives

$$\eta = -\frac{\partial \epsilon}{\partial t} - \nabla \cdot \mathbf{J} + \sigma. \tag{23.30}$$

Substituting the RHS of (23.28) for the divergence of the flux density yields

$$\eta = -\frac{\partial \epsilon}{\partial t} + (\mathbf{K}\nabla h) \cdot \nabla(\phi_e \rho_w g h) - \phi_e \rho_w g(\rho_w g \phi_e \alpha - S_s) h \frac{\partial h}{\partial t} + \sigma. \tag{23.31}$$

23.6. CHAPTER SUMMARY

In this chapter we began our examination of the transport of fluid mechanical energy by groundwater flow under isothermal conditions. We developed differential and integral equations describing the instantaneous conservation of fluid mechanical energy. Based on assumptions that are consistent with those that underlie the groundwater flow equation (see Chapter 2), we derived expressions for the fluid mechanical energy density and its rate of change, flux density, and dissipation rate.

24
Mechanical Energy: Time Averages

In this chapter we examine the *time-averaged* transport of fluid mechanical energy by time-periodic groundwater flow under isothermal conditions by applying time averaging to the results presented in Chapter 23. We develop differential and integral equations for the time-averaged conservation of fluid mechanical energy, and we derive expressions for time averages of the following: the fluid mechanical energy density and its rate of change, the flux density, and the dissipation rate. We assume that the hydraulic head field consists of multiple harmonic constituents.

Assume the hydraulic head field consists of multiple harmonic constituents; then it is given by the trigonometric series in (4.1a):

$$h(\mathbf{x}, t) = \sum_{m=1}^{\infty} h'(\mathbf{x}, t; \omega_m). \tag{24.1}$$

24.1. TIME AVERAGES

Define the *time average* of a function $f(\mathbf{x}, t)$ as

$$\langle f \rangle(\mathbf{x}) \equiv \lim_{T\to\infty} \frac{1}{T} \int_{t_0}^{t_0+T} f(\mathbf{x}, t)\, dt$$

where t_0 is a real but otherwise arbitrary time value. Thus, the time average of the function $f(\mathbf{x}, t)$ is the inner product of $f(\mathbf{x}, t)$ and unity:

$$\langle f \rangle(\mathbf{x}) = \langle f, 1 \rangle. \tag{24.2}$$

For a definition of the inner product, see (5.2).

In the special case where $f(\mathbf{x}, t)$ is time periodic with period T_1, its time average satisfies

$$\langle f \rangle(\mathbf{x}) = \frac{1}{T_1} \int_{t_0}^{t_0+T_1} f(\mathbf{x}, t)\, dt. \tag{24.3}$$

That is, the time average is equivalent to the average over a single period.

24.1.1. Results for Single-Constituent Fields

In this section we investigate time averages of fields consisting of single harmonic constituents.

Combining (24.2) and (B.11) we obtain

$$\langle c \rangle = c \; (c \text{ constant})$$
$$\langle \cos(\omega_m t) \rangle = \langle \sin(\omega_m t) \rangle = 0 \quad \forall\, |\omega_m| > 0.$$

Exercise 24.1 Time Averages of Products of Sinusoids

Problem: By transforming the variable of integration, show that

$$\left\langle \left[\cos \xi_h(\mathbf{x}, t; \omega_m)\right]^r \left[\sin \xi_h(\mathbf{x}, t; \omega_m)\right]^s \right\rangle = \left\langle \left[\cos(\omega_m t)\right]^r \left[\sin(\omega_m t)\right]^s \right\rangle \tag{24.4}$$

where r and s are real numbers.

Solution

$$\left\langle \left[\cos \xi_h(\mathbf{x}, t; \omega_m)\right]^r \left[\sin \xi_h(\mathbf{x}, t; \omega_m)\right]^s \right\rangle = \lim_{T\to\infty} \frac{1}{T} \int_{t_0}^{t_0+T} \left\{\cos\left[\omega_m t + \theta_h(\mathbf{x}; \omega_m)\right]\right\}^r \times \left\{\sin\left[\omega_m t + \theta_h(\mathbf{x}; \omega_m)\right]\right\}^s dt. \tag{24.5}$$

Hydrodynamics of Time-Periodic Groundwater Flow: Diffusion Waves in Porous Media, Geophysical Monograph 224, First Edition. Joe S. Depner and Todd C. Rasmussen.

Let $t' \equiv t + \theta_h(\mathbf{x};\omega_m)/\omega_m$. Then $dt = dt'$ and the previous equation becomes

$$\left\langle [\cos\xi_h(\mathbf{x},t;\omega_m)]^r [\sin\xi_h(\mathbf{x},t;\omega_m)]^s \right\rangle = \lim_{T\to\infty}\frac{1}{T}\int_{t_0'}^{t_0'+T} [\cos(\omega_m t')]^r [\sin(\omega_m t')]^s \, dt'$$

where

$$t_0' \equiv t_0 + \frac{\theta_h(\mathbf{x};\omega_m)}{\omega_m}.$$

Equation (24.4) immediately follows. □

Note 24.1 Time Averages of Some Sinusoids

The following relations [*Beyer*, 1984] are useful for evaluating various time averages:

$$\int \cos^{2n}(au)\, du = \frac{\sin(au)}{a}\sum_{r=0}^{n-1}\frac{(2n)!(r!)^2}{4^{n-r}(2r+1)!(n!)^2} \times \cos^{2r+1}(au) + \frac{(2n)!}{4^n(n!)^2}u$$

$$\int \cos^{2n+1}(au)\, du = \frac{\sin(au)}{a}\sum_{r=0}^{n}\frac{4^{n-r}(n!)^2(2r)!}{(2n+1)!(r!)^2} \times \cos^{2r}(au)$$

$$\int \cos^{n}(au)\sin(au)\, du = -\frac{\cos^{n+1}(au)}{(n+1)a}$$

$$n \in \mathbb{Z},\ n > 0.$$

These lead to

$$\langle \cos^{2n}(\omega_m t)\rangle = \frac{(2n)!}{4^n(n!)^2} \tag{24.6a}$$

$$\langle \cos^{2n+1}(\omega_m t)\rangle = 0 \tag{24.6b}$$

$$\langle \cos^{n}(\omega_m t)\sin(\omega_m t)\rangle = 0 \tag{24.6c}$$

$$n \in \mathbb{Z},\ n > 0.$$

In all three of the above cases the time average is independent of the sinusoid's frequency. Specific examples of nonzero averages involving even powers include

$$\langle \cos^2(\omega_m t)\rangle = 1/2 \tag{24.7a}$$

$$\langle \cos^4(\omega_m t)\rangle = 3/8 \tag{24.7b}$$

$$\langle \cos^6(\omega_m t)\rangle = 5/16. \tag{24.7c}$$

△

Substituting $s = 0$ in (24.4) and then combining the result with (24.6a) and (24.6b) lead to

$$\left\langle [h'(\mathbf{x},t;\omega_m)]^r \right\rangle = \begin{cases} \dfrac{r!}{4^{r/2}[(r/2)!]^2}[M_h(\mathbf{x};\omega_m)]^r, & r \text{ even} \\ 0, & r \text{ odd} \end{cases} \tag{24.8}$$

where

$$r \in \mathbb{Z}, r > 0.$$

For some specific values of r, these give

$$\langle h'(\mathbf{x},t;\omega_m)\rangle = \left\langle [h'(\mathbf{x},t;\omega_m)]^3\right\rangle = \left\langle [h'(\mathbf{x},t;\omega_m)]^5\right\rangle = \cdots = 0 \tag{24.9a}$$

$$\left\langle [h'(\mathbf{x},t;\omega_m)]^2\right\rangle = 1/2[M_h(\mathbf{x};\omega_m)]^2 \tag{24.9b}$$

$$\left\langle [h'(\mathbf{x},t;\omega_m)]^4\right\rangle = 3/8[M_h(\mathbf{x};\omega_m)]^4 \tag{24.9c}$$

$$\left\langle [h'(\mathbf{x},t;\omega_m)]^6\right\rangle = 5/16[M_h(\mathbf{x};\omega_m)]^6. \tag{24.9d}$$

Exercise 24.2 Time Averages of Products of Sinusoids

Problem: Evaluate the time average of the product $\cos^r \xi_h \sin \xi_h$, where r is a nonnegative integer.

Solution: Substituting $s = 1$ in (24.4) yields

$$\langle \cos^r \xi_h \sin\xi_h\rangle = \langle \cos^r(\omega_m t)\sin(\omega_m t)\rangle.$$

Using (24.6c) to evaluate the time average on the RHS gives

$$\langle \cos^r \xi_h(\mathbf{x},t;\omega_m)\sin\xi_h(\mathbf{x},t;\omega_m)\rangle = 0 \quad (r\in\mathbb{Z},\ r\ge 0). \tag{24.10}$$

□

Exercise 24.3 Time Averages of Products

Problem: Evaluate the time average of $h'^r \partial h'/\partial t$, where r is a nonnegative integer.

Solution:

$$\left\langle h'^r \frac{\partial h'}{\partial t}\right\rangle = -\omega_m M_h^{r+1}\langle \cos^r\xi_h \sin\xi_h\rangle.$$

Using (24.10) to evaluate the time average on the RHS yields

$$\left\langle h'^r \frac{\partial h'}{\partial t}\right\rangle = 0 \quad (r\in\mathbb{Z},\ r\ge 0).$$

□

Exercise 24.4 Time Averages of Polynomial Functions

Problem: Show that if $f(h')$ is a polynomial in h', then

$$\langle f(h')\nabla h'\rangle = \langle f(h')h'\rangle M_{\mathrm{h}}^{-1}\nabla M_{\mathrm{h}}. \tag{24.11}$$

This relation is sometimes useful for rewriting the time average of a vector in terms of the time average of a scalar.

Solution: Using (14.19) and (14.20) to write the hydraulic gradient in terms of the corresponding amplitude, phase, and wave phase gives

$$\begin{aligned}\langle f(h')\nabla h'\rangle &= \langle f(h')[\nabla M_{\mathrm{h}}\cos\xi_{\mathrm{h}} - M_{\mathrm{h}}\nabla\theta_{\mathrm{h}}\sin\xi_{\mathrm{h}}]\rangle\\ &= \langle f(h')\cos\xi_{\mathrm{h}}\rangle\nabla M_{\mathrm{h}} - \langle f(h')\sin\xi_{\mathrm{h}}\rangle M_{\mathrm{h}}\nabla\theta_{\mathrm{h}}.\end{aligned}$$

Writing $\cos\xi_{\mathrm{h}}$ as $M_{\mathrm{h}}^{-1}h'$ and defining $F(\cos\xi_{\mathrm{h}}) \equiv f(h')$, where F denotes a polynomial in $\cos\xi_{\mathrm{h}}$, lead to

$$\langle f(h')\nabla h'\rangle = \langle f(h')h'\rangle M_{\mathrm{h}}^{-1}\nabla M_{\mathrm{h}} - \langle F(\cos\xi_{\mathrm{h}})\sin\xi_{\mathrm{h}}\rangle M_{\mathrm{h}}\nabla\theta_{\mathrm{h}}.$$

Using (24.10) to eliminate the time average in the second term on the RHS yields (24.11). □

Exercise 24.5 Time Averages of Products

Problem: Use (24.8) to express the time average, $\langle h'^r\nabla h'\rangle$, where r is a nonnegative integer, in terms of M_{h}.

Solution: From calculus,

$$\begin{aligned}\langle h'^r\nabla h'\rangle &= \left\langle \frac{1}{r+1}\nabla h'^{r+1}\right\rangle\\ &= \frac{1}{r+1}\nabla\langle h'^{r+1}\rangle.\end{aligned}$$

Using (24.8) to express the time average on the RHS in terms of M_{h} gives

$$\langle h'^r\nabla h'\rangle = \frac{1}{r+1}\nabla\begin{cases}\dfrac{(r+1)!}{4^{(r+1)/2}[(r+1)/2!]^2}M_{\mathrm{h}}^{r+1}, & r+1 \text{ even}\\ 0, & r+1 \text{ odd.}\end{cases}$$

Evaluating the gradient and then simplifying yield

$$\langle h'^r\nabla h'\rangle = \begin{cases}\dfrac{(r+1)!}{4^{(r+1)/2}[(r+1)/2!]^2}M_{\mathrm{h}}^{r}\nabla M_{\mathrm{h}}, & r \text{ odd}\\ 0, & r \text{ even}\end{cases} \tag{24.12}$$

$$r \in \mathbb{Z},\ r > 0.$$

□

24.1.2. Results for Multiple-Constituent Fields

In this section we investigate time averages of fields consisting of multiple harmonic constituents.

Exercise 24.6 Time Averages in Terms of Amplitudes

Problem: Consider a hydraulic head field consisting of multiple harmonic constituents. Express the time-averaged values of h, ∇h, and $\mathbf{q}$ in terms of the constituent amplitudes.

Solution: Substituting the RHS of (24.1) for h and taking the time average give

$$\langle h\rangle = \left\langle\sum_{m=1}^{\infty} h'(\mathbf{x}, t; \omega_m)\right\rangle = \sum_{m=1}^{\infty}\langle h'(\mathbf{x}, t; \omega_m)\rangle.$$

Using (24.9) to evaluate the time average on the RHS yields

$$\langle h\rangle = 0. \tag{24.13}$$

Next, we note that

$$\begin{aligned}\langle\nabla h\rangle &= \nabla\langle h\rangle\\ \langle\mathbf{q}\rangle &= -\mathbf{K}\nabla\langle h\rangle.\end{aligned}$$

Combining these and (24.13) leads to

$$\langle\nabla h\rangle = \mathbf{0} \tag{24.14}$$

$$\langle\mathbf{q}\rangle = \mathbf{0}. \tag{24.15}$$

□

Exercise 24.7 Time Average of Product of Sinusoids

Problem: Evaluate the time average $\langle\cos\xi_{\mathrm{h}}(\mathbf{x}, t; \omega_m)\cos\xi_{\mathrm{h}}(\mathbf{x}, t; \omega_n)\rangle$.

Solution

$$\begin{aligned}&\langle\cos\xi_{\mathrm{h}}(\mathbf{x}, t; \omega_m)\cos\xi_{\mathrm{h}}(\mathbf{x}, t; \omega_n)\rangle\\ &\quad= \langle\cos[\omega_m t + \theta_{\mathrm{h}}(\mathbf{x}; \omega_m)]\cos[\omega_n t + \theta_{\mathrm{h}}(\mathbf{x}; \omega_n)]\rangle.\end{aligned}$$

Using trigonometric identity (B.2a) to rewrite the cosine terms on the RHS gives

$$\begin{aligned}&\langle\cos\xi_{\mathrm{h}}(\mathbf{x}, t; \omega_m)\cos\xi_{\mathrm{h}}(\mathbf{x}, t; \omega_n)\rangle\\ &\quad= \langle[\cos(\omega_m t)\cos\theta_{\mathrm{h}}(\mathbf{x}; \omega_m) - \sin(\omega_m t)\sin\theta_{\mathrm{h}}(\mathbf{x}; \omega_m)]\\ &\qquad\times[\cos(\omega_n t)\cos\theta_{\mathrm{h}}(\mathbf{x}; \omega_n) - \sin(\omega_n t)\sin\theta_{\mathrm{h}}(\mathbf{x}; \omega_n)]\rangle.\end{aligned}$$

Carrying out the multiplication gives

$$\begin{aligned}\Big\langle \cos\xi_h(\mathbf{x},t;\omega_m)\cos\xi_h(\mathbf{x},t;\omega_n)\Big\rangle &= \cos\theta_h(\mathbf{x};\omega_m)\cos\theta_h(\mathbf{x};\omega_n)\Big\langle \cos(\omega_m t)\cos(\omega_n t)\Big\rangle \\ &\quad - \cos\theta_h(\mathbf{x};\omega_m)\sin\theta_h(\mathbf{x};\omega_n)\Big\langle \cos(\omega_m t)\sin(\omega_n t)\Big\rangle \\ &\quad - \sin\theta_h(\mathbf{x};\omega_m)\cos\theta_h(\mathbf{x};\omega_n)\Big\langle \sin(\omega_m t)\cos(\omega_n t)\Big\rangle \\ &\quad + \sin\theta_h(\mathbf{x};\omega_m)\sin\theta_h(\mathbf{x};\omega_n)\Big\langle \sin(\omega_m t)\sin(\omega_n t)\Big\rangle.\end{aligned}$$

Using identities (B.12) to evaluate the time averages on the RHS yields

$$\Big\langle \cos\xi_h(\mathbf{x},t;\omega_m)\cos\xi_h(\mathbf{x},t;\omega_n)\Big\rangle = \delta_{mn}/2. \tag{24.16}$$

where δ_{mn} is the Kronecker delta [see equation (2.17)]. □

Exercise 24.8 Time Average of Product of Sinusoids

Problem: Evaluate the time average $\big\langle \cos\xi_h(\mathbf{x},t;\omega_m)\sin\xi_h(\mathbf{x},t;\omega_n)\big\rangle$, where $m \neq n$.

Solution:

$$\Big\langle \cos\xi_h(\mathbf{x},t;\omega_m)\sin\xi_h(\mathbf{x},t;\omega_n)\Big\rangle = \Big\langle \cos\big[\omega_m t + \theta_h(\mathbf{x};\omega_m)\big]\sin\big[\omega_n t + \theta_h(\mathbf{x};\omega_n)\big]\Big\rangle.$$

Using trigonometric identites (B.2a) and (B.2c) to rewrite the cosine and sine terms, respectively, on the RHS gives

$$\begin{aligned}\Big\langle \cos\xi_h(\mathbf{x},t;\omega_m)\sin\xi_h(\mathbf{x},t;\omega_n)\Big\rangle &= \Big\langle\big[\cos(\omega_m t)\cos\theta_h(\mathbf{x};\omega_m) - \sin(\omega_m t)\sin\theta_h(\mathbf{x};\omega_m)\big] \\ &\quad \times \big[\sin(\omega_n t)\cos\theta_h(\mathbf{x};\omega_n) + \cos(\omega_n t)\sin\theta_h(\mathbf{x};\omega_n)\big]\Big\rangle.\end{aligned}$$

Carrying out the multiplication gives

$$\begin{aligned}\Big\langle \cos\xi_h(\mathbf{x},t;\omega_m)\cos\xi_h(\mathbf{x},t;\omega_n)\Big\rangle &= \cos\theta_h(\mathbf{x};\omega_m)\cos\theta_h(\mathbf{x};\omega_n)\Big\langle \cos(\omega_m t)\sin(\omega_n t)\Big\rangle \\ &\quad + \cos\theta_h(\mathbf{x};\omega_m)\sin\theta_h(\mathbf{x};\omega_n)\Big\langle \cos(\omega_m t)\cos(\omega_n t)\Big\rangle \\ &\quad - \sin\theta_h(\mathbf{x};\omega_m)\cos\theta_h(\mathbf{x};\omega_n)\Big\langle \sin(\omega_m t)\sin(\omega_n t)\Big\rangle \\ &\quad - \sin\theta_h(\mathbf{x};\omega_m)\sin\theta_h(\mathbf{x};\omega_n)\Big\langle \sin(\omega_m t)\cos(\omega_n t)\Big\rangle.\end{aligned}$$

Using identities (B.12) to evaluate the time averages on the RHS yields

$$\Big\langle \cos\xi_h(\mathbf{x},t;\omega_m)\sin\xi_h(\mathbf{x},t;\omega_n)\Big\rangle = 0 \quad (m \neq n). \tag{24.17}$$

□

Exercise 24.9 Time Average in Terms of Amplitudes

Problem: Consider a hydraulic head field consisting of multiple harmonic constituents. Express the time average $\langle h^2\rangle$ in terms of the constituent amplitudes.

Solution: Substituting the RHS of (24.1) for h and taking the time average give the following:

$$\begin{aligned}\langle h^2\rangle &= \left\langle\left[\sum_{m=1}^{\infty} h'(\mathbf{x},t;\omega_m)\right]^2\right\rangle \\ &= \sum_{m=1}^{\infty}\sum_{n=1}^{\infty}\langle h'(\mathbf{x},t;\omega_m)h'(\mathbf{x},t;\omega_n)\rangle \\ &= \sum_{m=1}^{\infty}\sum_{n=1}^{\infty} M_h(\mathbf{x};\omega_m)M_h(\mathbf{x};\omega_n) \\ &\quad \times \big\langle \cos\xi_h(\mathbf{x},t;\omega_m)\cos\xi_h(\mathbf{x},t;\omega_n)\big\rangle.\end{aligned} \tag{24.18}$$

Using (24.16) to evaluate the time average on the RHS of (24.18) yields

$$\langle h^2\rangle = \frac{1}{2}\sum_{m=1}^{\infty} M_h^2(\mathbf{x};\omega_m). \tag{24.19}$$

□

Consider the time average of the cube of the hydraulic head field, $\langle h^3\rangle$, in the general case where the hydraulic head field consists of multiple harmonic constituents. Substituting the RHS of (24.1) for h and taking the time average give

$$\begin{aligned}\langle h^3\rangle &= \left\langle\left[\sum_{m=1}^{\infty} h'(\mathbf{x},t;\omega_m)\right]^3\right\rangle \\ &= \sum_{m=1}^{\infty}\sum_{n=1}^{\infty}\sum_{r=1}^{\infty}\Big\langle h'(\mathbf{x},t;\omega_m)h'(\mathbf{x},t;\omega_n)h'(\mathbf{x},t;\omega_r)\Big\rangle.\end{aligned}$$

We can also write this as

$$\begin{aligned}\langle h^3\rangle &= \sum_{m=1}^{\infty}\Big\langle\big[h'(\mathbf{x},t;\omega_m)\big]^3\Big\rangle \\ &\quad + \sum_{m=1}^{\infty}\sum_{n=1}^{\infty}\sum_{r=1}^{\infty}\Big\langle h'(\mathbf{x},t;\omega_m)h'(\mathbf{x},t;\omega_n)h'(\mathbf{x},t;\omega_r)\Big\rangle \\ &\qquad \text{(excluding the combination } m = n = r\text{)}.\end{aligned}$$

Using (24.9) to eliminate the first term on the RHS, expressing each constituent as the product of the corresponding amplitude and phase, and noting that the absolute value of the cosine function is bounded above by unity lead to

$$\left|\langle h^3\rangle\right| \le \sum_{m=1}^{\infty}\sum_{n=1}^{\infty}\sum_{r=1}^{\infty} M_{\text{h}}(\mathbf{x};\omega_m) \times M_{\text{h}}(\mathbf{x};\omega_n)M_{\text{h}}(\mathbf{x};\omega_r) \text{ (excluding } m = n = r).$$

We can express this as

$$\left|\langle h^3\rangle\right| \le \left[\sum_{m=1}^{\infty} M_{\text{h}}(\mathbf{x};\omega_m)\right]^3 - \sum_{m=1}^{\infty} M_{\text{h}}^3(\mathbf{x};\omega_m). \quad (24.20)$$

Exercise 24.10 Time Average of Cube of Hydraulic Head Field

Problem: Under what conditions, if any, does the RHS of (24.20) vanish at all locations and for all times? Ignore the trivial case $h = 0$.

Solution: All of the harmonic constituent amplitudes $M_{\text{h}}(\mathbf{x};\omega_m)$ $(m = 1, 2, 3, \ldots)$ are nonnegative, so the RHS vanishes if and only if the hydraulic head field consists of a single harmonic constituent. □

Exercise 24.11 Vanishing of Time Averages

Problem: By integrating, show that

$$\left\langle h^r \frac{\partial h}{\partial t}\right\rangle = 0 \quad (r \in \mathbb{Z},\ r \ge 0). \quad (24.21)$$

Solution

$$\left\langle h^r \frac{\partial h}{\partial t}\right\rangle = \lim_{T\to\infty}\left[\frac{1}{T}\int_{t_0}^{t_0+T}\left(h^r\frac{\partial h}{\partial t}\right)dt\right]$$

$$= \frac{1}{r+1}\lim_{T\to\infty}\left[\frac{1}{T}\, h^{r+1}\Big|_{t=t_0}^{t=t_0+T}\right]. \quad (24.22)$$

But because h is a cosine series it satisfies

$$\left|h^{r+1}\Big|_{t=t_0}^{t=t_0+T}\right| \le 2\left(\sum_{m=1}^{\infty} M_{\text{h}}(\mathbf{x};\omega_m)\right)^{r+1}.$$

We assume that the sum on the RHS is finite so that h is well defined. Therefore the RHS is finite. Using this to evaluate the limit on the RHS of (24.22) yields (24.21). □

24.2. TIME VARIATION OF WATER DENSITY AND POROSITY

Consider time-periodic flow in an elastically deformable porous medium. At any fixed location, temporal variations of hydraulic head correspond to temporal variations of water pressure and effective stress. These in turn cause temporal variations of water density and effective porosity.

The compressibility of water is relatively low, and for a wide range of porous materials the bulk compressibility is relatively low. Thus, if the temporal variation of hydraulic head is not too great, then we expect the deviations of the water density and the effective porosity from their respective time-averaged values to be relatively small. Under these conditions the water density and the effective porosity depend only weakly on the water pressure and hence only weakly on the hydraulic head.

Additionally, we assume that if the constituent frequencies are not too high, then temporal variations of the water density and the effective porosity are approximately in phase with the temporal variation of the hydraulic head. That is, we assume that the porous medium responds elastically and instantaneously to temporal changes in hydraulic head. This is consistent with the assumptions underlying the groundwater flow equation.

In keeping with the assumptions described above, we will assume the deviations of the water density and effective porosity from their time-averaged values are approximately proportional to the corresponding simultaneous deviations of the hydraulic head about its time-averaged value:

$$\phi_{\text{e}}(\mathbf{x},t) \approx \phi_{\text{e}0}(\mathbf{x}) + \left.\frac{\partial \phi_{\text{e}}}{\partial h}\right|_{\mathbf{x}0}(h - h_0) \quad (24.23\text{a})$$

$$\rho_{\text{w}}(\mathbf{x},t) \approx \rho_{\text{w}0} + \left.\frac{\partial \rho_{\text{w}}}{\partial h}\right|_{\mathbf{x}0}(h - h_0) \quad (24.23\text{b})$$

where

$$h_0 \equiv \langle h\rangle \quad (24.24\text{a})$$

$$\phi_{\text{e}0}(\mathbf{x}) \equiv \phi_{\text{e}}(\mathbf{x},t)|_{h=h_0} \quad (24.24\text{b})$$

$$\rho_{\text{w}0} \equiv \rho_{\text{w}}(\mathbf{x},t)|_{h=h_0} \quad (24.24\text{c})$$

and

$$\left.\frac{\partial \phi_{\text{e}}}{\partial h}\right|_{\mathbf{x}0} \equiv \frac{\partial \phi_{\text{e}}}{\partial h}\ (d\mathbf{x} = \mathbf{0}, h = h_0) \quad (24.25\text{a})$$

$$\left.\frac{\partial \rho_{\text{w}}}{\partial h}\right|_{\mathbf{x}0} \equiv \frac{\partial \rho_{\text{w}}}{\partial h}\ (d\mathbf{x} = \mathbf{0}, h = h_0) \quad (24.25\text{b})$$

The time-averaged values of effective porosity and density, respectively, satisfy the following:

$$\langle \phi_{\text{e}}\rangle = \phi_{\text{e}0}(\mathbf{x}) \quad (24.26\text{a})$$

$$\langle \rho_{\text{w}}\rangle = \rho_{\text{w}0}. \quad (24.26\text{b})$$

Exercise 24.12 Dependence of Effective Porosity on Hydraulic Head

Problem: Derive a compact expression for the first partial derivative of effective porosity with respect to hydraulic head, with position held constant, evaluated at $h = h_0$.

Solution: Using the chain rule, we obtain

$$\left.\frac{\partial \phi_{\text{e}}}{\partial h}\right|_{\mathbf{x}0} = \left.\left[\frac{\partial \phi_{\text{e}}}{\partial p}\frac{\partial p}{\partial h}\right]\right|_{\mathbf{x}0}.$$

Substituting the RHS of (23.15) for $\partial\phi_e/\partial p$ and using (2.3) with $d\mathbf{x} = \mathbf{0}$ to evaluate $\partial p/\partial h$ yield

$$\left.\frac{\partial\phi_e}{\partial h}\right|_{\mathbf{x}0} = (1 - \phi_{e0})\rho_{w0}g\alpha. \tag{24.27}$$

□

Exercise 24.13 Dependence of Water Density on Hydraulic Head

Problem: Derive a compact expression for the first partial derivative of water density with respect to hydraulic head, with position held constant, evaluated at $h = h_0$.

Solution: Using the chain rule, we obtain

$$\left.\frac{\partial\rho_w}{\partial h}\right|_{\mathbf{x}0} = \left.\left[\frac{\partial\rho_w}{\partial p}\frac{\partial p}{\partial h}\right]\right|_{\mathbf{x}0}.$$

Using (23.18) to evaluate $\partial\rho_w/\partial p$ and using (2.3) with $d\mathbf{x} = \mathbf{0}$ to evaluate $\partial p/\partial h$ yield

$$\left.\frac{\partial\rho_w}{\partial h}\right|_{\mathbf{x}0} = \rho_{w0}^2 g\beta_w. \tag{24.28}$$

□

We assume the hydraulic head field is time periodic (or almost periodic), so its time-averaged value is zero (i.e., $h_0 = 0$). Substituting the RHSs of (24.27) and (24.28) for the corresponding derivatives on the RHSs of (24.23) yields

$$\phi_e(\mathbf{x}, t) = \phi_{e0}(\mathbf{x}) + [1 - \phi_{e0}(\mathbf{x})]\rho_{w0}g\alpha(\mathbf{x})h(\mathbf{x}, t) \tag{24.29a}$$

$$\rho_w(\mathbf{x}, t) = \rho_{w0}\big[1 + \rho_{w0}g\beta_w h(\mathbf{x}, t)\big]. \tag{24.29b}$$

We will use these equations below to derive expressions for the time averages of various quantities related to the transport of fluid mechanical energy.

Exercise 24.14 Gradients of Effective Porosity and Water Density

Problem: Express the gradients of the effective porosity and the water density in terms of the hydraulic gradient.

Solution: Taking the gradients of (24.29) gives

$$\nabla\phi_e = (1 - \rho_{w0}g\alpha h)\nabla\phi_{e0} + (1 - \phi_{e0})\rho_{w0}gh\nabla\alpha + (1 - \phi_{e0})\rho_{w0}g\alpha\nabla h \tag{24.30a}$$

$$\nabla\rho_w = \rho_{w0}^2 g\beta_w\nabla h. \tag{24.30b}$$

□

We have assumed that the bulk compressibility and the time-averaged effective porosity generally depend on the space coordinates but not on the time. These are consistent with the definition of the specific storage,

$$S_s(\mathbf{x}) \equiv \rho_{w0}g[\alpha(\mathbf{x}) + \phi_{e0}(\mathbf{x})\beta_w].$$

and the assumption that it generally depends on the space coordinates but not on the time (see Chapter 2).

Exercise 24.15 Derivation: Upper Bounds

Problem: Derive approximate upper bounds for the derivatives $(\partial\phi_e/\partial h)|_{\mathbf{x}0}$ and $(\partial\rho_w/\partial h)|_{\mathbf{x}0}$ for a clay aquitard. Assume the following [*Freeze and Cherry*, 1979]:

For fresh groundwater:

$$\beta_w \approx 4.4 \times 10^{-10}\ \text{m}^2\ \text{N}^{-1}$$

$$\rho_{w0} \approx 1.0 \times 10^3\ \text{kg m}^{-3}$$

$$\rho_{w0}g \approx 9.8 \times 10^3\ \text{N m}^{-3}.$$

For clay:

$$0.40 \le \phi_{e0} \le 0.70$$

$$10^{-8}\ \text{m}^2\ \text{N}^{-1} \le \alpha \le 10^{-6}\ \text{m}^2\ \text{N}^{-1}.$$

Solution: Equations (24.27) and (24.28) give the following upper bounds on the derivatives:

$$\left.\frac{\partial\phi_e}{\partial h}\right|_{\mathbf{x}0} \le \left[1 - (\phi_{e0})_{\min}\right]\alpha_{\max}\rho_{w0}g \tag{24.31a}$$

$$\left.\frac{\partial\rho_w}{\partial h}\right|_{\mathbf{x}0} \le \beta_w\rho_{w0}^2 g. \tag{24.31b}$$

Substituting the numerical values listed above for the corresponding terms on the RHSs of these expressions yields

$$\left.\frac{\partial\phi_e}{\partial h}\right|_{\mathbf{x}0} \le 6 \times 10^{-3}\ \text{m}^{-1} \tag{24.32a}$$

$$\left.\frac{\partial\rho_w}{\partial h}\right|_{\mathbf{x}0} \le 4 \times 10^{-3}\ \text{kg m}^{-2}. \tag{24.32b}$$

Thus, for clay we expect the effective porosity and water density to undergo relative changes no greater than about 0.02 (= 0.006/0.40) and 4×10^{-6} (= 0.004/1000), respectively, per meter change in hydraulic head. □

24.3. TIME-AVERAGED ENERGY DENSITY AND ITS RATE OF CHANGE

24.3.1. Time-Averaged Energy Density

Taking the time average of (23.1) gives

$$\langle\epsilon\rangle = \langle\phi_e\rho_w gh\rangle.$$

Substituting the RHSs of (24.29a) and (24.29b) for ϕ_e and ρ_w, respectively, and simplifying lead to

$$\langle\epsilon\rangle = \rho_{w0}g\phi_{e0}\langle h\rangle + \rho_{w0}^2 g^2\left[\phi_{e0}\beta_w + (1-\phi_{e0})\alpha\right]\langle h^2\rangle + \rho_{w0}^3 g^3(1-\phi_{e0})\alpha\beta_w\langle h^3\rangle.$$

Using (24.13) to eliminate the first term on the RHS leaves

$$\langle\epsilon\rangle = \rho_{w0}^2 g^2\Big\{\left[\phi_{e0}\beta_w + (1-\phi_{e0})\alpha\right]\langle h^2\rangle + \rho_{w0}g(1-\phi_{e0})\alpha\beta_w\langle h^3\rangle\Big\}. \quad (24.33)$$

The RHS is of a sum of two terms, the first and second of which are quadratic and cubic, respectively, in $\rho_{w0}g$. We now examine the conditions under which the second term is negligible. The following dimensionless ratio measures the relative absolute magnitude of the second term in comparison to the first:

$$\nu \equiv \frac{\rho_{w0}g(1-\phi_{e0})\alpha\beta_w}{\phi_{e0}\beta_w + (1-\phi_{e0})\alpha}\frac{|\langle h^3\rangle|}{\langle h^2\rangle}. \quad (24.34)$$

If $\nu \ll 1$, then the magnitude of the second term is much less than that of the first term, so the second term is negligible.

Note 24.2 Material Properties for Shallow Geologic Environments
We expect that in most shallow geologic environments the following conditions will be satisfied (values from *Freeze and Cherry* [1979]):

$$0 \le \phi_{e0} \le 0.70 \quad (24.35a)$$

$$10^{-11}\ \mathrm{m^2\,N^{-1}} \le \alpha \le 10^{-6}\ \mathrm{m^2\,N^{-1}} \quad (24.35b)$$

$$\beta_w \approx 4.4\times10^{-10}\ \mathrm{m^2\,N^{-1}} \quad (24.35c)$$

$$\rho_{w0}g \approx 9.8\times10^{3}\ \mathrm{N\,m^{-3}}. \quad (24.35d)$$

△

It follows from relations (24.35) that $\phi_{e0}\beta_w \ge 0$, so the denominator of (24.34) satisfies the following:

$$\phi_{e0}\beta_w + (1-\phi_{e0})\alpha \ge (1-\phi_{e0})\alpha.$$

Combining this result and (24.19) and (24.20) with (24.34), we infer that $\nu \le \nu_{max}$, where

$$\nu_{max} \equiv 2\rho_{w0}g\beta_w M_{char} \quad (24.36)$$

$$M_{char}(\mathbf{x}) \equiv \frac{h_{max}^3(\mathbf{x}) - \sum_{m=1}^{\infty} M_h^3(\mathbf{x};\omega_m)}{\sum_{m=1}^{\infty} M_h^2(\mathbf{x};\omega_m)} \quad (24.37)$$

$$h_{max}(\mathbf{x}) \equiv \sum_{m=1}^{\infty} M_h(\mathbf{x};\omega_m). \quad (24.38)$$

For the purpose of illustration we assume that $M_{char} \le 10^3$ m. Substituting the numerical values given in relations (24.35) for the corresponding terms on the RHS of (24.36) yields $\nu_{max} = 10^{-2}$. Therefore, under the conditions described above we can approximate (24.33) as

$$\langle\epsilon\rangle = \frac{1}{2}\rho_{w0}^2 g^2\Big\{\phi_{e0}(\mathbf{x})\beta_w + \left[1-\phi_{e0}(\mathbf{x})\right]\alpha(\mathbf{x})\Big\} \times \sum_{m=1}^{\infty} M_h^2(\mathbf{x};\omega_m) \quad (24.39)$$

where we have used (24.19) to express $\langle h^2\rangle$ in terms of the sum of squared amplitudes. Equation (24.39) indicates that the time-averaged energy density is proportional to the sum of the squared amplitude functions $M_h(\mathbf{x};\omega_m)$ but independent of the phase functions $\theta_h(\mathbf{x};\omega_m)$.

24.3.2. Rate of Change of Energy Density

Taking the time average of (23.20) gives

$$\left\langle\frac{\partial\epsilon}{\partial t}\right\rangle = \left\langle(\alpha - \phi_e\alpha + \phi_e\beta_w)\rho_w^2 g^2 h\frac{\partial h}{\partial t}\right\rangle + \left\langle\phi_e\rho_w g\frac{\partial h}{\partial t}\right\rangle.$$

Substituting the RHSs of (24.29a) and (24.29b) for ϕ and ρ_w, respectively, gives

$$\left\langle\frac{\partial\epsilon}{\partial t}\right\rangle = \Big\langle\Big(\alpha + \left[\phi_{e0} + (1-\phi_{e0})\alpha\rho_{w0}gh\right](\beta_w - \alpha)\Big) \times \left[\rho_{w0} + \beta_w\rho_{w0}^2 gh\right]^2 g^2 h\frac{\partial h}{\partial t}\Big\rangle + \Big\langle\left[\phi_{e0} + (1-\phi_{e0})\alpha\rho_{w0}gh\right]\left[\rho_{w0}g + \beta_w\rho_{w0}^2 g^2 h\right]\frac{\partial h}{\partial t}\Big\rangle.$$

The RHS is the sum of two terms, each of which is the time average of the product of two factors. The first factor is a polynomial in h, and the second factor is $\partial h/\partial t$. By (24.21), each time-averaged term vanishes, yielding

$$\left\langle\frac{\partial\epsilon}{\partial t}\right\rangle = 0. \quad (24.40)$$

Exercise 24.16 Effect of Power Series Truncation

Problem: Suppose the linear approximations (24.29) were replaced with the full (nontruncated) power series

$$\phi_e = \phi_{e0} + (1-\phi_{e0})\alpha\rho_{w0}gh + O(h^2)$$

$$\rho_w = \rho_{w0} + \beta_w\rho_{w0}^2 gh + O(h^2).$$

Would this change the result obtained for the time average $\langle\partial\epsilon/\partial t\rangle$ in (24.40)? If so, how? What does your conclusion imply about the robustness of (24.40)?

Solution: No, it would not change the result. The first factor in each product would be a Maclaurin series in h rather than a polynomial in h, so (24.21) would still apply. Therefore, (24.40) is robust with regard to the order of the approximations for the water density and effective porosity. □

24.4. TIME-AVERAGED ENERGY CONSERVATION

Taking the time average of the differential equation for the conservation of fluid mechanical energy (23.10) and then rearranging give

$$\langle \eta \rangle + \langle \nabla \cdot \mathbf{J} \rangle = -\left\langle \frac{\partial \epsilon}{\partial t} \right\rangle + \langle \sigma \rangle.$$

Using (24.40) to eliminate the time average $\langle \partial \epsilon / \partial t \rangle$, we obtain

$$\langle \eta \rangle + \langle \nabla \cdot \mathbf{J} \rangle = \langle \sigma \rangle. \tag{24.41}$$

Consider the flux divergence term on the LHS. Reversing the order of the differentiation (with respect to the space coordinates) and the integration (with respect to time) produces the following time-averaged, differential equation for the conservation of fluid mechanical energy:

$$\langle \eta \rangle + \nabla \cdot \langle \mathbf{J} \rangle = \langle \sigma \rangle. \tag{24.42}$$

One way to understand (24.42) is to interpret it in the context of a stationary, nondeforming, finite control volume, V. Integrating (24.42) over the control volume yields

$$\int_V \langle \eta \rangle \, dV + \int_V \nabla \cdot \langle \mathbf{J} \rangle \, dV = \int_V \langle \sigma \rangle \, dV.$$

Applying the divergence theorem to the flux divergence term on the LHS gives the following time-average integral equation for the conservation of fluid mechanical energy:

$$\int_V \langle \eta \rangle \, dV + \int_S \langle \mathbf{J} \rangle \cdot \hat{\mathbf{u}} \, dS = \int_V \langle \sigma \rangle \, dV \tag{24.43}$$

where S denotes the control surface and $\hat{\mathbf{u}}$ represents the outward unit-normal vector for the control volume. Equation (24.43) states that in a time-averaged sense the sum of the energy dissipated within the control volume and the net energy outflow across the control surface balances the energy generated by sources within the control volume. Equations (24.42) and (24.43) are mathematically equivalent.

24.5. TIME-AVERAGED ENERGY FLUX DENSITY

Taking the time average of (23.23) gives

$$\langle \mathbf{J} \rangle = -\langle \epsilon \mathbf{K} \nabla h \rangle + \langle \phi_e \epsilon \mathbf{v}^{(s)} \rangle. \tag{24.44}$$

The solid matrix velocity appears in the second term on the RHS. Thus, evaluation of the flux density and its time average generally requires a conceptual model that incorporates some level of mechanical deformation. The development of a fully coupled hydromechanical conceptual model is beyond the scope of this book, but interested readers might refer to *Neuzil* [2003] for more information.

24.5.1. Slow-Deformation Approximation

In this section we consider the case where the time-averaged flux component associated with the deformation velocity is negligible compared to that associated with the flow relative to the deforming solid matrix:

$$\left| \langle \phi_e \epsilon \mathbf{v}^{(s)} \rangle \right| \ll \left| \langle \epsilon \mathbf{K} \nabla h' \rangle \right|. \tag{24.45}$$

We will refer to this case as the *slow-deformation approximation*.

Using the approximation (24.45) to simplify (24.44) gives

$$\langle \mathbf{J} \rangle = -\langle \phi_e \rho_w g h \mathbf{K} \nabla h \rangle.$$

Substituting the RHSs of (24.29a) and (24.29b) for the effective porosity and the water density, respectively, and grouping terms yield

$$\langle \mathbf{J} \rangle = -\rho_{w0} g \mathbf{K} \Big\{ \phi_{e0} \langle h \nabla h \rangle + \rho_{w0} g \big[\beta_w + (1 - \phi_{e0}) \alpha \big] \langle h^2 \nabla h \rangle + \rho_{w0}^2 g^2 (1 - \phi_{e0}) \alpha \beta_w \langle h^3 \nabla h \rangle \Big\}. \tag{24.46}$$

In evaluating the time average, we have assumed that over a relatively wide range of groundwater pressures the hydraulic conductivity generally is independent of the groundwater pressure (and therefore the hydraulic head). Consequently, the hydraulic conductivity is time invariant. We have also assumed that the medium may be spatially variable with respect to hydraulic conductivity and the acceleration of gravity is both space and time invariant.

24.6. TIME-AVERAGED ENERGY DISSIPATION

24.6.1. Basic Results

Taking the time average of (23.29) yields

$$\langle \eta \rangle \geq 0 \quad \forall \, \mathbf{x} \in D, \forall \, t. \tag{24.47}$$

Exercise 24.17 Energy Dissipation in a Source-Free Domain

Problem: Show that the net influx of fluid mechanical energy to any fixed control volume in a source-free domain is positive or zero.

Solution: Combining (24.47) and (24.42) leads to

$$\nabla \cdot \langle \mathbf{J} \rangle \leq 0.$$

Integrating the inequality over the control volume and applying the divergence theorem give

$$\int_S \langle \mathbf{J} \rangle \cdot \hat{\mathbf{u}} \, dS \leq 0.$$

That is, the net outflow of fluid mechanical energy is negative or zero; therefore, the net influx is positive or zero. □

24.6.2. Expressions

Taking the time average of (23.31) and using (24.40) to eliminate the time average $\langle \partial\epsilon/\partial t\rangle$ yield

$$\langle\eta\rangle = \Big\langle (\mathbf{K}\nabla h)\cdot\nabla(\phi_e\rho_w gh) - \phi_e\rho_w g(\rho_w g\phi_e\alpha - S_s)h\frac{\partial h}{\partial t}\Big\rangle + \langle\sigma\rangle. \quad (24.48)$$

Exercise 24.18 Time-Averaged, Per-Volume, Energy Dissipation Rate

Problem: Consider time-periodic flow. Show that the time-averaged, per-volume, mechanical energy dissipation rate is given by

$$\langle\eta\rangle = \big\langle(\mathbf{K}\nabla h)\cdot\nabla(\phi_e\rho_w gh)\big\rangle + \langle\sigma\rangle. \quad (24.49)$$

Solution: Comparing (24.48) and (24.49), clearly we need only show that

$$\Big\langle\phi_e\rho_w g(\rho_w g\phi_e\alpha - S_s)h\frac{\partial h}{\partial t}\Big\rangle = 0. \quad (24.50)$$

As yet this is unproven. By (24.29) we know that the effective porosity and the water density are linear in the hydraulic head. Therefore, we can write

$$\Big\langle\phi_e\rho_w g(\rho_w g\phi_e\alpha - S_s)h\frac{\partial h}{\partial t}\Big\rangle = \Big\langle f(h)\frac{\partial h}{\partial t}\Big\rangle. \quad (24.51)$$

where $f(h)$ is a polynomial function of h with nonnegative integer powers. Applying (24.21) to the RHS then gives (24.50), and (24.49) immediately follows. □

Exercise 24.19 Effect of Power Series Truncation

Problem: Suppose the linear approximations of (24.29) were replaced with the full (nontruncated) power series

$$\phi_e = \phi_{e0} + (1-\phi_{e0})\alpha\rho_{w0}gh + O(h^2)$$
$$\rho_w = \rho_{w0} + \beta_w\rho_{w0}^2 gh + O(h^2).$$

Would this change the result in (24.49)? If so, how? What does your conclusion imply about the robustness of (24.49)?

Solution: No, it would not change the result. The function $f(h)$ would be a Maclaurin series in h rather than a polynomial in h, so (24.21) would still apply. Therefore, (24.49) is robust with regard to the order of the approximations for the water density and effective porosity. □

Exercise 24.20 Derivation: Gradient

Problem: Express the gradient $\nabla(\phi_e\rho_w gh)$ in terms of the hydraulic head and its gradient.

Solution: Expanding the gradient gives

$$\nabla(\phi_e\rho_w gh) = \phi_e\rho_w g\nabla h + \phi_e gh\nabla\rho_w + \rho_w gh\nabla\phi.$$

Substituting the RHSs of (24.30a) and (24.30b) for the gradients of water density and effective porosity, respectively, and grouping terms with similar gradient factors yield

$$\nabla(\phi_e\rho_w gh) = \big[\phi_e\rho_w g + \phi_e\rho_{w0}^2 g^2\beta_w h + (1-\phi_{e0})\rho_w\rho_{w0}g^2\alpha h\big]\nabla h + \rho_w gh(1-\rho_{w0}g\alpha h)\nabla\phi_{e0} + (1-\phi_{e0})\rho_w\rho_{w0}g^2h^2\nabla\alpha.$$

Substituting the RHSs of (24.29a) and (24.29b) for the effective porosity and water density, respectively, and simplifying give

$$\nabla(\phi_e\rho_w gh) = \rho_{w0}g\Big\{\phi_{e0} + 2\rho_{w0}g\big[\phi_{e0}\beta_w + (1-\phi_{e0})\alpha\big]h + 3(1-\phi_{e0})\rho_{w0}^2g^2\alpha\beta_w h^2\Big\}\nabla h + \rho_{w0}g\big[1+\rho_{w0}g(\beta_w-\alpha)h - \rho_{w0}^2g^2\alpha\beta_w h^2\big]h\nabla\phi_{e0} + (1-\phi_{e0})\rho_{w0}^2g^2(1+\rho_{w0}g\beta_w h)h^2\nabla\alpha. \quad (24.52)$$

□

Substituting the RHS of (24.52) for $\nabla(\phi_e\rho_w gh)$ in (24.49) and exploiting the symmetry of $\mathbf{K}$ give

$$\langle\eta\rangle = \langle\eta\rangle_1 + \langle\eta\rangle_2 + \langle\eta\rangle_3 + \langle\sigma\rangle \quad (24.53)$$

where

$$\langle\eta\rangle_1 = \rho_{w0}g\Big\langle\Big\{\phi_{e0} + 2\rho_{w0}g\big[\phi_{e0}\beta_w + (1-\phi_{e0})\alpha\big]h + 3(1-\phi_{e0})\rho_{w0}^2g^2\alpha\beta_w h^2\Big\}\nabla h\cdot(\mathbf{K}\nabla h)\Big\rangle \quad (24.54a)$$

$$\langle\eta\rangle_2 = \rho_{w0}g(\mathbf{K}\nabla\phi_{e0})\cdot\Big\langle\big[1+\rho_{w0}g(\beta_w-\alpha)h - \rho_{w0}^2g^2\alpha\beta_w h^2\big]h\nabla h\Big\rangle \quad (24.54b)$$

$$\langle\eta\rangle_3 = (1-\phi_{e0})\rho_{w0}^2g^2(\mathbf{K}\nabla\alpha)\cdot\big\langle(1+\rho_{w0}g\beta_w h)h^2\nabla h\big\rangle. \quad (24.54c)$$

The results presented in this section are *not* based on the slow-deformation approximation and therefore do not require that approximation to be valid.

24.7. CHAPTER SUMMARY

In this chapter we examined the time-averaged transport of fluid mechanical energy by time-periodic groundwater flow under isothermal conditions. Our analysis is based on assumptions that are consistent with those of the groundwater flow equation (see Chapter 2). We derived differential and integral equations for the time-averaged conservation of fluid mechanical energy and expressions for time averages of the following: the fluid mechanical energy density and its rate of change, the flux density, and the dissipation rate. Our results are valid for the general case where the hydraulic head field consists of multiple harmonic constituents.

25

Mechanical Energy of Single-Constituent Fields

In this chapter we examine the time-averaged transport of fluid mechanical energy by a single-constituent wave propagating under isothermal conditions. We do this by applying the results presented in Chapter 24 to a hydraulic head field consisting of a single harmonic constituent with frequency ω_m (m fixed). Then (4.1a) gives $h(\mathbf{x}, t) = h'(\mathbf{x}, t; \omega_m)$, the hydraulic head is strictly periodic, and the corresponding time averages are equivalent to averages over a single-constituent period (see Section 24.1).

We use the time-averaged, integral equation for the conservation of fluid mechanical energy and the expression we derived for the time-averaged flux density to show that amplitude maxima are prohibited in source-free domains. Finally, we use the expression we derived for the time-averaged flux density to show that strictly true point sources and true line sources are nonphysical because they are energetically impossible to maintain.

25.1. TIME-AVERAGED ENERGY DENSITY

Substituting $h = h'$ in (24.39) and (24.40), respectively, gives for the time-averaged, fluid mechanical energy density and its rate of change:

$$\langle \epsilon \rangle = \frac{1}{2}\rho_{\mathrm{w0}}^2 g^2 \left\{ \phi_{\mathrm{e0}}(\mathbf{x})\beta_{\mathrm{w}} + \left[1 - \phi_{\mathrm{e0}}(\mathbf{x})\right]\alpha(\mathbf{x}) \right\} M_{\mathrm{h}}^2(\mathbf{x}; \omega_m) \tag{25.1a}$$

$$\left\langle \frac{\partial \epsilon}{\partial t} \right\rangle = 0. \tag{25.1b}$$

25.2. TIME-AVERAGED MECHANICAL ENERGY FLUX DENSITY

25.2.1. Slow-Deformation Approximation

In this section we derive expressions for the time-averaged, fluid mechanical energy flux density (i.e., the *average energy flux density*) by applying the slow-deformation approximation (see Section 24.5.1) to the case where the hydraulic head field consists of a single harmonic constituent.

Exercise 25.1 Expression for Time-Averaged Energy Flux Density

Problem: Consider a hydraulic head field consisting of a single harmonic constituent. Evaluate the average energy flux density, assuming that the slow-deformation approximation is valid.

Solution: Substituting $h = h'$ in (24.46) gives

$$\langle \mathbf{J} \rangle = -\rho_{\mathrm{w0}} g \mathbf{K} \Big\{ \phi_{\mathrm{e0}} \langle h' \nabla h' \rangle + \rho_{\mathrm{w0}} g \left[\beta_{\mathrm{w}} + (1 - \phi_{\mathrm{e0}})\alpha\right] \langle h'^2 \nabla h' \rangle + \rho_{\mathrm{w0}}^2 g^2 (1 - \phi_{\mathrm{e0}})\alpha\beta_{\mathrm{w}} \langle h'^3 \nabla h' \rangle \Big\}.$$

Using (24.12) to evaluate the time averages on the RHS and then simplifying yield

$$\langle \mathbf{J} \rangle = -\frac{\rho_{\mathrm{w0}} g}{2} \left\{ \phi_{\mathrm{e0}} + \frac{3}{4}\rho_{\mathrm{w0}}^2 g^2 (1 - \phi_{\mathrm{e0}})\alpha\beta_{\mathrm{w}} M_{\mathrm{h}}^2 \right\} M_{\mathrm{h}} \mathbf{K} \nabla M_{\mathrm{h}}. \tag{25.2}$$

□

Hydrodynamics of Time-Periodic Groundwater Flow: Diffusion Waves in Porous Media, Geophysical Monograph 224, First Edition. Joe S. Depner and Todd C. Rasmussen.

Equation (25.2) indicates that the average energy flux density is approximately parallel to the in-phase component of the specific-discharge vector (i.e., $\mathbf{q}'_{\mathrm{Ip}}$; see Exercise 14.7); it generally is *not* parallel to the phase velocity, the hydraulic gradient, or the specific discharge. In an anisotropic medium the average energy flux density and the direction of maximum attenuation generally are non-parallel, with the angle between the two being acute. In an isotropic medium the average energy flux density is approximately opposite the amplitude gradient, parallel to the direction of maximum attenuation.

Equation (25.2) also shows that the average flux density vanishes at those locations, if any, where either the amplitude or its gradient approaches zero.

25.2.1.1. Low-Order Approximation

Consider the expression for the average energy flux density (25.2). The RHS is the sum of two nonnegative terms, the first and second of which are linear and cubic, respectively, in $\rho_{w0}gM_h$. We now examine the conditions under which the second term is negligible. The following dimensionless ratio measures the magnitude of the second term relative to that of the first term:

$$\nu \equiv \frac{3\rho_{w0}^2 g^2 (1-\phi_{e0})\alpha\beta_w M_h^2}{4\phi_{e0}}.$$

Therefore, $\nu \le \nu_{max}$, where

$$\nu_{max} \equiv \frac{\rho_{w0}^2 g^2 \alpha_{max}\beta_w (M_h)_{max}^2}{(\phi_{e0})_{min}}. \tag{25.3}$$

and the subscripts max and min, respectively, indicate upper and lower bounds for the indicated variable. If $\nu_{max} \ll 1$, then the second term is negligible.

We will assume the material properties satisfy (24.35). To establish a point of comparison, we will assume that $(M_h)_{max} = 30\,\mathrm{m}$. For some perspective on the numerical values of constituent amplitudes, see Note 25.1.

Note 25.1 Numerical Values of Constituent Amplitudes

The mean spring range for the semidiurnal tides in the Minas Basin in the Bay of Fundy is 12.9 m [*Pugh*, 2004]. The Bay of Fundy is known for its extreme tidal range, possibly the greatest worldwide. Thus, for boundary forcing associated with oceanic tides the amplitude of any single harmonic constituent would not exceed half this, or about 6.5 m.

However, constituent amplitudes associated with artificial sources such as sinusoidal injection/pumping easily could exceed 6.5 m and in some cases might exceed 30 m. For example, *Song and Renner* [2007] report pressure amplitudes of 4–5 MPa for a series of laboratory hydraulic tests they conducted on rock core samples; equivalent hydraulic head amplitudes are approximately 400–500 m. △

To establish a point of comparison, we will assume that $(\phi_{e0})_{min} = 0.01$ (i.e., an effective porosity of 1%). Substituting the numerical values given above for the corresponding terms on the RHS of (25.3) yields

$$\nu_{max} = 4 \times 10^{-7}.$$

This result (i.e., $\nu \ll 1$) indicates that under a relatively wide range of natural conditions the second term of (25.2) is negligible compared to the first term. Under these conditions we can approximate (25.2) as

$$\langle \mathbf{J} \rangle = -\frac{\phi_{e0}\rho_{w0}g}{2} M_h \mathbf{K} \nabla M_h. \tag{25.4}$$

We can also write this in terms of the squared amplitude function:

$$\langle \mathbf{J} \rangle = -\frac{\phi_{e0}\rho_{w0}g}{4} \mathbf{K} \nabla M_h^2. \tag{25.5}$$

Comparing (25.5) to the generalized Darcy's law (2.4), apparently $-\nabla M_h^2$ and $\mathbf{K}$ take the roles of potential gradient and "steering"tensor, respectively, somewhat analogous to the way that $-\nabla h$ and $\mathbf{K}$ take these roles in the generalized Darcy's law.

Exercise 25.2 Time-Averaged, Mechanical Energy Flux Density

Problem: Write (25.5) in explicit notation, showing clearly how each variable depends on the space coordinates and the constituent frequency.

Solution

$$\langle \mathbf{J} \rangle (\mathbf{x};\omega_m) = -\frac{\phi_{e0}(\mathbf{x})\rho_{w0}g}{4} \mathbf{K}(\mathbf{x}) \nabla M_h^2(\mathbf{x};\omega_m).$$

□

Exercise 25.3 Point Source in Infinite, 3D, Ideal Medium

Problem: Consider a single-constituent (i.e., ω_m fixed) point source located at the origin of a 3D, infinite domain filled with an ideal medium (see Example 12.1). Express the time-averaged, fluid mechanical energy flux density in terms of the space coordinates and the constituent frequency. Assume that the slow-deformation approximation is valid.

Solution: Recall the pertinent expressions for the amplitude (Exercise 20.12) and the amplitude gradient (Exercise 20.15). Substituting these expressions for the corresponding terms, with $\mathbf{K} = K\mathbf{I}$, in (25.4) gives

$$\langle \mathbf{J} \rangle (\mathbf{x};\omega_m) = \frac{\phi_{e0}\rho_{w0}g|Q_0(\omega_m)|^2}{32\pi^2 K} \times \frac{\exp\left(-\sqrt{2\alpha_m}r\right)}{r^2}\left(\frac{1}{r} + \sqrt{\frac{\alpha_m}{2}}\right)\hat{\mathbf{r}}, \quad r > 0. \tag{25.6}$$

Thus, fluid mechanical energy is transported away from the point source. Furthermore, the magnitude of the flux density is unbounded at the source and decreases as the distance from the source increases. □

Exercise 25.4 Line Source in Infinite, 3D, Ideal Medium

Problem: Consider a single-constituent (i.e., ω_m fixed), infinitely long, uniform line source located at $x_1 = x_2 = 0$ in a 3D, infinite domain filled with an ideal medium (see Section 12.4.2). Express the time-averaged, energy flux density in terms of the space coordinates and the constituent frequency. Assume that the slow-deformation approximation is valid.

Solution: Recall the pertinent expressions for the amplitude (20.25a) and the amplitude gradient (20.31a). Substituting these expressions for the corresponding terms, with $\mathbf{K} = K\mathbf{I}$, in (25.4) gives

$$\langle \mathbf{J} \rangle(\mathbf{x}; \omega_m) = \frac{-\phi_{e0}\rho_{w0}g|\Delta Q_0(\omega_m)|^2\sqrt{\alpha_m}}{8\pi^2 K} \times N_0(\sqrt{\alpha_m}r)N_1(\sqrt{\alpha_m}r)\cos\left[\phi_1(\sqrt{\alpha_m}r) - \phi_0(\sqrt{\alpha_m}r) - \frac{\pi}{4}\right]\hat{\mathbf{r}}, \quad r > 0. \tag{25.7}$$

□

Consider the uniform line source of Exercise 25.4. Using (25.7), we briefly examine the behavior of the time-averaged energy flux density in the immediate vicinity of the line source. Equation (E.14) gives $\cos\left[\phi_1(0) - \phi_0(0) - \pi/4\right] = -1$, so $\langle \mathbf{J} \rangle$ points in the direction of $\hat{\mathbf{r}}$. Thus, fluid mechanical energy is transported away from the line source. Additionally, as $R \to 0$, we have $N_0(R) \to \infty$ and $N_1(R) \to \infty$, so the magnitude of the flux density is unbounded at the source.

Exercise 25.5 Plane Source in Infinite, 3D, Ideal Medium

Problem: Consider a single-constituent (i.e., ω_m fixed), uniform, vertically oriented plane source of infinite extent located at $x_1 = 0$ in a 3D, infinite domain filled with an ideal medium (see Example 9.3). Express the time-averaged, energy flux density in terms of the space coordinates. Assume that the slow-deformation approximation is valid.

Solution: Recall the pertinent expressions for the amplitude (Exercise 20.14) and the amplitude gradient (Exercise 20.19). Substituting these expressions for the corresponding terms, with $\mathbf{K} = K\mathbf{I}$, in (25.4) gives

$$\langle \mathbf{J} \rangle(\mathbf{x}; \omega_m) = \frac{\phi_{e0}\rho_{w0}g|\Delta Q_0(\omega_m)|^2}{8\sqrt{2}K\sqrt{\alpha_m}}\operatorname{sign}(x_1) \times \exp\left(-2\sqrt{\frac{\alpha_m}{2}}|x_1|\right)\hat{\mathbf{x}}, \quad |x_1| > 0 \tag{25.8}$$

where sign() denotes the *sign function* [see equation (D.7)]. Thus, fluid mechanical energy is transported away from the plane source. Furthermore, the magnitude of the flux density is bounded at the source and decreases as the distance from the source increases. □

Exercise 25.6 Effective Energy Transport Velocity

Problem: Define the *effective energy transport velocity* (for fluid mechanical energy) as

$$\mathbf{v}_\epsilon(\mathbf{x}; \omega_m) \equiv \frac{\langle \mathbf{J} \rangle(\mathbf{x}; \omega_m)}{\langle \epsilon \rangle(\mathbf{x}; \omega_m)}. \tag{25.9}$$

Derive a compact expression for $\mathbf{v}_\epsilon$ that is consistent with the slow-deformation approximation. What are the dimensions of $\mathbf{v}_\epsilon$?

Solution: Substituting the RHSs of (25.4) and (25.1a) for $\langle \mathbf{J} \rangle$ and $\langle \epsilon \rangle$, respectively, on the RHS of (25.9) and then simplifying give

$$\mathbf{v}_\epsilon(\mathbf{x}; \omega_m) = \left[\frac{-\phi_{e0}}{\phi_{e0}\beta_w + (1-\phi_{e0})\alpha}\right]\frac{\mathbf{K}}{\rho_{w0}g}\nabla Y_h$$

where $Y_h \equiv \ln M_h$. The dimensions of $\mathbf{v}_\epsilon$ are Lt^{-1}. □

25.2.2. Implications for Amplitude Maxima

In Section 17.3.1 we presented a mathematical argument for the prohibition of amplitude maxima in source-free domains. In this section we present a less abstract, more physical argument for that prohibition. Our argument, which uses the method of contradiction, is based on the conservation of fluid mechanical energy in isothermal flows.

Suppose a maximum of the amplitude function exists in a source-free domain at the point whose coordinates are $\mathbf{x} = \mathbf{x}_s$. For the purposes of this section only, define a control volume V that satisfies the following conditions:

- The control volume V is stationary; it does not move or deform.
- The control volume V is fully enclosed by the space domain D; it does not intersect the domain boundary Γ.
- The control volume V encloses the amplitude maximum.
- The control volume V has positive volume.
- The control surface S coincides with one of the amplitude level surfaces for which the amplitude is positive.

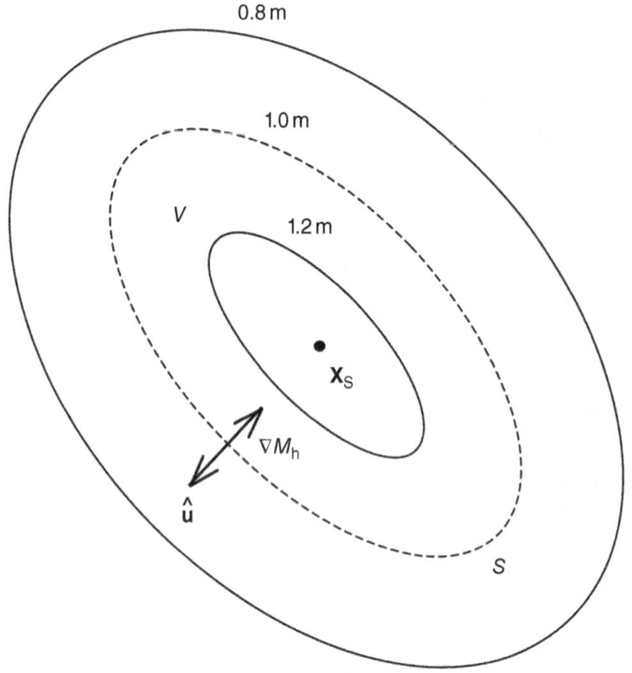

Figure 25.1 Hypothetical control volume for analysis of amplitude maximum.

- At every point of S, the amplitude gradient is nonzero.
- At every point of S, the outward unit-normal vector $\hat{\mathbf{u}}$ and the amplitude gradient point in opposite directions:

$$\nabla M_h(\mathbf{x}; \omega_m) = -|\nabla M_h(\mathbf{x}; \omega_m)|\hat{\mathbf{u}}(\mathbf{x}; \omega_m) \quad \forall\, \mathbf{x} \in S. \tag{25.10}$$

Figure 25.1 shows a hypothetical 2D example for which the control volume V satisfies the conditions specified above. In this figure the oval curves represent level surfaces of the amplitude function, the numerical labels give the corresponding amplitude values, and the dashed curve is the control surface, S. The amplitude maximum at $\mathbf{x} = \mathbf{x}_s$ exceeds 1.2 m.

Substituting $\sigma = 0$ in the time-averaged, integral equation for the conservation of fluid mechanical energy (24.43) produces

$$\int_V \langle \eta \rangle \, dV = -\int_S \langle \mathbf{J} \rangle \cdot \hat{\mathbf{u}} \, dS.$$

Using (25.4) and (25.10) to evaluate the integrand of the surface integral on the RHS yields

$$\int_V \langle \eta \rangle \, \mathrm{d}V = \frac{-\rho_{w0} g}{2} \int_S \phi_{e0} M_h |\nabla M_h| (\mathbf{K}\hat{\mathbf{u}}) \cdot \hat{\mathbf{u}} \, dS. \tag{25.11}$$

By assumption the hydraulic conductivity tensor is positive definite, so $(\mathbf{K}\hat{\mathbf{u}}) \cdot \hat{\mathbf{u}} > 0$. Our definition of the control volume in this section requires that the amplitude and its gradient are nonzero on S. The time-averaged effective porosity, ϕ_{e0}, is positive everywhere on S because zero points of ϕ_{e0} are zero points of the hydraulic conductivity, which by assumption we exclude from the space domain (see Section 2.5.1). Therefore, the integrand on the RHS of (25.11) is positive, giving

$$\int_V \langle \eta \rangle \, dV < 0.$$

This inequality, which contradicts (24.47), indicates that in a time-averaged sense the control volume as a whole undergoes negative dissipation of fluid mechanical energy. In the case of groundwater flowing through real porous media, water viscosity causes fluid mechanical energy to dissipate (i.e., $\eta \geq 0$), so this result is nonphysical. Therefore our initial supposition about the existence of an amplitude maximum must be false. That is, amplitude maxima are prohibited in source-free domains.

Exercise 25.7 Control Surface Contains No Stagnation Points

Problem: Consider time-periodic flow with a single harmonic constituent (i.e., ω_m fixed). Show formally that the control surface defined in Section 25.2.2 contains no stagnation points. *Hint*: Review Exercise 14.8.

Solution: We will use the method of contradiction. Suppose the control surface contains a stagnation point at $\mathbf{x} = \mathbf{x}_{nf}$. Applying the result (14.24a) and noting that $\hat{\mathbf{n}}(\mathbf{x}_{nf}) = \hat{\mathbf{u}}$ give

$$\left[\mathbf{K}(\mathbf{x}_{nf})\nabla M_h(\mathbf{x}_{nf}; \omega_m)\right] \cdot \hat{\mathbf{u}}(\mathbf{x}_{nf}; \omega_m) = 0 \quad \text{where} \quad \mathbf{x}_{nf} \in S.$$

Substituting the RHS of (25.10) for the amplitude gradient and momentarily dropping the explicit notation produce

$$-|\nabla M_h|(\mathbf{K}\hat{\mathbf{u}}) \cdot \hat{\mathbf{u}} = 0 \quad \text{where} \quad \mathbf{x}_{nf} \in S.$$

The quadratic form $(\mathbf{K}\hat{\mathbf{u}}) \cdot \hat{\mathbf{u}}$ is positive definite, so we can divide by $-(\mathbf{K}\hat{\mathbf{u}}) \cdot \hat{\mathbf{u}}$ to obtain

$$|\nabla M_h(\mathbf{x}_{nf}; \omega_m)| = 0 \quad \text{where} \quad \mathbf{x}_{nf} \in S.$$

Our definition of the control volume in this section requires the amplitude gradient to be nonzero on S, so this equation presents a contradiction. Therefore, our initial supposition must be false. That is, the control surface cannot contain stagnation points. □

25.2.3. Physical Interpretation of Vector Constituents

Using (19.9) and (25.2), we can now provide physical interpretations of the in-phase and quadrature components of the vector harmonic constituents.

25.2.3.1. Hydraulic Gradient Constituent

Equations (14.19) and (14.20) give the hydraulic gradient harmonic constituent, $\nabla h'(\mathbf{x}, t; \omega_m)$, as the sum of two component vectors, an in-phase component and a quadrature component. Referring to (14.20a), the in-phase component oscillates with time, and except at those two times in every constituent cycle when it vanishes, its direction alternates between those of $\pm\nabla M_{\rm h}(\mathbf{x}; \omega_m)$. That is, the direction of the in-phase component alternates between those of the dimensionless spatial attenuation vector $(-\nabla M_{\rm h})$ and its opposite.

Consider those points $\mathbf{x}$ (if any) where the amplitude gradient parallels a principal direction of the hydraulic conductivity tensor. At such points the time-averaged, fluid mechanical energy flux density (25.4) reduces to

$$\langle \mathbf{J} \rangle = -\frac{\phi_{\rm e0}\rho_{\rm w0}g}{2} M_{\rm h} K^{(\rm p)} \nabla M_{\rm h}$$

where $K^{(\rm p)}$ is the principal value (scalar) of the hydraulic conductivity in the direction of $\nabla M_{\rm h}(\mathbf{x}; \omega_m)$. Consequently, at such points the direction of the in-phase component alternates between those of $\pm\langle \mathbf{J} \rangle(\mathbf{x}; \omega_m)$. That is, the direction of the in-phase component alternates between those of the time-averaged, fluid mechanical energy flux density and its opposite.

Exercise 25.8 Interpreting the Quadrature Hydraulic Gradient Component

Problem: Interpret the quadrature component of the hydraulic gradient harmonic constituent in terms of the corresponding hydraulic head diffusion wave propagation velocity.

Solution: Referring to (14.20b), the quadrature component oscillates with time, and except at those two times in every constituent cycle when it vanishes, its direction alternates between those of $\pm\nabla\theta_{\rm h}(\mathbf{x}; \omega_m)$. Referring to (19.9), the direction of the quadrature component therefore alternates between those of $\pm\mathbf{v}(\mathbf{x}; \omega_m)$. That is, the direction of the quadrature component alternates between those of the hydraulic head diffusion wave propagation velocity and its opposite. □

25.2.3.2. Specific-Discharge Constituent

Equations (14.21) and (14.22) express the specific-discharge harmonic constituent, $\mathbf{q}'(\mathbf{x}, t; \omega_m)$, as the sum of two component vectors, an in-phase component and a quadrature component. Referring to (14.22a), the in-phase component oscillates with time, and except at those two times in every constituent cycle when it vanishes, its direction alternates between those of $\pm\mathbf{K}(\mathbf{x})\nabla M_{\rm h}(\mathbf{x}; \omega_m)$. Referring to (25.2), the direction of the in-phase component therefore alternates between those of $\pm\langle \mathbf{J} \rangle(\mathbf{x}; \omega_m)$. That is, the direction of the in-phase component alternates between those of the time-averaged, fluid mechanical energy flux density and its opposite.

The quadrature component does not lend itself to such a straightforward physical interpretation, except in special cases (see Exercise 25.9). Referring to (14.22b), the quadrature component oscillates with time, and except at those two times in every constituent cycle when it vanishes, its direction alternates between those of $\pm\mathbf{K}(\mathbf{x})\nabla\theta_{\rm h}(\mathbf{x}; \omega_m)$.

Exercise 25.9 Interpreting the Quadrature Specific-Discharge Component

Problem: Consider the specific-discharge harmonic constituent, $\mathbf{q}'(\mathbf{x}, t; \omega_m)$, at those points $\mathbf{x}$ (if any) where the phase gradient parallels a principal direction of the hydraulic conductivity tensor. Interpret the quadrature component in terms of the hydraulic head diffusion wave propagation velocity.

Solution: At such points the following holds:

$$\mathbf{K}(\mathbf{x})\nabla\theta_{\rm h}(\mathbf{x}; \omega_m) = K^{(\rm p)}\nabla\theta_{\rm h}(\mathbf{x}; \omega_m)$$

where $K^{(\rm p)}$ is the principal value (scalar) of the hydraulic conductivity in the direction of $\nabla\theta_{\rm h}(\mathbf{x}; \omega_m)$. Consequently, except at those two times in every constituent cycle when the quadrature component vanishes, its direction alternates between those of $\pm\nabla\theta_{\rm h}(\mathbf{x}; \omega_m)$. Referring to (19.9), the direction of the quadrature component therefore alternates between those of $\pm\mathbf{v}(\mathbf{x}; \omega_m)$, where $\mathbf{v}$ denotes the hydraulic head diffusion wave propagation velocity. □

25.3. TIME AVERAGED ENERGY DISSIPATION

25.3.1. General Results

As in the case where the hydraulic head field consists of multiple harmonic constituents, (24.53) gives the per-volume, time-averaged dissipation rate. Substituting $h = h'$ in (24.54) and using (24.11) to rewrite the time averages on the RHS give

$$\langle \eta \rangle_1 = \langle W \rangle_{\rm a} M_{\rm h}^{-2} \nabla M_{\rm h} \cdot (\mathbf{K}\nabla M_{\rm h}) + \langle W \rangle_{\rm b} \nabla\theta_{\rm h} \cdot (\mathbf{K}\nabla\theta_{\rm h}) \tag{25.12a}$$

$$\langle \eta \rangle_2 = \rho_{\rm w0} g \Big\langle \big[h'^2 + \rho_{\rm w0} g (\beta_{\rm w} - \alpha) h'^3 - \rho_{\rm w0}^2 g^2 \alpha \beta_{\rm w} h'^4 \big] \Big\rangle M_{\rm h}^{-1} (\mathbf{K}\nabla\phi_{\rm e0}) \cdot \nabla M_{\rm h} \tag{25.12b}$$

$$\langle \eta \rangle_3 = (1 - \phi_{\rm e0}) \rho_{\rm w0}^2 g^2 \langle h'^2 + \rho_{\rm w0} g \beta_{\rm w} h'^3 \rangle M_{\rm h}^{-1} (\mathbf{K}\nabla\alpha) \cdot \nabla M_{\rm h} \tag{25.12c}$$

where

$$\langle W\rangle_a \equiv \rho_{w0}g \times \left\langle \left\{\phi_{e0} + 2\rho_{w0}g\left[\phi_{e0}\beta_w + (1-\phi_{e0})\alpha\right]h' + 3(1-\phi_{e0})\rho_{w0}^2 g^2 \alpha\beta_w h'^2\right\} h'^2 \right\rangle \quad (25.13a)$$

$$\langle W\rangle_b \equiv \rho_{w0}g \times \left\langle \left\{\phi_{e0} + 2\rho_{w0}g\left[\phi_{e0}\beta_w + (1-\phi_{e0})\alpha\right]h' + 3(1-\phi_{e0})\rho_{w0}^2 g^2 \alpha\beta_w h'^2\right\} \left(M_h^2 - h'^2\right) \right\rangle. \quad (25.13b)$$

Exercise 25.10 Derivation: Time-Averaged Energy Dissipation

Problem: Evaluate $\langle W\rangle_a$ and $\langle W\rangle_b$.

Solution: Each of equations (25.13) is of the form

$$\langle W\rangle = \rho_{w0}g \times \left\langle \left\{\phi_{e0} + 2\rho_{w0}g\left[\phi_{e0}\beta_w + (1-\phi_{e0})\alpha\right]h' + 3(1-\phi_{e0})\rho_{w0}^2 g^2 \alpha\beta_w h'^2\right\} f(h') \right\rangle$$

where either $f(h') = h'^2$ or $f(h') = M_h^2 - h'^2$. Therefore $f(h')$ is a polynomial in even, positive powers of h'. Using (24.9) to evaluate the time averages on the RHSs of (25.13) leads to

$$\langle W\rangle_a = \frac{\rho_{w0}g}{2}\left[\phi_{e0} + 9(1-\phi_{e0})\left(\frac{\rho_{w0}g}{2}\right)^2 \alpha\beta_w M_h^2\right] M_h^2 \quad (25.14a)$$

$$\langle W\rangle_b = \frac{\rho_{w0}g}{2}\left[\phi_{e0} + 3(1-\phi_{e0})\left(\frac{\rho_{w0}g}{2}\right)^2 \alpha\beta_w M_h^2\right] M_h^2. \quad (25.14b)$$

□

Substituting the RHSs of (25.14) for the corresponding time averages on the RHS of (25.12a) gives

$$\langle\eta\rangle_1 = \frac{\rho_{w0}g}{2}\left\{\left[\phi_{e0} + 9(1-\phi_{e0})\left(\frac{\rho_{w0}g}{2}\right)^2 \alpha\beta_w M_h^2\right] \times \nabla M_h \cdot (\mathbf{K}\nabla M_h) + \left[\phi_{e0} + 3(1-\phi_{e0}) \times \left(\frac{\rho_{w0}g}{2}\right)^2 \alpha\beta_w M_h^2\right] M_h^2 \nabla\theta_h \cdot (\mathbf{K}\nabla\theta_h)\right\}. \quad (25.15)$$

Exercise 25.11 Derivation of Intermediate Result

Problem: Use the governing equation (6.7) to verify that in source-free domains

$$(\mathbf{K}\nabla\theta_h)\cdot\nabla\theta_h = (\mathbf{K}\nabla Y_h)\cdot\nabla Y_h + \nabla\cdot(\mathbf{K}\nabla Y_h) \quad (25.16)$$

where $Y_h \equiv \ln M_h$.

Solution: Substituting $M_u = 0$ in (6.7), dividing by M_h, and rearranging give

$$(\mathbf{K}\nabla\theta_h)\cdot\nabla\theta_h = \frac{1}{M_h}\nabla\cdot(\mathbf{K}\nabla M_h).$$

Substituting $M_h = \exp Y_h$ on the RHS, expanding the derivative terms, and then simplifying the result yield (25.16). □

We can use (25.16) to eliminate θ_h from the RHS of (25.15):

$$\langle\eta\rangle_1 = \frac{\rho_{w0}g}{2} M_h^2\left\{\left[2\phi_{e0} + 12(1-\phi_{e0})\left(\frac{\rho_{w0}g}{2}\right)^2 \alpha\beta_w M_h^2\right] \times \nabla Y_h \cdot (\mathbf{K}\nabla Y_h) + \left[\phi_{e0} + 3(1-\phi_{e0}) \times \left(\frac{\rho_{w0}g}{2}\right)^2 \alpha\beta_w M_h^2\right] \times \nabla\cdot(\mathbf{K}\nabla Y_h)\right\}. \quad (25.17)$$

Exercise 25.12 Derivation: Time-Averaged Energy Dissipation

Problem: Evaluate $\langle\eta\rangle_2$.

Solution: Using (24.9) to evaluate the time average on the RHS of (25.12b) and simplifying yield

$$\langle\eta\rangle_2 = \frac{\rho_{w0}g}{2}\left[1 - 3\left(\frac{\rho_{w0}g}{2}\right)^2 \alpha\beta_w M_h^2\right] M_h(\mathbf{K}\nabla M_h)\cdot\nabla\phi_{e0}.$$

Writing the gradient terms as gradients of logarithms gives

$$\langle\eta\rangle_2 = \frac{\phi_{e0}\rho_{w0}g}{2}\left[1 - 3\left(\frac{\rho_{w0}g}{2}\right)^2 \alpha\beta_w M_h^2\right] \times M_h^2(\mathbf{K}\nabla Y_h)\cdot\nabla\Phi_0 \quad (25.18)$$

where $\Phi \equiv \ln\phi$. □

Exercise 25.13 Derivation: Time-Averaged Energy Dissipation

Problem: Evaluate $\langle\eta\rangle_3$.

Solution: Using (24.9) to evaluate the time average on the RHS of (25.12c) and simplifying yield

$$\langle\eta\rangle_3 = 3(1-\phi_{e0})\left(\frac{\rho_{w0}g}{2}\right)^3 \beta_w M_h^3(\mathbf{K}\nabla M_h)\cdot\nabla\alpha.$$

Writing the gradient terms as gradients of logarithms gives

$$\langle\eta\rangle_3 = 3(1-\phi_{e0})\left(\frac{\rho_{w0}g}{2}\right)^3 \alpha\beta_w M_h^2(\mathbf{K}\nabla Y_h)\cdot\nabla\Lambda \tag{25.19}$$

where $\Lambda \equiv \ln\alpha$. □

25.3.2. Low-Order Approximations

Consider (25.14a), (25.14b), and (25.18). In each case the RHS is of a sum of two terms, the first of which is linear and the second of which is cubic, in $\rho_{w0}g/2$. We now examine the conditions under which the second term is negligible. Let z_1 and z_2 denote the dimensionless coefficients of the first and second terms, respectively; assume z_1 and z_2 are either numerical constants or functions of the effective porosity. Then the following dimensionless ratio measures the magnitude of the second term relative to the magnitude of the first term:

$$\nu \equiv \frac{|z_2|}{|z_1|}\left(\frac{\rho_{w0}g}{2}\right)^2 \alpha\beta_w M_h^2. \tag{25.20}$$

Therefore, $\nu \leq \nu_{max}$, where

$$\nu_{max} \equiv \frac{|z_2|_{max}}{|z_1|_{min}}\left(\frac{\rho_{w0}g}{2}\right)^2 \alpha_{max}\beta_w(M_h)^2_{max} \tag{25.21}$$

and the subscripts max and min, respectively, indicate upper and lower bounds for the indicated variable. If $\nu_{max} \ll 1$, then the second term is negligible.

We will assume the material properties satisfy (24.35). To establish a point of comparison, we will assume that $(M_h)_{max} = 30$ m. For some perspective on the numerical values of constituent amplitudes, see Note 25.1.

Substituting the numerical values given above for the corresponding terms on the RHS of (25.21) yields

$$\nu_{max} = 10^{-5}\frac{|z_2|_{max}}{|z_1|_{min}}. \tag{25.22}$$

Exercise 25.14 Derivation: Time-Averaged Energy Dissipation

Problem: Consider the two terms in the expression derived for $\langle\eta\rangle_2$, i.e., the RHS of (25.18). Use (25.22) to numerically evaluate ν_{max}.

Solution: In this case we have $z_1 = 1$ and $z_2 = -3$. Therefore, $|z_1|_{min} = 1$ and $|z_2|_{max} = 3$. Substituting these in (25.22) gives $\nu_{max} = 3 \times 10^{-5} \ll 1$. This suggests that under a wide range of conditions the second term of (25.18) is negligible in comparison to the first, so that we can approximate (25.18) as

$$\langle\eta\rangle_2 = \frac{\phi_{e0}\rho_{w0}g}{2} M_h^2(\mathbf{K}\nabla Y_h)\cdot\nabla\Phi_0. \tag{25.23}$$

□

A similar analysis of the relative magnitude of the second term on the RHS of each of (25.14a) and (25.14b), with the assumed minimum effective porosity $(\phi_{e0})_{min} = 0.01$, yields $\nu \leq 0.003$ and $\nu \leq 0.009$, respectively. Thus, (25.17) is approximated as

$$\langle\eta\rangle_1 = \frac{\phi_{e0}\rho_{w0}g}{2} M_h^2\Big[2(\mathbf{K}\nabla Y_h)\cdot\nabla Y_h + \nabla\cdot(\mathbf{K}\nabla Y_h)\Big]. \tag{25.24}$$

Substituting the RHSs of (25.24), (25.23), and (25.19) for $\langle\eta\rangle_1$, $\langle\eta\rangle_2$, and $\langle\eta\rangle_3$, respectively, in (24.53) yields

$$\langle\eta\rangle = \frac{\phi_{e0}\rho_{w0}g}{2} M_h^2\Big[2(\mathbf{K}\nabla Y_h)\cdot\nabla Y_h + \nabla\cdot(\mathbf{K}\nabla Y_h) + (\mathbf{K}\nabla Y_h)\cdot\nabla\Phi_0 + 3\frac{1-\phi_{e0}}{\phi_{e0}}\left(\frac{\rho_{w0}g}{2}\right)^2 \times \alpha\beta_w(\mathbf{K}\nabla Y_h)\cdot\nabla\Lambda\Big] + \langle\sigma\rangle. \tag{25.25}$$

Consider the sum within the square brackets on the RHS of (25.25). We will examine the conditions under which the fourth term is negligible relative to one or more of the other terms. Define

$$R_{1,2,3} \equiv \max\Big\{\big|2(\mathbf{K}\nabla Y_h)\cdot\nabla Y_h\big|, \big|\nabla\cdot(\mathbf{K}\nabla Y_h)\big|, \big|(\mathbf{K}\nabla Y_h)\cdot\nabla\Phi_0\big|\Big\}$$

$$R_4 \equiv \big|(\mathbf{K}\nabla Y_h)\cdot\nabla\Lambda\big|$$

where $R_{1,2,3}$ and R_4 have the same dimensions. The ratio of the absolute value of the fourth term to $R_{1,2,3}$ is then

$$\nu \equiv \frac{3(1-\phi_{e0})}{\phi_{e0}}\left(\frac{\rho_{w0}g}{2}\right)^2 \alpha\beta_w\frac{R_4}{R_{1,2,3}}. \tag{25.26}$$

To establish a point of comparison, initially assume that $R_4 \approx R_{1,2,3}$. Then the ratio ν is given by (25.20) with $z_1 = \phi_{e0}$ and $z_2 = 3(1-\phi_{e0})$. Assuming the same numerical limits as those used to derive (25.22) then gives $\nu \leq 0.003$. In fact, for R_4 as high as 30 times $R_{1,2,3}$ we get $\nu \leq 0.09$ (i.e., 30×0.003). These results suggest that under a relatively wide range of conditions the fourth term is negligible, so that we can approximate (25.25) as

$$\langle\eta\rangle = \frac{\phi_{e0}\rho_{w0}g}{2} M_h^2\Big[2(\mathbf{K}\nabla Y_h)\cdot\nabla Y_h + \nabla\cdot(\mathbf{K}\nabla Y_h) + (\mathbf{K}\nabla Y_h)\cdot\nabla\Phi_0\Big] + \langle\sigma\rangle. \tag{25.27}$$

In the special case where the porous medium is homogeneous with respect to bulk compressibility, (25.25) reduces to (25.27).

25.3.3. Slow-Deformation Approximation

In this section we examine the time-averaged, per-volume, fluid mechanical energy dissipation rate in the case where both (a) the slow-deformation approximation (see Section 24.5.1) is valid and (b) the hydraulic head field consists of a single harmonic constituent.

Exercise 25.15 Time-Averaged Divergence of Energy Flux Density

Problem: Consider a hydraulic head field consisting of a single harmonic constituent. Suppose (25.4) is valid (see Exercise 25.1). Evaluate the time average of the divergence of the fluid mechanical energy flux density.

Solution: By changing the order of differentiation (with respect to space coordinates) and integration (with respect to time) we have

$$\langle \nabla \cdot \mathbf{J} \rangle = \nabla \cdot \langle \mathbf{J} \rangle.$$

Implementing the approximation (25.4) gives

$$\langle \nabla \cdot \mathbf{J} \rangle = -\frac{\rho_{w0} g}{2}\Big[\nabla(\phi_{e0} M_h) \cdot (\mathbf{K}\nabla M_h) + \phi_{e0} M_h \nabla \cdot (\mathbf{K}\nabla M_h)\Big].$$

Expanding the gradient in the first term within square brackets produces

$$\langle \nabla \cdot \mathbf{J} \rangle = -\frac{\rho_{w0} g}{2}\Big[\phi_{e0}(\mathbf{K}\nabla M_h) \cdot \nabla M_h + M_h(\mathbf{K}\nabla M_h) \cdot \nabla \phi_{e0} + \phi_{e0} M_h \nabla \cdot (\mathbf{K}\nabla M_h)\Big].$$

Expressing the amplitude and effective porosity in terms of their corresponding logarithms and then simplifying lead to

$$\langle \nabla \cdot \mathbf{J} \rangle = -\frac{\phi_{e0}\rho_{w0} g}{2} M_h^2\Big[2(\mathbf{K}\nabla Y_h) \cdot \nabla Y_h + (\mathbf{K}\nabla Y_h) \cdot \nabla \Phi_0 + \nabla \cdot (\mathbf{K}\nabla Y_h)\Big]. \tag{25.28}$$

□

Exercise 25.16 Time-Averaged, Per-Volume, Energy Dissipation Rate

Problem: Consider a hydraulic head field consisting of a single harmonic constituent. Suppose (25.4) is valid (see Exercise 25.1). Evaluate the time-averaged, per-volume, fluid mechanical energy dissipation rate.

Solution: Taking the time average of (23.30) gives

$$\langle \eta \rangle = -\left\langle \frac{\partial \epsilon}{\partial t} \right\rangle - \langle \nabla \cdot \mathbf{J} \rangle + \langle \sigma \rangle.$$

Substituting the RHSs of (24.40) and (25.28) for $\langle \partial\epsilon/\partial t \rangle$ and $\langle \nabla \cdot \mathbf{J} \rangle$, respectively, leads to

$$\langle \eta \rangle = \frac{\phi_{e0}\rho_{w0} g}{2} M_h^2\Big[2(\mathbf{K}\nabla Y_h) \cdot \nabla Y_h + \nabla \cdot (\mathbf{K}\nabla Y_h) + (\mathbf{K}\nabla Y_h) \cdot \nabla \Phi_0\Big] + \langle \sigma \rangle. \tag{25.29}$$

□

The RHS of (25.29) is identical to that of (25.27).

Exercise 25.17 Time-Averaged, Per-Volume, Energy Dissipation Rate

Problem: Consider time-periodic flow in a source-free domain. Suppose the hydraulic head field consists of a single harmonic constituent and the porous medium is homogeneous with respect to effective porosity. Assume (25.4) is valid (see Exercise 25.1). Give an expression for the time-averaged, per-volume, fluid mechanical energy dissipation rate. What, if anything, can you conclude about its sign in this case?

Solution: Substituting $\sigma = 0$ and $\nabla \Phi_0 = \mathbf{0}$ in (25.29) yields

$$\langle \eta \rangle = \frac{\phi_{e0}\rho_{w0} g}{2} M_h^2\Big[2(\mathbf{K}\nabla Y_h) \cdot \nabla Y_h + \nabla \cdot (\mathbf{K}\nabla Y_h)\Big].$$

Using the governing equation (6.7) (with $M_u = 0$) to write the divergence term as a difference of quadratic forms in the amplitude and the phase, we obtain

$$\langle \eta \rangle = \frac{\phi_{e0}\rho_{w0} g}{2} M_h^2\Big[(\mathbf{K}\nabla Y_h) \cdot \nabla Y_h + (\mathbf{K}\nabla \theta_h) \cdot \nabla \theta_h\Big].$$

The effective porosity, water density, acceleration of gravity, and squared amplitude are nonnegative. The quantity within the square brackets is the sum of two positive-semidefinite quadratic forms, so it is nonnegative. It follows that the RHS is nonnegative. Consequently, the time-averaged dissipation rate is nonnegative. □

25.4. MECHANICAL POWER TRANSMISSION ACROSS A SURFACE

Consider 3D, time-periodic flow in which the hydraulic head field consists of a single harmonic constituent (i.e., ω_m fixed). The domain may be finite or infinite, and the resident porous medium is generally nonhomogeneous and anisotropic. Throughout this section we assume that the slow-deformation approximation is valid.

25.4.1. General Results

Consider a more or less arbitrary surface S in connection with the space domain. The surface may coincide with the domain boundary ($S = \Gamma$) or some portion thereof ($S \in \Gamma$) or it may be some other continuous surface within the space domain ($S \in D$). Also, the surface may coincide with the boundary of a closed volume but it need not.

Generally, in a time-averaged sense fluid mechanical energy is transmitted across the surface S. The time-averaged rate of energy transmission (i.e., *mechanical power transmission*) across the surface is

$$\langle P(\omega_m)\rangle = \int_S \langle \mathbf{J}(\mathbf{x};\omega_m)\rangle \cdot \hat{\mathbf{u}}\; dA \tag{25.30}$$

where $\hat{\mathbf{u}}$ is the outward unit-normal vector for the surface S. If the surface S is the boundary of a closed volume, then positive values for $\langle P(\omega_m)\rangle$ indicate that there is a net transmission of fluid mechanical power across the surface *out of* the enclosed volume and into the adjacent medium. Similarly, negative values for $\langle P(\omega_m)\rangle$ indicate that there is a net transmission of fluid mechanical power across the surface, from the adjacent medium *into* the enclosed volume.

25.4.2. Mechanical Power Transmission by Fluid Sources

Consider time-periodic flow in the presence of an isolated source, i.e., a point source, line source, or plane source. In a time-averaged sense, the fluid source transmits fluid mechanical energy to the adjacent medium. That is, in time-periodic flow, every fluid source is also a mechanical energy source. In this section we derive expressions for the source's time-averaged energy transmission rate, which we call the time-averaged, *mechanical power* of the source. Our derivations employ the *Gaussian surface* concept from classical electrodynamics [e.g., see *Jackson*, 1998].

25.4.2.1. Time-Averaged Power of an Isolated Point Source

Suppose the space domain D contains an isolated point source at the point $\mathbf{x} = \mathbf{x}_s$. Define a control volume, V, that contains the point $\mathbf{x} = \mathbf{x}_s$ and is simply connected. Then the time-averaged power transmission by the point source, if it exists, is given by the following limiting form of (25.30):

$$\langle P(\omega_m)\rangle = \lim_{V\to 0} \int_S \langle \mathbf{J}(\mathbf{x};\omega_m)\rangle \cdot \hat{\mathbf{u}}\; dA \tag{25.31}$$

where S denotes the control surface and $\hat{\mathbf{u}}$ is its outward unit-normal vector. For instance, it suffices to define V as the region enclosed by a sphere (i.e., a *Gaussian sphere*) centered at $\mathbf{x} = \mathbf{x}_s$. As the sphere's radius decreases in the limit, eventually the control volume will contain no other sources and will not intersect the domain boundary (i.e., Γ).

Exercise 25.18 Average Power of Point Source in Ideal Medium

Problem: Consider a point source in an infinite, 3D domain filled with an ideal medium. Express the time-averaged power delivered to the medium by the point source, at nonzero distance r from the point source, as a function of r. Is the time-averaged power delivered by the point source, at nonzero distance r, finite?

Solution: Define

$$\mathbf{r} \equiv \mathbf{x} - \mathbf{x}_s$$

$$r \equiv |\mathbf{r}|.$$

Also, let ϕ be the angle measured anticlockwise from the positive x_1 axis in the $x_1 - x_2$ plane, and let θ be the angle measured from the positive x_3 axis. Then $\hat{\mathbf{u}} = \hat{\mathbf{r}}$, and

$$\int_S dA = \int_0^{2\pi}\int_0^{\pi} r^2 \sin\theta \; d\theta \; d\phi.$$

Substituting these expressions for the appropriate quantities in (25.30) gives

$$\langle P(r;\omega_m)\rangle = \int_0^{2\pi}\int_0^{\pi} \langle \mathbf{J}(\mathbf{x};\omega_m)\rangle \cdot \hat{\mathbf{r}}\; r^2 \sin\theta \; d\theta \; d\phi.$$

Equation (25.6) indicates that $\langle \mathbf{J}(\mathbf{x};\omega_m)\rangle$ is independent of the angular coordinates θ and ϕ, so after evaluating the integrals we obtain

$$\langle P(r;\omega_m)\rangle = 4\pi r^2 \langle \mathbf{J}(\mathbf{x};\omega_m)\rangle \cdot \hat{\mathbf{r}}. \tag{25.32}$$

Substituting the RHS of (25.6) for $\langle \mathbf{J}(\mathbf{x};\omega_m)\rangle$ gives

$$\langle P(r;\omega_m)\rangle = \frac{\phi_{e0}\rho_{w0} g |Q_0(\omega_m)|^2}{8\pi K} \times \left[e^{-\sqrt{2\alpha_m} r} \left(\frac{1}{r} + \sqrt{\frac{\alpha_m}{2}} \right) \right], \quad r > 0. \tag{25.33}$$

The RHS is finite, so the time-averaged power delivered by the point source, at nonzero distance r from the source, is finite. □

Example 25.1 Average Power of Point Source in Ideal Medium

Consider a point source in an infinite, 3D domain filled with an ideal medium. In accordance with (25.31), the total, time-averaged power delivered to the medium by the point source, when it exists, is given by the limit

$$\langle P(\omega_m)\rangle = \lim_{r\to 0} \langle P(r;\omega_m)\rangle$$

where (25.33) gives $\langle P(r;\omega_m)\rangle$. Substituting the RHS of (25.33) for $\langle P(r;\omega_m)\rangle$ leads to

$$\langle P(\omega_m)\rangle = \frac{\phi_{e0}\rho_{w0}g|Q_0(\omega_m)|^2}{8\pi K}\lim_{r\to 0}\left[e^{-\sqrt{2\alpha_m}r}\left(\frac{1}{r}+\sqrt{\frac{\alpha_m}{2}}\right)\right] \to \infty.$$

That is, the RHS diverges. Apparently, in the limit as the radius r of the sphere centered at the point source vanishes, the effective cross-sectional area for flow vanishes. At the same time, the source amplitude $|Q_0(\omega_m)|$ remains fixed and finite. This causes the flow resistance to become infinite and hence to dissipate infinite energy. Strictly, then, a true point source in a 3D medium is nonphysical because it is energetically impossible to maintain. ◊

Although the result obtained in Example 25.1 indicates that strictly a point source in time-periodic flow is nonphysical, the result obtained in Exercise 25.18 suggests that under some conditions the point source concept may be useful for approximating the flow at nonzero distance r from the source. At nonzero distance r the cross-sectional area for flow is nonvanishing; therefore, the flow resistance is finite, resulting in finite energy dissipation.

25.4.2.2. Time-Averaged Power of an Isolated Line Source

Suppose the domain D contains an isolated line source whose strength generally is nonuniform (i.e., it may vary with position on the line source). Because the line source may be infinitely long, we expect that the time-averaged, fluid mechanical power it transmits to the adjacent medium may be infinite. However, the *per-length*, time-averaged, fluid mechanical power transmitted by the source may be finite, so we may be able to calculate it.

Define the control surface, $S(a)$, as a right circular cylinder (i.e., a *Gaussian cylinder*) of length l ($l > 0$) and radius a ($a > 0$) whose axis coincides with the line source. Define the control volume as the region enclosed by this cylindrical control surface. Also define a local cylindrical coordinate system (r, x_3) in which (a) $r = 0$ at the cylinder axis and r increases linearly with distance in the direction perpendicular to the axis and (b) the x_3 axis coincides with the cylinder axis.

Then the time-averaged power transmitted by this finite-length segment of the line source, if it exists, is given by the following limiting form of (25.30):

$$\langle P(\omega_m)\rangle = \lim_{a\to 0}\int_{S(a)} \langle \mathbf{J}(\mathbf{x};\omega_m)\rangle \cdot \hat{\mathbf{u}}\; dA \qquad (25.34)$$

where $\hat{\mathbf{u}}$ is the outward unit-normal vector of the control surface. As the cylinder radius decreases in the limit, eventually the control volume will contain no other sources and will not intersect the domain boundary (i.e., Γ).

Exercise 25.19 Average Power of Line Source in Ideal Medium

Problem: Consider an infinitely long, uniform line source in an infinite, 3D domain filled with an ideal medium. Express the time-averaged power delivered to the medium by a segment of the line source having length l, at nonzero distance r from the line source, as a function of r. Is the time-averaged power delivered by the line source, at nonzero distance r, finite?

Solution: Define a right-handed Cartesian coordinate system in which the x_3 axis coincides with the line source. Let ϕ be the angle in the x_1–x_2 plane measured clockwise from the positive x_1 axis and let r be the radial distance from the line source as measured in the x_1–x_2 plane:

$$r \equiv \sqrt{x_1^2 + x_2^2}.$$

In this case the time-averaged, fluid mechanical energy flux density is given by (25.7), which indicates that $\langle \mathbf{J}(\mathbf{x};\omega_m)\rangle$ is independent of the coordinates ϕ and x_3 and perpendicular to the x_3 axis. Therefore, $\hat{\mathbf{u}} = \hat{\mathbf{r}}$ and the contribution to the surface integral [on the RHS of (25.34)] from the two ends of the cylinder is zero; only the contribution from the curved side is nonzero:

$$\int_S dA = \int_0^l \int_0^{2\pi} r\; d\phi\; dx_3.$$

Substituting these expressions for the appropriate quantities in (25.30) leads to

$$\langle P(r;\omega_m)\rangle = \int_0^l \int_0^{2\pi} \langle \mathbf{J}(\mathbf{x};\omega_m)\rangle \cdot \hat{\mathbf{r}}\; r\; d\phi\; dx_3.$$

Evaluating the integrals gives

$$\langle P(r;\omega_m)\rangle = 2\pi\; r\; l\,\langle \mathbf{J}(\mathbf{x};\omega_m)\rangle \cdot \hat{\mathbf{r}}.$$

Substituting the RHS of (25.7) for $\langle \mathbf{J}(\mathbf{x};\omega_m)\rangle$ yields

$$\langle P(r;\omega_m)\rangle = \frac{-\phi_{e0}\rho_{w0}g|\Delta Q_0(\omega_m)|^2 l}{4\pi K} RN_0(R)N_1(R) \times \cos\left[\phi_1(R) - \phi_0(R) - \frac{\pi}{4}\right], \quad R > 0 \qquad (25.35)$$

where $R \equiv \sqrt{\alpha_m}r$. The RHS is finite, so the time-averaged power delivered by the line source, at nonzero distance r from the source, is finite. □

Example 25.2 Average Power of Line Source in Ideal Medium

Consider an infinitely long, uniform line source in a 3D domain filled with an ideal medium. As per (25.34), the total, time-averaged power delivered to the medium by a length-l segment of the line source, when it exists, is given by the limit

$$\langle P(\omega_m)\rangle = \lim_{r\to 0}\langle P(r;\omega_m)\rangle$$

where (25.35) gives $\langle P(r;\omega_m)\rangle$. Substituting the RHS of (25.35) for $\langle P(r;\omega_m)\rangle$ leads to

$$\langle P(\omega_m)\rangle = \frac{-\phi_{e0}\rho_{w0}g|\Delta Q_0(\omega_m)|^2 l}{4\pi K} \times \lim_{R\to 0}\left\{RN_0(R)N_1(R)\cos \times\left[\phi_1(R)-\phi_0(R)-\frac{\pi}{4}\right]\right\}.$$

Using (20.35) to evaluate the cosine term yields

$$\langle P(\omega_m)\rangle = \frac{\phi_{e0}\rho_{w0}g|\Delta Q_0(\omega_m)|^2 l}{4\pi K}\lim_{R\to 0}\left[RN_0(R)N_1(R)\right].$$

Noting that $N_\nu(R) = |K_\nu(e^{i\pi/4}R)|$ and using equations 10.30.2 (with $\nu = 1$) and 10.30.3 of *Olver and Maximon* [2016], we obtain

$$\langle P(\omega_m)\rangle \to \infty \quad \text{as} \quad r\to 0.$$

That is, the RHS diverges. Apparently, in the limit as the radius r of the cylinder centered at the line source vanishes, the effective cross-sectional area for flow vanishes. At the same time, the source amplitude $|\Delta Q_0(\omega_m)|$ remains fixed and finite. This causes the flow resistance to become infinite and hence to dissipate infinite energy. Strictly, then, a true line source in a 3D medium is nonphysical because it is energetically impossible to maintain. ◊

Exercise 25.20 Equivalence of Point/Line Sources in Two/Three Dimensions

Problem: Consider time-periodic flow in an infinite, 2D domain filled with an ideal medium. Is it possible to maintain a true point source in this setting?

Solution: Although the conclusion of Example 25.2 applies specifically to a line source in a 3D medium, because of dimensional equivalence, the same is true of a point source in a 2D medium. That is, strictly a true point source in a 2D medium is nonphysical because it is energetically impossible to maintain. □

Although the result obtained in Example 25.2 indicates that strictly a line source in time-periodic flow is nonphysical, the result obtained in Exercise 25.19 suggests that under some conditions the line source concept may be useful for approximating the flow at nonzero distance r from the source. At nonzero distance r the cross-sectional area for flow, for any finite-length segment of the line source, is nonvanishing; therefore, the flow resistance is finite, resulting in finite energy dissipation.

25.4.2.3. Time-Averaged Power of an Isolated Plane Source

Suppose the domain D contains an isolated plane source whose strength generally is nonuniform (i.e., it may vary with position on the plane source). Because the plane source may be infinite in extent, we expect that the time-averaged, fluid mechanical power it transmits to the adjacent medium may or may not be finite. However, the *per-area*, time-averaged, fluid mechanical power transmitted by the source should be finite, so we can calculate it.

Define the control surface, S, as a right circular cylinder of length $2b$ ($b > 0$) and radius a ($a > 0$) whose axis is perpendicular to and bisected by the plane source (i.e., a *Gaussian pillbox*). Define the control volume as the region enclosed by this cylindrical control surface. Also define a local cylindrical coordinate system (r, x_1) in which (a) $r = 0$ at the cylinder axis and r increases linearly with distance in the direction perpendicular to the cylinder axis and (b) the x_1 axis coincides with the cylinder axis and $x_1 = 0$ at the plane source.

Then the time-averaged, mechanical power transmitted by this disc-shaped portion of the plane source, if it exists, is given by the following limiting form of (25.30):

$$\langle P(\omega_m)\rangle = \lim_{b\to 0}\int_S \langle \mathbf{J}(\mathbf{x};\omega_m)\rangle \cdot \hat{\mathbf{u}}\, dA \qquad (25.36)$$

where $\hat{\mathbf{u}}$ is its outward unit-normal vector of the control surface. As the cylinder length decreases in the limit, eventually the control volume will contain no other sources and will not intersect the domain boundary (i.e., Γ).

Example 25.3 Average Power of Plane Source in Ideal Medium

Consider an infinite, uniform plane source in a 3D domain filled with an ideal medium. We will calculate the per-area, time-averaged, fluid mechanical power transmitted to the adjacent medium by the plane source. We begin by calculating the time-averaged, fluid mechanical power transmitted to the adjacent medium by the disc-shaped portion of the plane source with radius a.

Recall from Exercise 25.5 that in this case the time-averaged, fluid mechanical energy flux density is given by (25.8):

$$\langle \mathbf{J}(\mathbf{x};\omega_m)\rangle = \frac{\phi_{e0}\rho_{w0}g|\Delta Q_0(\omega_m)|^2}{8\sqrt{2}K\sqrt{\alpha_m}} \times \exp\left(-2\sqrt{\frac{\alpha_m}{2}}|x_1|\right)\operatorname{sign}(x_1)\hat{\mathbf{x}}_1, \quad |x_1| > 0. \qquad (25.37)$$

Here, $\langle \mathbf{J}\rangle$ is parallel to the x_1 axis, so the contribution to the surface integral [on the RHS of equation (25.36)] from the curved side of the cylinder is zero; only the contribution from the two ends is nonzero:

$$\langle P(\omega_m)\rangle = \lim_{b\to 0}\int_0^{2\pi}\int_0^{a}\left\{\left[\langle \mathbf{J}\rangle\cdot\hat{\mathbf{u}}\right]\Big|_{x_1=-b} + \left[\langle \mathbf{J}\rangle\cdot\hat{\mathbf{u}}\right]\Big|_{x_1=b}\right\} r\,dr\,d\theta.$$

The outward unit-normal vectors for the two opposite ends of the cylinder are given by $\hat{\mathbf{u}}(\mathbf{x}) = \mathrm{sign}(x_1)\hat{\mathbf{x}}_1$. Combining this result and (25.37) with the previous equation yields

$$\langle P(\omega_m)\rangle = \pi a^2 \frac{\phi_{e0}\rho_{w0}g|\Delta Q_0(\omega_m)|^2}{4\sqrt{2}K\sqrt{\alpha_m}}.$$

Dividing by the area (πa^2) gives the per-area, time-averaged, fluid mechanical power transmitted by the plane source to the adjacent medium:

$$\langle \Delta P(\omega_m)\rangle = \frac{\phi_{e0}\rho_{w0}g|\Delta Q_0(\omega_m)|^2}{4\sqrt{2}K\sqrt{\alpha_m}}. \tag{25.38}$$

◊

25.5. CHAPTER SUMMARY

In this chapter we examined the time-averaged transport of fluid mechanical energy by a single-constituent wave propagating under isothermal conditions. We derived expressions for time averages of the following: the fluid mechanical energy density and its rate of change, the flux density, and the dissipation rate. Our results indicate that the time-averaged flux density is parallel to the in-phase component of the corresponding specific-discharge vector harmonic constituent and the relationship between the time-averaged flux density and the square of the amplitude function [i.e., equation (25.5)], is similar to the relationship between the specific discharge and the hydraulic head (i.e., Darcy's law).

We also presented an informal, physical argument for the prohibition of amplitude maxima in source-free domains based on the conservation of fluid mechanical energy. This argument complements the more abstract, mathematical argument presented in Section 17.3.1.

Finally, we examined the time-averaged, fluid mechanical power transmitted to an adjacent medium by isolated fluid sources: point, line, and plane sources. We concluded that point and line sources provide useful approximations for the flow field at non-zero distances from the sources. This despite the fact that true point and line sources are not physically possible because they are energetically impossible to maintain. Lastly, we derived an expression (25.38) for the per-area, time-averaged, fluid mechanical power transmitted by a plane source to an infinite domain filled with an ideal medium.

Part VIII
Conclusion

26
Conclusion

26.1. OVERVIEW

26.1.1. Overview of Part I: Introduction

In Chapter 1 we introduced basic terminology on the time behavior of periodically forced groundwater flow systems, suggested criteria for the classification of time-periodic forcing, and briefly listed several areas of application for the theory of time-periodic groundwater flow.

26.1.2. Overview of Part II: Problem Definition

In Chapter 2 we derived a general IBVP to describe the transient flow of homogeneous groundwater through a multidimensional, nonhomogeneous, anisotropic, elastically deformable, porous medium for which the domain boundary is motionless. We described the major elements of the IBVP, including the space-time domain, governing equation, initial condition, boundary conditions, and other parameters.

In Chapter 3 we showed how, conceptually, one could use superposition to resolve the hydraulic head, source/sink strength, and boundary value into steady, nonperiodic transient, and periodic transient components. We also showed how, conceptually, one could use superposition to resolve the initial value into corresponding nonperiodic and periodic components. We defined the BVP (or IBVP) corresponding to each of the steady, nonperiodic transient, and periodic transient hydraulic head components. Finally, we concluded that we can express each periodic transient component of hydraulic head, source strength, and boundary value globally in time by a trigonometric series.

In Chapter 4 we expressed each of the periodic transient components of hydraulic head, source-sink strength, and boundary value as a sum of time-harmonic constituents (i.e., as a trigonometric series) whose amplitudes and phases are functions of position. We inferred basic relations between the constituent frequencies, and we defined a general BVP for the hydraulic head harmonic constituents. Finally, we showed how the periodic transient components of the hydraulic gradient and specific-discharge vectors could be expressed as sums of vector harmonic constituents and that these harmonic constituents are directly related to the hydraulic head harmonic constituents via the spatial gradient operator and Darcy's law.

In Chapter 5 we derived two rectangular forms (coupled and uncoupled) of the space BVP for the hydraulic head harmonic constituents. We also derived two forms (rectangular and general) of the space-time BVP for hydraulic head harmonic constituents.

In Chapter 6 we derived the polar form of the space BVP for the hydraulic head harmonic constituents and used the results to infer basic properties of the amplitude and phase functions. We also showed that for time-periodic flow in a source-free, 1D medium, under some conditions, it is possible to uncouple the polar form governing equations.

In Chapter 7 we defined complex-valued harmonic constituents of hydraulic head, source strength, and boundary value. We defined the complex-valued frequency response function and real-valued amplitude response and phase response functions for hydraulic head. We showed that the hydraulic head FRF satisfies the complex-variable form of the space BVP for hydraulic head harmonic constituents. We presented exercises that show the complex-variable form of the space BVP is mathematically equivalent to the rectangular and polar forms. Finally, we derived expressions for the FRFs of the corresponding hydraulic gradient and specific-discharge vectors.

Hydrodynamics of Time-Periodic Groundwater Flow: Diffusion Waves in Porous Media, Geophysical Monograph 224,
First Edition. Joe S. Depner and Todd C. Rasmussen.

In Chapter 8 we summarized and compared the three forms of the space BVP that we presented earlier in Part II.

26.1.3. Overview of Part III: Elementary Examples

In Chapter 9 we presented some elementary solutions of the complex-variable form of the space BVP for time-periodic, 1D flow in homogeneous media. We also related these solutions to those of point, line, and plane sources in multidimensional media.

In Chapter 10 we presented some elementary solutions of the complex-variable form of the space BVP for time-periodic, 1D flow in exponential media. The hydraulic conductivity and specific storage vary exponentially with distance in the direction parallel to the flow and at the same relative rate (μ) so that the hydraulic diffusivity is constant. The FRF solutions presented in this chapter are generalizations of those presented in Chapter 9.

In Chapter 11 we presented some elementary solutions of the complex-variable form of the space BVP for time-periodic, 1D flow in power law media. The hydraulic conductivity and specific storage vary continuously with distance in the direction parallel to the flow according to simple power laws. Unlike exponential media, in power law media the hydraulic diffusivity varies with the square of the space coordinate.

In Chapter 12 we presented some elementary solutions of the complex-variable form of the space BVP for multidimensional, time-periodic flow in ideal media. These include convergent/divergent flows with radial or axial symmetry.

In Chapter 13 we presented a particular family of FRFs for time periodic, multidimensional flow in a source-free domain filled with an exponential medium. However, unlike the solutions presented in Chapter 10, here the gradients generally do not parallel the direction of flow.

26.1.4. Overview of Part IV: Essential Concepts

In Chapter 14 we explained the concepts of attenuation and delay in the context of time-periodic flow. We defined scalar and vector measures of attenuation, and we related the concept of delay to those of in-phase versus quadrature components of the harmonic constituents. We also discussed collinearity of the harmonic constituent's amplitude and phase gradients. We defined local and global collinearity and listed examples of elementary FRF solutions in which the constituent amplitude and phase gradients are globally collinear, globally noncollinear, or neither. Finally, we showed that the question of whether the in-phase and quadrature components of a particular specific-discharge constituent are collinear is mathematically equivalent to the question of whether the corresponding hydraulic head constituent's amplitude and phase gradients are collinear.

In Chapter 15 we explored the time variation of the specific-discharge vector harmonic constituent, $\mathbf{q}'(\mathbf{x}, t; \omega_m)$. We showed that under rather general conditions (i.e., 3D flow through nonhomogeneous, anisotropic media) one of the following cases describes the time variation:

1. $\nabla M_{\rm h}(\mathbf{x}; \omega_m) = \mathbf{0}$ and $\nabla \theta_{\rm h}(\mathbf{x}; \omega_m) = \mathbf{0}$. Then $\mathbf{q}'(\mathbf{x}, t; \omega_m)$ is time invariant because it vanishes for all times. The envelope is a point.
2. $\nabla M_{\rm h}(\mathbf{x}; \omega_m)$ and $\nabla \theta_{\rm h}(\mathbf{x}; \omega_m)$ are locally collinear or one of them (but not both) vanishes locally. Then $\mathbf{q}'(\mathbf{x}, t; \omega_m)$ alternates direction and magnitude with time but does not rotate. The envelope is a line segment.
3. $\nabla M_{\rm h}(\mathbf{x}; \omega_m)$ and $\nabla \theta_{\rm h}(\mathbf{x}; \omega_m)$ are locally nonzero and locally noncollinear. Then $\mathbf{q}'(\mathbf{x}, t; \omega_m)$ rotates within the plane of oscillation, and its magnitude also may vary periodically with time. The rotation rate generally varies with time, and the rotation satisfies a Kepler-type law (see Exercise 15.13). The envelope is an ellipse.

In all cases the envelope of the specific-discharge constituent is symmetric about the origin in discharge space.

26.1.5. Overview of Part V: Stationary Points

In Chapter 16 we defined and described elliptic and uniformly elliptic operators and presented two maximum principles for differential inequalities involving elliptic operators. We showed how these principles could be applied to the analysis of stationary points of the rectangular form coefficient functions $A_{\rm h}(\mathbf{x}; \omega_m)$ and $B_{\rm h}(\mathbf{x}; \omega_m)$ of time-periodic flow. We concluded that the results so obtained are not very useful because they generally require one to know the signs of $A_{\rm h}(\mathbf{x}; \omega_m)$ and $B_{\rm h}(\mathbf{x}; \omega_m)$ in D.

In Chapter 17 we conducted a theoretical analysis of amplitude and phase extrema for time-periodic flow in source-free domains to infer general principles governing their possible existence and nature. For that analysis we used the polar forms of the governing equations (see Chapter 6) and the maximum principles that were introduced in Chapter 16. We showed that if the amplitude and phase functions vary smoothly over a region in which the amplitude is positive, then the following conditions are satisfied within that region:

- Maxima of the amplitude and phase functions are prohibited.
- Minima of the amplitude and phase functions are allowed but are not required.

In Chapter 18 we described the concept of flow stagnation in the context of time-periodic groundwater flow. We defined stagnation points and stagnation sets (isolated stagnation points, stagnation divides, and stagnation zones). Then we related these physical features to mathematical properties of the flow field, such as the stationary points of the amplitude and phase functions for hydraulic head harmonic constituents.

26.1.6. Overview of Part VI: Wave Propagation

In Chapter 19 we interpreted time-periodic groundwater flow as the steady-state propagation of spatially modulated (damped), traveling diffusion waves, i.e., harmonic, hydraulic head waves. The wave interpretation provides a means of conceptually organizing the space-time behavior of time-periodic flow. We presented and discussed two distinct wave interpretations—the constituent wave interpretation and the component wave interpretation. We gave rationale for preferring the component wave interpretation, which views each hydraulic head harmonic constituent field as an interference pattern that results from the superposition of the corresponding component waves.

In Chapter 20 we discussed dispersive and nondispersive distortion. We showed that under some conditions the concept of group velocity is valid for traveling wave packets. We listed examples from earlier chapters showing that constituent waves sometimes exhibit negative attenuation, and we presented several measures of wave attenuation. We also explained that harmonic, hydraulic head waves traveling through nonhomogeneous media generally undergo wave dilation and/or contraction.

In Chapter 21 we explored the conditions under which one can interpret time-periodic groundwater flow in one dimension as the steady-state propagation of traveling, harmonic waves. We explored the conditions under which waves travel at space-invariant velocities in a 1D medium. We also gave a hypothetical example, with an analytical solution, of a 1D medium through which waves propagate at nonuniform velocities.

In Chapter 22 we showed that each generic hydraulic head wave field satisfies a generalized wave equation (22.7), which is a homogenous, second-order PDE with real-valued coefficients that depend on the space coordinates and the constituent frequency. We expressed the coefficients explicitly in terms of the amplitude and phase functions and the frequency of the generic wave. That the coefficients of the generalized wave equation are known functions of the amplitude and phase has the following important consequences:

1. The coefficients generally are unknown unless the amplitude and phase functions are known, that is, unless the solution of the space BVP [e.g., the system of equations (7.17)] is known.
2. The coefficients depend implicitly not only on the material hydrologic properties **K** and S_s but also on the boundary conditions and the source distribution (see Exercise 22.6). Therefore, the coefficients do not directly represent intrinsic physical properties of the medium.
3. Superposition is not valid for the generalized wave equation 22.7.

These consequences limit the usefulness of the generalized wave equation.

26.1.7. Overview of Part VII: Energy Transport

In Chapter 23 we examined the transport of fluid mechanical energy by groundwater flow under isothermal conditions. We presented differential and integral equations describing the instantaneous conservation of fluid mechanical energy. Based on assumptions that are consistent with those that underlie the groundwater flow equation (see Chapter 2), we derived expressions for the density, flux density, and dissipation rate of groundwater mechanical energy.

In Chapter 24 we examined the time-averaged transport of fluid mechanical energy by time-periodic groundwater flow under isothermal conditions. Our analysis is based on assumptions that are consistent with those that underlie the groundwater flow equation (see Chapter 2). We derived differential and integral equations for the time-averaged conservation of fluid mechanical energy and expressions for time averages of the following: fluid mechanical energy density and its rate of change, flux density, and dissipation rate. Our results are valid for the general case where the hydraulic head field consists of multiple harmonic constituents.

In Chapter 25 we examined the time-averaged transport of fluid mechanical energy by a single harmonic constituent wave propagating under isothermal conditions. We found that the time-averaged, fluid mechanical energy flux density is parallel to the in-phase component of the corresponding specific-discharge vector harmonic constituent and the square of the amplitude function plays a role similar to that of a potential function in determining the time-averaged flux density. We also presented an informal, physical argument for the prohibition of amplitude maxima in source-free domains based on the conservation of fluid mechanical energy. This argument complements the more abstract, mathematical argument presented in Section 17.3.1. Finally, we examined the time-averaged, fluid mechanical power transmitted to an adjacent medium by isolated fluid sources—point, line, and plane sources. We concluded that, strictly, true point sources and true line sources are nonphysical because they are energetically impossible to maintain.

26.2. UNRESOLVED ISSUES

Because this book is introductory in nature, its scope is limited to the explanation of basic concepts. Consequently the discussion avoided a number of significant issues, some of which are listed below.

In Chapter 3 we assumed that in principle one could resolve the hydraulic head, source/sink strength, initial value, and boundary value into their steady, nonperiodic transient, and periodic transient components. We did not determine how, or whether it is necessary, to do so in practice. For instance, we did not discuss data analysis procedures or algorithms that might be required and/or useful for this purpose.

In Chapters 5–7 we assumed that the BVPs are well-posed and have unique solutions. We did not establish criteria for well-posedness, nor did we examine conditions for the existence or uniqueness of solutions.

In Chapter 15 we showed that if the hydraulic head amplitude and phase gradients are noncollinear, then the local specific-discharge vector harmonic constituent rotates in the plane of oscillation. We did not explore possible implications of this rotation, such as its potential impact on groundwater mixing.

In Chapter 18 we did not explore the topology of connected stagnation sets in great depth because we believe the topic requires a higher level of mathematical sophistication than the other topics in this book. However, a systematic study of the topology of connected stagnation sets may be worth further exploration.

In Chapter 21 we introduced various hypotheses (through travel, positive attenuation, opposing velocity) pertaining to 1D flows in which we can represent the hydraulic head harmonic constituent as the superposition of two component waves. Although all of the examples of 1D flow given in this book support these hypotheses, we did not prove the hypotheses generally. Effort toward proving these hypotheses may be justified, because such proofs would be a valuable contribution to the wave theory of time-periodic groundwater flow.

In Chapter 22 we showed that for time-periodic flow in source-free domains, under some conditions, the coefficients γ and β of the generalized wave equation are nonnegative (see Exercises 22.3–22.5). We did not systematically analyze the conditions under which the coefficients are nonnegative. Such analysis likely would provide a valuable contribution to the wave theory of time-periodic groundwater flow.

26.3. UNEXPLORED TOPICS

The following are some important topics related to time-periodic flow that this book has not explored.

Unconfined Flows Consider time-periodic flow in unconfined groundwater systems due to tidal forcing. If in the vicinity of the seaward boundary the position of the water table varies greatly over time, then the conceptual model presented in this book is invalid there because it fails to accommodate moving boundaries. An example is a water table that intersects a sloping beach [e.g., *Ataie-Ashtiani*, 1999a,b; *Urish and Ozbilgin*, 1989; *Urish and McKenna*, 2004].

Variable-Density Flows Consider time-periodic flow in coastal groundwater systems due to tidal forcing. In some cases the existence of a density gradient (e.g., near a sharp freshwater-saltwater interface) may require variable-density fluid modeling. For example, see *Cheng et al.*, [2004]. In this book we assumed that groundwater is a homogeneous fluid; that is, we neglected spatial variations of fluid density.

Uncertainty Limited spatiotemporal sampling of input/output fields and parameters and measurement errors create uncertainty, which one can treat within a stochastic framework. Although the basic conceptual model presented here is deterministic rather than stochastic, we believe it could serve as the foundation for a deterministic-stochastic conceptual model. For an example of a stochastic analysis of parameter uncertainty in the context of time-periodic flow, see *Trefry et al*, [2010, 2011].

Inverse Methods Much of the research completed on time-periodic flow has focused on the application of inverse methods to estimate physical properties such as hydraulic conductivity (or permeability) and specific storage. This book has not systematically addressed parameter estimation or inverse methods for time-periodic flow; interested readers should review the literature [e.g., *Alcolea et al.*, 2007].

Chemical Mass Transport Although we have not discussed chemical mass transport by time-periodic groundwater flows in this book, others have researched this topic [e.g., *Pool et al.*, 2014; *Smith et al.*, 2005; *Smith and Townley*, 2016; *Sposito*, 2006; *Yim and Mohsen*, 1992].

Numerical Methods Although we have not discussed them in this book, researchers have developed or tailored numerical methods specifically for modeling time-periodic groundwater flow. Examples include the finite-element method [*Smith et al.*, 2005] and the analytic element method [see *Bakker*, 2004, 2009].

26.4. FINAL NOTE

The goal of this book has been to present a clear and accessible mathematical introduction to the basic theory of time-periodic groundwater flow. Our emphasis on the mathematical description of the theory has necessarily led to a somewhat abstract approach, which has its limitations. For instance, we neglected to discuss physical mechanisms and natural spatial variation of material hydrologic properties in depth. Readers who seek to understand the complexities of time-periodic flow in natural environments can and should compensate for this neglect of real-world case studies and empirical data by studying the corresponding literature, which is relatively abundant. One can find peer-reviewed literature on time-periodic flow mainly in the subject areas listed in Section 1.4. Those who have completed this book should be exceptionally well prepared to read and understand the theory as they encounter it in the literature.

Part IX
Appendices

A
Hydraulic Head Components

A.1. STEADY VERSUS TRANSIENT COMPONENTS

The material in this appendix requires readers to have completed the material through Chapter 3.

A.1.1. Introduction

Suppose we are given a solution of the hydraulic head IBVP (see Chapter 2), h, the associated fields u and ψ, and all of the other parameters relevant to the IBVP. In addition, suppose that we are able to resolve h, u, and ψ into their corresponding steady and transient components. Can we then conclude that the steady and transient components of the hydraulic head satisfy the corresponding BVPs described in Chapter 3? The following sections address this question.

A.1.2. Analysis of the Governing Equation

Substituting the corresponding sums of steady and transient components for the hydraulic head and source/sink strength in the governing equation (2.6) yields

$$L\left[\bar{h}(\mathbf{x})\right] + \bar{u}(\mathbf{x}) + L\left[\tilde{h}(\mathbf{x}, t)\right] + \tilde{u}(\mathbf{x}, t) = 0 \qquad \text{(A.1)}$$
$$\forall\, t > 0,\ \mathbf{x} \in D.$$

The steady terms are independent of time, while the transient terms depend on time. All of the terms on the LHS generally depend on position as well. Define

$$c(\mathbf{x}, t) \equiv L\left[\tilde{h}(\mathbf{x}, t)\right] + \tilde{u}(\mathbf{x}, t). \qquad \text{(A.2)}$$

Then (A.1) is equivalent to the pair of simultaneous equations:

$$L\left[\bar{h}(\mathbf{x})\right] + \bar{u}(\mathbf{x}) = -c(\mathbf{x}, t)$$
$$L\left[\tilde{h}(\mathbf{x}, t)\right] + \tilde{u}(\mathbf{x}, t) = c(\mathbf{x}, t)$$
$$\forall\, t > 0,\ \mathbf{x} \in D.$$

The LHS of the first equation, being a linear combination of purely time-independent terms, is time independent. Therefore, the field $c(\mathbf{x}, t)$ is time independent; hence we write $c(\mathbf{x}, t) = c(\mathbf{x})$:

$$L\left[\bar{h}(\mathbf{x})\right] + \bar{u}(\mathbf{x}) = -c(\mathbf{x}) \qquad \text{(A.3a)}$$
$$L\left[\tilde{h}(\mathbf{x}, t)\right] + \tilde{u}(\mathbf{x}, t) = c(\mathbf{x}) \qquad \text{(A.3b)}$$
$$\forall\, t > 0,\ \mathbf{x} \in D.$$

In the special case where the hydraulic head field is purely steady, the source term also must be steady. Therefore, in the steady case both $\tilde{h}(\mathbf{x}, t) = 0$ and $\tilde{u}(\mathbf{x}, t) = 0$. Substituting these in (A.3b) produces $c(\mathbf{x}) = 0$. Similarly, in the special case where the hydraulic head field is purely transient, we obtain the same result using (A.3a). Therefore, $c(\mathbf{x})$ must vanish:

$$L\left[\bar{h}(\mathbf{x})\right] + \bar{u}(\mathbf{x}) = 0 \qquad \text{(A.4a)}$$
$$L\left[\tilde{h}(\mathbf{x}, t)\right] + \tilde{u}(\mathbf{x}, t) = 0 \qquad \text{(A.4b)}$$
$$\forall\, t > 0,\ \mathbf{x} \in D.$$

A.1.3. Analysis of the BC Equation

A similar analysis of the BC equation of the hydraulic head IBVP implies that the steady and transient hydraulic head components, respectively, satisfy

$$W\left[\bar{h}(\mathbf{x})\right] = \bar{\psi}(\mathbf{x}) \qquad \text{(A.5a)}$$
$$W\left[\tilde{h}(\mathbf{x}, t)\right] = \tilde{\psi}(\mathbf{x}, t) \qquad \text{(A.5b)}$$
$$\forall\, t > 0,\ \mathbf{x} \in \Gamma.$$

Hydrodynamics of Time-Periodic Groundwater Flow: Diffusion Waves in Porous Media, Geophysical Monograph 224,
First Edition. Joe S. Depner and Todd C. Rasmussen.

A.2. NONPERIODIC VERSUS PERIODIC COMPONENTS

A.2.1. Introduction

Suppose we are given a solution of the transient hydraulic head component IBVP (see Chapter 3), $\tilde{h}$, the associated fields $\tilde{u}$ and $\tilde{\psi}$, and all of the other parameters relevant to the IBVP. In addition, suppose that we are able to resolve $\tilde{h}$, $\tilde{u}$, and $\tilde{\psi}$ into their corresponding nonperiodic and periodic components. Can we then conclude that each of the nonperiodic and periodic hydraulic head components satisfies the corresponding IBVP described in Chapter 3? The following sections address this question.

A.2.2. Analysis of the Governing Equation

Substituting the corresponding sums of nonperiodic and periodic transient components for the hydraulic head and source/sink strength in the transient-component governing equation (3.4a) yields

$$L\left[\hat{h}(\mathbf{x},t)\right] + \hat{u}(\mathbf{x},t) + L\left[\mathring{h}(\mathbf{x},t)\right] + \mathring{u}(\mathbf{x},t) = 0 \qquad \text{(A.6)}$$

$$\forall\, t > 0,\ \mathbf{x} \in D.$$

All of the terms on the LHS generally are functions of the space coordinates and time. Define

$$g(\mathbf{x},t) \equiv L\left[\tilde{h}(\mathbf{x},t)\right] + \hat{u}(\mathbf{x},t).$$

Then (A.6) is equivalent to the pair of simultaneous equations

$$L\left[\tilde{h}(\mathbf{x},t)\right] + \hat{u}(\mathbf{x},t) = g(\mathbf{x},t) \qquad \text{(A.7a)}$$

$$L\left[\mathring{h}(\mathbf{x},t)\right] + \mathring{u}(\mathbf{x},t) = -g(\mathbf{x},t) \qquad \text{(A.7b)}$$

$$\forall\, t > 0,\ \mathbf{x} \in D.$$

The LHS of (A.7a), being a linear combination of nonperiodic terms, is either nonperiodic or zero. Similarly, the LHS of (A.7b), being a linear combination of periodic terms, is either periodic or zero. The function $g(\mathbf{x}, t)$ cannot be both nonperiodic and periodic. Therefore, $g(\mathbf{x}, t)$ must vanish:

$$L\left[\hat{h}(\mathbf{x},t)\right] + \hat{u}(\mathbf{x},t) = 0 \qquad \text{(A.8a)}$$

$$L\left[\mathring{h}(\mathbf{x},t)\right] + \mathring{u}(\mathbf{x},t) = 0 \qquad \text{(A.8b)}$$

$$\forall\, t > 0,\ \mathbf{x} \in D.$$

A.2.3. Analysis of the BC Equation

A similar analysis of the BC equation for the transient hydraulic head component IBVP implies that the nonperiodic and periodic transient hydraulic head components, respectively, satisfy the following BC equations:

$$W\left[\hat{h}(\mathbf{x},t)\right] = \hat{\psi}(\mathbf{x},t) \qquad \text{(A.9a)}$$

$$W\left[\mathring{h}(\mathbf{x},t)\right] = \mathring{\psi}(\mathbf{x},t) \qquad \text{(A.9b)}$$

$$\forall\, t > 0,\ \mathbf{x} \in \Gamma.$$

B
Useful Results from Trigonometry

B.1. TRIGONOMETRIC IDENTITIES

All of the identities listed here are from *Roy and Olver* [2016].
Pythagorean relation:

$$\cos^2\alpha + \sin^2\alpha = 1 \tag{B.1}$$

Angle-sum and angle-difference relations:

$$\cos(\alpha+\beta) = \cos\alpha\cos\beta - \sin\alpha\sin\beta \tag{B.2a}$$

$$\cos(\alpha-\beta) = \cos\alpha\cos\beta + \sin\alpha\sin\beta \tag{B.2b}$$

$$\sin(\alpha+\beta) = \sin\alpha\cos\beta + \cos\alpha\sin\beta \tag{B.2c}$$

$$\sin(\alpha-\beta) = \sin\alpha\cos\beta - \cos\alpha\sin\beta \tag{B.2d}$$

Cosine-sum relation:

$$\cos\alpha + \cos\beta = 2\cos\left(\frac{\alpha+\beta}{2}\right)\cos\left(\frac{\alpha-\beta}{2}\right) \tag{B.3}$$

Half-angle relations:

$$\cos\left(\frac{\alpha}{2}\right) = \pm\sqrt{\frac{1+\cos\alpha}{2}} \tag{B.4a}$$

$$\sin\left(\frac{\alpha}{2}\right) = \pm\sqrt{\frac{1-\cos\alpha}{2}} \tag{B.4b}$$

B.2. TRIGONOMETRIC SERIES

B.2.1. Introduction

Let $f(t)$ denote a real-valued function of time, represented globally by the trigonometric series

$$f(t) = \sum_{m=1}^{\infty} M_m \cos(\omega_m t + \theta_m) \tag{B.5}$$

where

$$m \in \mathbb{Z},\ m > 0$$

$$M_m \in \mathbb{R},\ M_m > 0,\ \frac{\partial M_m}{\partial t} = 0 \quad \forall\, m$$

$$\omega_m \in \mathbb{R},\ \frac{\partial \omega_m}{\partial t} = 0 \quad \forall\, m$$

$$0 < \omega_1 < \omega_2 < \omega_3 \cdots$$

$$t \in \mathbb{R}$$

$$\theta_m \in \mathbb{R},\ \frac{\partial \theta_m}{\partial t} = 0 \quad \forall\, m.$$

The frequencies (ω_m) may or may not be related to one another by rational factors, so $f(t)$ may or may not be strictly periodic in time. That is, $f(t)$ may be either purely periodic or almost periodic. In this appendix we derive equations for calculating the amplitudes M_m and the phases θ_m from $f(t)$.

Using the angle-sum relation for the cosine (B.2a), we can write the mth term of the series on the RHS of (B.5) as

$$M_m \cos(\omega_m t + \theta_m) = A_m \cos(\omega_m t) - B_m \sin(\omega_m t) \tag{B.6}$$

where

$$A_m = M_m \cos\theta_m \tag{B.7a}$$

$$B_m = M_m \sin\theta_m. \tag{B.7b}$$

Solving for M_m and θ_m gives the relations

$$M_m = \sqrt{A_m^2 + B_m^2} \tag{B.8a}$$

$$\tan\theta_m = \frac{B_m}{A_m}. \tag{B.8b}$$

Thus, knowledge of the coefficients A_m and B_m is equivalent to knowledge of the coefficients M_m and θ_m.

Hydrodynamics of Time-Periodic Groundwater Flow: Diffusion Waves in Porous Media, Geophysical Monograph 224,
First Edition. Joe S. Depner and Todd C. Rasmussen.

Substituting the RHS of (B.6) for the mth term of the series in (B.5) gives

$$f(t) = \sum_{m=1}^{\infty} \left[A_m \cos(\omega_m t) - B_m \sin(\omega_m t)\right]. \tag{B.9}$$

Formally, the trigonometric series in (B.5), or equivalently (B.9), is a *Fourier series*, and the coefficients A_m and B_m (or equivalently M_m and θ_m) are *Fourier coefficients.*

B.2.2. Inner Products and Time Averages

Define the *inner product* of two functions $f(t)$ and $g(t)$ as the following *mean*, or *time average*, of their product [*Mathematical Society of Japan*, 2000]:

$$\langle f(t), g(t)\rangle \equiv \lim_{T\to\infty} \frac{1}{T} \int_t^{t+T} f(v)g(v)dv. \tag{B.10}$$

Then the means of the sinusoids are

$$\langle \cos(\lambda t), 1\rangle = \begin{cases} 1, & \lambda = 0 \\ 0, & \lambda \neq 0 \end{cases} \tag{B.11a}$$

$$\langle \sin(\lambda t), 1\rangle = 0 \tag{B.11b}$$

and the means of the products of sinusoids are

$$\langle \cos(\omega t), \cos(\lambda t)\rangle = \begin{cases} 1, & \lambda = \omega = 0 \\ \frac{1}{2}, & |\lambda| = |\omega| > 0 \\ 0, & |\lambda| \neq |\omega| \end{cases} \tag{B.12a}$$

$$\langle \sin(\omega t), \sin(\lambda t)\rangle = \begin{cases} -\frac{1}{2}, & \lambda = -\omega \neq 0 \\ 0, & |\lambda| \neq |\omega| \text{ or } \lambda = \omega = 0 \\ \frac{1}{2}, & \lambda = \omega \neq 0 \end{cases} \tag{B.12b}$$

$$\langle \sin(\omega t), \cos(\lambda t)\rangle = \langle \cos(\omega t), \sin(\lambda t)\rangle = 0. \tag{B.12c}$$

Exercise B.1 Derivation: Time Averages of Sinusoids

Problem: Using the definition (B.10), prove relations (B.11). □

Exercise B.2 Derivation: Time Averages of Products of Sinusoids

Problem: Using the definition (B.10), prove relations (B.12). □

B.2.2.1. Special Case: Positive Frequencies

Suppose all of the frequencies ω_m ($m = 1, 2, 3, \ldots$) are positive. Then (B.9), (B.10), and (B.11) give

$$\langle f(t), 1\rangle = 0.$$

Next, consider the time averages of the products, $f(t)\cos(\omega_m t)$ and $f(t)\sin(\omega_m t)$. Multiplying (B.9) by $\cos(\omega_m t)$ and $\sin(\omega_m t)$, respectively, and then taking the time averages of the products give

$$\langle f(t), \cos(\omega_m t)\rangle = \left\langle \left\{ \sum_{n=1}^{\infty} \left[A_n \cos(\omega_n t) - B_n \sin(\omega_n t)\right]\right\}, \cos(\omega_m t)\right\rangle$$

$$\langle f(t), \sin(\omega_m t)\rangle = \left\langle \left\{ \sum_{n=1}^{\infty} \left[A_n \cos(\omega_n t) - B_n \sin(\omega_n t)\right]\right\}, \sin(\omega_m t)\right\rangle.$$

After some manipulation these become

$$\langle f(t), \cos(\omega_m t)\rangle = \sum_{n=1}^{\infty} A_n\langle \cos(\omega_n t), \cos(\omega_m t)\rangle - \sum_{n=1}^{\infty} B_n\langle \sin(\omega_n t), \cos(\omega_m t)\rangle$$

$$\langle f(t), \sin(\omega_m t)\rangle = \sum_{n=1}^{\infty} A_n\langle \cos(\omega_n t), \sin(\omega_m t)\rangle - \sum_{n=1}^{\infty} B_n\langle \sin(\omega_n t), \sin(\omega_m t)\rangle.$$

Applying the results in (B.12) leads to

$$\langle f(t), \cos(\omega_m t)\rangle = \frac{A_m}{2}$$

$$\langle f(t), \sin(\omega_m t)\rangle = -\frac{B_m}{2}.$$

Solving for the coefficients then yields

$$A_m = 2\langle f(t), \cos(\omega_m t)\rangle$$

$$B_m = -2\langle f(t), \sin(\omega_m t)\rangle$$

$$m = 1, 2, 3, \ldots.$$

C
Linear Transformation of Space Coordinates

C.1. INTRODUCTION

Consider the general linear coordinate transformation

$$\breve{\mathbf{x}} = \mathbf{C}\mathbf{x} \tag{C.1}$$

where

$\mathbf{x}$ is the N-vector of space coordinates in the original (untransformed) coordinate system;

$\breve{\mathbf{x}}$ is the N-vector of space coordinates in the transformed coordinate system; and

$\mathbf{C}$ is an $N \times N$ matrix of real-valued constants (dimensionless).

Throughout this appendix we use the breve accent (i.e., "˘") to denote a quantity expressed in terms of the transformed space coordinates.

We assume that the transformation (C.1) is invertible, i.e., the transformation matrix $\mathbf{C}$ is nonsingular. We denote its inverse by $\mathbf{C}^{-1}$.

We define the *Jacobian matrix* of the coordinate transformation (C.1) as

$$J_{mn} \equiv \frac{\partial \breve{x}_m}{\partial x_n} = C_{mn}.$$

That is, the Jacobian matrix is equal to the transformation matrix.

C.2. SCALAR FUNCTIONS

Let $f(\mathbf{x})$ and $g(\mathbf{x})$ denote general scalar functions of the space coordinates. Scalar functions are invariant to coordinate transformations:

$$\breve{f}(\breve{\mathbf{x}}) = f(\mathbf{x}).$$

C.3. GRADIENT VECTOR

By application of the chain rule we obtain the following relation between the gradients in the transformed and untransformed systems:

$$\nabla f(\mathbf{x}) = \mathbf{C}^{\mathrm{T}} \breve{\nabla} \breve{f}(\breve{\mathbf{x}}). \tag{C.2}$$

For instance, letting $f(\mathbf{x}) = h(\mathbf{x})$ in (C.2) yields the following relation between the hydraulic gradients in the transformed and untransformed coordinate systems:

$$\nabla h(\mathbf{x}) = \mathbf{C}^{\mathrm{T}} \breve{\nabla} \breve{h}(\breve{\mathbf{x}}). \tag{C.3}$$

C.4. SPECIFIC-DISCHARGE VECTOR

We assume the specific-discharge vector transforms as a velocity vector. Then the specific-discharge vectors in the transformed and untransformed systems are related as

$$\breve{\mathbf{q}}(\breve{\mathbf{x}}) = \mathbf{C}\mathbf{q}(\mathbf{x}). \tag{C.4}$$

If $\mathbf{C}$ is invertible, then we can solve the previous equation for $\mathbf{q}(\mathbf{x})$:

$$\mathbf{q}(\mathbf{x}) = \mathbf{C}^{-1} \breve{\mathbf{q}}(\breve{\mathbf{x}}). \tag{C.5}$$

Using Darcy's law to express the transformed and untransformed specific-discharge vectors in terms of the corresponding hydraulic conductivity tensors and hydraulic gradients gives

$$\mathbf{q}(\mathbf{x}) = -\mathbf{K}(\mathbf{x}) \nabla h(\mathbf{x}) \tag{C.6a}$$

$$\breve{\mathbf{q}}(\breve{\mathbf{x}}) = -\breve{\mathbf{K}}(\breve{\mathbf{x}}) \breve{\nabla} \breve{h}(\breve{\mathbf{x}}). \tag{C.6b}$$

Hydrodynamics of Time-Periodic Groundwater Flow: Diffusion Waves in Porous Media, Geophysical Monograph 224,
First Edition. Joe S. Depner and Todd C. Rasmussen.

C.5. DIVERGENCE OF SPECIFIC-DISCHARGE VECTOR

We define the divergence of a Cartesian N-vector $\mathbf{v}(\mathbf{x})$ as the scalar

$$\nabla \cdot \mathbf{v}(\mathbf{x}) = \frac{\partial v_p(\mathbf{x})}{\partial x_p}$$

where we have used Einstein notation (expressions are summed over all repeated indices). Using the chain rule, we can write this in terms of the transformed coordinates as

$$\begin{aligned}\nabla \cdot \mathbf{v}(\mathbf{x}) &= \frac{\partial v_p(\mathbf{x})}{\partial \breve{x}_m}\frac{\partial \breve{x}_m}{\partial x_p} \\ &= \frac{\partial}{\partial \breve{x}_m}\left[C_{mp} v_p(\mathbf{x})\right] \\ &= \breve{\nabla} \cdot [\mathbf{C}\mathbf{v}(\mathbf{x})].\end{aligned}$$

Substituting $\mathbf{q}(\mathbf{x})$ for $\mathbf{v}(\mathbf{x})$ and using (C.5) give

$$\nabla \cdot \mathbf{q}(\mathbf{x}) = \breve{\nabla} \cdot \breve{\mathbf{q}}(\breve{\mathbf{x}}).$$

C.6. HYDRAULIC CONDUCTIVITY TENSOR

Substituting the RHSs of (C.6) for the corresponding specific-discharge vectors in (C.4) yields

$$\breve{\mathbf{K}}(\breve{\mathbf{x}})\breve{\nabla}\breve{h}(\breve{\mathbf{x}}) = \mathbf{C}\mathbf{K}(\mathbf{x})\nabla h(\mathbf{x}).$$

Substituting the RHS of (C.3) for $\nabla h(\mathbf{x})$ in the previous equation leads to

$$\breve{\mathbf{K}}(\breve{\mathbf{x}})\breve{\nabla}\breve{h}(\breve{\mathbf{x}}) = \mathbf{C}\mathbf{K}(\mathbf{x})\mathbf{C}^{\mathrm{T}}\breve{\nabla}\breve{h}(\breve{\mathbf{x}}).$$

Since this must be satisfied for all possible hydraulic gradient vectors $\breve{\nabla}\breve{h}(\breve{\mathbf{x}})$, it must be true that

$$\breve{\mathbf{K}}(\breve{\mathbf{x}}) = \mathbf{C}\mathbf{K}(\mathbf{x})\mathbf{C}^{\mathrm{T}}.$$

Alternatively, solving the previous equation for $\mathbf{K}(\mathbf{x})$ gives

$$\mathbf{K}(\mathbf{x}) = \mathbf{C}^{-1}\breve{\mathbf{K}}(\breve{\mathbf{x}})(\mathbf{C}^{\mathrm{T}})^{-1} \tag{C.7}$$

where we have assumed that $\mathbf{C}$ is invertible.

C.7. HESSIAN MATRIX

We define the Hessian matrix of the function $f(\mathbf{x})$, denoted by $\mathbf{H}[f(\mathbf{x})]$, as the $N \times N$ matrix

$$H_{mn}[f(\mathbf{x})] \equiv \frac{\partial^2 f(\mathbf{x})}{\partial x_m \partial x_n} \quad (m = 1, 2, 3, \ldots, N;\ n = 1, 2, 3, \ldots, N).$$

Assuming the function $f(\mathbf{x})$ is C^2 continuous, corresponding mixed second partial derivatives are equal:

$$\frac{\partial^2 f(\mathbf{x})}{\partial x_m \partial x_n} = \frac{\partial^2 f(\mathbf{x})}{\partial x_n \partial x_m} \quad (m = 1, 2, 3, \ldots, M;\ n = 1, 2, 3, \ldots, N)$$

so that the Hessian matrix is symmetric:

$$H_{mn}[f(\mathbf{x})] = H_{nm}[f(\mathbf{x})] \quad (m = 1, 2, 3, \ldots, M;\ n = 1, 2, 3, \ldots, N).$$

We define the transformed Hessian matrix of the function $f(\mathbf{x})$, denoted by $\breve{\mathbf{H}}[\breve{f}(\breve{\mathbf{x}})]$, as the $N \times N$ matrix

$$\breve{H}_{mn}[\breve{f}(\breve{\mathbf{x}})] \equiv \frac{\partial^2 \breve{f}(\breve{\mathbf{x}})}{\partial \breve{x}_m \partial \breve{x}_n} \quad (m = 1, 2, 3, \ldots, M;\ n = 1, 2, 3, \ldots, N).$$

By applying the chain rule twice, we obtain the following relation between the Hessian matrices in the transformed and untransformed coordinate systems:

$$\mathbf{H}[f(\mathbf{x})] = \mathbf{C}^{\mathrm{T}}\breve{\mathbf{H}}[\breve{f}(\breve{\mathbf{x}})]\mathbf{C}. \tag{C.8}$$

Alternatively, solving the previous equation for $\breve{\mathbf{H}}[\breve{f}(\breve{\mathbf{x}})]$ gives

$$\breve{\mathbf{H}}[\breve{f}(\breve{\mathbf{x}})] = (\mathbf{C}^{\mathrm{T}})^{-1}\mathbf{H}[f(\mathbf{x})]\mathbf{C}^{-1}. \tag{C.9}$$

From (C.8), the determinants of the untransformed and transformed Hessian matrices are related as

$$\det \mathbf{H}[f(\mathbf{x})] = (\det \mathbf{C})^2 \det \breve{\mathbf{H}}[\breve{f}(\breve{\mathbf{x}})].$$

If the transformation is invertible, then $\mathbf{C}^{-1}$ exists and $|\det \mathbf{C}| > 0$; thus, the determinants of the untransformed and transformed Hessian matrices have the same sign.

C.8. INNER PRODUCT

Consider the inner product of the two vectors $(\mathbf{K}\nabla f) \cdot \nabla g$. We can write this as

$$\begin{aligned}(\mathbf{K}\nabla f) \cdot \nabla g &= (\mathbf{K}\nabla f)^{\mathrm{T}}\nabla g \\ &= (\nabla f)^{\mathrm{T}}\mathbf{K}\nabla g\end{aligned}$$

where we have exploited the symmetry of $\mathbf{K}$. Next, express the gradients and $\mathbf{K}$, in the previous equation, in terms of the corresponding transformed quantities [see equations (C.2) and (C.7)], to give

$$\begin{aligned}(\mathbf{K}\nabla f) \cdot \nabla g &= (\mathbf{C}^{\mathrm{T}}\breve{\nabla}\breve{f})^{\mathrm{T}}[\mathbf{C}^{-1}\breve{\mathbf{K}}(\mathbf{C}^{\mathrm{T}})^{-1}](\mathbf{C}^{\mathrm{T}}\breve{\nabla}\breve{g}) \\ &= (\breve{\nabla}\breve{f})^{\mathrm{T}}\mathbf{C}[\mathbf{C}^{-1}\breve{\mathbf{K}}(\mathbf{C}^{\mathrm{T}})^{-1}](\mathbf{C}^{\mathrm{T}}\breve{\nabla}\breve{g}) \\ &= (\breve{\nabla}\breve{f})^{\mathrm{T}}\breve{\mathbf{K}}\breve{\nabla}\breve{g}.\end{aligned}$$

Writing this in dot product notation gives the final result:

$$(\mathbf{K}\nabla f) \cdot \nabla g = (\breve{\mathbf{K}}\breve{\nabla}\breve{f}) \cdot \breve{\nabla}\breve{g}.$$

C.9. DOUBLE INNER PRODUCT

We define the double inner product of two $M \times N$ matrices **A** and **B** as

$$\mathbf{A} : \mathbf{B} \equiv A_{mn}B_{mn}$$

where we have used Einstein notation. For example, letting $\mathbf{A} = \mathbf{K}(\mathbf{x})$ and $\mathbf{B} = \mathbf{H}[f(\mathbf{x})]$ gives the double inner product of the $N \times N$ matrices **K** and $\mathbf{H}[f]$:

$$\mathbf{K}(\mathbf{x}) : \mathbf{H}[f(\mathbf{x})] = K_{mn}(\mathbf{x})H_{mn}[f(\mathbf{x})].$$

Exercise C.1 Derivation: Linear Algebra

Problem: Show that if **A**, **B**, **G**, and **D** are $M \times N$ matrices, then

$$(\mathbf{AB}) : (\mathbf{GD}) = (\mathbf{A}^{\mathrm{T}}\mathbf{G}) : (\mathbf{BD}^{\mathrm{T}}). \tag{C.10}$$

Solution: Using Einstein notation, we obtain

$$\begin{aligned}(\mathbf{AB}) : (\mathbf{GD}) &= (A_{mn}B_{np})(G_{ml}D_{lp}) \\ &= A^{\mathrm{T}}_{nm}G_{ml}D_{lp}B^{\mathrm{T}}_{pn} \\ &= (A^{\mathrm{T}}_{nm}G_{ml})(D_{lp}B^{\mathrm{T}}_{pn}) \\ &= (\mathbf{A}^{\mathrm{T}}\mathbf{G})_{nl}(\mathbf{DB}^{\mathrm{T}})_{ln} \\ &= (\mathbf{A}^{\mathrm{T}}\mathbf{G})_{nl}[(\mathbf{DB}^{\mathrm{T}})^{\mathrm{T}}]_{nl} \\ &= (\mathbf{A}^{\mathrm{T}}\mathbf{G}) : (\mathbf{BD}^{\mathrm{T}}).\end{aligned}$$

□

Letting $\mathbf{B} = \mathbf{D} = \mathbf{I}$ in identity (C.10) yields the identity

$$\mathbf{A} : \mathbf{G} = \mathrm{tr}(\mathbf{A}^{\mathrm{T}}\mathbf{G}).$$

This yields

$$\mathbf{K}(\mathbf{x}) : \mathbf{H}[f(\mathbf{x})] = \mathrm{tr}\{\mathbf{K}(\mathbf{x})\mathbf{H}[f(\mathbf{x})]\} \tag{C.11a}$$

$$\breve{\mathbf{K}}(\breve{\mathbf{x}}) : \breve{\mathbf{H}}[\breve{f}(\breve{\mathbf{x}})] = \mathrm{tr}\{\breve{\mathbf{K}}(\breve{\mathbf{x}})\breve{\mathbf{H}}[\breve{f}(\breve{\mathbf{x}})]\} \tag{C.11b}$$

where tr denotes the trace of the matrix.

Using (C.7) and (C.9), we can write the double inner product $\mathbf{K} : \mathbf{H}[f]$ in terms of transformed quantities as

$$\mathbf{K}(\mathbf{x}) : \mathbf{H}[f(\mathbf{x})] = (\mathbf{AB}) : (\mathbf{GD}) \tag{C.12}$$

where

$$\begin{aligned}\mathbf{A} &= \mathbf{C}^{-1}\breve{\mathbf{K}}(\breve{\mathbf{x}}) \\ \mathbf{B} &= (\mathbf{C}^{\mathrm{T}})^{-1} \\ \mathbf{G} &= \mathbf{C}^{\mathrm{T}}\breve{\mathbf{H}}[\breve{f}(\breve{\mathbf{x}})] \\ \mathbf{D} &= \mathbf{C}.\end{aligned}$$

Using identity (C.10) to evaluate the RHS of (C.12) leads to

$$\mathbf{K}(\mathbf{x}) : \mathbf{H}[f(\mathbf{x})] = \mathrm{tr}\{\breve{\mathbf{K}}(\breve{\mathbf{x}})\breve{\mathbf{H}}[\breve{f}(\breve{\mathbf{x}})]\}.$$

Substituting the LHS of (C.11b) for the RHS of the previous equation yields

$$\mathbf{K}(\mathbf{x}) : \mathbf{H}[f(\mathbf{x})] = \breve{\mathbf{K}}(\breve{\mathbf{x}}) : \breve{\mathbf{H}}[\breve{f}(\breve{\mathbf{x}})]. \tag{C.13}$$

C.10. DIVERGENCE OF A MATRIX

We define the divergence of an $N \times N$ matrix (e.g., a second-rank tensor) $\mathbf{F}(\mathbf{x})$ as the vector

$$\nabla \cdot \mathbf{F}(\mathbf{x}) \equiv \mathbf{e_n}\frac{\partial F_{pn}(\mathbf{x})}{\partial x_p}$$

where $\mathbf{e_n}$ denotes the basis vector for the Cartesian coordinate x_n and where we have used Einstein notation. That is, the nth element of the vector $\nabla \cdot \mathbf{F}$ is the divergence of the nth column vector of the matrix **F**. Using the chain rule, we can write this in terms of the transformed coordinates as

$$\begin{aligned}\nabla \cdot \mathbf{F}(\mathbf{x}) &= \mathbf{e_n}\frac{\partial F_{pn}(\mathbf{x})}{\partial \breve{x}_m}\frac{\partial \breve{x}_m}{\partial x_p} \\ &= \mathbf{e_n}\frac{\partial}{\partial \breve{x}_m}\left[C_{mp}F_{pn}(\mathbf{x})\right] \\ &= \breve{\nabla} \cdot [\mathbf{CF}(\mathbf{x})].\end{aligned} \tag{C.14}$$

Consider the case where $\mathbf{F}(\mathbf{x}) = \mathbf{K}(\mathbf{x})$. Equation (C.14) gives

$$\nabla \cdot \mathbf{K}(\mathbf{x}) = \breve{\nabla} \cdot [\mathbf{CK}(\mathbf{x})].$$

Substituting the RHS of (C.7) for $\mathbf{K}(\mathbf{x})$ on the RHS of the previous equation yields

$$\begin{aligned}\nabla \cdot \mathbf{K}(\mathbf{x}) &= \breve{\nabla} \cdot \{\mathbf{C}[\mathbf{C}^{-1}\breve{\mathbf{K}}(\breve{\mathbf{x}})(\mathbf{C}^{\mathrm{T}})^{-1}]\} \\ &= \breve{\nabla} \cdot [\breve{\mathbf{K}}(\breve{\mathbf{x}})(\mathbf{C}^{\mathrm{T}})^{-1}].\end{aligned} \tag{C.15}$$

Exercise C.2 Divergence of Matrix Product

Problem: Suppose that $\mathbf{A}(\mathbf{x})$ and $\mathbf{B}(\mathbf{x})$ are matrix fields. Show that

$$\nabla \cdot [\mathbf{A}(\mathbf{x})\mathbf{B}(\mathbf{x})] = [\nabla \cdot \mathbf{A}(\mathbf{x})] \cdot \mathbf{B}(\mathbf{x}) + [\mathbf{A}(\mathbf{x}) \cdot \nabla] \cdot \mathbf{B}(\mathbf{x}). \tag{C.16}$$

Solution: Using Einstein notation, we obtain

$$
\begin{aligned}
\nabla \cdot (\mathbf{AB}) &= \mathbf{e_n}\frac{\partial}{\partial x_p}\left(A_{pq}B_{qn}\right) \\
&= \mathbf{e_n}\frac{\partial A_{pq}}{\partial x_p}B_{qn} + \mathbf{e_n}A_{pq}\frac{\partial B_{qn}}{\partial x_p} \\
&= \mathbf{e_n}\frac{\partial A_{pq}}{\partial x_p}B_{qn} + \mathbf{e_n}A^{\mathrm{T}}_{qp}\frac{\partial}{\partial x_p}B_{qn} \\
&= (\nabla \cdot \mathbf{A}) \cdot \mathbf{B} + (\mathbf{A} \cdot \nabla) \cdot \mathbf{B}.
\end{aligned}
$$

□

Using identity (C.16), we can write (C.15) as

$$
\begin{aligned}
\nabla \cdot \mathbf{K}(\mathbf{x}) &= \left[\check{\nabla} \cdot \check{\mathbf{K}}(\check{\mathbf{x}})\right] \cdot (\mathbf{C}^{\mathrm{T}})^{-1} + \left[\check{\mathbf{K}}(\check{\mathbf{x}}) \cdot \check{\nabla}\right] \cdot (\mathbf{C}^{\mathrm{T}})^{-1} \\
&= \mathbf{C}^{-1}\check{\nabla} \cdot \check{\mathbf{K}}(\check{\mathbf{x}}).
\end{aligned}
$$

Combining the previous equation and (C.2) leads to

$$
\begin{aligned}
\left[\nabla \cdot \mathbf{K}(\mathbf{x})\right] \cdot \nabla f &= \left[\mathbf{C}^{-1}\check{\nabla} \cdot \check{\mathbf{K}}(\check{\mathbf{x}})\right] \cdot \left[\mathbf{C}^{\mathrm{T}}\check{\nabla}\check{f}(\check{\mathbf{x}})\right] \\
&= \left[\check{\nabla} \cdot \check{\mathbf{K}}(\check{\mathbf{x}})\right]^{\mathrm{T}}(\mathbf{C}^{\mathrm{T}})^{-1}\mathbf{C}^{\mathrm{T}}\check{\nabla}\check{f}(\check{\mathbf{x}}) \\
&= \left[\check{\nabla} \cdot \check{\mathbf{K}}(\check{\mathbf{x}})\right] \cdot \check{\nabla}\check{f}(\check{\mathbf{x}}).
\end{aligned}
$$

C.11. CROSS PRODUCT

Suppose $\mathbf{C}$ is a real-valued, 3×3, nonsingular matrix and $\mathbf{a}$ and $\mathbf{b}$ are real-valued, column 3-vectors. Then

$$
(\mathbf{Ca}) \times (\mathbf{Cb}) = (\det \mathbf{C})\mathbf{C}^{-\mathrm{T}}(\mathbf{a} \times \mathbf{b}). \tag{C.17}
$$

The following proof closely follows that sketched by jostpuur [2010, Mar 16].

First, express $\mathbf{C}$ in terms of its column vectors as

$$
\mathbf{C} = \left[\mathbf{C}_{\cdot 1}, \mathbf{C}_{\cdot 2}, \mathbf{C}_{\cdot 3}\right]
$$

where we define the column vectors as

$$
\mathbf{C}_{\cdot k} \equiv \begin{bmatrix} C_{1k} \\ C_{2k} \\ C_{3k} \end{bmatrix} \quad (k = 1, 2, 3).
$$

The inverse of a nonsingular matrix $\mathbf{C}$ is given by [*Noble and Daniel*, 1977]

$$
\mathbf{C}^{-1} = \frac{1}{\det \mathbf{C}}\operatorname{adj}\mathbf{C} \tag{C.18}
$$

where $\operatorname{adj}\mathbf{C}$ denotes the adjugate matrix of $\mathbf{C}$. We can express $\operatorname{adj}\mathbf{C}$ in terms of cross products of the column vectors as (see Exercise C.5)

$$
\operatorname{adj}\mathbf{C} \equiv \left[\mathbf{C}_{\cdot 2} \times \mathbf{C}_{\cdot 3}, \mathbf{C}_{\cdot 3} \times \mathbf{C}_{\cdot 1}, \mathbf{C}_{\cdot 1} \times \mathbf{C}_{\cdot 2}\right]^{\mathrm{T}}. \tag{C.19}
$$

Exercise C.3 Matrix Determinant

Problem: Express $\det \mathbf{C}$ *explicitly* in terms of the matrix elements C_{mn} $(m, n = 1, 2, 3)$.

Solution: The determinant is given by

$$
\begin{aligned}
\det \mathbf{C} &= \det \begin{bmatrix} C_{11} & C_{12} & C_{13} \\ C_{21} & C_{22} & C_{23} \\ C_{31} & C_{32} & C_{33} \end{bmatrix} \\
&= C_{11} \det \begin{bmatrix} C_{22} & C_{23} \\ C_{32} & C_{33} \end{bmatrix} - C_{12} \det \begin{bmatrix} C_{21} & C_{23} \\ C_{31} & C_{33} \end{bmatrix} \\
&\quad + C_{13} \det \begin{bmatrix} C_{21} & C_{22} \\ C_{31} & C_{32} \end{bmatrix}.
\end{aligned}
$$

Evaluating the determinants on the RHS, we obtain

$$
\begin{aligned}
\det \mathbf{C} = {} & C_{11}\left(C_{22}C_{33} - C_{32}C_{23}\right) - C_{12}\left(C_{21}C_{33} - C_{31}C_{23}\right) \\
& + C_{13}\left(C_{21}C_{32} - C_{31}C_{22}\right).
\end{aligned} \tag{C.20}
$$

□

Exercise C.4 Adjugate Matrix

Problem: Express the adjugate matrix *explicitly* in terms of the matrix elements C_{mn} $(m, n = 1, 2, 3)$.

Solution: Evaluating the transpose on the RHS of (C.19) leads to

$$
\operatorname{adj}\mathbf{C} = \begin{bmatrix} (\mathbf{C}_{\cdot 2} \times \mathbf{C}_{\cdot 3})^{\mathrm{T}} \\ (\mathbf{C}_{\cdot 3} \times \mathbf{C}_{\cdot 1})^{\mathrm{T}} \\ (\mathbf{C}_{\cdot 1} \times \mathbf{C}_{\cdot 2})^{\mathrm{T}} \end{bmatrix}. \tag{C.21}
$$

We also have

$$
\mathbf{C}_{\cdot 2} \times \mathbf{C}_{\cdot 3} = \begin{bmatrix} C_{12} \\ C_{22} \\ C_{32} \end{bmatrix} \times \begin{bmatrix} C_{13} \\ C_{23} \\ C_{33} \end{bmatrix} = \begin{bmatrix} C_{22}C_{33} - C_{32}C_{23} \\ C_{32}C_{13} - C_{12}C_{33} \\ C_{12}C_{23} - C_{22}C_{13} \end{bmatrix} \tag{C.22a}
$$

$$
\mathbf{C}_{\cdot 3} \times \mathbf{C}_{\cdot 1} = \begin{bmatrix} C_{13} \\ C_{23} \\ C_{33} \end{bmatrix} \times \begin{bmatrix} C_{11} \\ C_{21} \\ C_{31} \end{bmatrix} = \begin{bmatrix} C_{23}C_{31} - C_{33}C_{21} \\ C_{33}C_{11} - C_{13}C_{31} \\ C_{13}C_{21} - C_{23}C_{11} \end{bmatrix} \tag{C.22b}
$$

$$
\mathbf{C}_{\cdot 1} \times \mathbf{C}_{\cdot 2} = \begin{bmatrix} C_{11} \\ C_{21} \\ C_{31} \end{bmatrix} \times \begin{bmatrix} C_{12} \\ C_{22} \\ C_{32} \end{bmatrix} = \begin{bmatrix} C_{21}C_{32} - C_{31}C_{22} \\ C_{31}C_{12} - C_{11}C_{32} \\ C_{11}C_{22} - C_{21}C_{22} \end{bmatrix}. \tag{C.22c}
$$

Combining these with (C.21) yields

$$
\operatorname{adj}\mathbf{C} = \begin{bmatrix} C_{22}C_{33} - C_{32}C_{23} & C_{32}C_{13} - C_{12}C_{33} & C_{12}C_{23} - C_{22}C_{13} \\ C_{23}C_{31} - C_{33}C_{21} & C_{33}C_{11} - C_{13}C_{31} & C_{13}C_{21} - C_{23}C_{11} \\ C_{21}C_{32} - C_{31}C_{22} & C_{31}C_{12} - C_{11}C_{32} & C_{11}C_{22} - C_{21}C_{12} \end{bmatrix}. \tag{C.23}
$$

□

Exercise C.5 Derivation of Intermediate Result

Problem: Show that (C.18) is true, where the adjugate matrix is given by (C.19).

Solution: Premultiplying the RHS of (C.23) by $\mathbf{C}$ yields, after some laborious but straightforward algebra,

$$\mathbf{C}\,\mathrm{adj}\,\mathbf{C} = (\det \mathbf{C})\,\mathbf{I}$$

where $\mathbf{I}$ is the 3×3 identity matrix. Premultiplying by $\mathbf{C}^{-1}$ and dividing by $\det \mathbf{C}$ produce (C.18). □

Next, express the cross product of the vectors $\mathbf{Ca}$ and $\mathbf{Cb}$ as

$$\begin{aligned}(\mathbf{Ca}) \times (\mathbf{Cb}) &= \epsilon_{ijk}(\mathbf{Ca})_i(\mathbf{Cb})_j\hat{\mathbf{x}}_k \\ &= \epsilon_{ijk}\left(C_{im}a_m\right)\left(C_{jn}b_n\right)\hat{\mathbf{x}}_k \\ &= \left(\epsilon_{ijk}C_{im}C_{jn}\hat{\mathbf{x}}_k\right)a_m b_n \\ &= \left(\mathbf{C}_{\cdot m} \times \mathbf{C}_{\cdot n}\right)a_m b_n\end{aligned}$$

where we have used Einstein notation and ϵ_{ijk} is the Levi-Civita symbol. Writing out the terms of the double sum on the RHS and grouping terms, we obtain

$$\begin{aligned}(\mathbf{Ca}) \times (\mathbf{Cb}) &= \left(\mathbf{C}_{\cdot 1} \times \mathbf{C}_{\cdot 2}\right)\left(a_1b_2 - a_2b_1\right) \\ &\quad + \left(\mathbf{C}_{\cdot 2} \times \mathbf{C}_{\cdot 3}\right)\left(a_2b_3 - a_3b_2\right) \\ &\quad + \left(\mathbf{C}_{\cdot 3} \times \mathbf{C}_{\cdot 1}\right)\left(a_3b_1 - a_1b_3\right)\end{aligned}$$

which we can write more compactly as

$$(\mathbf{Ca}) \times (\mathbf{Cb}) = \left[\mathbf{C}_{\cdot 2} \times \mathbf{C}_{\cdot 3}\,, \mathbf{C}_{\cdot 3} \times \mathbf{C}_{\cdot 1}\,, \mathbf{C}_{\cdot 1} \times \mathbf{C}_{\cdot 2}\right] \begin{bmatrix}(\mathbf{a} \times \mathbf{b})_1 \\ (\mathbf{a} \times \mathbf{b})_2 \\ (\mathbf{a} \times \mathbf{b})_3\end{bmatrix}.$$

Using (C.19), we can write this as

$$(\mathbf{Ca}) \times (\mathbf{Cb}) = (\mathrm{adj}\,\mathbf{C})^{\mathrm{T}}(\mathbf{a} \times \mathbf{b}).$$

Finally, using (C.18) to express the adjugate matrix in terms of the determinant and inverse of $\mathbf{C}$ yields (C.17).

D
Complex Variables

D.1. BASIC FORMULAS

Consider a *complex variable* z:

$$z = x + iy \quad \text{(D.1)}$$

where both x and y are real-valued variables and i is the *imaginary unit* (i.e., $i \equiv \sqrt{-1}$). We say that x and y, respectively, are the real and imaginary parts of z. Equivalently, we write

$$x = \text{Re } z$$
$$y = \text{Im } z.$$

When written in the form (D.1), we say that z is expressed in *rectangular form*. *Euler's formula* is

$$e^{i\theta} = \cos\theta + i\sin\theta. \quad \text{(D.2)}$$

Using Euler's formula we can write z in *polar form* as

$$z = re^{i\theta} \quad \text{(D.3)}$$

where

$$r = |z| \equiv \sqrt{x^2 + y^2}$$
$$\theta \equiv \arctan\left(\frac{y}{x}\right) \quad (-\pi < \theta \leq \pi \text{ rad}).$$

We say that r and θ, respectively, are the *modulus* and *phase* of the complex variable z. Then $e^{i\theta}$ is a *unit-modulus* complex variable (i.e., $|e^{i\theta}| = 1$). We define the *complex conjugate* of z as

$$z^* \equiv x - iy = re^{-i\theta}. \quad \text{(D.4)}$$

Using the complex conjugate we can write

$$r = \sqrt{zz^*}$$
$$\theta = \frac{1}{2i}\ln\left(\frac{z}{z^*}\right).$$

D.2. ROOTS

If n is a positive integer, then every nonzero complex variable z has n distinct nth *roots*. The roots are given by [*Abramowitz*, 1972]

$$z^{1/n} = r^{1/n}\exp\left[i\left(\frac{\theta + k2\pi}{n}\right)\right] \quad (0 \leq k \leq n-1,\ k \in \mathbb{Z}) \quad \text{(D.5)}$$

where $r^{1/n}$ denotes the positive, real nth root of r. For given z and n, we call the root that corresponds to $k = 0$ the *principal root*; we call the other $n - 1$ roots the *auxiliary roots*. For $n > 1$ the principal root lies either in the right half of the complex plane or on the nonnegative part of the imaginary axis.

D.2.1. Square Roots

Applying (D.5) to the case $n = 2$ gives the two square roots [*Abramowitz*, 1972]

$$z^{1/2} = \pm\sqrt{r}\left[\cos\left(\frac{\theta}{2}\right) + i\sin\left(\frac{\theta}{2}\right)\right].$$

Using the half-angle relations (B.4) we can rewrite this as

$$z^{1/2} = \pm\sqrt{r}\left[\pm\sqrt{\frac{1+\cos\theta}{2}} \pm i\sqrt{\frac{1-\cos\theta}{2}}\right].$$

Noting that $x = r\cos\theta$, $y = r\sin\theta$, and $-\pi < \theta \leq \pi$ rad, we can express this as

$$z^{1/2} = \pm\left[\sqrt{\frac{r+x}{2}} + i\ \text{sign}(y)\sqrt{\frac{r-x}{2}}\right] \quad \text{(D.6)}$$

where we define the *sign function* as

$$\text{sign}(y) \equiv \begin{cases} -1, & y < 0 \\ 0, & y = 0 \\ +1, & y > 0. \end{cases} \quad \text{(D.7)}$$

Hydrodynamics of Time-Periodic Groundwater Flow: Diffusion Waves in Porous Media, Geophysical Monograph 224, First Edition. Joe S. Depner and Todd C. Rasmussen.

In (D.6), the principal square root is the one with the positive sign in front of the square brackets; the auxiliary square root is the one with the negative sign.

Exercise D.1 Square Roots of the Imaginary Unit

Problem: Express the square roots of the imaginary unit in both exponential and rectangular forms. Which is the principal square root?

Solution: In this case $x = 0$ and $y = 1$, so $r = 1$ and $\theta = \pi/2$ rad. Substituting these values for r and θ and $n = 2$ in the RHS of (D.5) yields

$$i^{1/2} = e^{i\pi/4},\ e^{i5\pi/4}$$

for $k = 0$ and $k = 1$, respectively. Similarly, substituting these values for x and y, respectively, in the RHS of (D.6) gives the corresponding rectangular forms:

$$i^{1/2} = \pm\frac{(1 + i)}{\sqrt{2}}.$$

The principal square root corresponds to $k = 0$:

$$\sqrt{i} = e^{i\pi/4} = \frac{1 + i}{\sqrt{2}}. \tag{D.8}$$

□

E
Kelvin Functions

Kelvin functions frequently arise in the analysis of 2D, axisymmetric flows in ideal media. For examples of such flows, see Section 12.4.

This appendix is based primarily on the material of *Olver and Maximon* [2016]. The notation used by *Olver and Maximon* [2016] to represent the general and principal-value inverse trigonometric functions (see Section 4.23 of *Roy and Olver* [2016]) is opposite to that used in this book. In this appendix we assume that all angles are measured in units of radians.

E.1. DEFINITIONS

Olver and Maximon [2016]. define the order-ν, *b-type Kelvin functions* as

$$\mathrm{ber}_\nu u \equiv \mathrm{Re}\left[e^{i\nu\pi/2} I_\nu\left(e^{i\pi/4}u\right)\right] \tag{E.1a}$$

$$\mathrm{bei}_\nu u \equiv \mathrm{Im}\left[e^{i\nu\pi/2} I_\nu\left(e^{i\pi/4}u\right)\right] \tag{E.1b}$$

and the order-ν, *k-type Kelvin functions* as

$$\mathrm{ker}_\nu u \equiv \mathrm{Re}\left[e^{-i\nu\pi/2} K_\nu\left(e^{i\pi/4}u\right)\right] \tag{E.2a}$$

$$\mathrm{kei}_\nu u \equiv \mathrm{Im}\left[e^{-i\nu\pi/2} K_\nu\left(e^{i\pi/4}u\right)\right] \tag{E.2b}$$

where $u \in \mathbb{R}$, $u \geq 0$, $\nu \in \mathbb{R}$, $I_\nu(z)$, and $K_\nu(z)$ are the order-ν, modified Bessel functions of the first and second kinds, respectively. Note that $\mathrm{ber}_\nu u$, $\mathrm{bei}_\nu u$, $\mathrm{ker}_\nu u$, and $\mathrm{kei}_\nu u$ are real valued. Therefore,

$$I_\nu\left(e^{i\pi/4}u\right) = e^{-i\nu\pi/2}\left(\mathrm{ber}_\nu u + i\,\mathrm{bei}_\nu u\right) \tag{E.3a}$$

$$K_\nu\left(e^{i\pi/4}u\right) = e^{i\nu\pi/2}\left(\mathrm{ker}_\nu u + i\,\mathrm{kei}_\nu u\right). \tag{E.3b}$$

When we use Kelvin functions in this book, the order-zero ($\nu = 0$) case arises most frequently.

Exercise E.1 Order-Zero Kelvin Functions

Problem: Write out all of the equations of Section E.1 for the case $\nu = 0$.

Solution: Substituting $\nu = 0$ in the equations of Section E.1 yields

$$\mathrm{ber}_0 u = \mathrm{Re}\ I_0\left(e^{i\pi/4}u\right)$$

$$\mathrm{bei}_0 u = \mathrm{Im}\ I_0\left(e^{i\pi/4}u\right)$$

$$\mathrm{ker}_0 u = \mathrm{Re}\ K_0\left(e^{i\pi/4}u\right)$$

$$\mathrm{kei}_0 u = \mathrm{Im}\ K_0\left(e^{i\pi/4}u\right)$$

$$I_0\left(e^{i\pi/4}u\right) = \mathrm{ber}_0 u + i\,\mathrm{bei}_0 u \tag{E.4a}$$

$$K_0\left(e^{i\pi/4}u\right) = \mathrm{ker}_0 u + i\,\mathrm{kei}_0 u. \tag{E.4b}$$

□

E.2. MODULUS AND PHASE FUNCTIONS

E.2.1. Definitions

Olver and Maximon [2016] define the order-ν *Kelvin modulus and phase functions* by the following:

$$M_\nu(u) \exp\left[i\theta_\nu(u)\right] \equiv \mathrm{ber}_\nu u + i\,\mathrm{bei}_\nu u \tag{E.5a}$$

$$N_\nu(u) \exp\left[i\phi_\nu(u)\right] \equiv \mathrm{ker}_\nu u + i\,\mathrm{kei}_\nu u \tag{E.5b}$$

where u, $M_\nu(u)$, $\theta_\nu(u)$, $N_\nu(u)$, and $\phi_\nu(u)$ are real valued. Then

$$M_\nu(u) \equiv \sqrt{\mathrm{ber}_\nu^2 u + \mathrm{bei}_\nu^2 u} \tag{E.6a}$$

$$\theta_\nu(u) \equiv \arctan\frac{\mathrm{bei}_\nu u}{\mathrm{ber}_\nu u} \tag{E.6b}$$

Hydrodynamics of Time-Periodic Groundwater Flow: Diffusion Waves in Porous Media, Geophysical Monograph 224,
First Edition. Joe S. Depner and Todd C. Rasmussen.

and

$$N_\nu(u) \equiv \sqrt{\text{ker}_\nu^2 u + \text{kei}_\nu^2 u} \tag{E.7a}$$

$$\phi_\nu(u) \equiv \arctan \frac{\text{kei}_\nu u}{\text{ker}_\nu u}. \tag{E.7b}$$

Exercise E.2 Order-Zero Modulus and Phase Functions

Problem: Write out all of the equations of Section E.2 for the case $\nu = 0$.

Solution: Substituting $\nu = 0$ in the equations of Section E.2 yields

$$M_0(u) \exp\left[i\theta_0(u)\right] = \text{ber}_0 u + i\,\text{bei}_0 u \tag{E.8a}$$

$$N_0(u) \exp\left[i\phi_0(u)\right] = \text{ker}_0 u + i\,\text{kei}_0 u \tag{E.8b}$$

$$M_0(u) \equiv \sqrt{\text{ber}_0^2 u + \text{bei}_0^2 u} \tag{E.9a}$$

$$\theta_0(u) \equiv \arctan \frac{\text{bei}_0 u}{\text{ber}_0 u} \tag{E.9b}$$

$$N_0(u) \equiv \sqrt{\text{ker}_0^2 u + \text{kei}_0^2 u} \tag{E.10a}$$

$$\phi_0(u) \equiv \arctan \frac{\text{kei}_0 u}{\text{ker}_0 u}. \tag{E.10b}$$

□

Exercise E.3 Bessel and Kelvin Functions

Problem: Express the Bessel function $I_\nu(e^{i\pi/4}u)$ $(u, \nu \in \mathbb{R})$ and its modulus and argument compactly in terms of the order-ν Kelvin modulus and phase functions. Repeat for the Bessel function $K_\nu(e^{i\pi/4}u)$ and its modulus and argument.

Solution: By combining (E.4) and (E.8), we obtain

$$I_\nu(e^{i\pi/4}u) = M_\nu(u) \exp\left\{i\left[\theta_\nu(u) - \frac{\nu\pi}{2}\right]\right\} \tag{E.11a}$$

$$\left|I_\nu(e^{i\pi/4}u)\right| = M_\nu(u) \tag{E.11b}$$

$$\arg I_\nu(e^{i\pi/4}u) = \theta_\nu(u) - \frac{\nu\pi}{2} \tag{E.11c}$$

and

$$K_\nu(e^{i\pi/4}u) = N_\nu(u) \exp\left\{i\left[\phi_\nu(u) + \frac{\nu\pi}{2}\right]\right\} \tag{E.12a}$$

$$\left|K_\nu(e^{i\pi/4}u)\right| = N_\nu(u) \tag{E.12b}$$

$$\arg K_\nu(e^{i\pi/4}u) = \phi_\nu(u) + \frac{\nu\pi}{2}. \tag{E.12c}$$

□

E.2.2. Behavior for Very Small and Very Large Arguments

E.2.2.1. Modulus Functions

From the definitions of b-type Kelvin modulus functions (see Section E.2) and the relevant power series in Section 10.65(i) of *Olver and Maximon* [2016], we have

$$u \to 0 \quad \Rightarrow \quad \begin{cases} 0 < M_\nu(u) < \infty \\ \dfrac{dM_\nu}{du} > 0. \end{cases}$$

From the asymptotic expansion in equation 10.68.16 of *Olver and Maximon* [2016] we have

$$u \to \infty \quad \Rightarrow \quad \begin{cases} M_\nu(u) \to \infty \\ \dfrac{dM_\nu}{du} > 0. \end{cases}$$

Similarly, from the definitions of the k-type Kelvin modulus functions (see Section E.2) and the power series for the k-type Kelvin functions in Section 10.65(ii) of *Olver and Maximon* [2016], we have

$$u \to 0 \quad \Rightarrow \quad \begin{cases} N_\nu(u) \to \infty \\ \dfrac{dN_\nu}{du} < 0. \end{cases}$$

Also, from the asymptotic expansion in equation 10.68.19 of *Olver and Maximon* [2016] we have

$$u \to \infty \quad \Rightarrow \quad \begin{cases} N_\nu(u) \to 0 \\ \dfrac{dN_\nu}{du} < 0. \end{cases}$$

E.2.2.2. Phase Functions

From the asymptotic expansion in equation 10.68.18 of *Olver and Maximon* [2016] we have

$$u \to \infty \quad \Rightarrow \quad \begin{cases} \theta_\nu(u) \to \infty \\ \dfrac{d\theta_\nu}{du} \to \frac{1}{\sqrt{2}}. \end{cases}$$

Combining the results of *Philip* [1979] and the second of equations 10.68.7 of *Olver and Maximon* [2016] gives, for integer j,

$$\theta_j(0) = \begin{cases} 7j\pi/4, & j < 0 \\ 3j\pi/4, & j \geq 0. \end{cases} \tag{E.13}$$

Similarly, from the asymptotic expansion in equation 10.68.21 of *Olver and Maximon* [2016] we have

$$u \to \infty \quad \Rightarrow \quad \begin{cases} \phi_\nu(u) \to -\infty \\ \dfrac{d\phi_\nu}{du} \to \frac{-1}{\sqrt{2}}. \end{cases}$$

Also, combining the results of *Philip* [1979] and the second of equations 10.68.15 of *Olver and Maximon* [2016] gives, for integer j,

$$\phi_j(0) = \begin{cases} -7j\pi/4, & j < 0 \\ -3j\pi/4, & j \geq 0. \end{cases} \tag{E.14}$$

E.3. ZEROS OF MODULUS FUNCTIONS

In this section we show that for real, nonnegative values of the order ν, the Kelvin modulus functions $M_\nu(u)$ and $N_\nu(u)$ are positive for all positive values of u. That is, we show that $M_\nu(u)$ and $N_\nu(u)$ have no real zeros for $u > 0$.

E.3.1. Zeros of *b*-Type Modulus Functions

Olver and Maximon [2016] (see their equation 10.27.6) give, for $\nu \in \mathbb{R}$,

$$I_\nu(z) = e^{\mp i\nu\pi/2} J_\nu\left(ze^{\pm i\pi/2}\right), \quad -\pi \le \arg z \le \frac{\pi}{2}$$

where $J_\nu(z)$ denotes the order-ν Bessel function of the first kind. Taking the modulus then gives

$$\left|I_\nu(z)\right| = \left|J_\nu\left(ze^{\pm i\pi/2}\right)\right|, \quad -\pi \le \arg z \le \frac{\pi}{2}.$$

Substituting $z = e^{i\pi/4}u$, where $u \in \mathbb{R}$, produces

$$\left|I_\nu\left(e^{i\pi/4}u\right)\right| = \left|J_\nu\left(e^{i3\pi/4}u\right)\right|.$$

When $\nu \ge -1$ the zeros of $J_\nu(z)$ are all real (see Section 10.21(i) of *Olver and Maximon* [2016]). Therefore,

$$\left|I_\nu\left(e^{i\pi/4}u\right)\right| > 0 \quad (\nu \ge -1,\ u \in \mathbb{R}).$$

Substituting the RHS of (E.11b) for the LHS yields

$$M_\nu(u) > 0 \quad (\nu \in \mathbb{R},\ \nu \ge -1,\ u > 0). \tag{E.15}$$

E.3.2. Zeros of *k*-Type Modulus Functions

Olver and Maximon [2016] (see their Section 10.42) give, for $\nu \in \mathbb{R}$,

$$\left|K_\nu(z)\right| > 0, \quad |\arg z| \le \frac{\pi}{2}.$$

Substituting $z = e^{i\pi/4}u$, where $u \in \mathbb{R}$, then yields

$$\left|K_\nu\left(e^{i\pi/4}u\right)\right| > 0, \quad u \in \mathbb{R}.$$

Substituting the RHS of (E.12b) for the LHS yields

$$N_\nu(u) > 0 \quad (\nu \in \mathbb{R},\ u > 0). \tag{E.16}$$

E.4. MONOTONICITY OF MODULUS FUNCTIONS

In this section we show that the *b*-type and *k*-type Kelvin modulus functions are monotone increasing and monotone decreasing, respectively. We assume that the Kelvin modulus and phase functions, and their first and second derivatives with respect to the argument u, exist and are continuous.

E.4.1. Monotonicity of *b*-Type Modulus Functions

In this section we prove informally that

$$\frac{dM_\nu}{du} > 0 \quad (\nu \in \mathbb{R},\ \nu \ge -1,\ u > 0). \tag{E.17}$$

Our proof uses the method of contradiction. Suppose $dM_\nu/du \le 0$ at some value $u = u_2$, where $0 < u_2 < \infty$ (see Figure E.1). Then $M_\nu(u)$ must have a local maximum, or a saddle point, at some value $u = u_1$ where $0 < u_1 \le u_2$. Thus,

$$\left.\frac{dM_\nu}{du}\right|_{u=u_1} = 0. \tag{E.18}$$

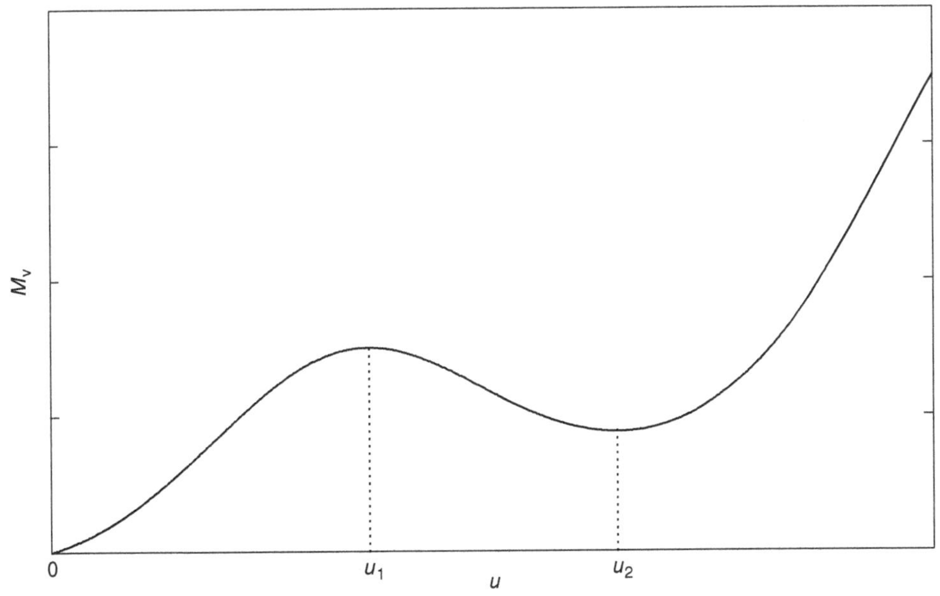

Figure E.1 Hypothetical, nonmonotonic, *b*-type modulus function.

Solving the second of equations 10.68.14 of *Olver and Maximon* [2016] for the second derivative of the amplitude function and then substituting $x = u_1$ and using (E.18) to simplify the result yield

$$\left.\frac{d^2 M_\nu}{du^2}\right|_{u=u_1} = \left[\left(\left.\frac{d\theta_\nu}{du}\right|_{u=u_1}\right)^2 + \left(\frac{\nu}{u_1}\right)^2\right] M_\nu(u_1), \quad (\nu \in \mathbb{R},\ \nu \geq -1). \tag{E.19}$$

There are two cases to consider:

Case 1 $\nu = 0$. Then equations 10.68.13 of *Olver and Maximon* [2016] (with $\nu = 0$) imply that zeros of the phase derivative, if any, do *not* coincide with zeros of the amplitude derivative. Consequently, the RHS of (E.19) is positive.

Case 2 $\nu \geq -1$ and $|\nu| > 0$. Then the RHS of (E.19) is positive regardless of the value of the phase derivative.

Combining the results from these two cases and (E.15) with (E.19) then gives

$$\left.\frac{d^2 M_\nu}{du^2}\right|_{u=u_1} > 0.$$

The second-derivative test then requires $M_\nu(u)$ to be a minimum at the critical point $u = u_1$. This contradicts our initial assumption that the point corresponds to a maximum or a saddle point. Therefore, $M_\nu(u)$ has no local maximum or saddle point, and (E.17) is valid.

Exercise E.4 Monotonicity of Integer-Order, *b*-Type Modulus Functions

Problem: Use the results obtained in this section and in Section 10.68(ii) of *Olver and Maximon* [2016] to show that

$$\frac{dM_\nu}{du} > 0 \quad (\nu \in \mathbb{Z},\ u > 0). \tag{E.20}$$

Solution: Equation (E.17) implies

$$\frac{dM_\nu}{du} > 0 \quad (\nu \in \mathbb{Z},\ \nu \geq -1,\ u > 0).$$

Equation 10.68.7 of *Olver and Maximon* [2016] allows us to extend this to corresponding negative integer values of the order ν, yielding (E.20). □

E.4.2. Monotonicity of *k*-Type Modulus Functions

In this section we prove informally that

$$\frac{dN_\nu}{du} < 0 \quad (\nu \in \mathbb{R},\ u > 0). \tag{E.21}$$

Our proof uses the method of contradiction. Suppose $dN_\nu/du \geq 0$ at some value $u = u_1$, where $0 < u_1 < \infty$ (see Figure E.2). Then $N_\nu(u)$ must have a local maximum, or a saddle point, at some value $u = u_2$ where $u_1 \leq u_2 < \infty$. Thus,

$$\left.\frac{dN_\nu}{du}\right|_{u=u_2} = 0. \tag{E.22}$$

Substituting $x = u_2$, $M = N$, and $\theta = \phi$ in the second of equations 10.68.14 of *Olver and Maximon* [2016] and then solving for the second derivative and using (E.22) to simplify the result yield

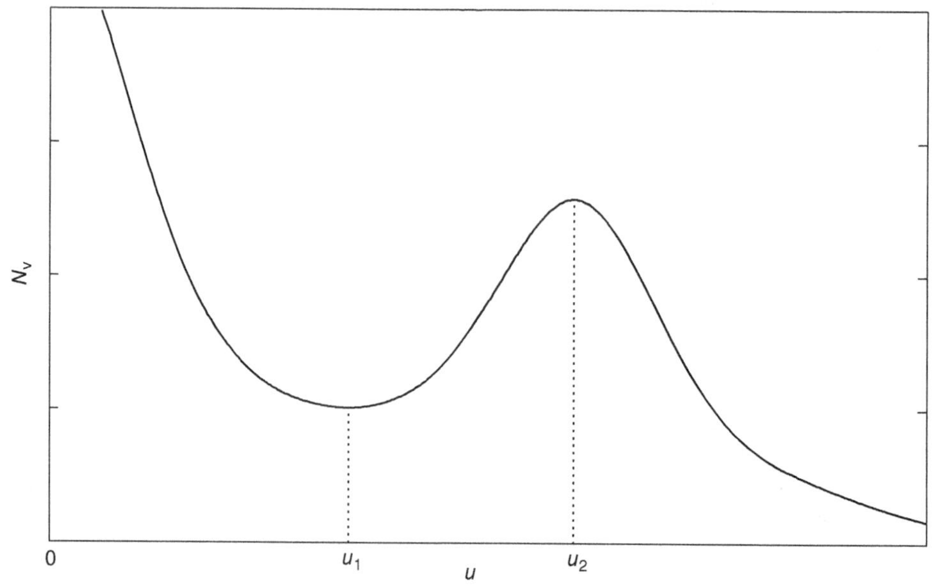

Figure E.2 Hypothetical, nonmonotonic, *k*-type modulus function.

$$\left.\frac{d^2N_\nu}{du^2}\right|_{u=u_2} = \left[\left(\left.\frac{d\phi_\nu}{du}\right|_{u=u_2}\right)^2 + \left(\frac{\nu}{u_2}\right)^2\right] N_\nu(u_2). \quad \text{(E.23)}$$

There are two cases to consider:

Case 1 $\nu = 0$. Then equations 10.68.13 of *Olver and Maximon* [2016] (with $M = N$, $\theta = \phi$, and $\nu = 0$) imply that zeros of the phase derivative, if any, do *not* coincide with zeros of the amplitude derivative. Consequently, the RHS of (E.23), with $\nu = 0$, is positive.

Case 2 $|\nu| > 0$. Then the RHS of (E.23), with $|\nu| > 0$, is positive regardless of the value of the phase derivative.

Combining the results from these two cases and (E.16) with (E.23) then gives

$$\left.\frac{d^2N_\nu}{du^2}\right|_{u=u_2} > 0.$$

The second-derivative test then requires $N_\nu(u)$ to be a minimum at the critical point $u = u_2$. This contradicts our initial assumption that the point corresponds to a maximum or a saddle point. Therefore, $N_\nu(u)$ has no local maximum or saddle point, and (E.21) is valid.

Exercise E.5 [$N_\nu(u)$ is Concave Upward]

Problem: Using the second of equations 10.68.14 of *Olver and Maximon* [2016], show that

$$\frac{d^2N_\nu}{du^2} > 0 \quad (\nu \in \mathbb{R},\ u > 0). \quad \text{(E.24)}$$

That is, using mathematics, show that the graph of $N_\nu(u)$ versus u is concave upward.

Solution: Substituting $x = u$, $M = N$, and $\theta = \phi$ in the second of equations 10.68.14 of *Olver and Maximon* [2016] and then solving for the second derivative yield

$$\frac{d^2N_\nu}{du^2} = \left[\left(\frac{d\phi_\nu}{du}\right)^2 + \left(\frac{\nu}{u}\right)^2\right] N_\nu(u) - \frac{1}{u}\frac{dN_\nu}{du}.$$

Combining this result with (E.16) and (E.21) yields (E.24). □

E.5. MONOTONICITY OF PHASE FUNCTIONS

In this section we show that the b-type and k-type Kelvin phase functions are monotone increasing and monotone decreasing, respectively. We assume that the Kelvin modulus and phase functions and their first and second derivatives with respect to the argument u exist and are continuous.

E.5.1. Monotonicity of b-Type Phase Functions

In this section we prove informally that

$$\frac{d\theta_\nu}{du} > 0 \quad (\nu \in \mathbb{R},\ u > 0). \quad \text{(E.25)}$$

Our proof uses the method of contradiction. Suppose $d\theta_\nu/du \le 0$ at some value $u = u_2$, where $0 < u_2 < \infty$ (see Figure E.3). Then $\theta_\nu(u)$ must have a local maximum, or a saddle point, at some value $u = u_1$, where $0 < u_1 \le u_2$. Thus,

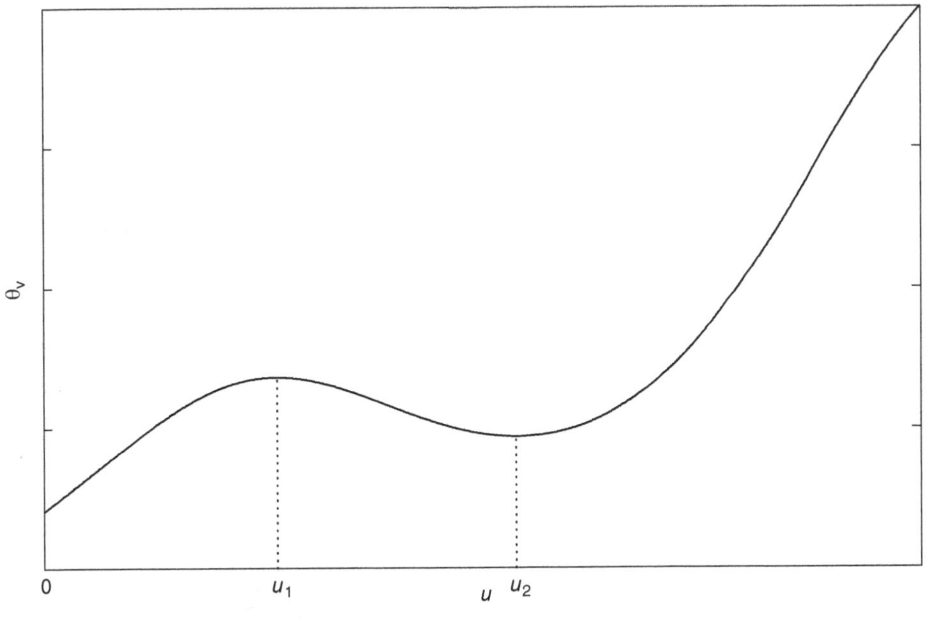

Figure E.3 Hypothetical, nonmonotonic, b-type phase function.

$$\left.\frac{d\theta_\nu}{du}\right|_{u=u_1} = 0. \tag{E.26}$$

Substituting $x = u$ in the first of equations 10.68.14 of *Olver and Maximon* [2016], expanding the derivative on the LHS, and then solving for the second derivative of the phase yield

$$\frac{d^2\theta_\nu}{du^2} = 1 - \left[\frac{1}{u} + \frac{2}{M_\nu(u)}\frac{dM_\nu}{du}\right]\frac{d\theta_\nu}{du}.$$

Substituting $u = u_1$ and using (E.26) then give

$$\left.\frac{d^2\theta_\nu}{du^2}\right|_{u=u_1} = 1. \tag{E.27}$$

The second-derivative test then requires $\theta_\nu(u)$ to be a minimum at the critical point $u = u_1$. This contradicts our initial assumption that $\theta_\nu(u)$ has a local maximum or a saddle point there. Therefore, $\theta_\nu(u)$ has no local maximum or saddle point and (E.25) is valid.

E.5.1.1. Alternate Proof for the Case $\nu \geq -1$

This section presents an alternate proof for the case $\nu \geq -1$. Substituting $x = u$ in equation 10.68.11 of *Olver and Maximon* [2016] produces

$$\frac{dM_\nu}{du} = \frac{-\nu}{u}M_\nu(u) - M_{\nu-1}(u)\cos\left[\theta_{\nu-1}(u) - \theta_\nu(u) - \frac{\pi}{4}\right].$$

Rearranging, and exploiting the evenness of the cosine function give

$$\frac{dM_\nu}{du} + \frac{\nu}{u}M_\nu(u) = -M_{\nu-1}(u)\cos\left[\theta_\nu(u) - \theta_{\nu-1}(u) + \frac{\pi}{4}\right]$$
$$= -M_{\nu-1}(u)\cos\left\{\left[\theta_\nu(u) - \theta_{\nu-1}(u) - \frac{\pi}{4}\right] + \frac{\pi}{2}\right\}.$$

Using the trigonometric identity (B.2a) with $\alpha = \theta_\nu(u) - \theta_{\nu-1}(u) - \pi/4$ and $\beta = \pi/2$ to rewrite the cosine term leads to

$$\frac{1}{M_{\nu-1}(u)}\left[\frac{dM_\nu}{du} + \frac{\nu}{u}M_\nu(u)\right] = \sin\left[\theta_\nu(u) - \theta_{\nu-1}(u) - \frac{\pi}{4}\right].$$

Substituting $w = \nu - 1$ gives

$$\frac{1}{M_w(u)}\left[\frac{dM_{w+1}}{du} + \frac{w+1}{u}M_{w+1}(u)\right]$$
$$= \sin\left[\theta_{w+1}(u) - \theta_w(u) - \frac{\pi}{4}\right].$$

Because of (E.15) and (E.17), the LHS is positive for $w \geq -1$; consequently

$$\sin\left[\theta_{w+1}(u) - \theta_w(u) - \frac{\pi}{4}\right] > 0 \quad (w \in \mathbb{R},\ w \geq -1,\ u > 0).$$

Combining this with equation 10.68.12 of *Olver and Maximon* [2016] yields

$$\frac{d\theta_\nu}{du} > 0 \quad (\nu \in \mathbb{R},\ \nu \geq -1,\ u > 0). \tag{E.28}$$

Exercise E.6 (Monotonicity of Integer-Order, *b*-Type Phase Functions)

Problem: Use the results obtained in this section and in Section 10.68(ii) of *Olver and Maximon* [2016] to show that

$$\frac{d\theta_\nu}{du} > 0 \quad (\nu \in \mathbb{Z},\ u > 0). \tag{E.29}$$

Solution: Equation (E.25) implies

$$\frac{d\theta_\nu}{du} > 0 \quad (\nu \in \mathbb{Z},\ \nu \geq -1,\ u > 0). \tag{E.30}$$

Substituting $x = u$ and $n = \nu$ in equation 10.68.7 of *Olver and Maximon* [2016], and then differentiating with respect to u give

$$\frac{d\theta_{-\nu}}{du} = \frac{d\theta_\nu}{du}, \qquad \nu \in \mathbb{Z}.$$

Combining this and (E.30) yields (E.29). □

E.5.2. Monotonicity of *k*-Type Phase Functions

In this section we prove informally that

$$\frac{d\phi_\nu}{du} < 0 \quad (\nu \in \mathbb{R},\ u > 0). \tag{E.31}$$

Our proof uses the method of contradiction. Suppose $d\phi_\nu/du \geq 0$ at some value $u = u_1$, where $0 < u_1 < \infty$ (see Figure E.4). Then $\phi_\nu(u)$ must have a local maximum, or a saddle point, at some value $u = u_2$, where $u_1 \leq u_2 < \infty$. Thus,

$$\left.\frac{d\phi_\nu}{du}\right|_{u=u_2} = 0. \tag{E.32}$$

Substituting $x = u$, $M = N$, and $\theta = \phi$ in the first of equations 10.68.14 of *Olver and Maximon* [2016], expanding the derivative on the LHS, and then solving for the second derivative of the phase, yields

$$\frac{d^2\phi_\nu}{du^2} = 1 - \left[\frac{1}{u} + \frac{2}{N_\nu(u)}\frac{dN_\nu}{du}\right]\frac{d\phi_\nu}{du}.$$

Substituting $u = u_2$ and using (E.32) then give

$$\left.\frac{d^2\phi_\nu}{du^2}\right|_{u=u_2} = 1. \tag{E.33}$$

The second-derivative test then requires $\phi_\nu(u)$ to be a minimum at the critical point $u = u_2$. This contradicts our initial assumption that $\phi_\nu(u)$ has a local maximum or a saddle point there. Therefore, $\phi_\nu(u)$ has no local maximum or saddle point, and (E.31) is valid.

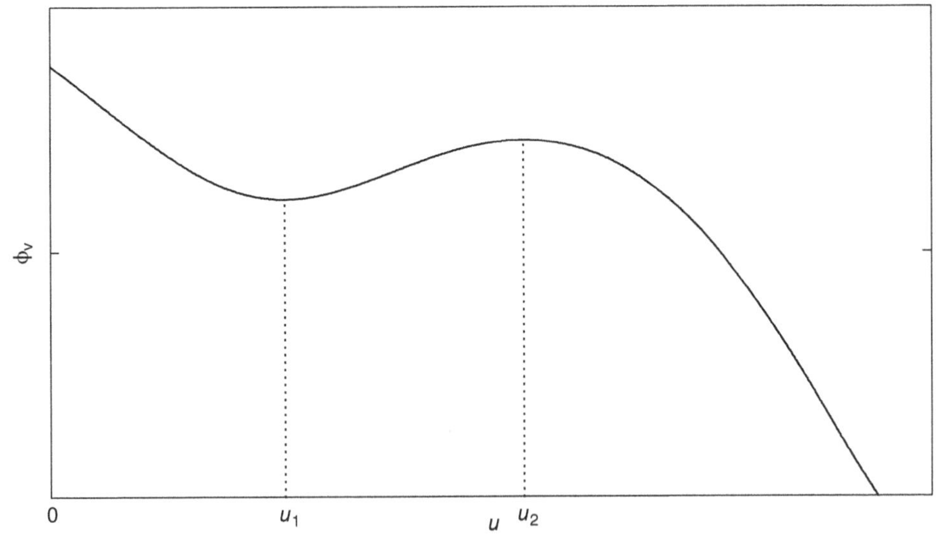

Figure E.4 Hypothetical, nonmonotonic, k-type phase function.

E.5.2.1. Alternate Proof for the Case $|\nu| \geq 1$

This section presents an alternate proof valid for the case $|\nu| \geq 1$. Substituting $x = u$, $M = N$, and $\theta = \phi$ in equation 10.68.11 of *Olver and Maximon* [2016] produces

$$\frac{dN_\nu}{du} = \frac{\nu}{u} N_\nu(u) + N_{\nu+1}(u) \cos\left[\phi_{\nu+1}(u) - \phi_\nu(u) - \frac{\pi}{4}\right].$$

Rearranging and exploiting the evenness of the cosine function give

$$\frac{dN_\nu}{du} - \frac{\nu}{u} N_\nu(u) = N_{\nu+1}(u) \cos\left[\phi_\nu(u) - \phi_{\nu+1}(u) + \frac{\pi}{4}\right]$$

$$= N_{\nu+1}(u) \cos\left\{\left[\phi_\nu(u) - \phi_{\nu+1}(u) - \frac{\pi}{4}\right] + \frac{\pi}{2}\right\}.$$

Using the trigonometric identity (B.2a) with $\alpha = \phi_\nu(u) - \phi_{\nu+1}(u) - \pi/4$ and $\beta = \pi/2$ to rewrite the cosine term leads to

$$\frac{1}{N_{\nu+1}(u)}\left[\frac{dN_\nu}{du} - \frac{\nu}{u} N_\nu(u)\right] = -\sin\left[\phi_\nu(u) - \phi_{\nu+1}(u) - \frac{\pi}{4}\right].$$

Substituting $w = \nu + 1$ gives

$$\frac{1}{N_w(u)}\left[\frac{dN_{w-1}}{du} - \frac{w-1}{u} N_{w-1}(u)\right]$$

$$= -\sin\left[\phi_{w-1}(u) - \phi_w(u) - \frac{\pi}{4}\right].$$

Because of (E.16) and (E.21), the LHS is negative for $w \geq 1$; consequently

$$\sin\left[\phi_{w-1}(u) - \phi_w(u) - \frac{\pi}{4}\right] > 0, \quad w \geq 1. \tag{E.34}$$

Substituting $x = u$, $M = N$, and $\theta = \phi$ in equation 10.68.12 of *Olver and Maximon* [2016] produces

$$\frac{d\phi_\nu}{du} = -\frac{N_{\nu-1}(u)}{N_\nu(u)} \sin\left[\phi_{\nu-1}(u) - \phi_\nu(u) - \frac{\pi}{4}\right].$$

Combining this result with (E.34) we obtain

$$\frac{d\phi_\nu}{du} < 0, \quad \nu \geq 1.$$

Equation 10.68.15 of *Olver and Maximon* [2016] allows us to extend this result to corresponding negative values of the order ν, yielding

$$\frac{d\phi_\nu}{du} < 0 \quad (\nu \in \mathbb{R},\ |\nu| \geq 1,\ u > 0). \tag{E.35}$$

BIBLIOGRAPHY

Abramowitz, M. (1972), Elementary analytical methods, *Handbook of Mathematical Functions*, edited by M. Abramowitz, and I. A. Stegun, Chap. 3, Dover, Mineola N. Y., available at http://files.eric.ed.gov/fulltext/ED250164.pdf

Adachi, J. I., and E. Detournay (1997), A poroelastic solution of the oscillating pore pressure method to measure permeabilities of "tight" rocks, Pap. 62, *Int. J. Rock. Mech. Min. Sci. Geomech. Abstr.*, *34*(3–4), 430.

Alcolea, A., E. Castro, M. Barbieri, J. Carrera, and S. Bea (2007), Inverse modeling of coastal aquifers using tidal response and hydraulic tests, *Ground Water*, *45*(6), 711–722, doi: 10.1111/j.1745-6584.2007.00356.x.

Aris, R. (1962), *Vectors, Tensors, and the Basic Equations of Fluid Mechanics*, Dover, New York.

Ataie-Ashtiani, B., R. E. Volker, and D. A. Lockington (1999a), Tidal effects on sea water intrusion in unconfined aquifers, *J. Hydrol.*, *216*(1–2), 17–31, doi: 10.1016/S0022-1694(98)00275-3.

Ataie-Ashtiani, B., R. E. Volker, and D. A. Lockington (1999b), Numerical and experimental study of seepage in unconfined aquifers with a periodic boundary condition, *J. Hydrol.*, *222*(1–4), 165–184, doi: 10.1016/S0022-1694(99)00105-5.

Bakker, M. (2004), Transient analytic elements for periodic Dupuit-Forcheimer flow, *Adv. Water Resour.*, *27*(1), 3–12, doi: 10.1016/j.advwatres.2003.10.001.

Bakker, M. (2009), Sinusoidal pumping of groundwater near cylindrical inhomogeneities, *J. Eng. Math.*, *64*(2), 131–143, doi: 10.1007/s10665-008-9244-0.

Bakker, M. (2015), Wigaem: An analytic element model for periodic groundwater flow, available at https://code.google.com/p/wigaem/, accessed 21 November, 2015.

Bear, J. (1972), *Dynamics of Fluids in Porous Media*, American Elsevier, New York.

Becker, M. W. and E. Guiltinan (2010), Cross-hole periodic hydraulic testing of inter-well connectivity, presented at 35th Workshop on Geothermal Reservoir Engineering, 1–3 February, SGP-TR-188, Stanford Univ. Stanford, Calif., available at https://pangea.stanford.edu/ERE/pdf/IGSstandard/SGW/2010/becker.pdf.

Bernabé, Y., U. Mok, and B. Evans (2006), A note on the oscillating flow method for measuring rock permeability, *Int. J. Rock Mech. Min. Sci.*, *43*(2), 311–316, doi: 10.1016/j.ijrmms.2005.04.013.

Beyer, W. H., Ed. in (1984), Section VIII: Calculus, *Standard Mathematical Tables*, 27th ed., pp.227–314. CRC Press, Boca Raton, Fla.

Black, J. H., and K. L. Kipp, Jr (1981), Determination of hydrogeological parameters using sinusoidal pressure tests: A theoretical appraisal, *Water Resour. Res.*, *17*(3), 686–692, doi: 10.1029/WR017i003p00686.

Bredehoeft, J. D. (1967), Response of well-aquifer systems to Earth tides, *J. Geophys. Res.*, *72*(12), 3075–3087, doi: 10.1029/JZ072i012p03075.

Bruggeman, G. A. (1999), *Analytical Solutions of Geohydrological Problems*, Elsevier, Amsterdam.

Burbey, T. J. and M. Zhang (2010), Assessing hydrofracing success from Earth tide and barometric response, *Ground Water*, *48*(6), 825–835, doi: 10.1111/j.1745-6584.2010.00704.x.

Burnett, W. C., P. K. Aggarwal, A. Aureli, H. Bokuniewicz, J. E. Cable, M. A. Charette, E. Kontar, S. Krupa, K. M. Kulkarni, A. Loveless, W. S. Moore, J. A. Oberdorfer, J. Oliveira, N. Ozyurt, P. Povinec, A. M. G. Privitera, R. Rajar, R. T. Ramessur, J. Scholten, T. Stieglitz, M. Taniguchi and J. V. Turner (2006), Quantifying submarine groundwater discharge in the coastal zone via multiple methods, *Sci. Total Environ.*, *367*(2–3), 498–543, doi: 10.1016/j.scitotenv.2006.05.009.

Butler, Jr., J. H., G. J. Kluitenberg, D. O. Whittemore, S. P. Loheide II, W. Jim, M. B. Billinger, and X. Zhan (2007), A field investigation of phreatophyte-induced fluctuations in the water table, *Water Resour. Res.*, *43*(W02404), doi: 10.1029/2005WR004627.

Cardiff, M., T. Bakhos, P. K. Kitanidis, and W. Barrash (2013), Aquifer heterogeneity characterization with oscillatory pumping: Sensitivity analysis and imaging potential, *Water Resour. Res.*, *49*(9), 5395–5410, doi: 10.1002/wrcr.20356.

Carr, P. A., and G. S. van der Kamp (1969), Determining aquifer characteristics by the tidal method, *Water Resour. Res.*, *5*(5) :1023–1031, doi: 10.1029/WR005i005p01023.

Carslaw, H. S., and J. C. Jaeger (1986), *Conduction of Heat in Solids*, 2nd ed., Oxford Univ. Press, New York.

Casimir, H. B. G. (1945), On Onsager's principle of microscopic reversibility, *Rev. Mod. Phys.*, *17*(2–3), 343–350, 10.1103/RevModPhys.17.343.

Chang, E., and A. Firoozabadi (2000), Gravitational potential variations of the sun and moon for estimation of reservoir compressibility, *Soc. Petrol. Eng. J.*, *5*(4), 456–465, doi: 10.2118/67952-PA.

Chapuis, R. P., C. Bélanger, and D. Chenaf (2006), Pumping test in a confined aquifer under tidal influence, *Ground Water*, *44*(2), 300–305, doi: 10.1111/j.1745-6584.2005.00139.x.

Cheng J., C. Chen, and M. Ji (2004), Determination of aquifer roof extending under the sea from variable-density flow modelling of groundwater response to tidal loading: Case study of the Jahe River Basin, Shandong Province, China, *Hydrogeology J.*, *12*, 408–423, doi: 10.1007/s10040-004-0347-z.

Cutillo, P. A., and J. D. Bredehoeft (2011), Estimating aquifer properties from the water level response to Earth tides, *Ground Water*, *49*(4), 600–610, doi: 10.1111/j.1745-6584.2010.00778.x.

Hydrodynamics of Time-Periodic Groundwater Flow: Diffusion Waves in Porous Media, Geophysical Monograph 224, First Edition. Joe S. Depner and Todd C. Rasmussen.

Davis, A. B., M. G. Trefry, N. Corngold, and A. Mandelis (2001), Letter: Many uses for diffusion waves, *Phys. Today (Lett.), 54*(3), 100–102, doi: 10.1063/1.4796266.

Erskine, A. D. (1991), The effect of tidal fluctuations on a coastal aquifer in the UK, *Ground Water, 29*(4), 556–562, doi: 10.1111/j.1745-6584.1991.tb00547.x.

Eves, H. (1984), Section V: Trigonometry, in *Standard Mathematical Tables*, 27th ed., edited by W. H. Beyer, pp. 132–155. CRC Press, Boca Raton, Fla.

Ferris, J. G. (1951), Cyclic fluctuations of water level as a basis for determining aquifer transmissibility, Bull. 33, Int. Geod. Geophys. Union, Assoc. Sci. Hydrol., General Assembly, Brussels, Int. Assoc. Hydrol Sci., available at http://pdw.hanford.gov/arpir/pdf.cfm?accession=D196025992.

Fischer, G. J. (1992), The determination of permeability and storage capacity: Pore pressure oscillation method, in *Fault Mechanics and Transport Properties of Rocks*, edited by B. Evans and T. F. Wong, pp. 187–211, Academic, New York, doi: 10.1016/S0074-6142(08)62823-5.

Freeze, R. A. and J. A. Cherry (1979), *Groundwater*, Prentice-Hall, Englewood Cliffs, N. J.

Furbish, D. J. (1991), The response of water level in a well to a time series of atmospheric loading under confined conditions, *Water Resour. Res., 27*(4), 557–568, doi: 10.1029/90WR02775.

Goldstein, H. (1980), *Classical Mechanics*, 2nd ed., Addison-Wesley, Reading, Mass.

Gradshteyn, I. S., and I. M. Ryzhik (1980), *Tables of Integrals, Series, and Products*, Academic New York.

Hanson, J. M. (1980) Reservoir response to tidal and barometric effects, *Geotherm. Resour. Counc. Trans. 4*, 337–340.

Hatch, C. E., A. T. Fisher, J. S. Revenaugh, J. Constantz, and C. Ruehl (2006), Quantifying surface water-groundwater interactions using time series analysis of streambed thermal records: Method development, *Water Resour. Res., 42*(W10410), doi: 10.1029/2005WR004787.

Hermance, J. F. (1998), *A Mathematical Primer on Groundwater Flow*, Prentice Hall, Upper Saddle River N. J.

Hobbs, P. J., and J. H. Fourie (2000), Earth-tide and barometric influences on the potentiometric head in a dolomite aquifer near the Vaal Range Barrage, South Africa, *Water SA, 26*(3), 353–360.

Hsieh, P. A., J. D. Bredehoeft, and J. M. Farr (1987), Determination of transmissivity from Earth tide analysis, *Water Resour. Res., 23*(10), 1824–1832, 10.1029/WR023i010p01824.

Hsieh, P. A., J. D. Bredehoeft, and S. A. Rojstaczer (1988), Response of well aquifer systems to Earth tides: Problem revisited, *Water Resour. Res., 24*(3), 468–472, doi: 10.1029/WR024i003p00468.

Hvorslev, M. J. (1951), Time lag and soil permeability in groundwater observations, Bull. 36, U.S. Army Corps of Engi., Waterways Experiment Station, Vicksburg, Miss., available at http://www.csus.edu/indiv/h/hornert/Geol%20500%20Spring%202014/Week_3_slug_tests/Hvorslev%201951.pdf.

Jackson, J. D. (1998), *Classical Electrodynamics*, 3rd ed., Wiley, New York.

Jacob, C. E. (1950), Flow of groundwater, in *Engineering Hydraulics*, edited by H. Rouse, pp. 321–386, Wiley, New York.

Jha, M. K., D. Namgial, Y. Kamii, and S. Peiffer (2008), Hydraulic parameters of coastal aquifer systems by direct methods and by an extended tide-aquifer interaction technique, *Water Resour. Mgmt., 22*(12), 1899–1923, doi: 10.1007/s11269-008-9259-3.

jostpuur (2010, Mar 16), Transformation of cross product [Online forum comment]. Message posted to https://www.physicsforums.com/threads/transformation-of-cross-product.384430/

Korn, G. A., and T. M. Korn (2000), *Mathematical Handbook for Scientists and Engineers*, 2nd ed., Dover, Mineola, N. Y.

Kranz, R. L., J. S. Saltzman, and J. D. Blacic (1990), Hydraulic diffusivity measurements on laboratory rock samples using an oscillating pore pressure method, *Int. J. Rock Mech. Min. Sci. Geomech. Abstr., 27*(5), 345–352, doi: 10.1016/0148-9062(90)92709-N.

Kruseman, G. P. and N. A. de Ridder (2000), Analysis and evaluation of pumping test data, Tech. Rep. 47, Int. Inst. for Land Reclamat. and Improv., Wageningen Netherlands, available at http://www2.alterra.wur.nl/Internet/webdocs/ilri-publicaties/publicaties/Pub47/Pub47.pdf.

Kümpel, H.-J., G. Grecksch, K. Lehmann, D. Rebscher, and K. C. Schulze (1999), Studies of *in situ* pore pressure fluctuations at various scales, *Oil Gas Technol. Rev. IFP*, 54(6), 679–688, doi: 10.2516/ogst:1999057.

Latinopoulos, P. (1984), Periodic recharge to finite aquifiers from rectangular areas, *Adv. Water Resour., 7*(3), 137–140, doi: 10.1016/0309-1708(84)90043-5.

Latinopoulos, P. (1985), Analytical solutions for periodic well recharge in rectangular aquifiers with third-kind boundary conditions, *J. Hydrol., 77*(1–4), 293–306, doi: 10.1016/0022-1694(85)90213-6.

Lautz, L. K. (2008a), Estimating groundwater evapotranspiration rates using diurnal water-table fluctuations in a semi-arid riparian zone, *Hydrogeol. J., 16*(3), 483–497, doi: 10.1007/s10040-007-0239-0.

Lautz, L. K. (2008b), Erratum: Estimating groundwater evapotranspiration rates using diurnal water-table fluctuations in a semi-arid riparian zone, *Hydrogeol. J., 16*(6), 1233–1235, doi: 10.1007/s10040-008-0338-6.

Maddock III, T., and L. B. Vionnet (1998), Groundwater capture processes under a seasonal variation in natural recharge and discharge, *Hydrogeol. J., 6*(1), 24–32, doi: 10.1007/s100400050131.

Mandelis, A. (2000), Diffusion waves and their uses, *Phys. Today, 53*(8), 29–34, doi: 10.1063/1.1310118.

Mandelis, A. (2001), *Diffusion-Wave Fields: Mathematical Methods and Green Functions*, Springer-Verlag, New York.

Mandelis, A., L. Nicolaides, and Y. Chen (2001), Structure and the reflectionless/refractionless nature of parabolic diffusion-wave fields, *Phys. Rev. Lett., 87*(020801), doi: 10.1103/PhysRevLett.87.020801.

Marine, I. W. (1975), Water level fluctuations due to Earth tides in a well pumping from slightly fractured crystalline rock, *Water Resour. Res., 11*(1), 165–173, doi: 10.1029/WR011i001p00165.

Mathematical Society of Japan (2000), Section 18: Almost periodic functions, in *Encyclopedic Dictionary of Mathematics*, Vol. 1, 2nd., edited by K. Ito, MIT Press, Cambridge, Mass., available at http:/mathsoc.jp/en/pamph/current/dictionary.html.

Mehnert, E., A. J. Valocchi, M. Heidari, S. G. Kapoor, and P. Kumar (1999), Estimating transmissivity from the water level fluctuations of a sinusoidally forced well, *Ground Water, 37*(6), 855–860, doi: 10.1111/j.1745-6584.1999.tb01184.x.

Merritt, M. L. (2004), Estimating hydraulic properties of the floridan aquifer system by analysis of Earth-tide, ocean-tide, and barometric effects, Collier and Hendry Counties, Florida. Water-Resources Investigat. Rep. 03-4267, U.S. Geol. Surv., available at http://pubs.usgs.gov/wri/wri034267/.

Monachesi, L. B., and L. Guarracino (2011), Exact and approximate analytical solutions of groundwater response to tidal fluctuations in a theoretical inhomogeneous coastal confined aquifer, *Hydrogeol. J., 7*, 1443–1449, doi: 10.1007/s10040-011-0761-y.

Morland, L. W., and E. C. Donaldson (1984), Correlation of porosity and permeability of reservoirs with well oscillations induced by Earth tides, *Geophys. J. R. Astron. Soc., 79*(3), 705–725, doi: 10.1111/j.1365-246X.1984.tb02864.x.

Narasimhan, T. N., B. Y. Kanehiro, and P. A. Witherspoon (1984), Interpretation of Earth tide response of three deep, confined aquifers, *J. Geophys. Res., 89*(B3), 1913–1924, doi: 10.1029/JB089iB03p01913.

Neeper, D. A. (2001), A model of oscillatory transport in granular soils, with application to barometric pumping and earth tides, *J. Contaminant Hydrol., 48*(3–4), 237–252, doi: 10.1016/S0169-7722(00)00181-9.

Neeper, D. A. (2002), Investigation of the vadose zone using barometric pressure cycles, *J. Contaminant Hydrol., 54*(1–2), 59–80, doi: 10.1016/S0169-7722(01)00146-2.

Neeper, D. A. (2003), Harmonic analysis of flow in open boreholes due to barometric pressure cycles, *J. Contaminant Hydrol., 60*(3–4), 135–162, doi: 10.1016/S0169-7722(02)00086-4.

Neuzil, C. E. (2003), Hydromechanical coupling in geologic processes, *Hydrogeol. J., 11*(1), 41–83, doi: 10.1007/s10040-002-0230-8.

Noble, B., and J. W. Daniel (1977), *Applied Linear Algebra*, 2nd ed., Prentice-Hall, Englewood Cliffs, N. J.

Nye, J. F. (1957), *Physical Properties of Crystals*, Oxford Univ. Press, London.

Olver, F. W. J., and L. C. Maximon (2016), Bessel functions, in *Digital Library of Mathemateical Functions*, Chap. 10, Nat. Insti. of Stand. Technol., available at http://dlmf.nist.gov/10, accessed 21 November 2015.

Onsager, L. (1931a), Reciprocal relations in irreversible processes, I, *Phys. Rev., 37*(4), 405–426, doi: 10.1103/PhysRev.37.405.

Onsager, L. (1931b), Reciprocal relations in irreversible processes, II, *Phys. Rev., 38*(12), 2265–2279, doi: 10.1103/PhysRev.38.2265.

Oosterbaan, R., J. Boonstra, and K. V. G. K. Rao (1996), The energy balance of groundwater flow, in *Int. Conf. Hydrology and Water Resources, 1993, New Delhi IN*, vol. 2 of *Subsurface-Water Hydrology*, edited by V. P. Singh and B. Kumar, pp. 153–160, Kluwer Academic, Dordrecht, Netherlands, available at http://www.waterlog.info/pdf/enerart.pdf.

Özişik, M. N. (1989), *Boundary Value Problems of Heat Conduction*, Dover, Mineola, N. Y.

Philip, J. R. (1979), Note on the Kelvin phase functions, *Math Comp., 33*(145), 337–341, doi: 10.1090/S0025-5718-1979-0514829-X.

Pool, M., V. E. A. Post, and C. T. Simmons (2014), Effects of tidal fluctuations on mixing and spreading in coastal aquifers: Homogeneous case, *Water Resour. Res., 50*(8), 6910–6926, doi: 10.1002/2014WR015534.

Protter, M. H., and H. F. Weinberger (1999), *Maximum Principles in Differential Equations*, 3rd ed., Springer Verlag, New York.

Pugh, D. (2004), *Changing Sea Levels*, Cambridge Univ. Press, Cambridge.

Rasmussen, T. C., and L. A. Crawford (1997), Identifying and removing barometric pressure effects in confined and unconfined aquifers, *Ground Water, 35*(3), 502–511, doi: 10.1111/j.1745-6584.1997.tb00111.x.

Rasmussen, T. C., and T. L. Mote (2007), Monitoring surface and subsurface water storage using confined aquifer water levels at the Savannah River Site, USA, *Vadose Zone J., 6*(2), 327–335, 10.2136/vzj2006.0049.

Rasmussen, T. C., K. G. Haborak, and M. H. Young (2003), Estimating aquifer hydraulic properties using sinusoidal pumping at the Savannah River Site, South Carolina, USA, *Hydrogeol. J., 11*(4), 466–482, doi: 10.1007/s10040-003-0255-7.

Renner, J., and M. Messar (2006), Periodic pumping tests, *Geophys J. Int., 167*(1), 479–493, doi: 10.1111/j.1365-246X.2006.02984.x.

Riddle, D. F. (1974), *Calculus and Analytic Geometry*, Wadsworth, 2nd ed., Belmont, Calif.

Rigord, P., Y. Caristan, and J. P. Hulin (1993), Analysis of porous media heterogeneities using the diffusion of pressure waves, *J. Geophys. Res., 98*(B6), 9781–9791, doi: 10.1029/92JB02695.

Rinehart, J. S. (1972), Fluctuations in geyser activity caused by variations in Earth tidal forces, barometric pressure and tectonic stresses, *J. Geophys. Res., 77*(2), 342–350, doi: 10.1029/JB077i002p00342.

Ritzi, Jr., R. W., S. Soorooshian, and P. A. Hsieh (1991), The estimation of fluid flow properties from the response of water levels in wells to the combined atmospheric and Earth tide forces, *Water Resour. Res., 27*(5), 883–893, doi: 10.1029/91WR00070.

Rojstaczer, S. A. (1988), Determination of fluid flow properties from the response of water levels in wells to atmospheric loading, *Water Resour. Res., 24*(11), 1927–1938, doi: 10.1029/WR024i011p01927.

Rojstaczer, S. A., and D. C. Agnew (1989), The influence of formation material properties on the reponse of water levels in wells to Earth tides and atmospheric loading, *J. Geophys. Res., 94*(B9), 12403–12411, doi: 10.1029/JB094iB09p12403.

Rojstaczer, S. A., and F. S. Riley (1990), Response of the water level in a well to Earth tides and atmospheric loading under unconfined conditions, *Water Resour. Res., 26*(8), 1803–1817, doi: 10.1029/90WR0021.

Rojstaczer, S. A., and F. S. Riley (1992), Correction: Response of the water level in a well to Earth tides and atmospheric loading under unconfined conditions, by S Rojstaczer and FS Riley, *Water Resour. Res., 28*(5), 1499, doi: 10.1029/92WR00362.

Roy, R., and F. W. J. Olver (2016), Elementary functions, in *Digital Library of Mathematical Functions*, chap. 4, Nat. Inst. of Stand. and Technol., available at http://dlmf.nist.gov/4, accessed 21 November 2015.

Roy, R., F. W. J. Olver, R. A. Askey, and R. Wong (2016), Algebraic and analytic methods, in *Digital Library of Mathematical Functions*, Nati. Inst. of Stand. and Technol., available at http://dlmf.nist.gov/1, accessed 21 November 2015.

Salazar, A. (2006), Energy propagation of thermal waves, *Eur. J. Phys., 27*(6), 1349–1355, doi: 10.1088/0143-0807/27/6/009.

Scales, J. A., and R. Snieder (1999), What is a wave? *Nature (Commentary), 401*, 739–740, doi: 10.1038/44453.

Seo, H. H. (2001), Modeling the influence of changes in barometric pressure on groundwater levels in wells, *Environ. Geol., 41*(1–2), 155–166, doi: 10.1007/s002540100361.

Serfes, M. E. (1991), Determining the mean hydraulic gradient of ground water affected by tidal fluctuations, *Ground Water, 29*(4), 549–555, doi: 10.1111/j.1745-6584.1991.tb00546.x.

Smith, A. J., and L. R. Townley (2015), Surface water-groundwater interaction: Flowthru-periodic, available at http://www.townley.com.au/streaklines/model_demonstrations_files/flowthru_periodic_files/flowthru_periodic.htm, accessed 21 November 2015.

Smith, A. J., L. R. Townley, and M. G. Trefry (2005), Visualization of aquifer response to periodic forcing, *Adv. Water Resour., 28*(8), 819–834, doi: 10.1016/j.advwatres.2005.02.001.

Song, I., and J. Renner (2006), Experimental investigation into the scale dependence of fluid transport in heterogeneous rocks, *Pure Appl. Geophys., 163*(10), 2103–2123, doi: 10.1007/s00024-006-0121-3.

Song, I., and J. Renner (2007), Analysis of oscillatory fluid flow through rock samples, *Geophys. J. Int., 170*(1), 195–204, doi: 10.1111/j.1365-246X.2007.03339.x.

Sposito, G. (2006), Chaotic solute advection by unsteady groundwater flow, *Water Resour. Res., 42*(W06D03), doi: 10.1029/2005WR004518.

Stallman, R. W. (1965), Steady one-dimensional fluid flow in a semi-infinite porous medium with sinusoidal surface temperature, *J. Geophys. Res., 70*(12), 2821–2827, doi: 10.1029/JZ070i012p02821.

Stewart, C. R., A. Lubinski, and K. A. Blenkarn (1961), The use of alternating flow to characterize porous media having storage pores, *J. Petrol. Technol., 13*(4), 383–389, doi: 10.2118/1650-G-PA.

Taniguchi, M. (2002), Tidal effects on submarine groundwater discharge into the ocean, *Geophys. Res. Lett., 29*(12), 2.1–2.3, doi: 10.1029/2002GL014987.

Toll, N. J., and T. C. Rasmussen (2007), Removal of barometric pressure effects and Earth tides from observed water levels, *Ground Water, 45*(1), 101–105, doi: 10.1111/j.1745-6584.2006.00254.x.

Townley, L. R. (1995), The response of aquifers to periodic forcing, *Adv. Water Resour., 18*(3), 125–146, doi: 10.1016/0309-1708(95)00008-7.

Trefry, M. G. (1999), Periodic forcing in composite aquifers, *Adv. Water Resour., 22*(6), 645–656, doi: 10.1016/S0309-1708(98)00037-2.

Trefry, M. G., and E. Bekele (2004), Structural characterization of an island aquifer via tidal methods, *Water Resour. Res., 40*(W01505), doi: 10.1029/2003WR002003.

Trefry, M. G., and C. D. Johnston (1998), Pumping test analysis for a tidally forced aquifer, *Ground Water, 36*(3), 427–433, doi: 10.1111/j.1745-6584.1998.tb02813.x.

Trefry, M. G., D. McLaughlin, D. R. Lester, G. Metcalfe, C. D. Johnston, and A. Ord (2011), Stochastic relationships for periodic responses in randomly heterogeneous aquifers, *Water Resour. Res., 47*(W08527), 18, doi: 10.1029/2011WR010444.

Trefry, M. G., D. McLaughlin, G. Metcalfe, D. Lester, A. Ord, K. Regenauer-Lieb, and B. Hobbs (2010), On oscillating flows in randomly heterogeneous porous media, *Philas. Trans. A, 368*(1910), 197–216, doi: 10.1098/rsta.2009.0186.

Urish, D. W., and T. E. McKenna (2004), Tidal effects on ground water discharge through a sandy marine beach, *Ground Water, 42*(7), 971–982, doi: 10.1111/j.1745-6584.2004.tb02636.x.

Urish, D. W., and M. M. Ozbilgin (1989), The coastal ground-water boundary, *Ground Water, 27*(3), 310–315, doi: 10.1111/j.1745-6584.1989.tb00454.x.

van der Kamp, G. S., and J. E. Gale (1983), Theory of Earth tide and barometric effects in porous formations with compressible grains, *Water Resour. Res., 19*(2), 538–544, doi: 10.1029/WR019i002p00538.

Weeks, E. P. (1979), Barometric fluctuations in wells tapping deep unconfined aquifers, *Water Resour. Res., 15*(5), 1167–1176, doi: 10.1029/WR015i005p01167.

Weisstein, E. (2016a) Euler differential equation, Wolfram Mathworld, http://mathworld.wolfram.com/EulerDifferentialEquation.html, accessed 15 August.

Weisstein, E. (2016b) Modified spherical Bessel differential equation, Wolfram Mathworld, http://mathworld.wolfram.com/ModifiedSphericalBesselDifferentialEquation.html, accessed 15 August.

Weisstein, E. (2016c) Modified spherical Bessel function of the first kind, Wolfram Mathworld, http://mathworld.wolfram.com/ModifiedSphericalBesselFunctionoftheFirstKind.html, accessed 15 August.

Weisstein, E. (2016d) Modified spherical Bessel function of the second kind, Wolfram Mathworld, http://mathworld.wolfram.com/ModifiedSphericalBesselFunctionoftheSecondKind.html, accessed 15 August.

Weisstein, E. (2016e) Modified Bessel differential equation, Wolfram Mathworld, http://mathworld.wolfram.com/ModifiedBesselDifferentialEquation.html, accessed 15 August.

Weisstein, E. (2016f) Lagrange's identity, Wolfram Mathworld, http://mathworld.wolfram.com/LagrangesIdentity.html, accessed 15 August.

Weisstein, E. (2016g) Linearly dependent functions, Wolfram Mathworld, http://mathworld.wolfram.com/LinearlyDependentFunctions.html, accessed 15 August.

Yano, Y., S. Nakao, K. Yasukawa, and T. Ishido (2000), Experimental and numerical investigation of sinusoidal pressure test, presented at 25th Workshop on Geothermal Reservoir Engineering, 24–26 January, SGP-TR-165, Stanford Univ. Stanford Calif., available at http://www.geothermal-energy.org/pdf/IGAstandard/SGW/2000/Yano.pdf.

Yim, C. S., and M. F. N. Mohsen (1992), Simulation of tidal effects on contaminant transport in porous media, *Ground Water, 30*(1), 78–86, doi: 10.1111/j.1745-6584.1992.tb00814.x.

Zawadzki, W., D. W. Chorley, and G. Patrick (2002), Capture-zone design in an aquifer influenced by cyclic fluctuations in hydraulic gradients, *Hydrogeol. J., 10*(6), 601–609, doi: 10.1007/s10040-002-0224-6.

Zucker, R. (1972), Elementary transcendental functions: Logarithmic, exponential, circular and hyperbolic functions, in *Handbook of Mathematical Functions*, edited by M. Abramowitz and I. A. Stegun, Chap. 4, Dover, Mineola, N. Y., available at http://files.eric.ed.gov/fulltext/ED250164.pdf.

INDEX

Hydrodynamics of Time-Periodic Groundwater Flow: Diffusion Waves in Porous Media, Geophysical Monograph 224, First Edition. Joe S. Depner and Todd C. Rasmussen.